Soil as World Heritage

David Dent
Editor

Soil as World Heritage

Springer

Editor
David Dent
Chestnut Tree Farmhouse, Forncett End
Norfolk, United Kingdom

Every effort has been made to contact the copyright holders of the figures and tables which have been reproduced from other sources. Anyone who has not been properly credited is requested to contact the publishers, so that due acknowledgment may be made in subsequent editions.

Disclaimer: The facts and opinions expressed in this work are those of the author(s) and not necessarily those of the publisher

ISBN 978-94-007-6186-5 ISBN 978-94-007-6187-2 (eBook)
DOI 10.1007/978-94-007-6187-2
Springer Dordrecht Heidelberg New York London

Library of Congress Control Number: 2013950419

Printed on acid-free paper

Springer is part of Springer Science+Business Media (www.springer.com)

Preface

The symposium *Soil as World Heritage* was held in the spring of 2012 to celebrate the half century of systematic field experiments at Balti, in the north of Moldova. The experiments monitor and evaluate the impact of crop rotations, monoculture, fallow, fertilization, tillage and irrigation on crop yields and soil fertility. The proceedings highlight the importance of such experiments for understanding the consequences of current farming practices, especially on the famous black earth or chernozem. But there is more.

Between 1965 and 1980, the *green revolution* increased crop yields two- to threefold, transcending differences in soils and climate. For a generation, food production was carried ahead of population growth, and political attention was turned away from land, food and agriculture. Policymakers today face new challenges and bigger challenges:

1. Burgeoning demand means that, by 2050, food production will need to be 70 % greater than now – double in developing countries. All this production must come from the same land and water resources or, if present trends continue, much less; there are no great reserves to draw on, the area under cereals peaked in the 1980s and diversion of arable to biofuel production intensifies the pressure.
2. We have passed peak soil. On top of historical land degradation, today's agricultural practices are driving land degradation, water shortage and contamination, loss of biodiversity and climate change. The last quarter century has witnessed degradation of one-quarter of the land surface; every continent and every biome is affected. The issue goes beyond mismanagement: tracts of the best land are being lost every year to cities and connecting infrastructure, and it appears inescapable that rising sea level will flood great cities and productive farmland.
3. The food system is unsustainable. The green revolution depended on cheap fuel, fertilizer and irrigation applied to new, responsive crop varieties. Fuel and fertilizer are no longer cheap, water resources are overcommitted and crop yields have levelled off – in some places they are falling.

4. Climate change is driven both by burning fossil fuels and by the insidious destruction of soil organic matter – yet the soil is the only buffer against climate change that we know how to manage. The symposium highlights the effects of drought on crop yields across southern and eastern Europe; yet more dramatically, by 2050, half of what is now India's high-potential wheat-growing land is likely to be heat-stressed, short-growing-season cropland.

One response is the international land grab where the power lies with the big players and which does nothing to help the global situation. Contributions to this symposium indicate a sustainable alternative that combines proven practices of conservation agriculture or ecological agriculture that retain and rebuild the soil with precision farming that tailors crops and operations to the natural variability of the landscape. This is high farming that demands high knowledge at the policy level and in the field – knowledge that depends on better information on land resources and relearning much that has been forgotten. We have to grow both the soil and the knowledge.

Chernozem is, simply, the best arable soil in the world. Historically, it has been the breadbasket of the Old World and the New. The chernozem of the Balti steppe was also at the heart of the foundation of soil science. Dokuchaev visited this very place, collecting material for *Russian chernozem,* and his first account is a concise statement of the principles of a new science. He wrote: 'The chernozem seemed to me, in 1877, so typical in its thickness, structure and humification that I called it first class. The analysis showed the content of humus was 5.718 %'. That soil now, under the plough, has nowhere more than 3.8 % humus and chernozem everywhere have lost 20–70 % of the humus that binds the soil together and created what appeared to be inexhaustible fertility. On present trends, by 2026, the humus content of chernozem across the country will be down to 2.5–3 %, and approaching a catastrophic shift to a different and unstable ecosystem; the black earth will turn to dust as it did in the prairies of America and Canada in the 1930s.

Agricultural practices are driving global warming, leaching of nutrients, pollution of water resources and diversion of rainfall away from replenishment of soil and groundwater to destructive runoff. These are pressing issues for our generation and will press harder on future generations. Long-term field experiments, and the scientific skills and experience that they nurture, will be increasingly valuable as a foundation and a focus for interdisciplinary teams of specialists studying the effects of farming practices on the soil and on both above- and below-ground components of flora, fauna and microorganisms. Experimental data built up over the last 50 years demonstrate the damage caused by human activity to the productivity and integrity of chernozem and, also, ways to restore its fertility.

For all these reasons, the chernozem of the Balti steppe under the long-term field experiments has been proposed as the first World Heritage Site for soil and soil science as *an outstanding example of human interaction with the environment that has become vulnerable under the impact of irreversible change, of significant ongoing ecological and biological processes in the evolution and development of terrestrial ecosystems and communities of plants and animals, and containing the most important natural habitats for in situ conservation of biological diversity, including threatened species of outstanding universal value*. By safeguarding this unique ecosystem and testimony to civilization, we may work towards sustainable development of society – and agriculture in particular. The ongoing scientific work is also a foundation for public appreciation of soils and soil science which is critical for wise policy and management.

These proceedings include contributions from 14 countries under headings: *The Soil and Environment, Soil Fertility: Lessons from Long-Term Field Experiments, Different Ways of Doing Things*, and *Soil Policy and Communications to Decision Makers*. On the last topic, there has been much wringing of hands by the scientific community about the lack of effective action to arrest land degradation, loss of biodiversity and climate change. Inaction is not due to lack of information: inaction stems from a lack of acceptable courses of action. If acceptable, and effective, courses of action are to be developed, the scientific community must involve itself in practical and political developments – even though this means venturing to the exposed frontiers of its own knowledge and experience. Therefore, at the request of the President and Government of Moldova, our communications to decision makers include recommendations of all the participants. These recommendations include definition of a new research thrust to support more sustainable land use through crop rotations that can be commercially viable, self-sufficient in energy, and which restore the stocks of soil organic matter; and a soil resolution that may serve as a basis for legislation. Important and achievable recommendations include:

1. Initiatives have to be within the framework of national policy for food and water security. Our first recommendation is to review this policy in the light of present knowledge of the land and develop *a national program for food and water security and safety worked out at local, regional and state level, including support for or creation of markets for the required production and services such as water management and carbon sequestration.*
2. Knowing what you want to achieve, it may be useful to set out ground rules in the form of a soil law. This is our second main recommendation: *adoption of a soil law to secure the services provided by the soil to society and the environment. This law should be the basis for allocation of payments or other incentives necessary to achieve the required protection of soil services.*

   Examples of incentives include *green water credits* paid to farmers for water management services (in the shape of approved soil water management and soil conservation practices). This does not require the government to find new money; credits are paid for by the direct beneficiaries of this service, the water users. Also, we may draw on EU experience of integrating soil protection within the Common Agricultural Policy; to receive support from the EU budget, farmers should respect standards set nationally to protect the soil against erosion, maintain soil organic matter and soil structure and avoid degradation of habitats and landscape features.
3. *Application of such policy and compliance with its conditions requires revitalized state services, working in partnership with land users, to elaborate*

whole-farm and community plans for rational land use, to provide on-farm support for the adoption of best practice and to monitor the state of soil and water resources. This recommendation does require new money!

4. The system of landholding goes against the requirement of sustainable land management, but this is not the time for another upheaval. We recommend evolution of the system towards something better fitted to the task. Possibilities include extension of the period of leasehold to, at least, the length of a sustainable crop rotation (say 7 or 8 years) or, better allow 99-year leases so that the leaseholder has incentive to take good care of the land. The final, easy-to-implement recommendation is: *support for cooperation between individual farmers for purchasing inputs, marketing produce and services, soil and water conservation at the landscape scale, and mutual exchange of know-how and support for services to cooperatives by contractors, especially for the purchase of new equipment needed for conservation farming.*

Norwich, January 2013 David Dent

Contents

Part I The Soil and Environment

Part III Different Ways of Doing Things

Role of Honour: those whose time and efforts have established and maintained the Selectia long-term field experiment in crop rotations and continuous monocultures

Years	Head of the Department	Years	Scientists in charge	Years	Technical workers	Years	Workers
1946–1949	Mihail Sidorov	1945–1960	Nicolae Lebedev	1960–1967	Lidia Necrasova	1956–1968	Andrei Doroftei
		1954–1961	Vasile Cazanji	1962–1996	Lidia Cuţulima	1961–	Ana Meacicova
1950–1956	Iosif Liberştein	1964–1979	Iurie Bondarenco	1968–1977	Vera Gulita	1965–1997	Nina Buţenco
		1963–1989	Gheorghe Şonţu	1970–2002	Dmitri Gasnaş	1992–1995	Alexandru Bodnari
1956–1984	Petru Chibasov	1970–	Lidia Bulat	1971–	Lidia Camennaia	2000–2006	Anatol Frasiniuc
		1984–	Boris Boincean	1964–1978	Maria Cozac		
1984–1992	Cozma Cibotari	1991–1994	Victor Garaşchiuc	1979–2002	Maria Revenco		
1992–	Boris Boincean	2002–	Mihail Bugaciuc	1993–	Tanea Bugaciuc		
		2002–	Vadim Cuzeac	1996–2000	Eudochia Labutina		
		2008–	Marin Cebotari	2011–	Andrei Negrescu		
				2011–	Marina Ilușca		

Those whose time and efforts have established and maintained the Selectia long-term field experiment on fertilization systems in crop rotations

Head of the Department	Scientists in charge	Years	Technical workers and years	Workers
Matina M.S. (1945–1970)	Nica L.T.	1957–1970	Popa T. (1970–2006)	Vaţetco R.P. (1967–2002)
Naconecinîi Zinaida I. (1964–1992)	Nichitchin B.G.	1954–1980	Ranga I. (1970–1978)	Ţurcan O. (1984–2005)
Boincean Boris P. (1991–)	Iachimov S.V.	1970–1976	Cîşlaru Z. (1970–)	Corobeinicov E. (1975–1996)
	Revenco E.I.	1972–1979	Barac L. (1970–1994)	Ivanov A. (1967–1981)
	Taran M.G.	1970–1976	Grigorenco O. (1968–1976)	Roşca D. (1983–1987)
	Barbuţă D.N.	1979–1990	Babin G. (1959–1976)	Anţeperovici S. (1975–1987)
	Crasnojan V.D.	1981–1993	Cucu L. (1968–1977)	Anţeperovici L. (1975–1987)
	Stadnic S.S.	1981–1997	Colibaba S. (1973–1979)	Calistru M. (1984–1987)
	Bolduma A.	2002	Cricsfelid C. (1972–1980)	Melnic N. (1976–1988)
	Zbanc E.	2007–2010	Chiaburu V. (1984–1988)	Moldovan V. (2001–2005)
		2009	Lupuleac G. (1975–1977)	Mereuţă T. (2001–2005)
			Dochienco L. (1972–1974)	Cozionova N. (2002–2004)
			Reaboi V. (1984–2000)	Damian R. (1985–1987)
			Botnariuc B. (1986–1992)	Şarîghina R. (1977–1992)
			Chilimova E. (1983–1988)	Micşhanscaia E. (1975–1987)
			Saico L. (1981–1984)	
			Biciuşca M. (1984–1985)	

Volcov A. (1996–1998)
Mahu L. (1975–1981)
Fediuc C. (2002–2008)
Gavriliuc N. (2000–2004)
Balan L. (1981–1983)
Ceaglîc S. (1975–2001)
Ghiţan G. (1956–1970)
Halfina P. (1973–1975)
Gutman F. (1982–1996)
Socoliuc I. (1984–1987)
Voitic Z. (1983–1985)
Ianuşchevici E. (1985–1991)
Danilchina I. (1978–1980)
Secrieru I. (2006–)
Bodiu N. (2011–)

Those whose time and efforts have contributed to the establishment and maintenance of the Selectia long-term experiment on irrigation

Years	Head of the Department	Years	Scientists in charge	Years	Technical workers	Years	Workers
1968–1996	Mihalcevschi V.D.	1968–1970	Perju V.E.	1968–1971	Buchinschi I.G.	1968–1975	Lupu M.M.
1997–2001	Hropotinschi P.	1968–1973	Geletchi V.S.	1968–1970	Urman P.M.	1968–1976	Ciolacu M.I.
2001–	Boincean B.P.	1970–1977	Urman P.M.	1969–1974	Poblotchi A.G.	1970–1976	Melnic N.N.
		1974–1976	Ilienco G.	1972–1977	Conicova N.A.	1970–1980	Movilă P.I.
				1973–1996	Martea M.P.		
		1974–2000	Pobloţchi A.G.	1974–1976	Bucatari I.I.	1972–1973	Martea M.P.
		1977–2000	Conicova N.A.	1976–1977	Ungureanu V.	1977–1982	Guţu I.I.
		1980–1989	Tcacenco V.I.	1976–1977	Chiaburu V.	1983–1990	Dranca A.A.
		1995–1997	Hropotinschi P.M.	1977–1981	Chitic I.V.	1983–1990	Dranca O.L.
		1995–	Ungurean A.I.	1977–1979	Mitrofanova L.I.	1985–2006	Golbur L.M.
		1996–	Martea M.P.	1980–1988	Rudniţchi D.N.	1995–2003	Surcov A.V.
		2001–2005	Hropotinschi P.M.	1980–1989	Tcacenco T.G.	1996–2004	Dolghier A.C.
				1981–1983	Bejenari V.N.	1996–2001	Rotari M.
				1984–1988	Moscalenco M.G.	1999–2006	Furtună T.G.
				1984–1989	Erenciuc I.V.	2004–2009	Aftanas I.
				1984–1990	Temciuc C.I.	2004–	Savca N.V.
				1984–1996	Ponomarenco V.M.	2008–2010	Tambur G.G.
				1995–2006	Statna M.N.	2008–2010	Fostei N.P.
				1997–1998	Punga V.		
				2009–2010	Rusnac S.		

Part I
The Soil and Environment

Chapter 1
Chernozem: Soil of the Steppe

A. Ursu, A. Overenco, I. Marcov, and S. Curcubăţ

Abstract Chernozem is the predominant soil of Moldova and the country's greatest natural treasure. Its profile is very thick, well humified and well structured – properties inherited from the steppe. Only grassland with its many-branched and deeply-ramified root system is able to produce abundant organic matter and humification throughout the solum. The underlying horizon, enriched in secondary carbonates, is a marker of the soil water regime that determines the different subtypes of chernozem. From north to south, less and less water percolates through the profile; in phase with the water regime, *Leached chernozem* gives way to *Typical chernozem* which, in turn, gives way to *Carbonate chernozem.* All chernozem share the thick, black, granular topsoil – remarkable for its fertility and resilience – but more than a century of cropping has degraded the chernozem; even where the soil profile is intact, it has lost half of its native humus and requires different and better treatment if its productivity is to be sustainable.

1.1 Introduction

Soils constitute the greatest natural treasure of the Republic of Moldova. The predominant soil is chernozem – described by the founder of the soil science as 'the king of soils'. Its remarkable fertility is determined by its rich composition, unique conservatism and resilience to degradation – even after being worked for centuries. Over millennia, chernozem accumulated and preserved a great store of energy and nutrients in the form of humus which stabilises the famous granular structure and endows the soil with great permeability and available water capacity. The soil is easily worked, maintains its structure and resists erosion by wind and

A. Ursu • A. Overenco • I. Marcov • S. Curcubăţ (✉)
ASM Institute of Ecology and Geography, Chisinau, Republic of Moldova
e-mail: stelacurcubat@rambler.ru

D. Dent (ed.), *Soil as World Heritage*, DOI 10.1007/978-94-007-6187-2_1,

water. These attributes were soon appreciated wherever the chernozem is found, encouraging unlimited exploitation for agriculture but, under intensive exploitation, even the chernozem has yielded to the processes of degradation.

1.2 Soil-Forming Processes

Chernozem was formed during a long period of soil genesis in the steppe and forest-steppe zones under herbaceous vegetation rich in grasses (Fig. 1.1). According to some mineralogical estimates, chernozem pedogenesis lasted for hundreds of thousands of years (Alekseev 2003).

Chernozem is the result of synthesis and accumulation of enormous quantities of organic matter subjected, partially, to mineralisation and, partially, to humification. This organic matter, conserved as humus and acting together with the mineral parent material, provided conditions for formation of complex, organo-mineral calcium humate that determines the essential character of chernozem – its blackish colour – and, together with the root system of the steppe vegetation, creates its strong granular structure. Humus is found throughout the solum, which may be 80–100 cm thick (Fig. 1.2), although the amount decreases with depth. Such a deeply humified and well-structured soil can be formed only by steppe vegetation, which develops a very deep and branching root system. Under forest, by contrast, herbaceous vegetation is ephemeral and the forest litter and perennial root systems do not create deeply humified soils.

The temperate seasons drive seasonal and annual rhythms of vegetation. Organic matter produced by plants (the primary producers) serves as food for animals and is

Fig. 1.1 Feather grass steppe

Fig. 1.2 *Typical chernozem*, moderately humified

further comminuted and decomposed by the soil fauna and microorganisms, ensuring the annual and multiyear cycling of organic matter (and carbon). The balance is conditioned by the amount of primary organic matter and amount of synthesised humus. Under natural steppe ecosystems, the balance is positive; humus and biologically sequestered mineral elements extracted from the regolith accumulate up to a point that may be called a *climax* state, when the amount of humus decomposed to its initial components (water + CO_2 + minerals) becomes equal to the newly synthesised humus.

All chernozem share the thick, black, granular topsoil. Within the chernozem zone, lower taxonomic units may be distinguished by attributes conditioned by variations of intrazonal pedogenetic factors. Climate plays a decisive role; the diagnostic horizon for division of chernozems into subtypes is the subsoil horizon of secondary carbonates (CSRM 1999) which is conditioned by the water regime – *percolative* in the north, *non-percolative* in the south. The depth of the carbonate horizon is an easily identified characteristic that indicates other soil attributes conditioned by geographical position and climate – notably humus content, structure and potential productivity. The more percolative the water regime, the more deeply carbonates are leached from the soil profile. The *Typical chernozem*[1] is the characteristic soil of the steppe, having all the specific characters of the type: a thick

[1] In the *World reference base for soil resources 2006* (IUSS 2006) Typical and Leached chernozem key out as *Haplic chernozem,* Carbonate chernozem *as Calcic chernozem. Luvic and Vertic chernozem* are as in WRB.

Table 1.1 Physicochemical characteristics of a Typical chernozem, moderately humified

	Humus	$CaCO_3$		Exchangeable cations		
				Ca^{++}	Mg^{++}	Σ
Depth, cm	%		pH	me/100 g		
0–10	3.9		7.1	29.7	5.3	34.9
30–40	3.7		7.2	30.4	5.1	35.4
50–60	2.3		7.5	28.5	4.9	33.3
70–80	1.4	3.8	8.2			
90–100	1.0	8.1	8.6	26.3	7.7	30.1
110–120		13.7	8.6			

solum, well humified, granular structure and with the carbonate horizon between the A and B layers. The solum of *Leached chernozem,* under a percolative regime, is leached of carbonates which occur only below the B horizon; the *Carbonate chernozem*, under a non-percolative regime, contains carbonates throughout the soil profile, including the topsoil.

Chernozem includes two further subtypes with a transitional character. *Luvic chernozem* is formed at the furthest limit of the type under mixed oak forest and borders with the *grey* soil type; it retains all the attributes conditioned by steppe vegetation but adds some specific characters – a powdering of silica on the structural elements of the upper horizon and clay accumulation in the lower part of the solum. Its genesis might be explained by forest invasion into the steppe. Another opinion is that this subtype is formed in rare cases where woodland has been succeeded by steppe grassland. Both may be correct. Within the chernozem zone, *Vertisols* occur as a lithomorphic soil type on heavy-textured, illite-montmorillonite parent material. Adjacent to Vertisols, on the same clay parent material, the transitional subtype of *Vertic chernozem* may be found. Its topsoil is typical for chernozem but the B horizon exhibits the vertic characteristics of coarse polyhedral structure with slickensides.

The chernozem of Moldova is almost entirely cultivated and has been for centuries. For all its resilience, over time the topsoil has lost its granular structure, the humus content has decreased, nutrient reserves accumulated over millennia have been depleted and the topsoil has been compacted and exposed to erosion by rain-splash and runoff. There is practically no virgin chernozem left in former steppe regions, so there is no way to check its original composition, except for Luvic chernozem that is still preserved under forest. Analyses made in the past show that the humus content of Typical chernozem exceeded 5–7 % (Dokuchaev 1883, 1900; Krupenicov 1967; Krupenicov et al. 1961; Ursu 2005). Currently, the arable Typical chernozem of the Balti Steppe contains 3.9 % humus (Table 1.1) and the thickness of the solum (with humus content >1 %) is about 80 cm.

The same soil under an oak shelterbelt, 60 years old, has 6.9 % of humus in the topsoil. It is difficult to explain such increase within 60 years; possibly, the soil sample collected in the shelterbelt contained organic residues that are difficult to remove, as Dokuchaev (1883) mentioned with reference to a soil sample collected from the forest in Cuhuresti. Nevertheless, the increase of organic matter content in

Table 1.2 Physicochemical characteristics of a Leached chernozem

	Humus	$CaCO_3$		Exchangeable cations		
				Ca++	Mg++	Σ
Depth, cm	%		pH	me/100 g		
0–20	4.6		6.6	26.8	9.6	36.4
30–40	4.5		6.8	26.4	9.6	36.0
45–55	3.6		6.9	25.2	10.8	36.0
60–70	3.1		7.3	24.8	10.4	35.2
70–80	2.3		7.3	23.6	11.6	35.2
80–90	1.8	2.8	7.4	23.2		33.2
110–120	1.2	11.9	7.8	22.8	9.6	32.4

the surface layer of chernozem under a grass sward and forest conditions suggests a real possibility of restoring the humus balance.

Regional variants of Typical chernozem may be distinguished according to humus content: in forest-steppe and the Balti Steppe, the humus content of the plough layer exceeds 3.3–3.5 %; in the Southern Plain the range is 2.5–3.2 % – this lower humus content is considered to be an indicator of the dryer climate (Krupenicov and Ursu 1985; Ursu 2006). So, within the Typical chernozem subtype, two groups are distinguished – *moderately humified* and *weakly humified*; the former borders with Leached chernozem, the latter with Carbonate chernozem. However, assigning quantitative indices of humus content to these two groups is problematic because humus content can be conditioned by texture; a clayey *Typical weakly humified chernozem* may have a higher humus content than a loamy *Typical moderately humified chernozem.* Virgin chernozem of practically all subtypes can be characterised as *humified*; likewise chernozem under forest plantations, etc. which may be recovering their original humus status.

On the Balti steppe and the hilly regions of forest-steppe, Leached chernozem occurs alongside Typical chernozem; there is no clear transition between the leached and typical subtypes so the boundary between them is arbitrary. The former occurs at higher elevation; it has a thick, well-humified and deeply structured solum (80–120 cm) with a high base saturation but without carbonates, which appear below 80–85 cm (Table 1.2). The leached subtype is more humified and with a thicker solum than the typical; the humus content of the arable layer is 4.6 %, decreasing gradually to 1.2 % at 110–120 cm.

The Carbonate chernozem subtype, formed under xerophytic steppe conditions, presents the southern boundary of the type.

1.3 Conclusions

The chernozem is the outstanding natural wealth of the country. It created itself over many thousands of years and accumulated enormous reserves of energy and nutrients in the form of humus – about 1 billion tons in Moldova alone. This ‘king of soils' has many attributes favourable to agriculture and great natural fertility. But it is not used properly; it is being degraded and destroyed. The chernozem deserves better; it deserves respect and requires good husbandry and efficient protection. Soil science has developed concepts and practical measures that enable sustainable use of chernozem – but these measures need to be implemented.

Each generation may and should be able to use the soil for food production. At the same time, it is obliged to maintain and pass on this treasure to future generations.

References

Alekseev VE (2003) On the age of soil and soil formation in the territory of Moldova. In: Andries SV (ed) Soil – one of the main problems of the 21st century. International scientific and practical conference on the 50th anniversary of the N Dimo Institute. Pontos, Chisinau, pp 41–51 (Romanian)

CSRM (1999) Classification of the soil of the Republic of Moldova. SMMSS, Chisinau (Romanian)

Dokuchaev VV (1883) Russian chernozem. Independent Society for Economics, St. Petersburg (Russian)

Dokuchaev VV (1900) On the question of the soils of Bessarabia. Pochvovedenie 1:2–23 (Russian)

IUSS Working Group (2006) World reference base for soil resources. A framework for international classification, correlation and communication, vol 103, World Soil Resource Report. FAO, Rome

Krupenicov IA (1967) The chernozem of Moldova. Cartea Moldoveneasca, Chisinau (Russian)

Krupenicov IA, Ursu AF (1985) The soils of Moldova, vol 2. Stiinta, Kishinev (Russian)

Krupenicov IA, Rodina AK, Strijova GP, Ursu AF (1961) The chernozem of the Northern part of Moldova. Bull Mold Branch Acad Sci USSR 85:24–32 (Russian)

Ursu AF (2005) On monitoring the dehumification of a Typical chernozem. In: Pedological factors and processes in the temperate zone, Ser. 9, vol. 6. Stiinta, Iasi (Romanian), pp 29–34

Ursu AF (2006) Rationele pedogeografice şi particularităţile regionale de utilizare şi protejare a solurilor. Tipografia ASM, Chisinau

Chapter 2
The Quality of Moldovan Soils: Issues and Solutions

S. Andrieş, V. Cerbari, and V. Filipciuc

Abstract From the economic point of view, soil is Moldova's most valuable natural resource. Maintaining the productive capacity of the soil over the long term and increasing its fertility to ensure food security should be primary goals of the whole nation. Soil investigations are aimed at solving the problem of maintaining the quality and production capacity of agricultural land in the face of high rates of soil degradation. Annual direct and indirect losses amount to 4.8 billion lei, including 1.85 billion lei from the irreversible loss of 26 million tons of fertile topsoil by erosion from slopes; 878 million lei from the complete destruction of soil cover by landslides, ravines and excavation for social needs; and 2.07 billion lei in lost agricultural production. Irreversible losses as a result of total destruction of soil cover over about 30 years amount to about 36.5 billion lei.

Problems and solutions for the protection, improvement and rational management of agricultural soils are listed in state programs elaborated with support of the Nicolae Dimo Institute.

2.1 Introduction

As support and living environment for people, plants and animals, the soil cover is the main part of Moldova's natural capital but it is finite and, on the human time scale, non-renewable. Soil is the wealth of the entire nation and should be used in the national interest in accordance with the law, regardless of land ownership. Proper management of soil resources is a primary social issue; the needed increase of agricultural production can be achieved only through rational use of soil resources.

S. Andrieş (✉) • V. Cerbari • V. Filipciuc
Nicolae Dimo Institute of Pedology, Agrochemistry and Soil Protection, Chisinau, Republic of Moldova
e-mail: ipaps_dimi@mtc.md

D. Dent (ed.), *Soil as World Heritage*, DOI 10.1007/978-94-007-6187-2_2,

We need to know the productive capacity of the arable land and its resilience under existing farming systems. Unfortunately, the quality of our arable is far from satisfactory; areas affected by erosion and landslides continue to expand; processes of humus loss, structure deterioration and compaction, salinity and sodicity, waterlogging and drought are intensifying – leading to breakdown of biophysical cycles, soil deterioration and loss of fertility.

Food and agriculture contribute 15–20 % of GDP. They depend on soil resources for agricultural and livestock production, environmental services and social welfare, especially in rural areas. Therefore, proper management of soil fertility is a primary social issue (National Program for Soil Fertility 2001, 2002, 2004).

State programs for soil conservation and sustainable use and management require land-use planning for multiple functions. Here we summarise the results of 60 years of soil investigation by the Nicolae Dimo Institute, the Design Institute for Land Management and other investigations in soil science conducted in scientific institutions within the country.

2.2 Condition of the Land Resource

The land resource of the Republic of Moldova comprises 3,384,600 ha of which farmland occupies 2,498,300 ha (73.8 %) including arable, 1,812,730 ha (72.6 %); perennial crops, 298,780 ha (12.0%); meadows, 352,550 ha (14.1 %); and fallow, 14,210 hectares (0.4 %). Farmland amounts to only 0.5 ha per caput including 0.4 ha of arable. About 1,877,100 ha is held by 1,310,000 private landowners with the average holding of 1.4 ha divided between two and five individual plots (Agency for Land Relation and Cadastre 2010).

Scientifically blind land reform fragmented holdings and created up-and-down slope alignment of plots that has not helped sustainability, fertility or farm production. The weighted average *bonitat* rating of farmland in 2010 was 63 points, yielding a modest 2.5 t/ha of winter wheat, whereas in the 1970s the rating was equal to 70 points. The current value of one *bonitat* point in terms of agricultural production is $8/ha/year, so recent losses as a result of depreciating soil quality amount to $56/ha/year, adding up to $105,420,000 per year for all farmland.

2.3 Land Evaluation and Soil Quality Information System

Land evaluation at the broad scale is the responsibility of soil scientists in the Design Institute for Land Management (DILM). Soil mapping still follows procedures drawn up prior to the land reform – which do not correspond to the current situation. A better identification of land quality problems and solutions and planning of measures to combat land degradation require periodic land evaluation and the creation of an electronic soil quality information system. A correct cadastre of soil quality requires, in turn: (1) improvement of the soil survey investigation system and regular re-evaluation of the entire country every 20 years and (2) improvement of soil classification and soil evaluation methods. A prototype methodology for soil investigation and electronic soil quality information system developed by the Nicolae Dimo Institute and the Agency for Land Relation and Cadastre of Moldova has been tested at the community level for 17 villages in Teleneshti district (Cerbari et al. 2010).

2.4 Soil Degradation

Krupenikov (2008) describes in detail five types and 40 forms of soil degradation that lower the productive capacity of the land and, in the worst case, completely destroy the soil cover. Erosion by runoff water affects 2.5 million ha, loss of soil nutrients also 2.5 million ha, salinity and sodicity affect 220,000 ha, landslides 81,000 ha and nearly all arable land suffers from loss of soil structure and secondary compaction (Tables 2.1 and 2.2).

Soil erosion is the main agent of land degradation and contamination of water resources (Nour 2001). The affected area increased from 594,000 ha in 1965 to 878,000 ha at present – an average annual increase of more than 7,000 ha involving a reduction of productivity of some 20 % in the *slightly eroded* category, 40 % for

Table 2.1 Soil quality indices

Quality indices	Index appreciation: Optimum/allowable limit	Index appreciation: Actual	Actual index condition
Erosional			
Soil loss t/ha	5	15–20	Extremely high
Humus loss kg/ha	70	700	Extremely high
Nutrient loss (NP) kg/ha	10–12	50	High
Agrophysical			
Soil structure (sum of aggregates 10–0.25 cm) %	60–80	40–45	Unsatisfactory
Bulk density g/cm^3	1.10–1.22	1.25–1.30	Moderately compacted
Porosity %	50–55	45–50	Low
Infiltration speed mm/h	42–70	20–30	Very low
Agrochemical			
Humus content % at 0–30 cm	>4	80% of arable land has less than 3% humus (low content)	
Humus balance t/ha/year	Steady or positive	−0.7	Negative
Optimum P content at 0–30 cm mg/100 g soil	3.0–4.0	60 % of arable has low soil P	
Nutrient balance (NPK) kg/ha	Steady or positive	−130 to −150	Extremely negative

Table 2.2 Land degradation in Moldova, after National Program for Soil Fertility (2002)

No.	Factors and forms of degradation	Affected farmland, 1,000 ha	Damage, US$1,000 Annual	By soil loss
1	Sheet and rill erosion	878	221,365	–
2	Gully erosion	8.8	7,622	370,594
3	Landslides	24.1	–	1,014,923
4	Complete destruction of soil by excavation	5	–	210,565
5	Secondary compaction	2,183	39,730	–
6	Salinity in gley soils on slopes and in depressions	20	3,640	–
7	Salinity in alluvial soils	99	5,405	–
8	Sodicity in steppe soils	25	1,820	–
9	Humus loss	1,037	18,873	–
10	Low/very low mobile phosphorus content	785	28,574	–
11	Salinity and compaction under irrigation	12.8	699	–
12	Other factors	1,258	108,751	1,722,422
Total		–	436,479	3,318,504

moderately eroded and more than 50 % for *highly eroded* soils. Sheet, rill and gully erosion are widespread; over the period 1911–1965, the area of gullies doubled from 14,434 ha to 24,230 ha. After 1965, some of the gullied land was afforested and some areas were levelled, reducing the gullied area to 8,800 ha in 1999, but in 2005 the recorded area was 11,800 ha thanks to cessation of controls and feckless management of the land over recent years.

Landslides: Some 800,000 ha of land is affected by inactive landslides that are prone to reactivation (*Ecopedological Monitoring* 1996; Cerbari 2011a); the area of farmland affected spread from 21,200 ha in 1970 to 24,500 ha in 2010. The main preventative and control measures are diversion of runoff water, drainage, land levelling and afforestation. These are costly but it is more costly to neglect and abandon affected areas.

Humus loss: Humus is a prime index of soil fertility, determining agrophysical, agrochemical and agrobiological soil attributes. It has been established by experiment that increasing humus content by 1 % yields 1.0 t/ha of maize or 0.8 t/ha of winter wheat (Andries 2007). Under the plough, Moldovan soils have lost about 40 % their original humus reserves. Over the last 15 years (1994–2009), the application of farmyard manure has decreased 60-fold; the area sown to perennial grasses decreased 4–5-fold; and over large areas crop residues are simply burnt in the fields. As a result, the soil's humus balance is negative (−0.7 t/ha/year and, with losses by erosion, −1.1tonnes); every year, our arable loses some 2.4 million tons of humus. Increased input of organic matter is entirely possible using crop rotations with more land under perennial and annual legumes, grasses and green manure and by application of farmyard manure.

Secondary compaction: Intensive tillage and damage to soil structure have led to loss of resilience against compaction and formation of a plough pan at a depth of 25–35 cm with a bulk density of 1.7–1.8 g/cm^3. Control measures include:

- Use of organic manure and compost, incorporation of plant residues, green manure, farmyard manure and treated sewage sludge
- In acid soils, addition of sugar-factory sludge at 4–6 t/ha along with organic manure to create a lime reserve which promotes soil aggregation
- Subsoiling at 35–40 cm once in 3 years to disrupt the plough pan
- Crop rotation with 20–30 % perennial grasses

Business-as-usual will lead to greater humus loss, structural damage and compaction – with serious consequences for farm production. It is necessary to implement gradually no-till and minimum-till soil conservation systems (Boincean 2011; Cerbari 2011b), but these need special machinery, skilled application of herbicides taking account of the peculiarities of crops, soil conditions and local climate and the use of varieties adapted to the improved soil conditions created by these technologies. Implementation of these new soil management systems requires testing and local adaptation in several places, and production or acquisition of the necessary machinery and other inputs.

Depletion of nutrients: In the years 1961–1965, 19 kg/ha of NPK and 1.3 t/ha of manure were applied annually; the soil nutrient balance was negative and the main crops yielded 1.6 t/ha of winter wheat, 2.8 t/ha of maize and 19.0 t/ha of sugar beet. During the era of chemical agriculture (1965–1990), the use of mineral fertilisers increased ninefold to 172 kg/ha NPK and manure application also increased to 6.6 t/ha. For the first time, there was a positive nutrient balance and, as a result, mobile phosphorus content doubled and exchangeable potassium increased by 2–3 mg/100 g of soil. Wheat yields increased to about 4t/ha. The better farms achieved on average 4.0–5.5 t/ha of wheat, 5.5–7.5 t/ha of corn and 45–50 t/ha of sugar beet.

But over the last 15 years, application of mineral fertilisers has fallen 15–20-fold. While crops extract 150–180 kg/ha NPK annually from the soil, only 15–20 kg/ha NPK is added as fertiliser; the nutrient balance is again profoundly negative (National Program for Soil Fertility 2002; Andries 2007; Boincean 2011). Rebuilding soil fertility requires a balanced increase in the amount of manure and fertiliser and their rational application based on soil maps and optimising crop rotations; and long-term field experiments with fertilisers have a crucial role to play.

Sodicity and salinity affect naturally wet and irrigated soils. In the period 1966–1990, major works were carried out on soil improvement with irrigation, drainage and gypsum amendment. Irrigation is very beneficial, especially in the south of Moldova where droughts occur once in 3 years and irrigation will increase yields 1.5–2-fold. However, local water resources have a high salt content and sodium adsorption ratio which promote soil salinity and sodicity.

In the 1990s, the irrigated area reached 308,000 ha but has now fallen back to 46,000 ha (Fig. 2.1). Currently, alluvial soils are rated as good −17 %, satisfactory −34 % and unsatisfactory −49 % (about 90,000 ha). Financial losses from salinity and waterlogging amount to about 50 million lei (Cerbari 1995).

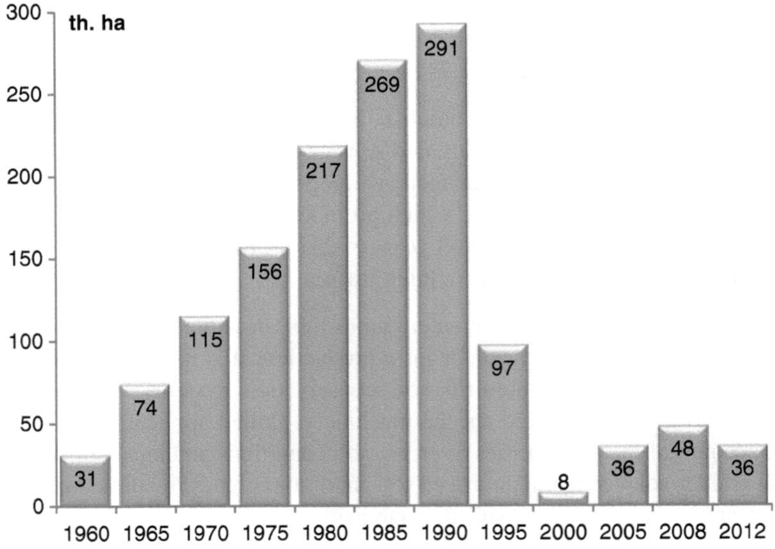

Fig. 2.1 Changes in the area of irrigated land

In the northern and central parts of the country, there are about 50,000 ha of poorly drained soils. In the 1970s to 1990s, more than 40,000 ha was improved by drainage but, over the last 14–16 years, the drains have not been maintained and no new work has been undertaken.

Management of the 25,000 ha of sodic (solonetz) soils poses special problems on account of their high exchangeable sodium content. In the years 1965–1990, the regular application of gypsum and ameliorative fertilisation was undertaken, using technology developed by the N Dimo Institute, but this has been discontinued and high sodium levels have returned.

The following State Programs have been developed and adopted to protect, improve and sustainably use the soil resources of Moldova:

- Complex National Program to increase soil fertility (2001)
- Recovery of Degraded Land and Increase Soil Fertility Complex Program, Parts 1 and 2 (2004)
- Conservation and improvement of soil fertility for the years 2011–2020 (Government decision no. 626 of 20.08.2011)

Work continues on all these programs.

2.5 Conclusions

1. Exploitative use of the soil, ignoring the need to maintain its biological regenerative capacity, has decreased soil fertility. Intensive farming in the years 1970–1990 increased agricultural production but many interventions damaged

the long-term quality of soil resources; about half of the county's soils are degraded – and still degrading.
2. Land reform did not create conditions for soil fertility improvement, rational soil use and increase of agricultural production. Instead it damaged the national economy and soil quality.
3. A strategy of step-by-step implementation of sustainable agriculture along with land consolidation will guarantee soil quality and maintain productivity in the long term.
4. Systems of soil tillage and fertilisation should ensure a steady-state or positive humus and nutrient balances and the formation of a biogenic topsoil with a good soil structure.
5. Field experiments confirm that the cardinal amelioration of the chernozem topsoil can be achieved only by complex measures which include soil conservation, radical increase of soil organic matter through the systematic use of green manure and other available organic fertilisers, and benign systems of mineral fertilisation and control of weeds and pests.
6. Planning for multiple uses of the land, considering all natural and man-made components, must go hand in hand with programs of protection, amelioration and sustainable use.

References

Agency for Land Relation and Cadastre (2010) Land cadastre of the Republic of Moldova, 1 January 2010, ALRC, Chisinau (Romanian)

Andries S (2007) Optimization of nutritional regimes and crops productivity. Pontos, Chisinau (Romanian)

Boincean B (2011) Soil tillage – trends and perspectives. In: Science and innovation, culture and arts. Akademos 22(3):61–67 (Romanian)

Cerbari V (1995) Bulletin of ecopedological monitoring (pedoamelioration), 2nd edn. Agroinformreclama, Chisinau (Romanian)

Cerbari V (1996) Bulletin of ecopedological monitoring (land degraded by erosion). Agroinformreclama, Chisinau (Romanian)

Cerbari V (2011a) Program of development and implementation of conservative technologies in agriculture. Moldovan Agric 4–5:7–9 (Romanian)

Cerbari V (2011b) No-till – agricultural system that protects the soil. Moldovan Agric 8–9:9–14 (Romanian)

Cerbari V et al. (2010) Soil quality monitoring (data base, conclusions, prognosis, recommendations). Pontos, Chisinau (Romanian)

Krupenikov IA (2008) Chernozem, genesis, perfection, tragedy of degradation, conservation and ways to regeneration. Pontos, Chisinau (Russian)

National Program for Soil Fertility (2001) National complex program for soil fertility increase. Pontos, Chisinau (Romanian)

National Program for Soil Fertility (2002) National complex program for recovery of degraded lands and soil fertility increase. Part I: Degraded soils amelioration. Pontos, Chisinau, Romanian

National Program for Soil Fertility (2004) National complex program for recovery of degraded lands and soil fertility increase. Part II: Soil fertility increase. Pontos, Chisinau (Romanian)

Nour DD (2001) Soil erosion. The essence of the process: consequence, minimalisation and stabilisation. Pontos, Chisinau (Russian)

Chapter 3
The State of Soil Erosion in the Republic of Moldova and the Need for Monitoring

E. Kuharuk and O. Crivova

Abstract Privatisation of land in Moldova, in 1992, created conditions ripe for soil erosion – and other problems in agricultural management. Systematic survey of soil erosion, completed in 2008, assessed more than 36 % of farmland as eroded in some degree, with substantial consequences for crop yields, water resources and the environment. The social and economic conditions that have favoured erosion in recent times have not been attended to, and there is pressing need for a national soil monitoring process and program for soil conservation and improvement.

3.1 Introduction

There is pressing need for soil protection in the Republic of Moldova. First, on account of the crucial role of soil in the biosphere: it is a powerful energy accumulator, it regulates the water cycle and the composition of the atmosphere and it is a reliable barrier to the migration of pollutants (Dobrovolski and Nikitin 2006). Secondly, Moldova has become, once again, an essentially agricultural country. The soil is its main capital asset but is in a critical condition – suffering significant degradation, especially from soil erosion. Soil erosion is a constant threat governed by relief, climate, land cover, soil type and management. In Moldova, beginning in 1987 but, especially from the beginning of land privatisation in 1992, all soil and water conservation activities came to a halt. At the same time, the conditions were created for unrestrained soil erosion. The new owners who came into possession – sometimes of less than hectare of sloping land – neither knew nor cared about soil and water conservation. Erosion-containment is not the immediate concern of poor

E. Kuharuk (✉) • O. Crivova
Nicolae Dimo Institute of Pedology, Agrochemistry and Soils Protection, Chisinau, Republic of Moldova
e-mail: icpadimo@agriculture.md; Ecaterina-ecostrategii@yahoo.com

D. Dent (ed.), *Soil as World Heritage*, DOI 10.1007/978-94-007-6187-2_3,

people in rural areas: much more important for them is to obtain crops from the gift and profit from it.

Although agriculture is the basis of local and export markets, it has been left to its own devices – and both soil fertility and productivity have declined. There are many reasons for this: demographic, economic and social (Kuharuk 2012). Even the reform of the fiscal system creates perverse incentives: in 2012, farms with an income not exceeding 25,200 lei (about US$2,100) will be assessed 7 % tax on income but, if their income exceeds this level, agricultural producers will be taxed at 18 % (Miheeva 2012).

Land policy should be informed by reliable information. The information presented here on the extent of soil erosion in the Republic of Moldova is the most up-to date available but is far from comprehensive.

3.2 Materials and Methods

Qualitative estimation of soil characteristics in the field, soil description and sampling were executed according to the methodology described by Rowell (1994/1998). Cartometric analysis of the degree of soil erosion was undertaken according to the procedures detailed by Kuharuk (1991, 2006) using soil maps at scales 1:10,000 and 1:200,000.

3.3 Results and Discussion

A full-profile chernozem with its topsoil intact contains, on average, 3.52 % of humus in the 0–50 cm layer. At the *low* level of erosion, average humus content decreases by 20 %; at the *medium* level of erosion, the average loss of humus is 42 %; at the high level of erosion, the loss is 64 %. Soil productivity declines along with soil fertility so that crop losses on highly eroded soils amount to 90 % and for the medium category up to 60 %! Table 3.1 summarises the condition of the soil

Table 3.1 Erosion of agricultural land in the Republic of Moldova as of 1 January 2008

				Erosion level:					
	Total	Total eroded:		Low		Medium		High	
Zones	area, ha	ha	%	ha	%	ha	%	ha	%
Northern	875,176	296,685	34.0	194,340	22.2	73,515	8.4	28,881	3.4
Central	696,345	314,748	45.2	165,730	23.8	102,363	14.7	46,655	6.7
Southern	732,411	285,640	39.0	157,468	21.5	91,551	12.5	36,621	5.0
South-eastern	234,753	51,880	22.1	36,152	15.4	12,442	5.3	3,287	1.4
Total	2,538,685	949,468	37.4	553,433	21.8	279,255	11.0	116,780	4.6

cover of all agricultural lands and their degree of erosion according to the natural climatic zones of the country (Kuharuk 1991).

According to the data on 1 January 2008, soil erosion afflicted more than one third of all agricultural land in the country. Annual economic losses from loss of crop yields, alone, amount to 3.61 billion lei (US$260 million dollars), not including Transnistria (Program of Land Management 2005). These losses are surely dwarfed by the costs to water supply, infrastructure and the environment which are, as yet, not properly assessed.

It has been proven by experiment in long-term field trials at Lebedenco, in Cahul Region (and discussed elsewhere in this symposium), that correct and competent deployment of erosion-containment measures has clear and unambiguous economic benefits: crop yields have increased by 30–40 % and there are big environmental benefits from keeping the topsoil on sloping land. But scientific investigation should not stop with the long-term field experiments; their practical results should be implemented in other places! When assessing the future perspectives of soil erosion, we should take account of not only present-day erosion processes but also likely changes of climatic and agricultural conditions that will affect the intensity of these processes. Fragmented land ownership and underfinanced management aggravates the problem of inadequate land cover. Consequently, if present trends continue, the area of land damaged by erosion will expand and the severity of erosion will increase.

In the present state of the country, and of the farming community in particular, erosion-containment measures have to be funded by the government. Otherwise, even the best policies and plans remain only on paper. At the same time, we know that without wholehearted and competent involvement of the people, in the first place by the entire rural population, no government and no authority is able to tackle this issue effectively. In August 2011 the Government of Moldova ratified a program for *Soil Fertility Conservation and Increase 2011–2020* which stipulates mitigation measures against soil erosion and landslides and for blocking of gullies. But the budget for scientific investigation of erosion-containment measures and new research was cut; development and improvement of technology, taking account of local natural and agricultural conditions, has been suspended.

3.4 Conclusions

Sound agricultural policy depends on sound information and this should include information on the status and trends of the soil cover. Equally it is important to be able to assess the effectiveness of land policy – which can only be done with accurate and up-to-date data on the ecological and economic effectiveness of specific erosion-containment measures, technologies and their combined application at local and regional level. Therefore, it is necessary to introduce the concept – and practice – of soil-erosion monitoring as part of a systematic state program of land quality monitoring.

References

Dobrovolski GV, Nikitin ED (2006) Soil ecology. Moscow University Edition, Science, Moscow (Russian)

Kuharuk ES (1991) Cartometric analysis of soil cover of Republic of Moldova and use of large-scale maps. Printed summary of PhD in Agriculture Dissertation, Baku (Russian)

Kuharuk ES (2006) Soil cover from LS Berg to the present. In: Collected volume – Academician LS Berg 130 years. Eco-Tiras, Bender (Russian)

Kuharuk ES (2012) Let's join forces in combating soil erosion. www.pan.md, N3(277), 13 January 2012 (Russian). Accessed May 2013

Miheeva L (2012) www.puls.md, N5(401) 17 February 2012, 11 p, (Russian). Accessed May 2013

Program of Land Management (2005) Program of land management for degraded soils and increasing their fertility. Part 1. Pontos, Chisinau (Romanian)

Rowell DL (1994/1998) Soil science: methods and applications. Longman Scientific and Technical/Kolos Edition, London/Moscow (Russian)

Chapter 4
Biota of Typical Chernozem Under Different Land Uses in Long-Term Field Experiments

I. Senikovskaya

Abstract The influence of land use and management on the biological properties of Typical chernozem in the north of the Republic of Moldova has been investigated. The values of most biological indices and the soil organic carbon content decrease in the sequence: 10–23-year-old lea grassland → arable → 30–43-year-old black fallow. A stable state of the biota is supported by a topsoil humus content of 4.9–5.1 %. Inclusion of areas with natural vegetation in a crop rotation restores the vital activities of the biota that are degraded by long-term arable use. A scale of biological parameters of Typical chernozem is proposed to assess the stability of the biota and to develop national soil quality standards.

4.1 Introduction

The soil cover of the Republic of Moldova is suffering intensified degradation and desertification (Andries et al. 2004). The combination of natural and anthropogenic factors has created disequilibrium between the accumulation of carbon in the soil and the loss of soil organic carbon (Banaru and Plamadeala 2010). The carbon budget is profoundly negative and this has a big impact on the soil biota (Senikovskaya et al. 2010). Biological degradation accompanies and, sometimes, exacerbates all kinds of soil deterioration. Therefore, there is a need for ecological certification of soils and national quality standards.

Soil quality and stability depend on many physical, chemical and biological attributes, and characterisation requires the selection of the properties most sensitive to changes in management practices (Cameran et al. 1998; Filip 2002; Paz Jimenez et al. 2002). Biological properties fulfil this requirement but they vary both seasonally

I. Senikovskaya (✉)
Nicolae Dimo Institute of Soil Science, Agrochemistry and Soil Protection, Chisinau, Republic of Moldova
e-mail: Irina-irina_sen@mail.ru

D. Dent (ed.), *Soil as World Heritage*, DOI 10.1007/978-94-007-6187-2_4,

and spatially (Trasar-Cepeda et al. 1998; Bending et al. 2004). Therefore, indices based on a combination of soil properties are needed to reflect the effects of the major soil processes on soil quality and resilience. An important issue to be resolved is the selection of reference criteria: in order to evaluate soil quality, it is necessary to develop individual parameters and scales for different soils and types of soil degradation. In this respect, monitoring and statistical analysis of the results of long-term field experiments make it possible to produce databases on status of the soil biota and to develop a system of evaluation of degradation and recovery processes. This research aims to determine the influence of different land use management on the biological properties of Typical chernozem[1] and to develop scale parameters of the stability of the soil biota.

4.2 Experimental Site and Methods

Research was conducted on the long-term field experiments of the Selectia Research Institute for Field Crops (Boincean 1999), located in the north of the Republic of Moldova, in District No. 3 of Typical and Leached Chernozems of the wooded steppe of the Balti ridges and plains (Ursu 2011). Within the long-term experiments, plots were selected from lea[2] established for 10–23 years, long-term arable and black fallow established for 30–43 years. The soil is Typical chernozem with humus content of 5.06 % under lea, 4.17 % under unfertilised arable crop rotation, and 3.54 % under black fallow. Soil samples were collected from the 0–30 cm layer of the experimental plots during 1994–1998 and 2006–2007.

Status of invertebrates: The status of invertebrates was identified from test cuts by manually sampling the soil layers to the depth of soil fauna occurrence (Gilyarov 1965). Diversity at the family level and classification according to feeding habits were categorised according to Gilyarov and Striganova (1987).

Microbiological properties: Microbial biomass carbon was measured on the 0–30 cm layer by the rehydration method (the difference between carbon extracted with 0.5 M K_2SO_4 from fresh soil samples and from soil dried at 65–70 °C for 24 h – Blagodatsky et al. 1987). K_2SO_4-extractable carbon (C) concentrations in the dried and fresh soil samples were measured simultaneously by dichromate oxidation; K_2SO_4-extractable C was determined at 590 nm using a Specol-221 spectrophotometer. The quantity of K_2SO_4-extractable C from fresh soil samples was used as a measure of labile organic C (Graham and Haynes 2006). Counts of culturable microorganisms (heterotrophic bacteria, humus-mineralising microorganisms, bacteria from the family of the *Azotobacter* and fungi) were obtained on agar plates (Zvyagintsev 1991).

Enzymatic activity: Dehydrogenase activity (potential) was determined colorimetrically by the presence of triphenylformazan from TTC (2, 3, 5-triphenyltetrazolium

[1] Haplic chernozem in *World reference base for soil resources 2006.*

[2] Arable land sown to grass or pasture.

chloride) added to air-dry soil (Haziev 2005). Polyphenol oxidase and peroxidase activities (potential) were determined colorimetrically using hydroquinone as a substrate (Karyagina and Mikhailovskaya 1986).

Soil chemical properties: Organic carbon was determined by dichromate oxidation; the humus content was estimated using the coefficient 1.724 (Arinushkina 1970).

Microbiological, enzymatic and chemical indices were evaluated by analysis of variance.

4.3 Results and Discussion

In Typical chernozem, soil organic carbon content and the values of most biological indices decreased from long-term lea to arable to black fallow (Table 4.1, Fig. 4.1). The highest numbers, biomass, activity and diversity of invertebrates and microorganisms, except the humus-mineralising microorganisms, are found under lea. This regularity is observed in the mean values of indicators as well as their confidence intervals.

The greatest number and diversity of invertebrates were found under lea; as many as 12 families were identified. The families of *Lumbricidae*, *Elateridae*, *Enchytraeidae*, *Chilopoda* and *Formicidae* were dominant under lea (Demchenko 2004). Earthworms comprised 33.8 % of the total invertebrates (and 63.1 % of saprophages), *Elateridae* 22.4 %, *Enchytraeidae* 19.0 % and *Chilopoda* 8.6 % without taking account of members of the ant family.

Long-term arable contributes to the degradation of soil biota. Simplification of the structure of faunal communities and diminished biodiversity were registered during the entire observation period. In the arable soil, the number of invertebrate families decreased to 8. The family of *Lumbricidae* occupies a dominant position but the number of saprophages diminished by a factor of 2.1, predatory species by 3.2 and phytophages by 5.8. This shows the destruction of trophic links in feed circuits.

Similarly, the structure of soil microbial communities was affected. There was a 2.5-fold increase in the number of humus-mineralising microorganisms. The ratio between bacteria and fungi increased from 190 to 342. Activities of soil enzymes also diminished: dehydrogenase by 1.4–1.6 and polyphenol oxidase to only 8 %.

The lowest values of biological parameters were observed in the soil under black fallow – the extreme variant of degradation under the plough. No invertebrates were found and *Azotobacter* was absent from the microbial community. Microbial biomass is 3.3 times lower than under lea and enzymatic activity was very low – in some years it stood at zero. The peroxidase activity is an exception that demonstrates the high degree of mineralisation of humus under black fallow.

The indicators of the edaphic fauna varied widely, both seasonally and in different sampling plots. The highest values of the variables were recorded under arable. The coefficients of variation increased from 22 to 31 % under lea to 40–52 % under arable. The largest variations were observed in the abundance and biomass of

Table 4.1 Soil invertebrates in Typical chernozem under different land management

Indices	Average values	Confidence intervals	V, %	n
10–23-year-old lea				
Number of invertebrates/m^2	356.0	317.0–395.0	22	11
Biomass of invertebrates, g/m^2	29.4	24.2–34.6	24	11
Number of *Lumbricidae*/m^2	141.7	112.5–171.0	28	11
Biomass of *Lumbricidae*, g/m^2	23.5	18.1–28.8	31	11
Arable				
Number of invertebrates/m^2	141.8	118.2–164.6	40	18
Biomass of invertebrates, g/m^2	17.8	12.3–23.4	48	18
Number of *Lumbricidae*/m^2	91.1	73.5–108.7	47	18
Biomass of *Lumbricidae*, g/m^2	16.5	11.3–21.7	52	18
30–43-year-old black fallow				
Number of invertebrates/m^2	0	0	0	12
Biomass of invertebrates, g/m^2	0	0	0	12

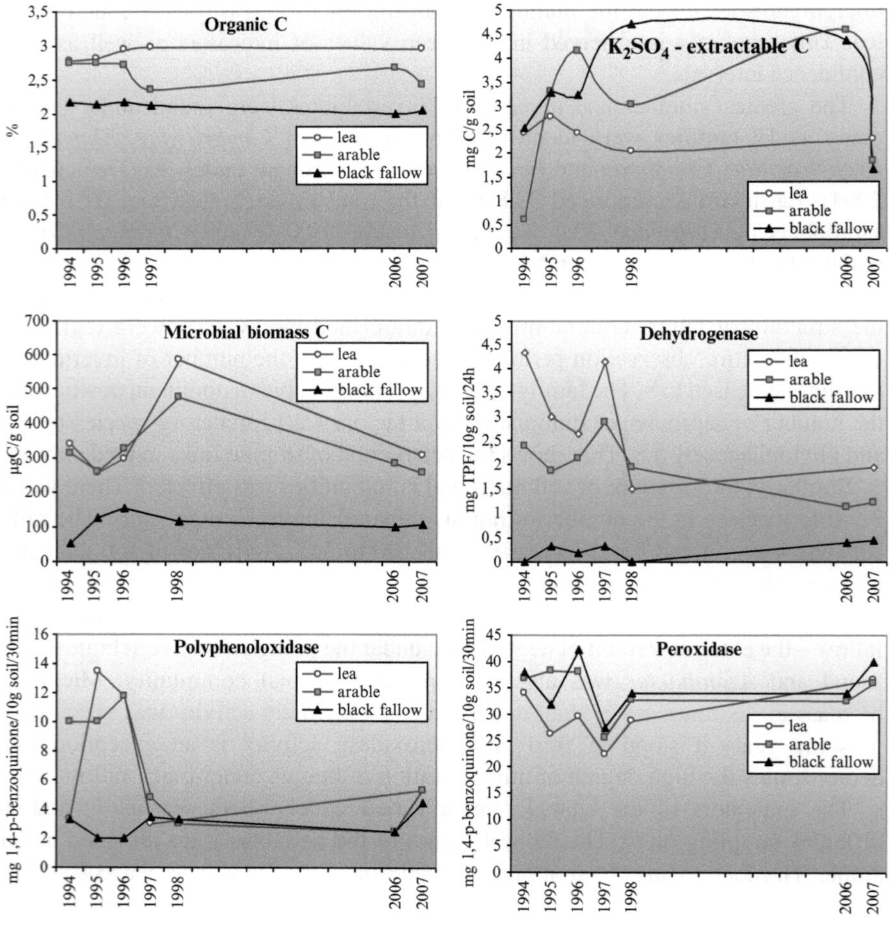

Fig. 4.1 Dynamics of organic C, K_2SO_4-extractable C, microbial biomass C and enzymatic activities of the Typical chernozem (0–30 cm) under different management

Table 4.2 Organic carbon, K_2SO_4-extractable C, microbial biomass C and ratios in Typical chernozem under different management, 0–30 cm (means, $n = 9–14$, $P \leq 0.05$)

Indices	Lea	Arable	Black fallow
Humus content, %	4.97	4.53	3.63
C_{org}, %	2.88	2.63	2.11
K_2SO_4-extractable C, mg C/g soil	2.38	2.91	3.29
Microbial biomass C, μg C/g soil	355.8	318.4	109.0
C_{MB}/C_{org}	1.23	1.21	0.52
C_{MB}/CK_2SO_4	14.94	10.94	3.32

MB microbial biomass = $C_{org} - K_2SO_4$-extractable C

Table 4.3 Scale of biological parameters for Typical chernozem (0–30 cm)

Degree of stability	Humus content, %	Invertebrates		MB, μg C/g soil	Humus-mineralising microorganisms, CFU/g soil*10^6	Dehydrogenase, mg TPF/10 g soil/24 h
		Number/m^2	Biomass g/m^2			
Very high	>5.1	>395	>35	>487	4.7–5.4	>3.6
High	4.9–5.1	317–395	24–35	275–487	5.4–7.5	2.2–3.6
Moderate	4.4–4.7	118–165	12–23	266–371	15.2–17.3	1.6–2.3
Low	3.5–3.7	0	0	82–136	2.6–4.7	0.1–0.4

earthworms. The increase of coefficients of variation in long-term arable chernozem indicates a decline in environmental sustainability. Although the three plots exhibited different levels of fertility, indicators of microbiological and enzymatic properties registered moderate and high variability. The coefficient of variation for microbial biomass C was 26–39 %; for numbers of microorganisms, 14–62 %; and for enzymatic activities, 14–91 %.

Soil organic carbon enters the soil through exudates from live roots, crop residues and carbon fixation by autotrophic microorganisms (Boincean 1999; Graham and Haynes 2006). Root mortality is generally considered to contribute most of the soil organic matter. As a result, contents of organic C, microbial biomass C and dehydrogenase activity were much greater under lea than under arable or black fallow (Fig. 4.1, Table 4.2). The absolute content of K_2SO_4-extractable C varied widely during the observation period: it was most constant under lea (2.03–2.76 mg C/g soil), but there was a tendency to increase under arable and under black fallow – which may be a result of the activation of autotrophic carbon fixation by soil algae. However, the ratio C_{MB}/CK_2SO_4 decreased from 14.94 under lea to 10.94 under arable and 3.32 under black fallow. A similar decrease was apparent in the share of the microbial C in the total organic carbon; the ratio C_{MB}/C_{org} decreased from 1.23 under lea to 0.52 under black fallow (Table 4.2).

Table 4.3 presents a scale to estimate the stability and quality of Typical chernozem, grouping the biological parameters of soils according to the humus content. The stability of the soil biota is greatest under seminatural lea – and least under black fallow. Arable chernozem under crop rotation, without fertiliser, occupies an intermediate position.

4.4 Conclusions

Degradation of chernozem under long-term arable is reflected in the decline of the humus content and soil biological properties in the sequence: 10–23-year-old lea > arable >30–43-year-old black fallow. The number and biomass of invertebrates diminished by a factor of 1.7–2.5. In the extreme conditions of black fallow land, microbial biomass carbon decreased by 3.3 compared with the soil under natural vegetation. In long-term arable chernozem, dehydrogenase activity went down by 1.5 and humus-destroying microorganisms dominate.

Multiannual fallow under natural vegetation promotes the conservation and regeneration of invertebrates and microorganisms; they have a high level of biomass and enzyme activity. A stable state of the biota is provided by a topsoil humus content of 4.9–5.1 %. The incorporation of areas with natural vegetation in a crop rotation created conditions for the improvement of the vital activity of the soil biota which had been degraded by long-term arable use.

A scale of biological parameters is proposed for Typical chernozem to assess the stability of biota and to establish the national soil quality standards.

References

Andries S, Banaru A, Filipciuc V et al. (2004) Complex program of recovery of degraded lands and increase of soil fertility. Part II. Pontos, Chisinau (Romanian)

Arinushkina EV (1970) Guide for chemical analysis of soils. MSU, Moscow (Russian)

Banaru A, Plamadeala V (2010) Evolution of the humus balance in arable soils 1971–2005. In: Cerbari V (coordinator) Andries S (editor) Monitoring soil quality of the Republic of Moldova (database, conclusions, prognosis, recommendations). Pontos, Chisinau (Romanian), pp 399–305

Bending GD, Turner MK, Rayns F et al. (2004) Microbial and biochemical soil quality indicators and their potential for differentiating areas under contrasting agricultural management regimes. Soil Biol Biochem 36:1785–1792

Blagodatsky SA, Blagodatskaya EV, Gorbenko AJ, Panikov NS (1987) Rehydration method for the determining of the microbial biomass in the soil. Pochvovedenie 4:64–71 (Russian)

Boincean BP (1999) Organic agriculture in the Republic of Moldova (crop rotation and soil organic matter). Stiinta, Chisinau (Russian)

Cameran K, Beare M, Mcharen K, Hong DI (1998) Selecting physical, chemical and biological indicators of soil quality for degraded or polluted soils. In: XVIII World congress of soil science, vol. 2. Montpellier, p 679

Demchenko E (2004) Perspectives of biological methods to enhance the stability of soil invertebrate communities in agricultural lands. In: Boincean BP (ed) Field crops – results and prospects: proceedings of the international scientific and practical conference 2004. Selectia RIFC, Balti (Russian), pp 242–245

Filip Z (2002) International approach to assessing soil quality by ecologically-related biological parameters. Agric Ecosyst Environ 88:169–174

Gilyarov MS (1965) Zoological method of diagnostics of soils. Nauka, Moscow (Russian)

Gilyarov MS, Striganova BR (eds) (1987) Quantitative methods in soil zoology. Nauka, Moscow (Russian)

Graham MH, Haynes RJ (2006) Organic matter status and the size, activity and metabolic diversity of the soil microbial community in the row and inter-row of sugarcane under burning and trash retention. Soil Biol Biochem 38:21–31

Haziev FH (2005) Methods of soil enzymology. Russian Academy of Sciences, Ufa, Moscow (Russian)

Karyagina LA, Mikhailovskaya NA (1986) Determination of polyphenoloxidase and peroxidase activities in the soil. J Acad Sci BSSR 2:40–41 (Belorussian)

Paz Jimenez M, Horra AM, Pruzzo L, Palma RM (2002) Soil quality: a new index based on microbiological and biochemical parameters. Biol Fertil Soils 35(4):302–306

Senikovskaya I, Marinescu C, Balan T (2010) Impact of the agricultural use on the biota of Typical chernozem. In: Cerbari V (coordinator) Andries S (editor) Monitoring soil quality of the Republic of Moldova (database, conclusions, prognosis, recommendations). Pontos, Chisinau (Romanian), pp 164–171

Trasar-Cepeda C, Leirós C, Gil-Sotres F, Seoane S (1998) Towards a biochemical quality index for soils: an expression relating several biological and biochemical properties. Biol Fertil Soils 26:100–106

Ursu A (2011) Soils of Moldova. Stiinta, Chisinau, pp 198–206 (Romanian)

Zvyagintsev DG (ed.) (1991) Methods of soil microbiology and biochemistry. MSU, Moscow (Russian)

Chapter 5
Essential Mass and Structure of Soil Microorganisms in the Black Earth

N. Frunze

Abstract New data are presented on the microbial biomass in chernozem soil and its prokaryote and eukaryote components. The essential mass of soil microorganisms varies between 537 and 705 μgC/g soil under a crop rotation including lucerne and between 385 and 585 μgC/g under the rotation without lucerne. The highest values of this index were been registered in the soil receiving long-term applications of farmyard manure. Fungi accounted for 63–80 % of the microbial biomass. The soil's C:N ratio conditioned the fungi to bacteria ratio: 1.8–3.9 under the rotation with lucerne and 1.7–3.6 under the rotation without lucerne; i.e. fungi to bacteria ratios were inversely proportional to the C:N ratio in the soil.

5.1 Introduction

The mass of microorganisms is the most active component of soil, responsible for maintenance and cycling of carbon and nutrients. It is also the most unstable component in terms of its reserves and structure, or composition. In general, fungi are thought to predominate (Anderson and Domsch 1975; West 1986; Frey et al. 1999) although Ananieva et al. (2006, 2008) affirm a preponderance of bacteria. The fungi to bacteria ratio is the structural parameter reflecting the humidity gradient in, for instance, soil use and management and the decomposition of plant residues (Beare et al. 1990; Bailey et al. 2002; Baath and Anderson 2003). It is sensitive to modifications of the environment such as pH (Baath and Anderson 2003), substrate (Blagodatskaya and Anderson 1998) and soil organic carbon content (Frey et al. 1999; Bailey et al. 2002) so it is widely used to assess the impact of management on the soil's living component. This chapter evaluates

---

N. Frunze (✉)
Institute of Microbiology and Biotechnology of MAS, Chisinau, Republic of Moldova
e-mail: ninafrunze@mail.ru

D. Dent (ed.), *Soil as World Heritage*, DOI 10.1007/978-94-007-6187-2_5,

microbial biomass and structure, in particular the relationship between soil fungi and bacteria, and the contribution of prokaryotes and eukaryotes to the microbial biomass.

5.2 Research Site and Methods

Microbial communities were investigated in the topsoil of eight variants of Typical chernozem within a field experiment at the MAS Biotron Experimental Station, near Chisinau in Moldova. The variants were a control, soil under long-term application of mineral fertilisers, soil fertilised with farmyard manure and with mixed fertilisation (manure + mineral fertiliser + crop residues); apart from the

Table 5.1 Crop rotations and fertiliser application

Year	Crop	Mineral fertiliser	Organic fertiliser	Organic + mineral fertiliser
Crop rotation with lucerne				
1995, 2002	1st year lucerne	$N_{90}P_{56}K_{258}$	Manure 10 t/ ha = $N_{90}P_{27}K_{235}$	Manure 10 t/ha = $N_{90}P_{27}K_{235}$
1996, 2003	2nd year lucerne	$N_{160}P_{100}K_{458}$	Residual effect	Residual effect
1997, 2004	3rd year lucerne	$N_{100}P_{62}K_{286}$	"	Green manure
1998, 2005	Winter wheat	$N_{156}P_{68}K_{0}$	"	Residual effect = $N_{45}P_{10}K_{47}$ + wheat straw
1999, 2006	Maize for grain	$N_{185}P_{82}K_{0}$	Manure10 t/ ha = $N_{90}P_{27}K_{235}$	Manure 10 t/ha = $N_{90}P_{27}K_{235}$
2000, 2007	Maize silage	$N_{145}P_{68}K_{145}$	Residual effect = $N_{90}P_{13}K_{70}$	Residual effect = $N_{90}P_{13}K_{70}$
2001, 2008	Winter wheat	$N1_{56}P_{0}K_{0}$	Residual effect = $N_{45}P_{10}K_{47}$	Residual effect = $N_{45}P_{10}K_{47}$ + wheat straw
Crop rotation without lucerne				
1995, 2002	Fodder beet	$N_{143}P_{48}K_{285}$	Manure10 t/ ha = $N_{90}P_{27}K_{235}$	Manure 10 t/ha = $N_{90}P_{27}K_{235}$
1996, 2003	Maize silage	$N_{89}P_{42}K_{81}$	Residual effect = $N_{90}P_{13}K_{70}$	Residual effect = $N_{90}P_{13}K_{70}$
1997, 2004	Winter wheat	$N_{141}P_{62}K_{90}$	Residual effect = $N_{45}P_{10}K_{47}$	Residual effect = $N_{45}P_{10}K_{47}$+ catch crop of peas = $N_{27}P_{34}K_{22}$
1998, 2005	Soya beans	$N_{54}P_{36}K_{60}$	No residual effect	Green manure from soya = $N_{27}P_{34}K_{22}$
1999, 2006	Spring barley	$N_{120}P_{45}K_{60}$	No residual effect	Barley straw = $N_{27}P_{34}K_{22}$
2000, 2007	Maize for grain	$N_{156}P_{68}K_{110}$	Manure 10 t/ ha = $N_{90}P_{27}K_{235}$	Manure 10t/ha = $N_{90}P_{27}K_{235}$
2001, 2008	Winter wheat	$N_{141}P_{62}K_{99}$	Residual effect = $N_{90}P_{13}K_{70}$	Residual effect = $N_{90}P_{13}K_{70}$ + wheat straw = $N_{27}P_{34}K_{22}$

control, plots ("sample" in Figs. 5.1 and 5.2) received equivalent amounts of NPK. Each variant was sampled under crop rotations with and without a perennial legume. Over the period 2006–2008, soil samples were taken from the 0–20 cm layer, three times during the growing season, from plots under forage crops (Table 5.1). While still at field moisture content, the soil samples were passed through a 2 mm sieve.

The separate contribution of fungi and bacteria to the respiration induced by the substrate (SIR) was determined by selective inhibition by antibiotics which stopped growth without inhibiting the respiration of the microorganisms (Anderson and Domsch 1975). SIR was evaluated against the initial maximum rate of soil respiration after enrichment with glucose: Glucose solution (10 mg/g) was added to 5 g soil in a flask; the sample was incubated for 4 h at 22 °C then a 2 ml aliquot of air extracted and analysed in a chromatograph. Antibiotics (cycloheximide, a fungicide, and streptomycin, a bactericide) (2 mg/g), were introduced separately and together; glucose solution was added and SIR was determined after a further period of incubation. A soil sample to which only glucose was added served as a control. All measurements were made on five replicates.

The fungi to bacteria ratio contributing to SIR was determined according to the formulae: $C = (A - V)/(A - \mathrm{D}) \times 100\ \%$ and $B = (A - C)/(A - D) \times 100\ \%$ – where A is respiration (emission of CO_2) of the soil with glucose; B is respiration of the soil with glucose + fungicide; C is respiration of the soil with glucose + bactericide; and D is respiration of the soil with glucose + bactericide + fungicide. Microbial mass was calculated as C_{mic}, μg/g = μgC – CO_2/g dry soil/h; i.e. SIR, being multiplied by the experimentally established value 40.04 + 0.37, was transformed to microbial mass, expressed in carbon units (Anderson and Domsch 1975). The data were processed following the procedures given by Comarov et al. (2000).

5.3 Results and Discussion

The mass of the microorganisms varied between 536.97 and 705.07 μgC/g under the crop rotation with lucerne and between 384.75 and 584.95 μgC/g without lucerne (Fig. 5.1). The greatest biomass was in the rotation with lucerne under long-term manuring. There were distinct seasonal changes: Soil microorganisms metabolised more glucose and achieved their maximum mass in the spring; SIR slowed down considerably in summer but there was resurgence in autumn – though not to spring levels.

Differences in functional activity are evidence of metabolic diversity and, also, different structure within microbial communities. The biggest microbial biomass, under long-term application of farmyard manure, was conditioned not only by the high specific metabolic activity (which did not differ from the variant with mineral fertilisers) but by its great reserves of microbial biomass – confirmed by independent determination by the rehydration method (Frunze 2005).

The microbial community under mineral fertilisation was characterised by the development of microorganisms *following* addition of glucose – evidence of a

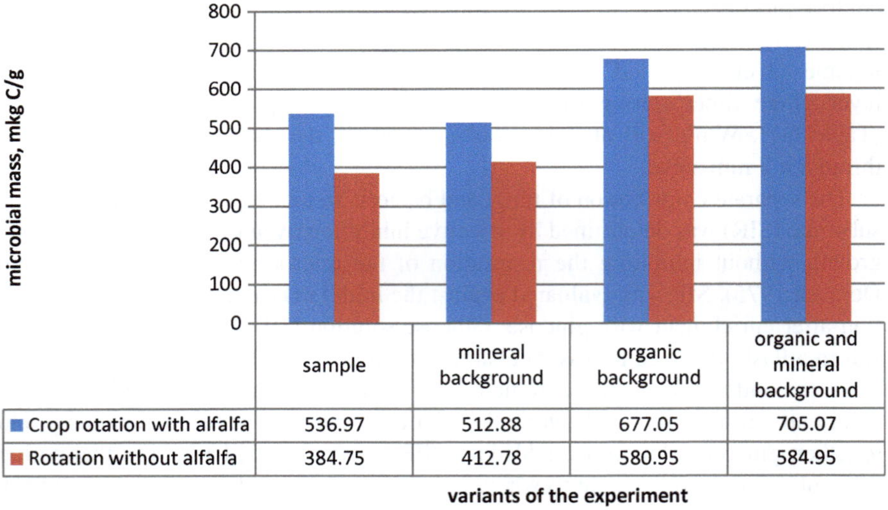

Fig. 5.1 Microbial biomass of the black earth

Table 5.2 Eukaryote and prokaryote biomass (μgC/g), fungi to bacteria ratio and physicochemical properties of the typical black earth (average for 3 years)

				Biomass, mkg C/g		
Variant	Humus %	C/N	pH	Fungi	Bacteria	Fungi/bacteria
Crop rotation with lucerne						
Control	3.0	8.60	8.2	340.71	176.55	1.93
Mineral fertiliser	3.0	7.03	8.2	328.70	180.55	1.82
Organic fertiliser	3.3	4.80	8.2	564.93	156.53	3.61
Organic + mineral	3.4	4.40	8.3	576.95	148.52	3.88
Crop rotation without lucerne						
Control	2.8	7.17	8.2	232.60	136.51	1.70
Mineral fertiliser	2.9	6.10	8.2	264.63	144.51	1.83
Organic fertiliser	3.2	4.49	8.2	420.79	116.49	3.61
Organic + mineral	3.1	4.20	8.3	416.79	140.51	2.97

predominance of slowly growing microorganisms (k-strategists) able to exploit the substrate to the utmost. In unfertilised soil, the microbial community is a complex of microorganisms, most of the time inert but able to grow rapidly to take advantage of any improvement (r-strategists).

Antibiotics induced differential changes in the structure of the soil microbial biomass (Table 5.2). The largest mass was registered following administration of bactericides: 340.71–576.95 μgC/g under the rotation with lucerne and 232.60–416.79 μgC/g without lucerne. This indicates that, in black earth, bacterial biomass is a minor component: 148.52–176.55 μgC/g in the crop rotation with lucerne and 116.49–144.51 μgC/g without lucerne. The fungi to bacteria ratio is

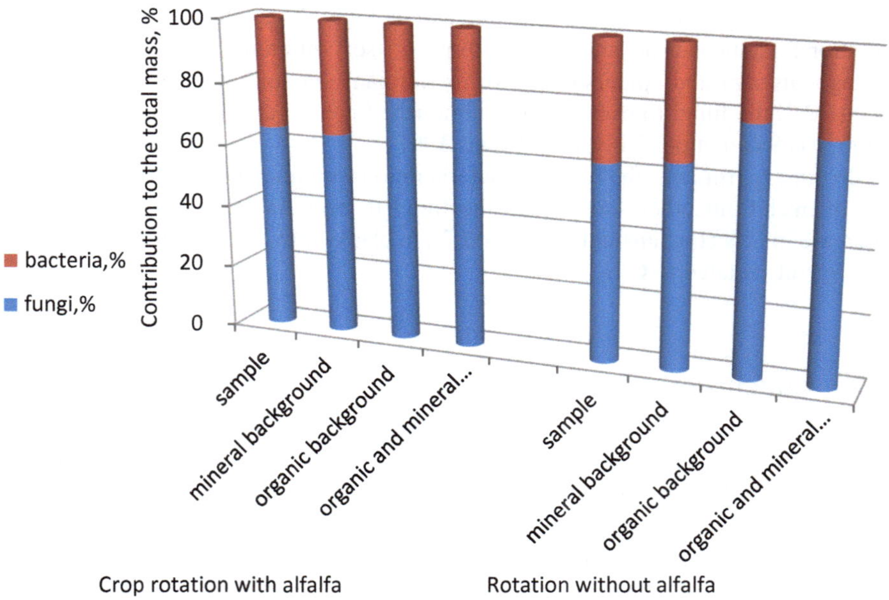

Fig. 5.2 The correlation fungi/bacteria in the microbial mass of typical black earth

1.82–3.88 under the rotation with lucerne and 1.70–3.61 without lucerne, with the highest ratios in the manured soil.

Comparison with the literature underlines the great reserves of microbial biomass in the black earth, which may be considered a factor of stability of the soil ecosystem (Polianskaia et al. 1995). Our data demonstrate that microbial biomass is diminished in arable soils; application of manure and fertiliser creates more favourable conditions but the microbial biomass remains less than in the natural ecosystem (Frunze 2005). Nevertheless, despite long agricultural usage, the black earth has preserved a pool of microorganisms with high potency for reactivation. According to Blagodatsky et al. (1994), it is typical of a small pool of potentially active microorganisms without pronounced r- or k-strategists. Blagodatskaya et al. (2003) consider that with the cultivation on glucose the microbial community possesses at least some of the features of k-strategists – corresponding to the rule of ecology concerning the interconnection of the habitat conditions with the characteristics of its biota and Odum's hypothesis that the speed of the growth of microorganisms should diminish in the more mature ecosystems.

In Typical chernozem under crop rotation, assay of microbial biomass by SIR shows a dominance of the fungi: 62.78–79.45 % of the total biomass of soil microorganisms (Fig. 5.2). These results are comparable with those obtained from black earths elsewhere.

It is noteworthy that low fungi to bacteria ratios (0.5–0.6) have been observed in arable soils (Velvis 1997) and under pasture (1.0) (Bailey et al. 2002), while the ratio is higher under coniferous forest (1.1) and much greater in prairie (13.5). On the basis of such results, Bailey suggested that there is better maintenance of soil organic carbon in the soils dominated by fungi. Vegetation and pH also have a strong influence on the composition of soil microorganism communities (Blagodatskaya and Anderson 1998); in Germany on soils with pH values from 3.0 to 7.2, the fungi to bacteria ratio diminished greatly at higher pH; at pH 3 the ratio was 9 and at pH 7 it was 2 (Velvis 1997).

In our experiments, the fungi to bacteria ratio is modified by the soil's C:N ratio. As distinct from other soils where the fungi to bacteria ratio increases with an increase in C:N (Ingham and Horton 1987), in black earth the fungi to bacteria ratio is inversely related to C:N.

5.4 Conclusions

1. The dominant contribution to the emission of CO_2 by soil microorganisms in typical black earth under forage crops is made by fungi (62.8–79.4 %). Therefore, we conclude that communities of microorganisms in the black earth preponderantly consist of fungi (eukaryotes).
2. The fungi to bacteria ratio was 1.82–3.88 under the crop rotation with lucerne and 1.70–3.61 under the rotation without lucerne. The highest values of this index have been registered in the soil regularly treated with organic fertilisers.
3. Values of the fungi to bacteria ratio were inversely average proportional to the C:N ratio in the soil.
4. Forage crops had a positive impact on the structural composition of the soil microorganism communities and on the biomass mass of fungi.

References

Ananieva ND, Susian EA, Chernova OV, Chernov II, Macarova OL (2006) The correlation of fungi and bacteria in the biomass of various types of soils, determined by selective inhibition. Microbiology 75(6):807–813 (Russian)

Ananieva ND, Poleanskaia LM, Susian EA et al (2008) A comparative evaluation of microbial biomass of the soils, determined by the methods of direct and substrate-induced respiration. Microbiology 77(3):404–412 (Russian)

Anderson JPE, Domsch KH (1975) Measurement of bacterial and fungal contribution to respiration of selected agricultural soils. Can J Microbiol 21:314–322

Baath E, Anderson TH (2003) Comparison of soil fungal/bacterial ratios in a pH gradient using physiological and PLFA-based techniques. Soil Biol Biochem 35(7):955–963

Bailey VL, Smith JI, Bolton H (2002) Fungal-to-bacterial biomass ratios in soils investigated for enhanced carbon sequestration. Soil Biol Biochem 34:997–1007

Beare MH, Neely CL, Coleman DC, Hargroove WL (1990) A substrate-induced respiration (SIR) method for measurement of fungal and bacterial biomass on plant residues. Soil Biol Biochem 22(5):585–594

Blagodatskaya EV, Anderson TH (1998) Interactive effects of pH and substrate quality on the fungal-bacterial ratio and qCO_2 of microbial communities in forest soils. Soil Biol Biochem 30(10/11):1269–1274

Blagodatskaya EV, Hohlova OS, Anderson TH, Blagodatsky SA (2003) The microbial DNA pool and microbiological activity of paleo-soil of the southern Ural Region. Microbiology 72(6):847–853 (Russian)
Blagodatsky SA, Blagodatskaia EV, Rozanova LN (1994) Kinetics of the strategy of growth of microorganisms in the soil after a long-term usage of various systems of fertilizers. Microbiology 63(2):298–307 (Russian)
Comarov AS, Grabarnik PA, Galitsky VV (2000) The analysis of the results of observations. Materials on mathematical support for computers, Pushkino (Russian)
Frey SD, Elliot ET, Paustian K (1999) Bacterial and fungal abundance and biomass in conventional and no tillage agroecosystems along two climatic gradients. Soil Biol Biochem 31(4):573–585
Frunze NI (2005) The soil microbial biomass as a reserve of biogenic elements. Agrochemistry 9:20–23 (Russian)
Ingham ER, Horton KA (1987) Bacterial, fungal and protozoan responses to chloroform fumigation in stored soil. Soil Biol Biochem 19(5):545–550
Polianskaia LM, Golovchenko AV, Zveaghintsev DG (1995) The microbial biomass in the soils. Rep Acad Sci 144:846–848 (Russian)
Velvis H (1997) Evaluation of the selective respiratory inhibition method for measuring the ratio of fungal:bacterial activity in acid agricultural soils. Biol Fertil Soils 25:354–360
West AW (1986) Improvement of the selective inhibition technique to measure eukaryote-prokaryote ratios in soils. Microbiol Methods 5:125–138

Chapter 6
Effects of Long-Term Fertility Management on the Soil Nematode Community and Cyst Nematode *Heterodera schachtii* Population in Experimental Sugar Beet Fields

L. Poiras, E. Iurcu-Straistraru, A. Bivol, N. Poiras, I. Toderas, M. Bugaciuc, and B.P. Boincean

Abstract Nematode faunal responses were studied under long-term, continuous sugar beet and crop rotations without fertilizer and with the combined application of manure with mineral fertilizers. Differences attributable to the different treatments were assessed in terms of abundance, species diversity of nematode communities, and functional guilds combining feeding groups and life strategy. Continuous sugar beet gave rise to an increasing abundance of nematodes, especially plant-parasitic nematodes and, relatively, bacterivores; and the number of *Heterodera schachtii* cysts exceeded the threshold of economic damage. Under crop rotations including grains, in combination with organic manure and nitrogen fertilizer, plant-parasitic nematodes decreased and the *H. schachtii* population was below the damage threshold. In the experimental plots with fertilizers, the nematode bacterivores guild Ba2 (Cephalobidae and Plectidae) and fungivores guild Fu2 (Aphelenchus and Aphelenchoides) were numerous; however, plant-parasitic nematodes from guilds PP3, Hoplolaimidae (Heteroderinae) and Pratylenchidae, and partly PP2, Anguinidae and Dolichodoridae, decreased.

6.1 Introduction

The cyst nematode *Heterodera schachtii* Schmidt is one of the most dangerous pests of sugar beet (*Beta vulgaris* L. ssp. *saccharifera*). Symptoms of nematode infestation include patchy wilting of the leaves, with yellowing and death of the outer leaves, a stunted taproot with an increased lateral root formation (root beard), and the presence of the pinhead-sized white females and brown cysts; field

L. Poiras (✉) • E. Iurcu-Straistraru • A. Bivol • N. Poiras • I. Toderas
ASM Institute of Zoology, Chisinau, Republic of Moldova
e-mail: poiras@yahoo.co.uk; poiras@gmail.com

M. Bugaciuc • B. Boincean
Selectia Research Institute of Field Crops, Balti, Republic of Moldova

D. Dent (ed.), *Soil as World Heritage*, DOI 10.1007/978-94-007-6187-2_6,

symptoms are patches of stunted plants with yellowing leaves and other symptoms of nutrient deficiency. With a nematode population of 300–400 eggs and second-stage juveniles per 100 g of dry soil, yield losses may be more than 50 % (Baldwin and Mundo-Ocampo 1991; Curto 2008; Okada et al. 2004; Sigareva 1977). In Moldova, the study of nematodes in sugar beet was begun by Nesterov in the second half of the last century. He found that damaging *H. schachtii* infestations are related to the weather; a long period of soil temperature above 10 °C increases the number of generations of the pest; the development cycle lasts 30–45 days and there may be 3–4 generations per year (Nesterov 1970, 1973).

Today, farmers can choose between several agronomic techniques that will control cyst nematodes if correctly and punctually applied. These include crop rotation with a long period between successive beet crops and long-term application of farmyard manure and nitrogen fertilizers (Boincean 1999; Neher 2001; Perry and Moens 2006; Poiras et al. 2010). Every year, the nematode larvae hatch spontaneously from cysts and, if they find no host, the population is reduced. Many studies of crop rotation for sugar beet propose the inclusion of non-susceptible plants for 3 or 4 years between sugar beet planting to reduce the nematode population (Curto 2008). The objectives of this study were to compare the long-term effects of a combination of manure and nitrogen fertilizers and crop rotation (maize – wheat – sugar beet) on the density of juveniles and cysts of *H. schachtii* in the long-term field experiments of Selectia RIFC at Balti and to evaluate the relationship between soil proprieties and nematode communities.

6.2 Material and Methods

The species diversity of nematode communities was studied in 24-year experimental plots of continuous sugar beet and in crop rotations with and without manure and nitrogen fertilizers. The experimental plots comprised (1) unfertilized continuous sugar beet (P); (2) continuous sugar beet with application of nitrogen fertilizer $N_{60}P_{30}K_{30}$ (N) and 40 t/ha of organic manure (M) (P + M + N); (3) unfertilized three-field rotation of maize, wheat, and sugar beet (CR); and (4) three-field rotation of maize, wheat, and sugar beet with application of 40 t/ha manure and fertilizer (CR + M + N). Soil samples were taken from 0–20 cm depth in each plot in autumn of 2010 (before harvest). Nematodes and cysts were extracted from 100 cm^3 of soil and 100 g of sugar beet roots by a modified Baermann and Fenwick procedure (Bezooijen 2006).

At least 150 nematodes were identified from each sample and assigned to trophic groups, characterized by feeding habits (Yeates et al. 1993): bacterivores (Ba), fungivores (Fu), omnivore-carnivores (Om-Ca), and plant parasites (PP). To analyze the community structures, nematode families were allocated to functional guilds (Ferris et al. 2001) defined as combinations of feeding groups and life strategy by *cp* values from extreme r-strategy to k-strategy using the maturity index $MI = \sum v(i) \times f(i)$, where $v(i)$ is cp value of taxon i according to their r and k characteristics, $f(i)$ is the frequency of taxon i in a sample and by the plant-parasitic index (PPI) which was determined according to the ratio of plant-parasitic genera and MI (Bongers 1990).

6.3 Results and Discussion

Altogether, 44 species of plant-parasitic and free-living nematodes were encountered with densities of 550–3,500 individuals/100 cm^3 soil and different numbers of cysts of *H. schachtii*, depending on the different agronomic treatments. According to trophic groups, plant parasites and bacterivores were characterized by species diversity and abundance, especially in continuous sugar beet culture (Table 6.1, Fig. 6.1).

Figure 6.2 illustrates the morphology, biology, and pathogenesis of *H. schachtii*. In the continuous sugar beet plots (P), the number of *H. schachtii* cysts (1,000–2,000 or more) exceeded the threshold of economic damage. Endoparasitic species of genus

Table 6.1 Species and abundance of nematodes in long-term experimental plots

	Treatments			
	P	P + M + N	CR	CR + M + N
Number of species in plots	34	31	28	29
Trophic groups: Plant parasites (PP)	10	9	9	9
Bacterivores (Ba)	11	10	9	10
Fungivores (Fu)	5	6	5	5
Omnivore-carnivores (Om-Ca)	8	6	5	5
Maturity index (MI)	2.65	2.55	2.5	2.2
Ratio indexes PPI/MI	0.86	0.84	0.84	0.95
No. individuals/100 cm^3 soil	1,720–3,500	950–1,680	640–1,120	550–1,320
No. cysts/100 cm^3 soil	1,450–2,000	750–1,500	0–120	0–68

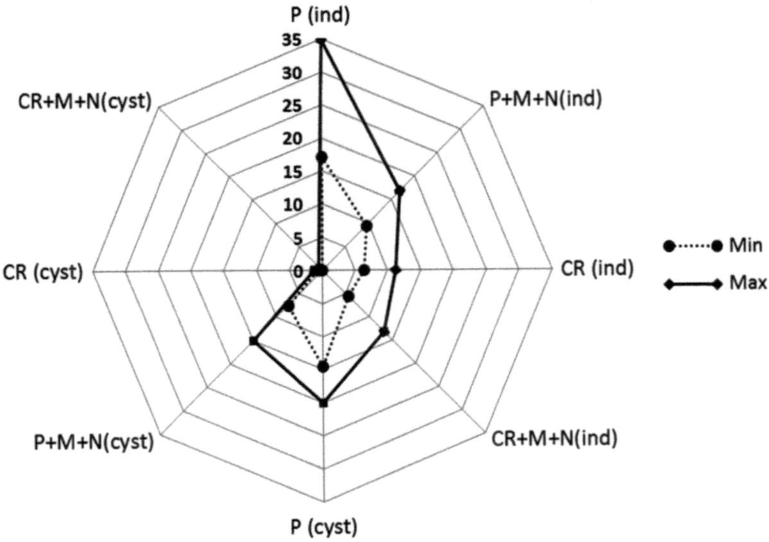

Fig. 6.1 Number of individuals of nematode communities and cysts of *H. schachtii* ($\times 10^2$/100 cm^3 soil) according to plot treatment

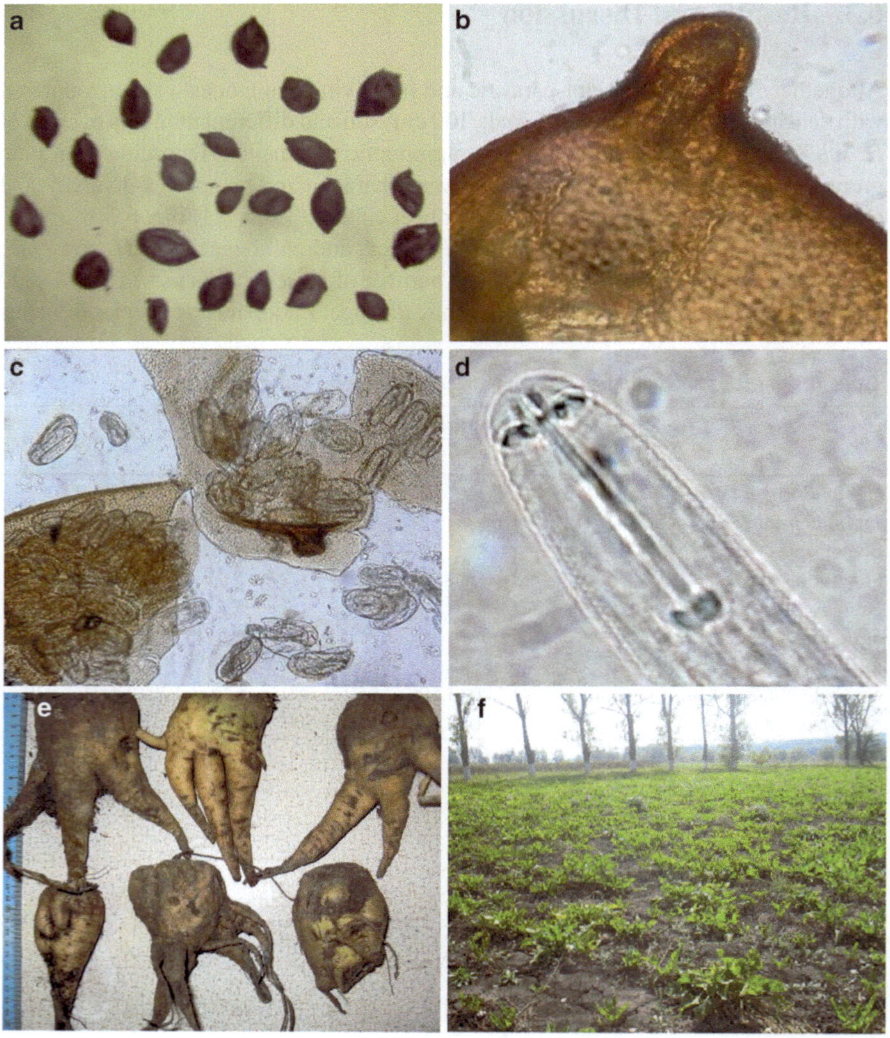

Fig. 6.2 *Heterodera schachtii* (**a**) cysts, (**b**) anterior cyst, (**c**) invasive second-stage juvenile (J2) coming from destroyed cyst, (**d**) head of J2 with strong stylet, (**e**) deformed sugar beet roots after *H.schachtii* infestation (**f**) typical field symptoms of infestation by cyst nematodes

Pratylenchus (*P. pratensis, P. subpenetrans*) and ectoparasite *Paratylenchus nanus* were also numerous. Crop rotations including small grains and maize in combination with organic manure and nitrogen fertilizers (CR + M + N) effectively reduced the numbers of *H. schachtii* to 15–20 cysts/100 cm^3 (well below the damage threshold).

The application of manure and nitrogen fertilizers changed the nematode community indices: values of MI (2.65) increased in continuous sugar beet and decreased in crop rotation with fertilizers (2.2), although mineral fertilizers slightly depressed omnivore-carnivore nematodes. The ratio PPI:MI varied from optimal use of nutrient sources (0.84) to slight nutrient disturbance (0.95).

Table 6.2 shows the assignment of nematode families and genera to functional guilds.

Table 6.2 Nematode assemblages under different fertilization treatments

		Treatments[b]			
Species of nematodes	Guild[a]	P	P + M + N	CR	CR + M + N
PP					
Tylenchus davainei	PP2	+	+	+	+
Filenchus filiformis	PP2	+	+	+	+
Nothotylenchus acris	PP2	+	+	+	+
Lelenchus minutus	PP2	+	–	–	–
Helicotylenchus dihystera	PP2	–	+	–	–
H. crenatus	PP2	+	–	–	–
Ditylenchus misselus	PP2	+	+	+	+
Pratylenchus brachyurus	PP3	+	–	–	–
P. pratensis	PP3	+	+	+	+
P. subpenetrans	PP3	–	+	+	+
Bitylenchus dubius	PP2	–	–	+	+
Heterodera schachtii	PP3	+	+	+	+
Paratylenchus nanus	PP2	+	+	+	+
Fu					
Aphelenchus avenae	Fu2	+	+	+	+
A. paramonovi	Fu2	–	–	+	+
Aphelenchoides composticola	Fu2	–	+	–	–
A. limberi	Fu2	+	+	+	+
A. parietinus	Fu2	+	+	+	+
A. subtenuis	Fu2	+	+	–	–
A. saprophilus	Fu2	+	+	+	+
Om-Ca					
Eudorylaimus maritus	Om4	+	+	+	+
E. papillatus	Om4	+	+	–	–
Ecumenicus monohystera	Om4	+	+	+	+
Mesodorylaimus centrocercus	Om4	+	–	+	+
Crassolabium ettersbergensis	Om4	–	+	+	+
Prodorylaimus sp.	Om4	+	+	–	–
Aporcelaimellus paraobtusicaudatus	Om5	+	–	+	+
Mononchus sp.	Ca4	+	+	–	–
Mylonchulus brachyuris	Ca4	+	–	–	–
Ba					
Cephalobus persegnis	Ba2	+	+	+	+
Chiloplacus propinquus	Ba2	+	–	–	–
Eucephalobus oxyuroides	Ba2	–	–	+	+
E. striatus	Ba2	+	+	+	+
Heterocephalobus elongatus	Ba2	+	+	–	+
A.buetschlii	Ba2	+	+	+	+
Acrobeloides nanus	Ba2	+	–	+	+

(continued)

Table 6.2 (continued)

Species of nematodes	Guild[a]	Treatments[b] P	P + M + N	CR	CR + M + N
Acrobeloides tricornis	Ba2	+	–	+	+
Alaimus primitivus	Ba2	+	+	–	–
Mesorhabditis signifera	Ba1	+	+	+	+
Aulolaimus oxycephalus	Ba2	–	+	–	–
Anaplectus granulosus	Ba2	+	+	–	–
Plectus cirratus	Ba2	–	+	+	+
Plectus parietinus	Ba2	–	+	+	+
Plectus rhizophilus	Ba2	+	–	–	–

[a]Functional guilds: *Ba* bacterivores, *Fu* fungivores, *PP* plant parasites, *Om* omnivores, *Ca* predators, number in functional groups indicate *cp* values
[b]Treatments: *P* permanent culture, P + M + N permanent culture + manure + N fertilizer, *CR* crop rotation, CR + M + N, crop rotation + manure + N fertilizer

The allotment of nematodes into guilds with the similar trophic function and life strategy provides a basis for applying nematode faunal analyses to an integrated assessment of the condition of the soil food web (Ferris et al. 2001; Liang et al. 2009; Yeates et al. 1993). To describe the long-term changes in soil conditions, we exclude opportunistic bacterivores (Ba1) that reproduce rapidly in response to nutrient conditions in the soil; in our study, Ba1 was represented only by the genus *Mesorhabditis* but replaced by numerous Ba2 (*Cephalobidae* and *Plectidae*) and Fu2 (*Aphelenchus* and *Aphelenchoides*) that are relatively tolerant to chemical stress or polluted soils and, so, may have survived fumigation deeper in the soil and recovered rapidly. *Cephalobidae* require fewer food bacteria to maintain their population and are less affected by food species than *Rhabditidae;* they tend to be most abundant in soils with fewer pores or more clay (Liang et al. 2009). Organic manure and nitrogen fertilizers greatly increased the abundance of Ba2 bacterivores, relatively reduced omnivore-carnivores (Om4, Om5 and Ca4), and reduced numbers of plant parasites (PP3 and, partly, PP2). Plant-parasitic nematodes are potentially more responsive to the presence of host plants than to soil amendment; in this case, crop species may have influenced nematode community structure more than management practices (Neher 2001). In these experimental plots, the nematode bacterivores Ba2, which are natural antagonists of the sugar beet cyst nematode, were numerous.

6.4 Conclusions

Compared with unfertilized, continuous sugar beet, long-term application of farmyard manure and nitrogen fertilizer and crop rotation change the structure of nematode communities by increasing the number of bacterivorous *Cepholabidae* (Ba2) and fungivorous *Aphelenchidae* (Fu2), thereby suppressing the plant-parasitic species especially the beet cyst nematode *Heterodera schachtii*.

References

Baldwin JG, Mundo-Ocampo M (1991) *Heteroderinae* cyst- and non-cyst-forming nematodes. In: Nickle WR (ed) Manual of agricultural nematology. Marcel Dekker, New York

Bezooijen JV (2006) Methods and techniques for nematology. Wageningen University, Wageningen

Boincean BP (1999) Ecological agriculture in the Republic of Moldova (crop rotation and soil organic matter). Stiinta, Chisinau (Russian)

Bongers T (1990) The maturity index: an ecological measure of enviromental disturbance based on nematode species composition. Oecologia 83:14–19

Curto G (2008) Sustainable methods for management of cyst nematodes. In: Ciancio A, Mukerji KG (eds) Integrated management and biocontrol of vegetable and grain crops nematodes. Springer, Dordrecht, pp 221–237

Ferris H, Bongers T, Goede RGM (2001) A framework for soil food web diagnostics: extension of the nematode faunal analysis concept. Appl Soil Ecol 18:13–29

Liang W, Lou Y, Li Q et al (2009) Nematode faunal responseto long-term application of nitrogen fertilizer and organic manure in Northeast China. Soil Biol Biochem 41:883–890

Neher D (2001) Role of nematodes in soil health and their use as indicators. J Nematol 33:161–168

Nesterov PI (1970) Plant-parasitic nematodes – pests of agricultural plants in the Republic of Moldova. ASMSSR, Chisinau (Russian)

Nesterov PI (1973) Sugar beet nematodes. Stiinta, Chisinau (Russian)

Okada H, Harada H, Kadota I (2004) Application of diversity indices and ecological indices to evaluate nematode community changes after fumigation. Jpn J Nematol 34(2):89–98

Perry RN, Moens M (2006) Plant nematology. CABI, Wallingford

Poiras L, Iurcu-Straistraru E, Antofica A et al (2010) Structure of phyto-nematode communities of sugar beet depending on crop rotation and continuous culture. In: Scientific papers in horticulture, viticulture, forestry & plant protection. State Agrarian University, Chisinau, pp 429–435 (Romanian)

Sigareva DD (1977) Infestation of plant parasitic nematodes of winter wheat, rye and sugar beet grown in rotation and monoculture. In: Collected papers: free-living, soil, entomopathogenic and plant-parasitic nematodes, L, Proceeding of the Zoological Institute, USSR Academy of Sciences, Leningrad, pp 61–67 (Russian)

Yeates GW, Bongers RG, Goede RGM et al (1993) Feeding habits in soil nematode families and genera – an outline for soil ecologists. J Nematol 25(3):315–331

References

Chapter 7
Biochemical Parameters of Arable Chernozem in Long-Term Field Experiments

E. Emnova

Abstract The biochemistry of cultivated chernozem was investigated using microbial eco-physiological indicators and specific enzymatic activities. Soil-specific biochemical properties (soil FDA hydrolase, urease and protease activities) responded to fertilizer application more than general biochemical attributes (microbial biomass carbon, basal soil respiration, microbial and metabolic quotients).

Compared with Typical chernozem in northern Moldova, Calcareous chernozem in the south-east of the country is more stressed but, despite lesser stocks of soil organic carbon, it provides more carbon for microbial growth and maintenance. Under maize, the microbial quotient of Calcareous chernozem (1.7–1.8 %) was twice that in Typical chernozem; basal respiration (2.33 μgCO_2-C/g dry soil/ha) was 3.0–4.5 times higher; and the metabolic quotient reached 9.2–9.4 mg CO_2-C/g C_{mic}/ha, 2.6–3.8 times higher. On Calcareous chernozem, ploughless cultivation contributed to increases in soil organic carbon, microbial biomass and microbial quotient – but did not increase crop yields.

7.1 Introduction

Biochemical processes drive the carbon and nutrient cycles, so soil biochemical attributes lend insights into soil quality (Leirós et al. 2000; Gil-Sotres et al. 2005). Nannipieri and others (1995, cited by Trasar-Cepeda et al. 2008) group soil biochemical properties into general parameters directly related to microbial activity (microbial biomass C and N, microbial respiration, N mineralization capacity, etc.) and specific parameters (which include the activities of extracellular hydrolytic enzymes involved in the C, N, S and P cycles) which are rather more independent of microbial activity. Soil microbial biomass is involved in many and various soil processes – including decomposition of organic residues, nutrient cycling,

E. Emnova (✉)
Institute of Genetics and Plant Physiology, MAS, Chisinau, Republic of Moldova
e-mail: kateemnova@mail.ru

D. Dent (ed.), *Soil as World Heritage*, DOI 10.1007/978-94-007-6187-2_7,

mobilization of nutrients (particularly phosphates), degradation of xenobiotic compounds and pollutants, soil structure, carbon storage and biological control and suppression of plant pathogens – so we may consider it to be an indicator of soil quality.

Microbial biomass carbon (C_{mic}) and its related parameters – microbial quotient (C_{mic}:C_{org}), basal soil respiration (CO_2-C rate) and metabolic quotient (qCO_2) – have been proposed as eco-physiological indicators (Anderson 2003; Anderson and Domsh 2010). They have been widely applied to investigation of microbial responses to soil management practices (Corcimaru et al. 2011; Emnova 2012; Emnova et al. 2010, 2011a; Kaschiuk et al. 2010; Leirós et al. 2000; Merenuic et al. 2008; Senicovscaia et al. 2008). Although soil quality is an amalgam of many properties, most workers choose just a few to assess it (Emnova et al. 2011b). Here, we use microbial eco-physiological indicators and some specific soil enzymatic activities to investigate the biochemistry of Moldavian chernozem under modern farm management, represented by long-term field experiments.

7.2 Experimental Sites and Methods

Moldova lies within the temperate continental climatic zone: mean annual temperature 9.3 °C, mean January temperature −4 °C and mean summer temperature about 20 °C; mean annual rainfall ranges from 500–600 mm in the north to 450–500 mm in the south. Chernozem soils comprise more than 78 % of the agricultural land (National Program for Soil Fertility 2005). One of the long-term field experiments on Typical chernozem[1] at RIFC Selectia, at Balti in the north of the country, is a six-field rotation established in 1991 (Boincean et al. 2004, 2007 and this symposium) that includes variants with organic (Org) and mixed mineral-organic (Min + Org) fertilization. These two were selected for comparative research on soil biochemical properties (Table 7.1). Both maintain a positive balance of soil organic matter (SOM) but achieve different increases in crop yields compared with the unfertilized control and conventional (Min) fertilization. The long-term field experiment on Calcareous chernozem at the Chetrosu Experimental Station of Agricultural State University of Moldova, established in 1969 at Anenii Noi in the south-east of the country, has applied various intensive management practices since 1989 (Bucur et al. 2004).

Soil samples were collected at Selectia on 14–16 June, 2010, (mid-growing season) and at Chetrosu on September, 2011 (before the maize harvest). Samples were collected from 0 to 20 cm from each of 3–4 replicates per treatment by combining 5 soil cores from each plot, 70 samples in all. After removing plant debris and stones, the soil was passed through a 2 mm sieve and stored at 4 °C for no longer than the one month necessary for biochemical analysis.

---

[1] *Typical chernozem* and *Leached chernozem* are equivalent to Haplic chernozem in *World reference base for soil resources 2006; Calcareous chernozem* is equivalent to Calcic chernozem.

Table 7.1 Details of experimental sites

	Typical chernozem, Balti	Calcareous chernozem, Chetrosu
Characteristics	Lat. 47°45′N; long. 27°55′E	Lat. 46°54′N; long. 29°02′E
pH_{water}	6.6–7.1	7.8–8.0
Soil organic matter	4.65 %	3.68 %
Plot size	242 m^2 (5.6 m × 43.2 m)	55 m^2 (5.5 m × 10 m)
Crops	6-field crop rotation: vetch and oats > winter wheat (*Triticum durum*) > sugar beet (*Beta vulgaris*) > maize-for-grain (*Zea mays* L.) > spring barley (*Hordeum vulgare* L.) > sunflower (*Helianthus annuus* L.)	5-field rotation: peas (*Pisum sativum*) > winter wheat (*Triticum durum*) > maize-for-grain (*Zea mays* L.) (1) > maize-for-grain (2) > lucerne (*Medicago sativa*)
Fertilization:		
Organic	90 t/ha farmyard manure twice per rotation: 60t/ha before sugar beet, 30 t/ha before sunflower	60 t/ha farmyard manure once per rotation, before maize
Mineral + organic	$N_{300}P_{255}K_{225}$ (kg/ha) + 90 t farmyard manure per rotation	$N_{120}P_{120}K_{120}$ (kg/ha) before winter wheat + green manure

Microbial biomass carbon (C_{mic}) assay employed the rehydratation method (Blagodatskii et al. 1987). For each treatment, two 5–10 g replicate samples were dried at 65–70 °C for 24 h then rehydrated with 0.5 M K_2SO_4 (ratio 1:2 w/v) to destroy the cells and release the microbial carbon into solution; two control samples were kept refrigerated prior to treatment with K_2SO_4 solution. K_2SO_4-extractable organic C was measured in both dried and fresh samples using dichromate oxidation: a 1.6 ml aliquot of filtered soil extract was mixed with 2.4 ml dichromate solution (1.28 g $K_2Cr_2O_7$ in 400 ml of deoinized water, dissolved in 2 l concentrated H_2SO_4), incubated at 140°C for 20 min, and the optical density was measured at 340 nm against a blank mixture. C_{mic} (μgC/g oven-dry soil) was calculated as $(C_d - C_f)/k_c$ where $(C_d - C_f)$ is the difference of carbon measured in dried and fresh soil samples; k_c (the portion of cell components released in solution after the drying-rehydratation procedure) was 0.25.

Basal soil respiration (BSR) was determined according to Isermeyer (1995): 25–50 g air-dry soil, adjusted to 40 % moisture, was incubated at 21 °C in the dark for 7 days in a sealed 0.75 l glass jar with a vessel containing 10 ml distilled water to maintain humidity and another with 20 ml M NaOH for trapping CO_2. The CO_2 released during incubation and absorbed by the NaOH was determined by titration and recorded as μgCO_2-C/g soil/h at 21 °C.

Nitrogen mineralization capacity (NMC) was determined simultaneously with BSR (Leirós et al. 2000) in soil extracts (0.05 M NaCl) from the difference between the values of ammoniacal and nitric N determined before and after incubation, respectively, with Nessler and disulfo-phenolic acid reagents (Arinushkina 1980). NMC was recorded as μg total inorganic N/g soil/7-d at 21 °C.

Metabolic quotient (qCO_2), the quantity of CO_2-C produced per unit of microbial biomass C per hour, was calculated as the ratio CO_2-C:C_{mic} and expressed in mgCO_2-C/gC_{mic}/h (Anderson 2003).

Total organic carbon (C_{org}) was assayed by acid dichromate oxidation and estimation of the excess of dichromate according to Tiurin's method (Arinushkina 1980).

Microbial quotient was calculated as a ratio C_{mic}:C_{org}, in per cent of total organic C (Anderson 2003).

Dehydrogenase (Dh) activity (EC 1.1.1.1) was determined using Galstean's procedure as described by Khaziev (1990), *urease (Ure) activity* (EC 3.5.1.5) by Khaziev's method (1990), *FDA hydrolysis rate (FDHR)* following Schnurer and Rosswall (1982) and *protease (Pro) activity* following Ladd and Butler (1972). All enzymatic activities were analysed at recommended optimal pH and temperature.

Statistical Analysis. Two-way ANOVA (StatSoft STATISTICA 7.0) was conducted with fertilizer and crop type as fixed factors. Differences between means within a crop types for each investigated parameter were identified using paired t-tests.

7.3 Results and Discussion

Most parameters measured in Typical chernozem under the 6-field rotation at Balti are significantly influenced by crop type (Emnova 2012); the exception was nitrogen mineralization capacity – which varied enormously (Table 7.2) although there was a trend towards lower NMC under Min + Org fertilization (Table 7.3). Two-way ANOVA revealed no significant influence of either organic or mixed Min + Org fertilization on microbial quotient, basal soil respiration, metabolic quotient or dehydrogenase activity. General biochemical properties related to microbial activity were less influenced by fertilization than were the extracellular enzymatic activities. This was confirmed by one-way ANOVA with the system of fertilization as one independent variable (Table 7.3); the activities of extracellular hydrolytic enzymes as well as C_{org} content were modified by the system of fertilization system, but C_{mic} and related microbial parameters were not.

The microbial biomass in the top of the cultivated layer was 176–240 (mean 212 $\pm$ 29) μgC/g dry soil under organic fertilization and 182–257 (mean 225 $\pm$ 31) μgC/g dry soil under mixed Min + Org fertilization. Similar values have been reported from the Selectia long-term field experiments by Corcimaru et al. (2011), by Merenuic et al. (2008, 2009) and by Senicovscaia et al. (2008). The microbial quotients and basal soil respiration parameters also are similar to those reported by Senicovscaia et al. (2008) and Corcimaru et al. (2011). In our study, the qCO_2 values for the 6-field rotation ranged from 1.5 to 3.6 mg CO_2-C/gC_{mic}/h (mean 2.82 $\pm$ 0.79) under organic fertilization and from 1.3 to 2.9 mgCO_2-C/gC_{mic}/h (mean 2.25 $\pm$ 0.57) under Min + Org fertilization. The least value of qCO_2 in the 6-field rotation (1.3 mgCO_2-C/gC_{mic}/h) was observed under sugar beet with Min + Org fertilization,

Table 7.2 Two-way analysis of variances for biochemical parameters of Typical chernozem at Balti

Dependent variables	Independent variables[a]	F^b	P-value[c]
Total organic carbon (0–20 cm), C_{org}	Fertilization system	15.91	<0.001***
	Crop types	7.95	<0.001***
	Interaction	2.21	0.07
Microbial biomass C, C_{mic}	Fertilization system	6.24	0.017*
	Crop types	20.79	<0.001***
	Interaction	0.32	0.89
Microbial quotient, C_{org}:C_{mic}	Fertilization system	0.90	0.35
	Crop types	20.21	<0.001***
	Interaction	0.66	0.66
Basal soil respiration, CO_2-C	Fertilization system	0.88	0.35
	Crop types	4.61	0.002**
	Interaction	2.98	0.023*
Metabolic quotient, qCO_2	Fertilization system	0.88	0.35
	Crop types	4.61	0.002**
	Interaction	2.98	0.024*
Nitrogen mineralization capacity, NMC (total inorganic N mineralized)	Fertilization system	0.09	0.75
	Crop types	0.57	0.72
	Interaction	0.37	0.87
Dehydrogenase activity, Dh	Fertilization system	0.61	0.44
	Crop types	7.63	<0.001***
	Interaction	0.16	0.99
FDA hydrolysis rate, FDHR	Fertilization system	10.69	0.002**
	Crop types	12.34	<0.001***
	Interaction	3.23	0.016**
Urease activity, Ure	Fertilization system	18.54	<0.001***
	Crop types	11.40	<0.001***
	Interaction	2.26	0.07
Protease activity, Pro1	Fertilization system	5.71	0.022*
	Crop types	17.43	<0.001***
	Interaction	0.42	0.83
Protease activity, Pro2	Fertilization system	6.19	0.018*
	Crop types	1.24	0.31
	Interaction	0.59	0.71
Crop productivity (Harvest, % of unfertilized control)	Fertilization system	8.14	0.007**
	Crop types	58.37	<0.001***
	Interaction	1.27	0.30

***$p < 0.001$; **$0.001 < p < 0.01$; *$0.01 < p < 0.05$; $n = 48$

[a]Fertilization system (Min + Org vs. Org) and crop types (six cereal and row crops in 6-field rotation) were the independent variables

[b]Fisher's criterion

[c]Confidence level

while the qCO_2 value was much higher with organic fertilization (3.5 mg CO_2-C/gC_{mic}/h). For comparison, qCO_2 for Typical chernozem under sugar beet in the nearby conventional 10-field crop rotation with Min + Org fertilization was 1.84 ± 0.15 mgCO_2-C/gC_{mic}/h (Corcimaru et al. 2011).

Table 7.3 Biochemical parameters for Typical chernozem topsoil under the 6-field rotation at Balti

Balti, northern Moldova	Organic fertilization			Mineral + organic fertilization			One-way ANOVA	
	Min.	Max.	Mean ± S.D.[a]	Min.	Max.	Mean ± S.D.	*F*	*p*
$C_{org,}$, %	2.36	2.59	2.47 ± 0.09	2.40	2.63	2.56 ± 0.08	8.43	0.005**
C_{mic}, μgC/g dry soil	176	240	212 ± 29	182	257	225 ± 31	2.03	0.16
C_{mic}:C_{org}, %	0.75	1.00	0.86 ± 0.12	0.75	0.99	0.88 ± 0.11	0.30	0.59
BSR, μgCO_2-C/g soil/h	0.35	0.61	0.58 ± 0.13	0.25	0.69	0.51 ± 0.17	0.55	0.46
qCO_2, mgCO_2-C/ gC_{mic}/h	1.50	3.60	2.82 ± 0.79	1.30	2.90	2.25 ± 0.57	2.70	0.11
NMC, μgN/g soil/7-d	0.36	1.34	0.66 ± 0.36	0.24	1.01	0.58 ± 0.33	0.14	0.74
Dh, μgTPF/g soil/h	1.96	3.10	2.66 ± 0.43	1.87	2.98	2.55 ± 0.43	0.38	0.54
FDHR, μgFl[b]/g soil/h	54.3	60.0	55.8 ± 3.6	53.2	65.7	58.4 ± 4.1	4.32	0.043*
Ure, μgNH_4/g soil/h	39.4	54.8	45.1 ± 5.2	41.2	63.2	51.2 ± 7.5	8.18	0.006**
Pro1, μgTyr[c]/g soil/h	18.6	36.5	25.5 ± 6.1	21.8	40.0	28.5 ± 6.6	2.10	0.15
Pro2[d], μgTyr/g soil/h[1]	26.6	35.4	31.0 ± 3.5	30.1	45.4	37.9 ± 5.5	6.31	0.016*

**$0.001 < p < 0.01$; *$0.01 < p < 0.05$

[a]S.D. (standard deviation, σ), $n = 6$

[b]Fluorescein

[c]Tyrosine

[d]Retest of soil Pro activity on the same soil samples, which have been incubated for NMC measurements; for other abbreviations see text

According to the criteria proposed by Anderson (2003), values below 2 % for the C_{mic}:C_{org} ratio or above 2.0 for qCO_2 may be considered critical for neutral soils. The estimation of microbial eco-physiological parameters of the arable Typical chernozem shows C_{mic}:C_{org} ratios below 2.0 %, which reflects a scarcity of soil carbon for growth and maintenance of microbial biomass. The metabolic quotient is also below 2.0 in the nearby long-term black fallow (Corcimaru et al. 2011); the microbial biomass under crop rotation eliminated CO_2-C more actively than under black fallow.

The parameter qCO_2 indicates the efficiency with which soil microorganisms use the carbon resources of the soil; we might expect that stressed soils will provide higher qCO_2 values than less-stressed soils (Anderson and Domsh 2010). It appears that the Balti chernozem is stressed – in spite of the crop rotation and organic fertilization – although both lead to much lower loss of organic carbon from the plough layer than the comparable soil under a conventional management (Boincean et al. 2007).

Soil dehydrogenase (Dh) activity is an intracellular enzymatic complex found in all microorganisms. In this investigation, Dh activity was found to be significantly influenced by crop type but not by fertilization system: values ranged from 1.87 to 3.10 μg triphenylformazan (TPF)/g soil/h at 30 °C (mean 2.60 ± 0.43). This value is close to published data on Dh activity for Leached chernozem in northern Moldova; under wheat and maize, means were 2.77 and 3.63 μgTPF/g soil/h, respectively (Merenuic et al. 1985).

The specific biochemical parameters – activities of hydrolytic enzymes – depended significantly on both crop type and fertilization (Table 7.2). FDA hydrolysis rate (FHDR) ranged from 54.3 to 60.0 μg fluorescein/g soil/h at 24 °C (mean 55.8 ± 3.6) with organic fertilization and from 53.2 to 65.7 μg fluorescein/g soil/h (mean 58.4 ± 4.1) with Min + Org fertilization. The highest FHDR was measured in soil under sugar beet under both systems of fertilization. Urease activity ranged from 39.4 to 54.8 μgNH_4/g soil/h at 37 °C (mean 45.1 ± 5.2) with organic fertilization and from 41.2 to 63.2 μgNH_3/g soil/h (mean 51.2 ± 7.5) with Min + Org fertilization. Under sugar beet, maize, spring barley (maximum value) and sunflower, Urease values were significantly higher with Min + Org compared with the same crops with only organic fertilizer (Emnova et al. 2011b). Caseine-protease activity ranged from 18.6 to 36.5 μg tyrosine/g soil/h at 50 °C (mean 25.5 ± 6.1) with organic fertilization and from 21.8 to 40.0 μg tyrosine/g soil/h (mean 28.5 ± 6.6) with Min + Org fertilization (Table 7.3). Pro activity with Min + Org fertilizers was generally higher than with organic fertilizer alone. The maximal soil Pro activity with both systems of fertilization was observed in soil under maize; the next highest (21 % less) was under sugar beet (the precursor of the maize crop).

No other values for Typical chernozem appear to be available for comparison of soil-specific biochemical properties. However, the values reported by Trasar-Cepeda et al. (2008) for arable *Umbrisols* and *Regosols* in the temperate-humid zone of Galicia, NW Spain, are within or close to ranges reported here.

Table 7.4 Values of biochemical parameters for Typical and Calcareous chernozem under maize in crop rotations

	Typical chernozem, Balti		Calcareous chernozem, Chetrosu			
	Conventional tillage				Ploughless tillage	
Parameter	Org	Min + Org	Org	Min + Org	Org	Min + Org
C_{org}	2.46 ± 0.03[a]	2.60 ± 0.04	1.47 ± 0.12	1.48 ± 0.04	1.59 ± 0.10	1.53 ± 0.10
C_{mic}	217 ± 24	220 ± 10	253 ± 24	252 ± 15	289 ± 21	284 ± 26
C_{mic}:C_{org}	0.88 ± 0.10	0.84 ± 0.05	1.72 ± 0.15	1.71 ± 0.13	1.82 ± 0.16	1.86 ± 0.27
CO_2-C rate	0.77 ± 0.08	0.52 ± 0.13	2.33 ± 0.53	2.33 ± 0.57	3.65 ± 0.56	1.85 ± 0.34
qCO_2	3.60 ± 0.25	2.40 ± 0.69	9.40 ± 2.6	9.20 ± 2.1	12.70 ± 2.3	6.60 ± 1.4

[a]Mean ± standard deviation, σ; $n = 4$

Table 7.4 provides a comparison of general biochemical parameters for Typical and Calcareous chernozems under maize in the long-term crop rotations. Despite its lower content of soil organic carbon, the Calcareous chernozem provides more carbon for microbial growth and maintenance than the Typical chernozem. In both soils, the microbial biomass (C_{mic}) was not influenced significantly by the system of fertilization, ranging between 252 and 253 μgC/g soil in Calcareous chernozem as against 217–220 μgC/g soil in Typical chernozem. The microbial quotient (C_{mic}:C_{org}) was twice higher in Calcareous chernozem (1.7–1.8 %) compared with Typical chernozem; basal respiration (2.33 μgCO_2-C/g dry soil/h) was 3–4.5 times greater; and qCO_2 (9.2–9.4mgCO_2-C/g C_{mic}/h) 2.6–3.8 was times greater than in Typical chernozem.

Compared with conventional ploughing, ploughless tillage raised soil organic carbon, microbial biomass and microbial quotient – in accord with other reports of the favourable effects of reduced tillage (Kandeler et al. 1999; van Groenigen et al. 2010). In the crop rotation under investigation, ploughless tillage interacted differently with organic and mixed fertilization – with different consequences for microbial respiration and metabolic quotient. With organic manure, CO_2-C emissions and qCO_2 were 3.65 ± 0.56 μg CO_2-C/g dry soil/h and 12.7 ± 2.3 mgCO_2-C/g C_{mic}/h, respectively, 1.6 and 1.4 times greater than under the plough. With Min + Org fertilization and ploughless tillage, values for both parameters were less than half the organic values, as well 1.2–1.3 times less than under the plough. Thus, Calcareous chernozem under maize with ploughless tillage and fertilized by farmyard manure eliminates an enormous amount of CO_2-C – which indicates a strongly stressed soil microbial biomass.

Boincean et al. (2007), reporting 26 years of data from Balti, noted that maize-for-grain responds poorly to fertilization: achieving only 8–14 % increase over the unfertilized crop. Even so, the mean yields were 6.88 and 6.65 t/ha for organic and mixed Min + Org fertilization. The multiyear (1990–2010) analysis of maize yield from ploughed Calcareous chernozem at Chetrosu shows mean values 5.23 t/ha with organic manure and 4.91 t/ha with mixed fertilization. Despite its favourable effect on microbial activity, ploughless tillage did not increase maize yields; the mean values were 4.81 and 4.41 t/ha, respectively.

7.4 Conclusions

It appears that maize yields in the two crop rotations on two types of chernozem correlate mainly with SOM content and annual rainfall. But soil biochemical parameters reveal the different intensity of biochemical processes which influence the stocks of soil organic carbon and release of nutrients from plant residues and organic fertilizers. In the north of the country, Typical chernozem under two advanced systems of fertilization is characterized by low microbial quotients (on average 1.0 %) and metabolic quotients only occasionally higher than 2.0mgCO_2-C/gC_{mic}/h. The inevitable loss of soil organic matter under arable management is compensated by manuring, but enhancement of fertilizer efficiency remains a challenge. In the dryer south-east of the country, Calcareous chernozem under current farm practice is characterized by almost double the microbial quotient (on average 1.7–1.8 %) and significantly higher than 2.0mgCO_2-C/gC_{mic}/h metabolic quotients. In other words, arable Calcareous chernozem is much more stressed than the comparable Typical chernozem.

In short:

1. *Under the plough*, Typical chernozem in northern Moldova is characterized by low microbial quotients (on average 1.0 %) and, as a rule, low metabolic quotients (but sometimes a little above 2.0mgCO_2-C/gC_{mic}/h). The values of biochemical parameters are generally within published ranges.
2. The soil-specific biochemical properties (soil FDA hydrolase, urease and protease activities) were more influenced by fertilization systems compared with general biochemical properties (microbial biomass C, basal soil respiration, microbial and metabolic quotients).
3. For ploughed Calcareous chernozem in south-eastern Moldova, the values of microbial quotient are almost twice higher (on average 1.7–1.8 %), and metabolic quotients are significantly higher (2.0mgCO_2-C/g C_{mic}/h).
4. Judging by the metabolic quotient (qCO_2), arable Calcareous chernozem in south-eastern Moldova is much more stressed than its counterpart Typical chernozem in the north of the country.
5. *Ploughless tillage* enhances soil organic matter, microbial biomass (C_{mic}) and microbial quotient in Calcareous chernozem but does not stimulate maize yields. With organic manure, the CO_2-C elimination and qCO_2 parameters were twice higher than with mixed Min + Org fertilization.

Acknowledgements I thank Prof B Boincean and Dr L Nica (Selectia RIFC at Balti) and Dr V Gnidiuc (Chetrosu, ASUM) for the opportunity to sample soils from the long-term experiments and also researchers Oxana Daraban, Iana Druta and Olga Ursachi and undergraduate student Anna Hanganu for experimental assistance. I am grateful to Academician S Toma for valuable advice on data analysis.

References

Anderson TH (2003) Microbial eco-physiological indicators to asses soil quality. Agric Ecosyst Environ 98:285–293

Anderson TH, Domsh KH (2010) Soil microbial biomass: the eco-physiological approach. Soil Biol Biochem 42:2039–2043

Arinushkina EV (1980) Agrochemical methods of soil analysis. Nauka, Moscow (Russian)

Blagodatskii SA, Blagodatskaia EV, Gorbenko AJ, Panikov NS (1987) Rehydratation method of soil microbial biomass determination. Pedology (Russia) 4:64–71 (Russian)

Boincean BP, Bugaciuc M, Nica L et al (2004) Agrotechnical research results from Selectia RIFC. In: Boincean B (ed) Field crops cultivation – results and prospects. University Press, Balti, pp 56–84 (Romanian)

Boincean B, Nica L, Bugaciuc M et al (2007) Long-term productivity and fertility of arable chernozem soils of Moldova. In: Boincean B et al (eds) Sustainable including, ecological agriculture – results, problems and perspectives. University Press, Balti, pp 22–30

Bucur G, Nicolaev N, Vasiliev M et al (2004) Development of a tillage system with high economic efficiency and enhanced soil fertility. Information Bulletin, Institute of Economics and Information, Chisinau (Romanian)

Corcimaru S, Merenuic G, Boincean BP, Bugaciuc M (2011) The influence of different technologies of sugar beet cultivation on the microbial activity in soil. J Moldovan Acad Sci Life Sci 2(314):166–170 (Russian)

Emnova E (2012) Biochemical processes in chernozem soil under different fertilization systems. Chem J Moldova Gen Ind Ecol Chem 7:45–51

Emnova E, Toma S, Boincean BP et al (2010) Biochemical parameters of Typical chernozem soil under sunflower and vetch + oats in crop rotation with different fertilization. In: Papers (Scientific works). Agronomy series, vol. 53, 1, University of Agricultural Sciences and Veterinary Medicine, Iasi, pp 9–14

Emnova E, Toma S, Daraban S et al (2011a) Biochemical properties of Typical chernozem soil under sugar beet in crop rotation. In: Vozian V et al (eds) Sugar beet – strategic crop for sustainable development of agriculture of Moldova Republic. Central Press, Chisinau, pp 175–183 (Russian)

Emnova E, Toma S, Daraban O, Druta I (2011b) Hydrolytic activity of Typical chernozem and mobilisation of plant nutrients. In: Increasing the impact of research and development of capacity for innovation. International conference on 65th anniversary of MSU. MSU, Chisinau (Russian), pp 230–234

Gil-Sotres F, Trasar-Cepeda C, Leirós MC, Seoane S (2005) Different approaches to evaluating soil quality using biochemical properties. Soil Biol Biochem 37:877–887

Isermeyer H (1995) Estimation of soil respiration in closed jars. In: Alef K, Nannipieri P (eds) Methods in applied soil microbiology and biochemistry. Academic Press, London, pp 214–216

Kandeler E, Palli S, Stemmer M, Gerzabek MH (1999) Tillage changes microbial biomass and enzymatic activities in particle-size fractions of a Haplic Chernozem. Soil Biol Biochem 31:1253–1264

Kaschiuk G, Alberton O, Hungria M (2010) Three decades of soil microbial biomass studies in Brazilian ecosystems: Lessons learned about soil quality and indications for improving sustainability. Soil Biol Biochem 42:1–13

Khaziev FX (1990) Methods of soil enzymology. Nauka, Moscow (Russian)

Ladd JN, Butler JHA (1972) Short-term assays of soil proteolytic enzyme activity using proteins and dipeptide derivatives as substrates. Soil Biol Biochem 37:1373–1385

Leirós MC, Trasar-Cepeda C, Seoane S, Gil-Sotres F (2000) Biochemical properties of acid soils under climax vegetation (Atlantic oakwood) in an area of the European temperate-humid zone (Galicia, NW Spain): general parameters. Soil Biol Biochem 32:733–745

Merenuic G, Iliinscaya S, Iscenco N (1985) Enzymatic activity of Moldavian soils. J Moldovan Acad Sci Biol Chem Sci 6:44–47 (Russian)

Merenuic G, Boincean BP, Corcimaru S et al (2008) Estimation of agricultural soil on the basis of soil microbiological parameters. Mediul Ambiant/Scientific. J Inf Ecol Cult 3(39):13–16 (Russian)

Merenuic G, Corcimaru S, Tanase A, Bogaciuc M (2009) Biological degradation of Typical chernozem in agro-ecosystems. Mediul Ambiant/Scientific. J Inf Ecol Cult 2(44):19–22 (Russian)

National Program for Soil Fertility (2005) Program of reclamation of degraded lands and improvement of soil fertility. Part 1. Reclamation of degraded lands. Pontos, Chisinau (Russian)

Schnurer J, Rosswall T (1982) Fluorescein diacetate hydrolysis as a measure of total microbial activity in soil and litter. Appl Environ Microbiol 43:1256–1261

Senicovscaia I, Marinescu C, Balan T, Cerbari V, Boincean BP (2008) Biota and its relationship with physical properties of typical chernozem at different condition of agricultural use. J Moldovan Acad Sci Life Sci 2(305):170–179 (Romanian)

Trasar-Cepeda C, Leirós MC, Seoane S, Gil-Sotres F (2008) Biochemical properties of soils under crop rotation. Appl Soil Ecol 39:133–143

van Groenigen KJ, Bloem J, Baath E et al (2010) Abundance, production and stabilization of microbial biomass under conventional and reduced tillage. Soil Biol Biochem 42:48–55

Chapter 8
Energy Status of Soil Agro-ecosystems

D. Girla (Dubit)

Abstract The activity and resilience of agro-ecosystems depend on soil – as a reservoir and biochemical reactor transforming substances and energy. But under cultivation, energy reserves have been much depleted. Protection of soil functions and solar gain in the form of soil organic matter is a priority for environmental protection.

Energy stocks under the plough with addition of green manure and mineral fertilizer, and with addition of farmyard manure, are compared with stocks under the same treatments under conservation tillage. In chernozem, there is no evidence of stabilization of the humus content at a new equilibrium value under arable farming; the energy balance of conventional crop rotations is profoundly negative. Over 38 years, a six-field rotation with application of 60 tonnes/ha farmyard manure every cycle lost 19–29 % of its initial organic matter; under mineral fertilizer with one catch crop of green manure, losses were 30–33 %.

8.1 Introduction

Energy stocks maintain ecosystems (Dobrovolski and Nikitin 1990), and soil organic matter is the largest stock of concentrated energy produced by plants and, subsequently, transformed by soil fauna and microorganisms (Skiopu 1988). The greater the soil organic matter content, the greater the energy of the ecosystem and the better plants are supplied with nutrients for growth and yield formation and (Kovda 1973).

D. Girla (Dubit) (✉)
State Agrarian University of Moldova, Chisinau, Republic of Moldova
e-mail: girla@mail.ru

D. Dent (ed.), *Soil as World Heritage*, DOI 10.1007/978-94-007-6187-2_8,

8.2 Experimental Site and Methods

The research was conducted on Calcareous chernozem[1] under the long-term field experiment with crop rotation and fertilization, established 40 years ago at the Chetrosu Experimental Station of the State Agrarian University of Moldova at Chisinau. The crop rotation is peas > winter wheat > maize silage > maize-for-grain > lucerne. Comparisons are made between conventional cultivation with the mouldboard plough and cultivation with the paraplough which does not invert the furrow and between two systems of fertilization – 60 tonnes/ha of farmyard manure applied once in the rotation cycle (in the autumn before sowing maize) and mineral fertilization ($N_{120}P_{120}K_{120}$) with ploughing-in of a catch crop of winter rape sown after the wheat harvest. The energy reserves in soil organic matter were calculated under the different management regimes based on the soil organic matter content of the entire 160 cm soil profile. Taking Kovda's estimate of the energy content of humus as 4–5 kcal/gram, a working value of 4.5 was used.

8.3 Results and Discussion

Table 8.1 summarizes the stocks and changes in soil organic matter and their energy equivalence under the different systems of tillage and fertilization. In every case, the energy budget is profoundly negative. Energy stocks were highest under the ploughing-with-farmyard manure treatment and lowest under the paraplough-with-NPK treatment. Losses over the period 1970–2008 were, in ascending order, mouldboard plough + farmyard manure < paraplough + farmyard manure < ploughed + NPK, paraplough + NPK.

Annual average losses of organic matter from the whole 160 cm soil profile were:

- Ploughed + farmyard manure – 1.70 t/ha (−0.5 %)

Table 8.1 Energy stocks and changes in soil organic matter under crop rotation with different tillage and fertilization

Management regime	Organic matter t/ha 1970	Organic matter t/ha 2008	Energy in soil organic matter 2008, KJ	± Changes since 1970, %
Ploughed, green manure + NPK	284.8	197.7	$3,725 \cdot 10^6$	−30
Ploughed, farmyard manure	284.8	231.7	$4,365 \cdot 10^6$	−19
Paraplough, green manure + NPK	284.8	189.5	$3,570 \cdot 10^6$	−33
Paraplough, farmyard manure	284.8	202.7	$3,819 \cdot 10^6$	−29

1 kcal = 4,187 J, KJ = 1,000 J

[1] Calcic chernozem in *World reference base for soil resources (2006).*

- Paraplough + farmyard manure – 2.16 t/ha (−0.8 %)
- Ploughed + NPK + green manure – 2.29 t/ha (−0.8 %)
- Paraplough + green manure + NPK – 2.51 t/ha (−0.9 %)

These high rates of loss under fairly good management support the contention of Ursu (2011), Krupenicov (2008) and Krupenikov et al. (2011) that in chernozem soils there is no evidence of stabilization of the humus content at a new equilibrium value under arable farming; losses are still high after a 100 years of cultivation.

Observations on-site suggest that even weeds are an important source of organic matter contributing to humus and energy stocks. Winter wheat is less weedy than maize, and most of the soil organic matter comes from the wheat crop itself. Under maize, the greater amount of weeds (particularly large perennial weeds with suckers, rhizomes and big root systems) makes for a closer C:N ratio than wheat straw and a significant contribution to the soil's energy stocks at some cost to the harvested yield of the maize crop.

References

Dobrovolski G, Nikitin E (1990) The ecological significance of soil. In: Functions of soils in the biosphere and ecosystems. Nauka, Moscow (Russian)

Kovda VA (1973) Fundamentals of soil, vol 1. Nauka, Moscow (Russian)

Krupenikov IA (2008) Chernozem. Morphology, perfection, the tragedy of degradation and the way of protection and regeneration. Pontos, Chisinau (Russian)

Krupenikov IA, Boincean BP, Dent DL (2011) The black earth. Ecological principles for sustainable agriculture on chernozem soils. Springer, Dordrecht

Skiopu D (1988) Agricultural ecosystems and the law of entropy. Science and Technology Publishing House, Bucharest (Romanian)

Ursu A (2011) Soils of Moldova. Stiinta, Chisinau (Romanian)

References

Chapter 9
Heavy Metals in the Anthropogenic Cycle of Elements

G. Jigau, M. Motelica, M. Lesanu, E. Tofan, L. Georgescu, C. Iticescu, V. Rogut, and S. Nedealcov

Abstract The increasing entry or unnatural concentration of man-made substances in soils affects every level of the biosphere. At the molecular-ionic level, it is apparent in significantly increased contents of heavy metals in agro-ecosystems and biogeochemical cycles. This may not, yet, amount to pollution but increased concentrations are being found in farm products – so it is prudent to control the heavy metal content of soils and develop agricultural technologies that can reduce their concentration. A management strategy should include: (a) systematic monitoring of the content of heavy metals and categorization of background levels according to the soil's buffering capacity, taking account of expected changes, current and possible soil degradation in particular; (b) ecological appraisal of agricultural technologies and prohibition of aggressive practices and chemicals; (c) adoption of biologically based practices and substitution of synthetic plant-protection products and fertilizers with biological products and manure; (d) augmenting the buffering capacity of the soil by increasing the content of organic matter and adjusting pH; and (e) special measures to reduce the impact of heavy metals, treatments that create geochemical barriers to the mobility of the elements (application of lime, gypsum, bentonite and farmyard manure) as well as cultivation of industrial crops.

G. Jigau (✉) • M. Motelica • M. Lesanu
State University of Moldova, Chisinau, Republic of Moldova
e-mail: pedologieaplicativa@gmail.com

E. Tofan
Al I Cuza University, Iasi, Romania

L. Georgescu • C. Iticescu
Dunărea de Jos University, Galati, Romania

V. Rogut • S. Nedealcov
Republican Centre for Applied Soil Science, Chisinau, Republic of Moldova

D. Dent (ed.), *Soil as World Heritage*, DOI 10.1007/978-94-007-6187-2_9,

9.1 Introduction

With the creation of the *technosphere*, soil formation entered a phase distinguished by new forms of energy and materials – we might call it the technogenic era. Technical developments and pressures continuously increase and diversify – and their consequences are hard to forecast. While they have released society from many constraints, the sustainability of current agro-ecosystems is in doubt. Sustainability requires identification of the biophysical laws and mechanisms of managed soils and a change of focus from the soil as a means of production towards soil as a functional system (Jigau 2009). This means reassessing relations between soil and environment, taking account of the flows of man-made materials and energy, and relations between plant and soil in the exchange of materials and energy under the new regime.

Agro-ecosystems are being changed by pollutants. These may be categorized by source as industrial, agricultural and household (Table 9.1). The impact of agricultural sources is mostly low, but almost the entire land surface is affected and their impact only increases. For instance, mineral fertilizers contain heavy metals, and, in Europe, the use of mineral fertilizers increased fivefold from the 1960s to 1990s. Based on the average values (Table 9.2), a dose of N_{100} P_{100} K_{100} fertilizer introduces the following quantities (g/ha) of heavy metals: Pb 10.4, Ni 23.5, Zn

Table 9.1 Sources of pollution of agricultural soils in the Carpatho-Danubian-Pontic region

Source of pollution	Type of pollution	Pollutants: Radioactive substances	Pesticides	Heavy metals	Incidental substances	Other	Impact
Agricultural	Fertilizers	+	–	+	+	+	Weak
	Plant protection	–	–	+	–	–	Weak
	Irrigation	–	+ ?	+ ?	+	+	Weak
	Livestock	–	–	+ ?	+	+	Weak
Industrial	Use of fossil fuels	–	–	+	+ +	+ +	Moderate local
	Manufacturing	–	–	–	+ +	+ +	Moderate local
	Transport	–	–	+	+	+	Moderate
	Mining	–	–	+	+ +	+ +	Moderate local
	Trans-boundary	+	–	+	+	+	Weak
Domestic	Manure storage	+ ?	+ ?	+	+ +	+ +	Moderate local
	Unauthorized dumps	+ ?	+ ?	+	+ +	+ +	Moderate local
	Sewage sludge	+ ?	+ ?	+ +	+ +	+ +	Moderate/ high local

Table 9.2 Average content of heavy metals and arsenic in mineral fertilizers

	Heavy metal content, mg/kg									
Fertilizer	Cd	Pb	Ni	Zn	Cu	Mn	Hg	As	Cr	Co
Potassic	0.3	8.0	14.0	23.0	16.0	10.1	–	1.4	5.7	1.5
Nitrogenous	0.3	0.2	19.0	30.0	26.0	76.0	–	2.5	42.0	1.3
Phosphatic	1.4	13.0	2.0	49.0	33.0	–	0.06	–	46.0	–
Complex	30.0	7.5	18.0	59.0	39.0	194.0	–	3.0	116.0	36.0

50.1, Cu 36.0, Cr 58.9 and Cd 1.4. Fertilizers applied at recommended rates are not polluting but they do increase the background levels of heavy metals and arsenic, which enter biogeochemical cycles. The same applies to organic fertilizers; application of 50 t/ha of farmyard manure introduces (g/ha) Pb 38, Ni 75 and Cd 2.3 (Jigau et al. 2005).

Limestone stockpiles are another source of heavy metals; each tonne of limestone applied to the land introduces 44.2 gPb, 3.4 gCd and 35.4gNi – so wind and water deposit significant amounts of these elements around quarries.

Plant-protection chemicals introduce mercury, arsenic, copper and other pollutants onto soil. And we should not neglect smoke and exhaust gases: along roads and railways, there has been a steady increase in concentrations of lead and other heavy metals; and downwind of chimneys and power stations, wherever waste is burned, there are concentrations of heavy metals that are subsequently carried downwind and downstream (Jigau et al. 2005).

9.2 Soil Vulnerability to Heavy Metal Pollution

In most cases, the risk of pollution by heavy metals is minimal, but their accumulation in soils and sediments increases their participation in biogeochemical cycles. As an illustration, Table 9.3 presents schematic information on the accumulation of heavy metals in vineyards.

From Table 9.3 we see that copper and nickel moved from moderate adsorption capacity to high adsorption capacity, chromium and lead moved from low to high, arsenic from strong to low, mercury from moderate to low, and cadmium moved from extremely low to moderate adsorption capacity. This situation suggests that the persistence of introduced chemical elements, although present in small quantities, acts on the components of the system and, over time, these elements enter biogeochemical cycles like any other.

The involvement of heavy metals in anthropogenic soil formation depends on external and intrinsic factors. Among the more important external factors are *location* in respect of industrial facilities, lines of communication and other pollutants; *lithology; landforms, hydrology and hydrogeology,* in particular natural drainage and erosion; and *land use*, especially the intensity and aggression of agricultural practices. Intrinsic factors include *soil texture*, especially the content of fine clay (<0.001 mm) and its distribution in the profile (Fig. 9.1); the *mineralogy*

Table 9.3 Orders of magnitude of heavy metal accumulation in vineyard soils

Attributes of elements	Biological adsorption	Biological adsorption coefficient					
		100*n*	10*n*	*n*	0.9*n*	0.0*n*	0.00*n*
Biological elements	Energetic						
	High		Zn, As	Cr, Cu, Pb, Ni			
Accessory accumulated chemicals	Moderate			Cu, Ni, Hg	Cd		
	Low					Hg, As	Cr, Pb
	Very low						

Kovda (1973)

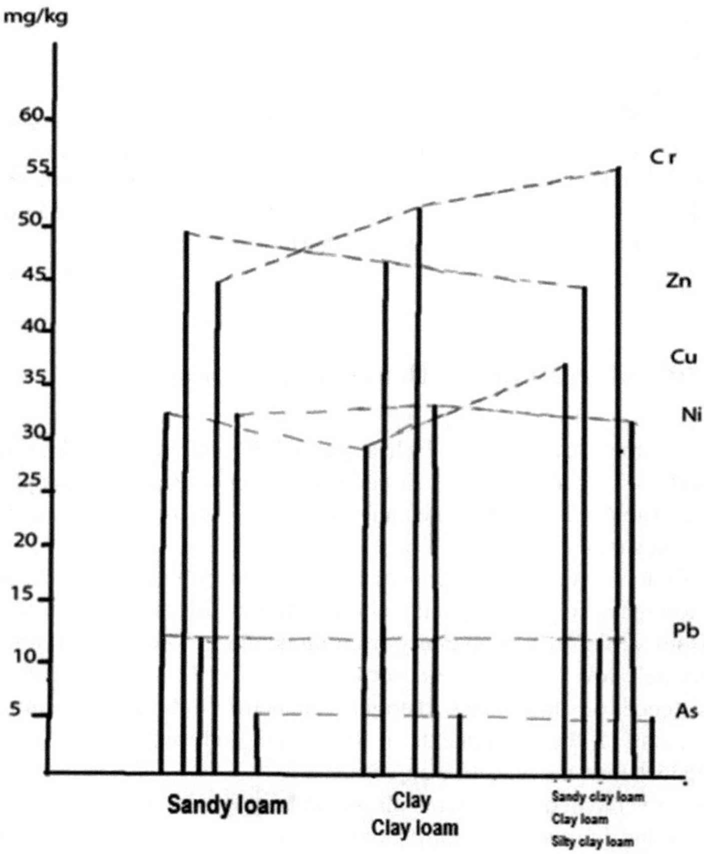

Fig. 9.1 Relationships between soil texture and heavy metal content

Table 9.4 Soil vulnerability to pollution with heavy metals

Class	Characteristics
Very low vulnerability	Soils with very high capacity to retain heavy metals: CEC > 60 meq/100 g, generally very fine textured, dominantly smectite clay, very thick or very rich in organic matter
Low vulnerability	Soils with a high capacity to retain heavy metals: CEC 35–60 meq/100 g, fine-textured but predominantly kandite clays, very thick
Moderate vulnerability	Soils with moderate capacity to retain heavy metals: CEC 12–35 meq/100 g, medium texture, moderate or low organic matter content, thick or very thick fine textured or moderately thick very fine textured
High vulnerability	Soils with low capacity to retain heavy metals: CEC 6–12 meq/100 g, mostly coarse texture, low organic matter content, moderately to strongly eroded, thin humus layer
Very high vulnerability	Soils with very low capacity to retain heavy metals: CEC <5 meq/100 g, coarse or very coarse texture, very low organic matter content, strongly eroded

Table 9.5 Implications of soil degradation for vulnerability to heavy metal pollution

Kind of degradation	Implications for vulnerability to heavy metals
Loss of humus	Lower specific surface, CEC and capacity to bind metals as organo-mineral compounds
Secondary compaction	Less active volume and loss of permeability, therefore accumulation of pollutants in the topsoil
Constriction of pore space	Impedance of root system and cycles of materials in the soil profile
Stratification of profile	Reduced permeability, hydraulic conductivity and translocation of materials lead to accumulation of heavy metals in the topsoil
Secondary sodicity	Increased pH, mobilization of organic colloids and heavy metals which migrate from the upper to the middle and lower parts of the profile
Waterlogging	Stagnation of water in the upper part of the profile. Subsequent evaporation increases concentrations of heavy metals in soil solution

Table 9.6 Dependence of heavy metals on soil organic matter content, global means

	Heavy metals, mg/kg		
Organic matter content %	Pb	Cu	Ni
<2.0	31.0	18.5	51.5
2.0–2.5	16.0–19.5	12.0–18.0	14.5–24.5
2.7–3.25	–	20.0–50.0	10.0–23.5
>3.5	22.0–28.0	–	–

of the fine clay; humus content; cation exchange capacity (CEC); soil reaction; pore volume and pore size; permeability; and hydraulic conductivity. Based on these characteristics, classes of soil vulnerability to pollution by heavy metals are drawn up in Table 9.4 (Jigau 2011).

Subclasses may be defined according to different factors of soil degradation; agriculture promotes various kinds of land degradation that increase soil vulnerability to pollution by heavy metals (Table 9.5). Table 9.6 presents relationships between heavy metals and humus content; it shows a clear trend of increasing copper and lead content as humus content increases and the opposite trend for nickel (since the accumulation of nickel is unrelated to soil texture, we suppose that its behaviour is determined by pH).

Taking into account the specified intrinsic and external factors, land may be categorized according to the risk of heavy metals entering biogeochemical cycles:

1. Polluted land (waste dumps, factories, road and rail corridors)
2. Land used for intensive, continuous cropping and heavily fertilized
3. Land irrigated with waste water and/or fertilized with sediment, sludge and industrial wastes
4. Flood plains and other depressions accumulating sediment
5. Poorly drained land with fine-textured soil and a high water table
6. Land of low humus content and moderate or high soil erosion
7. Degraded land (compacted, leached, silted up)
8. Land with heterogeneous ground cover

9.3 Plant and Soil

According to Kovda (1973), plants assimilate some 80 chemical elements from the soil, but only two heavy metals, copper and zinc, belong to the special group of biologically active chemical elements; a further six are accessory chemical elements or impurities (Table 9.7). Impurities may be ecological and absolute; they are assimilated automatically because they are present in soil solution; their content in plants is determined not by the needs of the plant but by their concentration in soil solution. There is no strict dependence between the concentration of heavy metals in soils and their content in plants. However, their background

Table 9.7 Heavy metals in the biological cycle

Category	Subcategory	Specifications	Elements
Biologically active	Absolute	Essential to life	O, H, C, N, Mg, K, P, S, I, Mn, Ca, Fe, Cu, Co
	Special	Necessary to many but not all organisms	Si, Na, Cl, F, Mn, Sr, B, Zn, Br
Accessory	Ecological	Not necessary for life but automatically assimilated by living organisms, especially plants, because they are present in solution	All the chemicals that go into solution as a result of rock weathering (except K, Na, Li and Rb)
	Absolute	Have no role in living material but always present in soils and rocks, from which they are absorbed by and pass through living organisms	Cl, Li, Rb, He, noble gases, Ra, Rn, all dispersed elements (e.g. Co, Ba, Th, Au, Hg)

Table 9.8 Ranges of heavy metal content in soil (mg/kg) and grapes (mg/kg dry matter)

	Soil		Grapes	
Element	Content	LCA	Content	LCA
Cu	13.2–117.9	25.0	0.68–2.97	5.0
Zn	21.1–82.0	68.0	0.02–3.23	10.0
Pb	7.4–29.6	20.0	0.01–0.35	0.4
Cd	0.03–0.77	0.24	0.01–0.04	0.03
Cr	17.3–78.2	90.0	0.01–1.25	0.1
Ni	17.9–78.0	45.0	0.15–0.42	0.5
As	3.7–10.0	5.6	<0.04	0.2
Hg	0.01–0.23	0.2	<0.05	0.2

After Jigau et al. (2005)

concentrations, as conditioned by progressive accumulation in the soil, determine the variability of the content heavy metals in farm produce. Table 9.8 shows that the chromium, nickel and cadmium content in grapes may exceed the allowable concentration limit (LCA).

9.4 Management Strategy for Heavy Metals in the Soil

The impact of heavy metals and arsenic on biogeochemical cycles, according to their natural background in soils, is in the order Cd > Pb > As > Zn > Ni. About 90 % of the gross amount of heavy metals that arrives on the soil accumulates there. The migration of heavy metals and arsenic in agro-ecosystems is determined by the particular physico-chemical characteristics of each element and the biological characteristics of plants; their degree of accumulation by the same species of plants varies 4–90-fold; between species it varies 2–30-fold in seed and 30–130-fold in the vegetative parts. The continuing concentration of these elements in biogeochemical cycles can only increase their absorption by plants. Moreover, the flow of heavy metals into soil is virtually irreversible; not being biologically useful, most heavy metals are extracted from the soil only in small quantities. A long-term management strategy is needed, which should include:

- *Systematic monitoring of the metal content in soils and categorization of background levels according to the soil's self-cleaning and buffering capacity.* Categorization should be varied according to expected changes – in particular current and possible soil degradation – taking account of the different vulnerability of soils to heavy metals.
- *Ecological appraisal of all agricultural technologies and prohibition of aggressive practices and chemicals.*
- *Adoption of biologically based practices and substitution of synthetic plant-protection products and fertilizers with biological products and organic fertilizers.*

- *Augmenting buffering capacity* by increasing soil organic matter content (manuring) and clay content (marling) and by adjusting pH.
- *Special measures to reduce the impact of sources of heavy metals in soil.* In the case of industrial sources of heavy metals, treatments that create geochemical barriers to the mobility of the elements can be recommended: lime, gypsum, bentonite clay and farmyard manure, as well as cultivation of industrial crops such as castor bean.
- *Measures to increase biological productivity* by practising the most advanced technologies and processes to intensify the process of extraction of heavy metals by crops.

9.5 Conclusions

1. Man-induced changes in soil genesis are affecting the functioning of all levels of the ecosystem. These changes include a significant increase of heavy metals in biogeochemical cycles.
2. Levels of heavy metals are increased in locations downwind and downstream of polluting sources, for instance, lead along corridors of communication and copper and arsenic in soils that are continuously and intensively cropped.
3. Increased background levels may not, yet, amount to pollution, but contents in farm produce do sometimes exceed allowable levels. Therefore, it is necessary to manage the content of heavy metals in soil through systematic monitoring, avoidance of aggressive practices and adoption of practices that reduce the share of technogenic elements in biogeochemical cycles.

References

Jigau G (2009) Soil genesis and soil physics. CEP Moldova State University, Chisinau (Romanian)

Jigau G (2011) Structural-functional levels of soil system integration, processes and evaluation indices. Sci Pap UASVM Bucharest Ser A 14:16–22

Jigau G, Grigheli G, Nedealcov S, Stasiev S (2005) Soil pollution with heavy metals and radionuclides in the Prut and Dniester landscape. Soil Form Factors Temp Zone New Ser 4:145–156 (Romanian)

Kovda VA (1973) The principles of pedology. General theory of soil formation, vol 2. Nauka, Moscow (Russian)

Chapter 10
Effects of Long-Term Application of Fertilizers on the Trace Element Content of Soils

T. Leah

Abstract Chemical fertilizers are a major source of soil contamination with heavy metals; their incidental minerals commonly contain nickel, cadmium, zinc, lead and other trace elements. Moreover, they contribute to soil acidification which increases the mobility of trace metals. Both aspects need to be considered when using mineral fertilizers, and soil and plant quality should be monitored. The trace element content in long-term fertilized Typical chernozem is categorized as *none* (i.e. not contaminated). Grey forest soils, Leached chernozem, Common chernozem and Calcareous chernozem were in the *low* category.

10.1 Introduction

The foundation of the Nicolae Dimo Institute for Pedology, Agrochemistry and Soil Protection in 1953 marked the beginning of a new stage of agronomic research in Moldova. Trials and long-term field experiments were set up to verify and generalize research results, establish indices of soil fertility, and develop systems of fertilization and new technologies. By Government Decision of July 11, 1994 but already founded on long experience, experimental stations coordinated by the N Dimo Institute were established in representative agro-climatic zones. The Experimental Station for Pedology, Agrochemistry and Ecology at Ivancea, in Orhei District, founded in 1964, works on systems of fertilization in crop rotations on Leached chernozem[1] and Grey forest soil. The Experimental Station for Soil

[1] *WRB* (IUSS 2006) equivalents of Moldovan soil taxonomic units: Calcareous chernozem – *Calcic chernozem;* Common, Typical and Leached chernozem – *Haplic chernozem;* Dark grey/ Grey forest – *Phaeozem/Albic luvisol.*

T. Leah (✉)
Nicolae Dimo Institute of Pedology, Agrochemistry and Soil Protection,
Chisinau, Republic of Moldova
e-mail: Tamara-tamaraleah09@gmail.com

D. Dent (ed.), *Soil as World Heritage*, DOI 10.1007/978-94-007-6187-2_10,

Table 10.1 Content of impurities in fertilizers, lime and gypsum

Impurity	Content %	Impurity	Content %
Boron	0.1–0.2	Strontium	0.5–2.1
Molybdenum	0.05–0.13	Fluorine	0.3–3.8
Manganese	1.0–1.5	Arsenic	10^{-3}–10^{-4}
Copper	0.01–0.5	Cadmium	10^{-4}
Zinc	0.05–1.5	Lead	10^{-4}
Superphosphate, mg/kg			
Arsenic	1.2–2.2	Lead	7–92
Cadmium	50–170	Nickel	7–32
Chromium	66–243	Selenium	0–4.5
Cobalt	0–9	Vanadium	20–180
Copper	4–79	Zinc	50–143

Science and Agrochemistry at Grigorevca, in Causheni District, founded in 1961, works on systems for Calcareous chernozem. Since August 2000, both have been part of the European Soil Organic Matter Network.

Nutrient regimes under the influence of fertilizers have been monitored – both to establish optimal levels of soil fertility and to protect the environment. In this context, the influence of fertilization on the content of trace elements in soils has been investigated. In Moldova, the most widely used fertilizers are superphosphate, potassium chloride and ammonium nitrate. Their raw materials contain significant amounts of strontium, uranium, zinc, lead, vanadium, cadmium, lanthanides and toxic elements (Table 10.1, after Kazak et al. 1987).

The accompanying elements in superphosphate, potassium chloride, ammonium nitrate and other fertilizers are widely distributed. Application of mineral fertilizers at an annual dose of 90kg NPK/ha deposits about 7 g/ha of copper, 10 g zinc, 0.2 g cadmium, 3 g lead, 4 g nickel and 5 g chromium (Leah 1984). In the long term, these trace elements accumulate in the soils, commonly in mobile forms that find their way into crops and groundwater. Their removal is not generally considered and, indeed, complicated by farming systems.

10.2 Experimental Sites and Methods

Samples of soils from the N Dimo long-term field experiments, which have been systematically fertilized for more than 40 years, and unfertilized control soils were analyzed for trace elements using atomic absorption spectrometry.

10.3 Results and Discussion

Determination of the total content of trace elements in the profile of Leached chernozem fertilized for more than 40 years at doses of $(NPK)_{240}$ revealed significantly greater amounts of trace elements than in the unfertilized control soil (Table 10.2).

Table 10.2 Total trace elements in the profile of Leached chernozem, mg/kg

	Ti	V	Cr	Mn	Co	Ni	Cu	Pb
Depth, cm	*Unfertilized*							
0–20	3,162	89	339	1,020	15.5	44.7	27.2	22.2
20–30	3,212	98	447	1,660	17.4	42.7	23.9	29.8
30–40	3,311	133	447	1,585	16.5	38.9	22.4	28.2
40–50	3,981	158	398	1,445	14.1	39.8	23.9	28.2
50–60	4,169	110	398	1,362	15.1	38.9	23.9	28.2
60–80	2,512	112	417	1,349	18.2	37.2	23.9	28.2
80–100	2,239	98	363	1,360	19.1	30.2	35.5	30.0
100–130	2,884	90	269	1,380	19.1	28.8	26.9	29.8
130–150	1,445	69	234	1,724	16.2	38.9	17.4	20.0
	Ti	V	Cr	Mn	Co	Ni	Cu	Pb
Depth, cm	*Fertilized – $(NPK)_{240}$*							
0–20	3,981	126	631	3,162	35.5	66.3	39.9	28.9
20–30	5,677	140	488	1,794	61.7	66.1	39.0	31.6
30–40	3,388	159	417	1,660	52.6	61.7	39.8	29.8
40–50	3,512	135	398	2,512	41.7	49.0	32.4	29.1
50–60	5,012	135	398	1,794	25.1	36.3	23.4	29.1
60–80	5,981	170	240	1,479	36.2	36.3	21.4	29.1
80–100	6,607	148	275	1,479	41.7	52.5	29.5	35.1
100–130	2,291	93	479	1,479	26.9	66.1	29.5	31.4
130–150	1,514	79	891	1,238	18	61.7	32.3	27.8

In unfertilized soil, there are accumulations of trace metals in the humose plough layer and, also, in the calcareous horizon where secondary carbonates have accumulated. In the fertilized soil, total trace elements increased by a factor of 1.3–3. Cr, Mn and Cu have accumulated in the top 20 cm and V, Co and Pb in the 20–30 cm layer, which indicates that these elements have affinity for soil organic matter; the sequence of accumulation in the topsoil is Mn > Co > Cr > Ni = Cu > V > Ti = Pb. However, the greater part of the trace elements migrates down profile and accumulates at some point within the more calcareous subsoil between 50 and 150 cm.

Over several years, the use of fertilizers such as ammonium nitrate and potassium chloride, and others, contributes to soil acidification. The optimum pH for most crops is in the range 6.0–6.5, but another negative effect of fertilizer application is to increase the mobility of some trace elements – which can lead to a deficiency of Zn, Cu and Mn in the topsoil. Most of the trace elements from fertilizers are in a weakly mobile state but, sometimes, they have a high mobility – depending on the type of fertilizers – and high mobility of trace elements in soils leads to toxicity for plants (Leah 1995, 2004, 1986).

Table 10.3 Content of mobile forms of microelements in Leached chernozem

	Unfertilized, mg/kg			$(NPK)_{240}$, mg/kg		
Depth cm	*Mn*	*Cu*	*Zn*	*Mn*	*Cu*	*Zn*
0–20	91.0	0.9	1.2	92.5	0.9	1.8
20–30	61.5	0.9	0.9	82.5	0.9	3.4
30–40	43.5	0.9	0.9	59.0	0.9	3.8
40–50	46.0	0.9	1.8	53.5	1.3	4.9
50–60	43.5	0.9	1.6	47.5	1.5	2.2
60–80	25.5	0.9	1.4	35.0	2.1	1.9
80–100	23.5	0.9	1.2	33.5	1.6	1.6
100–130	23.5	1.8	1.4	27.5	1.7	3.1
130–150	25.0	1.7	2.7	25.5	1.3	4.3

Table 10.4 Active pollution index of fertilized soils

		Index of active contamination		
Soil	Depth cm	*Mn*	*Cu*	*Zn*
Grey soil $N_{180}P_{120}K_{120}$	0–30	1.94	0.90	1.18
	30–40	0.84	1.33	0.45
Leached chernozem $N_{120}P_{120}K_{60}$	0–30	1.64	1.00	0.90
	30–40	1.00	1.52	0.56
Typical chernozem $N_{120}P_{120}K_{60}$	0–30	1.03	1.71	1.10
	30–40	1.05	1.00	1.30
Common chernozem $N_{120}P_{120}K_{60}$	0–30	1.70	1.10	1.30
	30–40	0.90	2.10	1.10
Calcareous chernozem $N_{90}P_{90}K_{60}$	0–30	1.00	1.04	1.12
	30–40	1.00	2.17	0.94

In fertilized Leached chernozem, the content of mobile forms is higher than in the control soil. The largest amount of mobile forms of trace elements is contained in the humus horizon (Table 10.3).

10.3.1 Active Pollution Index

Long and, sometimes, irrational application of high doses of mineral and organic fertilizers leads to high concentrations of heavy metals in soils and plants (Leah 2010). Ecological indicators of heavy-metal accumulation were derived from the analysis of the long-term fertilized and control soils in the field experiments. The *active pollution index* is the content of mobile forms of element in fertilized soil compared with the unfertilized control; it characterizes the degree of contamination of the soil by elements that can pass through the food chain in plants. Table 10.4 shows manganese pollution in the humus horizon (0–30cm) of Grey soil and Leached chernozem; zinc in Grey soil, Typical chernozem and Calcareous chernozem; and copper in Leached, Typical and Ordinary chernozem. In other cases, excessive accumulation occurred below the humus horizon.

Table 10.5 Index of trace elements accumulation in fertilized soils

Soil	Depth cm	Index of accumulation					
		Cu	*Zn*	*Pb*	*Ni*	*Mn*	*Cr*
Grey soil $N_{180}P_{120}K_{120}$	0–30	1.8	1.6	1.4	1.0	1.4	1.3
	30–40	0.8	0.9	0.6	1.5	0.9	0.7
Leached chernozem $N_{120}P_{120}K_{60}$	0–30	1.6	1.7	1.2	1.2	0.9	1.0
	30–40	0.9	0.8	1.0	1.0	1.0	1.0
Typical chernozem $N_{120}P_{120}K_{60}$	0–30	0.9	0.9	0.9	0.9	0.8	0.9
	30–40	1.1	1.0	0.9	1.0	1.2	0.9
Common chernozem $N_{120}P_{120}K_{60}$	0–30	1.1	1.1	1.1	2.0	1.2	1.0
	0–40	2.1	1.8	1.7	2.1	1.2	1.6
Calcareous chernozem $N_{90}P_{90}K_{60}$	0–30	1.1	1.2	1.1	1.0	0.9	1.0
	30–40	0.8	1.0	1.2	1.2	1.0	1.0

10.3.2 Index of Trace Element Accumulation

This is the content of total forms of contaminants relative to the content in the non-polluted control, showing the degree of accumulation of microelements in fertilized soils (Table 10.5).

The index of accumulation for the 0–30 cm layer is higher than average in Grey soil (1.0–1.8) and Common chernozem (1.0–2.0) and lower in Typical chernozem (0.8–0.9). The accumulation index of trace elements in fertilized soils is conditioned by the content of humus, clay minerals, carbonates, Fe-Mn oxides and soil reaction. Retained trace elements have different solubility and plant availability (Leah 1997, 1986). Acidity caused by chemical fertilizers favours mobilization of trace elements and absorption of these elements by plants: raising the reaction to pH 6–7 decreases solubility and causes symptoms of deficiency in plants.

From these two indices, we may classify soils according to the degree of contamination and activity of the pollutants as class I (0.1–1.0), unpolluted soils; class II (1.1–2.0), slightly polluted; class III (2.1–3.0), moderately polluted; class IV (3.1–4.0), polluted; and class V (4.1–5.0), highly polluted. The soils from the long-term fertilization experiments are slightly polluted, the only exception being Typical chernozem which remains in class I, unpolluted.

10.4 Conclusions

1. Forty years of systematic fertilization has brought about accumulation of trace elements in humus and calcareous horizons – to the extent that most soils may be classified as slightly polluted with trace elements. In this respect, as in numerous investigations of the evolution of soil properties, soil fertility and environmental quality, long-term field experiments have been invaluable.

2. Soils with a long history of application of mineral fertilizers accumulate trace elements that can then be taken up by plants and transfer elsewhere in the environment. Such soils occur beyond experimental plots, and it is prudent to undertake quality monitoring, including trace elements.

References

IUSS Working Group (2006) World reference base for soil resources, vol 103, World soil resources report. FAO, Rome

Kazak VG, Onishchenko TL, Gorbunov AV (1987) Impurities in phosphate fertilizers. Chemistry in Agriculture 3:61–62 (Russian)

Leah T (1984) Effect of fertilizers on the content of heavy metals in soils of Moldova. In: Contribution of young scientists and specialists in the intensification of agricultural production. First republican scientific-practical conference. Vilina Ukraina, Lvov (Russian), pp 16–17

Leah T (1995) Heavy metal pollution of agricultural soils as a result of application of mineral fertilizers. In: Report of the international scientific- practical conference on European year for nature conservation in Moldova: problems, achievements and perspectives. Institute of Ecology, Chisinau (Romanian) p 133

Leah T (1997) Criteria for identification and classification of soils polluted with heavy metals. In: N Dimo Institute (ed) Agrochemical problems in modern agriculture. Collected papers dedicated to the centenary of Academician I Dicusar. Agrarian State University of Moldova, Chisinau, pp 113–115 (Romanian)

Leah T (2004) Influence of systematic fertilization on the accumulation of heavy metals in soils of long-term field experiments. In: State Centre for Certification and Approval of Plant Protection Products and Fertilizers (ed) Agrochemical problems in modern agriculture. Abstracts of scientific-practical conference protection of plants, achievements and perspectives. Agrarian State University of Moldova, Chisinau (Romanian), pp 198–199

Leah T (2010) Impact of current farming systems on the content of heavy metals and pesticide residues in the soils. In: N Dimo Institute (ed) Monitoring soil quality in Moldova (data base, conclusions, forecasts, recommendations). Pontos, Chisinau, pp 393–399 (Romanian)

Leah T (1986) Quantitative indicators of changes in the content of trace elements in soils under the influence of fertilizers. In: Collected papers on fertilizers, soil fertility and crops productivity in Moldova. Nicolae Dimo Institute of Pedology, Agrochemistry and Soil Protection, Kishinev (Russian), pp 109–115

Chapter 11
Potassium in Brown Forest Soils

O. Lobova and V. Vakhnyak

Abstract The parameters that characterize the state of potassium in the soils of the Ukrainian Carpathians were determined and the dependence between different states of potassium and other soil parameters ascertained using statistical and mathematical analysis. Different kinds of soils have significantly different potassium regimes; Brown forest soils have much less total and exchangeable potassium than Chernozem. For Brown forest soils, the most significant factor is the mineralogy of the soil parent material. The lowest potassium values were found in the Pre-Carpathian province: measured potassium values in Brown forest soils in the province range from 940 to 3,510 mg/100 g total potassium, 21–61 mg/100 g non-exchangeable potassium and for exchangeable potassium, from 2 to 20 mg/100 g by the Kirsanov method and 8–40 mg/100 g by the Maslova method. The application of fertilizers in this province should be adjusted accordingly.

11.1 Introduction

The territory of the Ukrainian Carpathians (Outer and Pre-Carpathians) exhibits complexity of soil-forming conditions and a great variety of soils. Brown forest soil[1] is the main soil type in the mountains and Brown-podzolized soil[2] in the Pre-Carpathian region, but every subtype of Brown forest soils may be observed in the territory: typical, argillic, podzolized, brown-podzolized and soddy-brown

[1] Cambisols in *World Reference Base for Soil Resources 2006*, Burozems in the classification of the Russian Federation.

[2] Albeluvisols in *World Reference Base for Soil Resources 2006*.

O. Lobova (✉)
National University of Life and Environmental Sciences of Ukraine, Kiev, Ukraine
e-mail: oksana.varkhl@rambler.ru

V. Vakhnyak
Podolsky State Agricultural and Technical University, Kamenetz-Podolsk, Ukraine

D. Dent (ed.), *Soil as World Heritage*, DOI 10.1007/978-94-007-6187-2_11,

forest soil. All are genetically young by virtue of Pleistocene glaciation, a cool climate and, in the mountains, active erosion and deposition. Soil formation has proceeded mainly through weathering of the parent materials (to produce illite and smectite clays and hydrous iron oxides) and leaching. The active transformation of minerals in the soil is displayed in the forms, dynamics and thermodynamics of potassium, knowledge of which is essential for the proper and efficient use of potash fertilizers and which may be used as a diagnostic characteristic.

11.2 Materials and Methods

Various subtypes of Cambisols (*Brown forest soils* in Ukraine classification) and Albeluvisols (*Brownish-podzolic soils* in Ukraine) were investigated in the hilly Outer Carpathian and Pre-Carpathian provinces: *Gleyic Cambisols,* silty clay loam under coniferous forest; *Gleyic Cambisols,* clay loam in a mountain valley; *Gleyic Cambisols,* loam in hayfields; *Dystric Cambisols,* sandy loam under deciduous forest; *Eutric Albeluvisols,* silty clay loam under pasture; *Gleyic Albeluvisols,* clay loam under pasture; and *Gleyic Albeluvisols,* clay loam on arable land.

Soil indices were examined by conventional methods: particle size distribution following NA Kaczynski, soil organic matter content, reaction in water and saline solution, extractable acidity and extractable aluminium, hydrolytic acidity and the sum of adsorbed bases according to Kappen-Hilkovyts. Base saturation and adsorption capacity were calculated. Different forms of potassium were determined: water-soluble, exchangeable following Kirsanov (0.2 M HCl) and Maslova (1 M CH_3COONH_4), non-exchangeable following Pcholkin (2 M HCl), and total potassium by Smith's method.

11.3 Results and Discussion

11.3.1 Outer Carpathian Province (OCP)

Soil potassium depends on the mineralogy of the soil parent material and, also, environmental conditions such as the diversity of landforms and vegetation within small areas (Table 11.1).

The total content of potassium in the brown soils of the Carpathian province lies within the standard parameters for this type of soil. Exchangeable potassium determined by the Maslova procedure is in 2–3 times higher than the values determined by the Kirsanov procedure – and closer to the non-exchangeable values. This may be attributed to the natural acidity of leached mountain soils which are, therefore, unreactive with acid reagents.

Within the study area, the greatest total potassium content was observed in the mountain valley, probably related to the heavy soil texture and, consequently, the

Table 11.1 Statistical indices of potassium content in brown forest soils of the Outer Carpathians (*OCP*) and Pre-Carpathians (*PCP*) (mg/100 g soil of all soil horizons)

	Potassium				
Index	Total	Water-soluble	Kirsanov exchangeable	Maslova exchangeable	Non-exchangeable
OCP: (1) *Gleyic Cambisols,* silty clay loam, coniferous forest					
[a]$M \pm m_s$	2,481 ± 887	0.01 ± 0.008	13.7 ± 1.26	35.7 ± 4.64	51.8 ± 11.6
min–max	1,606–3,675	$3 \cdot 10^{-3}$–$6 \cdot 10^{-3}$	12.5–15.4	31.8–42.1	35.5–60.6
(2) *Gleyic Cambisols,* clay loam, mountain valley					
$M \pm m_s$	2,586 ± 214	0.02 ± 0.009	16.7 ± 3.97	48.9 ± 8.11	54.6 ± 2.73
min–max	2,336–2,767	$4 \cdot 10^{-3}$–$2.8 \cdot 10^{-2}$	12.2–21.3	37.0–54.1	52.5–58.6
(3) *Gleyic Cambisols,* loam, hayfields					
$M \pm m_s$	2,115 ± 1,093	0.01 ± 0.008	8.16 ± 2.02	25.3 ± 7.49	38.3 ± 15.8
min–max	938–3,150	$2 \cdot 10^{-3}$–$1.8 \cdot 10^{-2}$	5.34–10.1	20.0–35.9	24.7–53.6
PCP: (1) *Dystric Cambisols,* sandy loam, deciduous forest					
$M \pm m_s$	2,055 ± 130	0.05 ± 0.005	5.74 ± 1.17	19.8 ± 4.33	29.8 ± 8.04
min–max	1,909–2,264	$3 \cdot 10^{-3}$–$1.3 \cdot 10^{-1}$	4.43–7.14	15.0–24.2	21.1–38.8
(2) *Eutric Albeluvisols,* silty clay loam, pasture					
$M \pm m_s$	2,062 ± 1,423	0.014 ± 0.007	4.40 ± 1.21	12.2 ± 2.57	25.4 ± 4.70
min–max	1,394–5,283	$5 \cdot 10^{-3}$–$2.8 \cdot 10^{-2}$	2.33–5.77	9.06–15.8	18.5–34.2
(3) *Gleyic Albeluvisols,* clay loam, pasture					
$M \pm m_s$	1,918 ± 162	0.06 ± 0.05	7.99 ± 2.53	23.1 ± 8.99	32.6 ± 10.8
min–max	1,675–2,091	$2.8 \cdot 10^{-2}$–$1.4 \cdot 10^{-1}$	5.59–11.2	13.8–32.9	22.6–45.8
(4) *Gleyic Albeluvisols,* clay loam, arable land					
$M \pm m_s$	2,069 ± 133	0.04 ± 0.02	5.94 ± 1.85	28.3 ± 8.04	32.3 ± 3.75
min–max	1,918–2,239	$1.2 \cdot 10^{-2}$–$8.4 \cdot 10^{-2}$	4.77–8.68	18.7–38.0	28.2–37.3

[a]*Note: M* mean, m_s standard deviation from the mean

increased clay mineral content; water-soluble, exchangeable and non-exchangeable potassium were also highest at this site. The gross amount of potassium depends on the potassium content of the parent material and the soil texture; in general, total potassium followed soil texture but, also, decreased from strongly gleyed to weakly gleyed mid-loamy soils. Regression-correlation analysis of the various indices revealed certain trends in the distribution of different forms of potassium in relation to separate soil indices:

- A direct correlation between total and Maslova-exchangeable potassium.
- Variation in different forms of potassium is most frequently influenced by acidity (actual, potential, exchangeable). Correlation is generally inverse for extractable acidity (EA) whereas, for pH, correlations may be direct and inverse. Reaction and redox conditions determine the intensity of weathering. The dynamics of these parameters is greater than the changes of local climate conditions, but different vegetation cover creates microclimates manifested in different fluxes of heat and moisture through the soil profile. It is clear that pH depends on the composition of plant residues and transformation products that fall onto the soil surface and migrate through the profile. Thus, soil acidity is indirect evidence of environmental aggressivity.

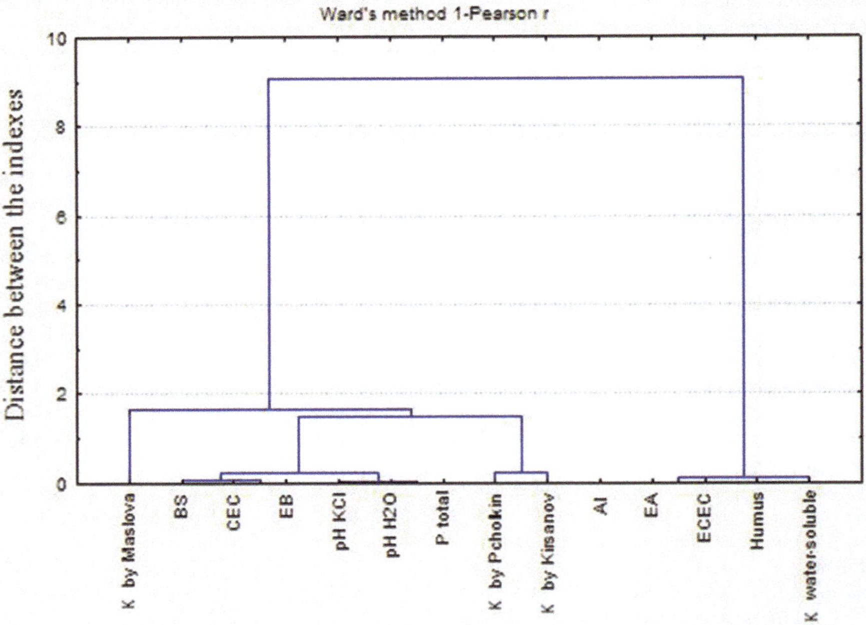

Fig. 11.1 Links between forms of potassium and other soil indices under coniferous forest

– The strongest correlations under the influence of soil adsorption complex (SAC) properties are seen in the Gleyic Cambisols in the hayfield, which exhibit the highest values of cation exchange capacity, total exchangeable bases and base saturation.

Cluster analysis was applied to elucidate multivariate correlations between different indices. Under coniferous forest (Fig. 11.1), total potassium was connected with the acidity of the soil solution; it is clear that changes of acidity shift the balance of potassium from the fixed state to other forms that may be cycled or leached from the profile. Exchangeable and non-exchangeable potassium are distanced from any other soil indices; the impact of these other indices on the states of potassium is probably indirect. The properties of SAC formed a separate cluster.

Gleyic Cambisols in the mountain valley show a similar relationship between different forms of acidity and Maslova-exchangeable potassium and between non-exchangeable potassium, actual acidity (pH_{H2O}) and humus content. A separate cluster of total potassium, Kirsanov-exchangeable potassium and extractable aluminium points to the actual reason for acidity.

The dendrogram of the Cambisols in the hayfield shows similarities between non-exchangeable and total potassium and potential acidity and, indirectly, with actual acidity. At the same time, Kirsanov-exchangeable and Maslova-exchangeable potassium form separate clusters, equidistant from the characterized indices.

In general, cluster analysis confirms the affinity between different forms of acidity and soil potassium content, regardless of the ecosystem, and also indicates the influence of several environmental conditions on the distribution of the different forms of potassium.

Total potassium varies down profile but is highest in the parent material, confirming the importance of parent materials for soils' potassium content. Profile variations of non-exchangeable potassium were similar; differences down profile were greater but, again, there was a clear maximum at the base of the profile so it is also dependent on the composition of the parent material.

Maslova-exchangeable and Kirsanov-exchangeable potassium often decrease in the transitional horizons and increase again down to the parent material, corresponding more closely with profile changes of soil acidity. Taking into account the results of cluster analysis, we may discern the higher dynamic of exchangeable potassium and its multilateral dependencies, connected with indices of the process of soil genesis.

11.3.2 Pre-Carpathian Province (PCP)

For comparative analysis of soils of the Pre-Carpathian province, we may take as a benchmark the Albeluvisol under deciduous (broad-leaved) forest which exhibits the typical features of its subtype although it is more acidic (in terms of pH, ECEC and extractable aluminium Al) than other PCP soils and has low cation exchange capacity (CEC), base saturation (BS) and exchangeable bases (EB) – all features generally determined by the parent material (Table 11.1). In comparison, Eutric Albeluvisols (silty clay loams) under pasture show greater acidity (actual and potential), accumulation of humus in the topsoil and abrupt decrease down profile. They exhibit high base saturation but low cation exchange capacity and exchangeable bases; compared with the soil under deciduous forest, these values are quite high, but compared with other Pre-Carpathian soils, quite low. Under arable, humus content was much reduced (again with abrupt decrease down profile).

Soils under arable and pasture show more variability of total and non-exchangeable potassium than soils under forest and in the mountain valley; however, there is a uniform distribution of Kirsanov-exchangeable and Maslova-exchangeable potassium (notwithstanding the much higher values of the latter). Only the Gleyic Albeluvisols (Burozem-podzolic soils) under pasture exhibited an increase in exchangeable potassium in the parent rock and non-exchangeable potassium decreased down profile. In general, the distribution of total potassium resembled that of the OCP in increasing down profile; only the Gleyic Albeluvisols under pasture exhibited a maximum in the topsoil; those under arable exhibited decreasing total potassium down to the parent material. It would appear that agricultural use affects potassium dynamics rather than its absolute content.

Given that there is dynamic equilibrium between the various forms of potassium, the balance between the exchangeable and non-exchangeable potassium is set for a

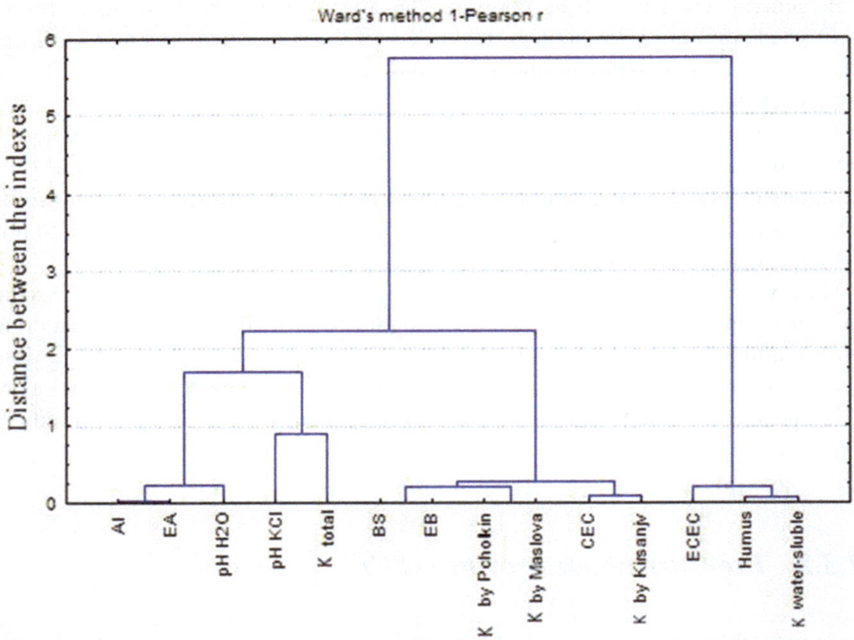

Fig. 11.2 Dendrogram of connections between forms of potassium and soil indices under deciduous forest

long time – depending on conversion to an exchangeable state of potassium ions released from the crystal lattice of micaceous minerals by chemical weathering. The activity of chemical weathering comes from dissolved CO_2 and, particularly, acidic root secretions – hydrogen displaces potassium from the lattice. Potassium is also released by aluminosilicate collapse as a result of the activity of silicate bacteria.

Our data confirm correlation between different forms of potassium in the soil, but, in the study area, correlation between soil physical and chemical properties and forms of potassium is weak; the most significant factor is soil mineralogy. The weak correlation between potassium and soil indices may be explained by the heterogeneous impact of the soil properties; in the case of Gleyic Albeluvisols under pasture and arable, physical and chemical properties clearly influenced total and water-soluble potassium (Fig. 11.2). Generally, acidity affected the content of exchangeable potassium.

Cluster analysis indicates that, under deciduous forest, water-soluble potassium was strongly associated with humus and ECEC; exchangeable and non-exchangeable potassium clusters with SAC properties. Investigation of other Pre-Carpathian soils shows a similar picture of connections between different potassium forms and the impact of acidity and SAC properties on its content.

11.4 Conclusions

Different kinds of soils have significantly different potassium regimes. Cambisols have much less total and exchangeable potassium than chernozem. The Pre-Carpathian area is characterized by the lowest contents of potassium: measured potassium values in Albeluvisols in the province range from 940 to 3,510 mg total potassium/100 g, 21–61 mg non-exchangeable potassium/100 g and for exchangeable potassium, from 2 to 20 mg/100 g by the Kirsanov method and 8–40 mg/100 g by the Maslova method. The application of fertilizers in this province should be adjusted accordingly.

In the soils under pasture and under cultivation, the variability of potassium content is much higher than in natural ecosystems.

Bibliography

Gorbunov NI, Voronin TV (1968) Weathering of potassium minerals in soils. Agrochemistry 5:45–51 (Russian)

Gorbunov N (1969) Genesis and transformation of minerals in soils. Pochvovedenie 3:106–118 (Russian)

Zharikova EA (2001) Potassium in soils of the Far East territories. Pochvovedenie 9:1060–1068 (Russian)

Zharikova EA (2004) The ability of soils to accumulate potassium (an example from the Amur region in the Far East of Russia). Pochvovedenie 7:819–827 (Russian)

Chapter 12
Pure and Applied Agrophysics

E.V. Shein, V.M. Goncharov, and M.A. Mazirov

Abstract Soil physical properties are conservative, many of them changing only slightly over time, but they vary significantly in space—a problem for soil physical investigations in long-term field experiments. The fundamental physical properties of soddy-podzolic soil (albeluvisol) under a long-term field experiment have hardly changed under the influence of a century of liming and chemical and organic fertilization. However, significant changes of *approximation parameters* of penetration resistance that depend on soil moisture indicate that the application of manure increased interparticle contacts within the normal range of soil moisture. These approximation parameters indicate a significant increase in interparticle forces with a decrease in moisture content in both the control and limed treatments that resulted from a greater sand fraction in the control and aggregation of particles due to liming.

12.1 Introduction

Many soil properties change under the conditions imposed by intensive farming, notably under the impact of mechanical compaction. The creation of a compacted layer at the plough sole affects the soil water regime and soil aeration. This is accompanied by changes in the soil biota and, often, in the character of transformation of soil organic matter. The application of fertilizer and manure also affects many soil properties including relatively stable (conservative) physical properties such as aggregate-size distribution and specific surface (Rachman et al. 2003; Dutartre et al. 1993). Depending on the particular farming practices, soil properties,

E.V. Shein (✉) • V.M. Goncharov
Soil Science Faculty, Lomonosov Moscow State University, Moscow, Russia
e-mail: evgeny.shein@gmail.com

M.A. Mazirov
Russian State Agrarian University-Timiryazev Agricultural Academy, Moscow, Russia

D. Dent (ed.), *Soil as World Heritage*, DOI 10.1007/978-94-007-6187-2_12,

and the weather, these changes may be clearly manifest; in other cases, they cannot be diagnosed by traditional methods. For instance, several studies have shown that the long-term application of fertilizers exerts a negligible effect on soil texture and bulk density (Munkholm et al. 2002). Many physical properties of soils are quite stable; their minor changes cannot be properly estimated by the routine methods so it is important to develop methods that will allow unambiguous judgements about changes in the physical properties of soils under the impact of agricultural management. To this end, we have studied a wide range of the physical properties of a soddy-podzolic soil and their changes under the impact of mineral fertilizers, lime, and manure.

12.2 Experimental Site and Methods

Field studies were undertaken on plots of the long-term experiment of the Timiryazev Agricultural Academy, established by Prof. AG Doyarenko in 1912. The site slopes at 1° towards the northwest within the southern part of Klin–Dmitrov Ridge; the soil is light loamy soddy medium podzolic.[1] The experimental area of 1.5 ha is divided into two parts with six rectangular fields in each. In the first part, continuous crops of winter rye, potatoes, barley, clover, flax, and bare fallow are cultivated; in the second part, a rotation of bare fallow–winter rye–potatoes–oats (barley) with clover–clover–flax is followed. Each field under continuous crops is split into eleven plots of 100m^2 on which different variants of fertilization have been applied: unfertilized control (two plots), N, P, K, NP, NK, PK, NPK, manure, and NPK + manure. Since the fall of 1949, half of each plot (50 m^2) has been limed once every 6 years (Kiryushin and Safonov 2002).

In April 2008, auger samples were taken from 0 to 10, 10 to 20, 20 to 30, and 30 to 40 cm layers from the continuously cropped plots under the control, lime, NPK, and NPK + manure treatments. Bulk density was determined on samples taken by a cylindrical (Pol'skii) auger.

Particle-size distribution was determined in two stages. First, the ground soil was sieved through 1mm and 0.25 mm screens to separate coarse soil particles (>0.25 mm). Then particle-size distribution in the fraction <0.25 mm was determined on a Fritsch Analysette 22 laser diffractometer after ultrasonic pretreatment. Thus, we obtained data on the content of coarse fractions (>0.25 mm) and the particle-size distribution for finer fractions. This procedure was necessary because the large coarse fraction in the bulk soil hampered the measurement of particle-size distribution curves for the finer fractions (Shein et al. 2006).

The soil's specific surface was determined by desorption equilibrium above saturated salt solutions: 3–5 g soil samples were wetted and stored for two weeks in desiccators above water to reach quasi-equilibrium saturation and then placed in desiccators with saturated salt solutions ensuring relative vapour pressures of 0.15,

[1] Albeluvisol in the *World reference base for soil resources 2006*.

0.332, 0.55, 0.86, and 0.98. The desorption of water from the samples continued for about three months, until equilibrium was attained; then the samples were dried at 105 °C and their water content determined. The specific surface was calculated according to the BET method (Shein 2005; Shein and Karpachevskii 2007).

Aggregate-size distribution in the upper horizons (0–10 and 10–20 cm) was determined by dry sieving using a Retsch device (Retsch 2005).

Organic carbon was determined with an AH7529 auto-analyzer at 900–1,000 °C in a flow of oxygen (Shein and Karpachevskii 2007).

Wetting heat (*WH* cal/g) was determined using an OX12K calorimeter and calculated as

$$WH = \frac{K_{\mathrm{k}} \cdot t_{\mathrm{n}}}{P_{\mathrm{d}}}, \tag{12.1}$$

where K_{k} is the heat capacity of the calorimeter, t_{n} is the real rise in temperature, and P_{d} is the mass of absolutely dry soil sample (Shein and Karpachevskii 2007).

The strength of dry aggregates of sizes 3–5 and 5–7 mm was determined in 20 replicates using a cone penetrometer developed by PA Rebinder:

$$P_{\mathrm{m}} = 1.108 \cdot \frac{F}{h^2}, \tag{12.2}$$

where F is the load, kg; h is the depth of the cone penetration, cm; and 1.108 is the coefficient for the cone of 30°. The penetration resistance (P_{m}), measured in kg/cm^2, was determined at various water contents (Khaidapova and Pestonova 2007).

12.3 Results and Discussion

Data on the soil physical properties are summarized in Table 12.1. According to their particle-size distribution, the soils generally have a medium loamy texture; in the surface horizons of the control plot, the texture is coarse loamy; a coarsening of soil texture to loamy sand with the high content of coarse (>0.25 mm) fractions is also observed in the 30–40 cm layer of the plot treated with NPK fertilizers. The two-peaked particle-size distribution curves observed in the experimental plots are typical of poorly sorted morainic deposits.

The control differs somewhat from other variants, having a higher content of coarse particles. The variant treated with manure is characterized by a higher content of the finest particles throughout the profile, which may be related to an initial textural difference between the plots; it is also possible that the increased content of fine particles in this variant is due to the formation of colloidal and fine-clay organo-mineral particles in the manured soil.

Specific surface values are fairly low and constant throughout the profile, except for the 30–40 cm layer of the control and, especially, the plot with NPK and manure

Table 12.1 Physical properties of soddy-podzolic soil

Variants	Depth cm	Particle-size fraction % <0.01 mm	<0.001 mm	>0.25 mm	Bulk density g/cm^3	Aggregates 10–0.25 mm %	$C_{org.}$ %	S_{HO} m^2/g	Plastic limit %	Liquid limit %
Control	0–10	27.61	3.92	24.64	1.56	71.89	1.04	22.01	14.49	19.24
	10–20	28.17	3.90	28.34	1.58	71.71	1.04	20.84	15.10	18.16
	20–30	29.27	4.59	34.18	1.80	–	0.65	18.67	12.60	15.29
	30–40	34.04	4.98	28.64	1.75	–	0.24	26.20	12.33	15.25
Lime	0–10	31.48	4.23	25.30	1.67	76.77	1.20	22.60	15.10	19.88
	10–20	29.54	3.95	24.34	1.65	82.58	1.20	21.80	15.56	19.43
	20–30	31.34	4.23	25.56	1.88	–	1.05	20.79	15.53	20.03
	30–40	25.82	3.41	32.32	1.97	–	0.75	14.10	12.61	14.02
NPK	0–10	32.53	4.77	24.88	1.58	72.43	1.16	21.77	15.16	19.92
	10–20	31.10	4.26	26.50	1.50	71.92	1.20	17.69	14.45	18.96
	20–30	29.13	3.88	35.30	1.67	–	0.67	19.02	12.87	16,59
	30–40	17.39	2.42	59.10	1.75	–	0.26	13.75	12.95	13,71
NPK + manure	0–10	32.12	4.23	20.6	1.44	81.13	1.77	27.83	12.82	26.45
	10–20	31.08	4.20	25.86	1.40	75.10	2.09	31.38	21.38	27.22
	20–30	27.42	3.76	29.02	1.75	–	0.57	20.40	13.20	17.13
	30–40	35.57	6.55	25.20	1.84	–	0.21	46.66	13.30	19.48

C_{org} content of soil organic matter, S_{H2O} soil specific surface

where this index increases. This may be due to an increased affinity for water of the organic substances in these soils and, probably, some difference in the mineralogical composition of the deeper layers (Shein 2005; Shein and Karpachevskii 2007). We may also note an increased water retention capacity of the soil of this plot, in particular, the increased values of the plastic and liquid limits. The wetting heat varies between 0.78 and 3.13 cal/g; according to this index, the soils are slightly hydrophilic. The maximum values of wetting heat are found in the variant with NPK and manure, which may be explained by the addition of hydrophilic organic matter.

It is known that bulk density determined using the Pol'skii auger is usually higher than when measured using rings. However, the auger method may be applied for comparative assessments of bulk density in different variants of the experiment. According to our data, bulk density increases down the soil profile in all variants; the lowest values are observed in the Ap horizon of the variant with NPK and manure, which we attribute to the manuring. The highest values are observed in the limed variant; they are significantly higher than those in the control. Increased bulk density in the limed variant was noted during the survey of 1996–1998, but it was concluded at that time that the difference between the control and the limed variant was within the experimental error.

Comparative analysis of these physical and chemical properties of the upper soil layers from different variants of the experiment shows relatively small differences between major physical characteristics of the soil solid phase. There is a lot of spatial heterogeneity in the values of the studied indices within particular soil profiles and between them. It may be supposed that a somewhat coarser soil texture in the control and a greater clay content in the variant with NPK and manure application are related to the initial heterogeneity in the soil properties rather than to the different agricultural loads on the soils; the degree of changes in the soil physical properties under the impact of different agricultural loads is relatively small. Only the variant with NPK and manure application differs significantly from other variants in having a higher content of the finest particles and, hence, higher water retention capacity at the plastic and liquid limits.

The fact of relatively small differences between the major physical properties of soils under different variants of the experiment prompted a search for other differentiating properties. In particular, the strength of soil aggregates and penetration resistance were studied at different water contents. These characteristics are indicative of the strength of interparticle bonds. The strength of air-dry aggregates in the variants with lime and with NPK and manure is higher than that in the control and NPK variants for both studied depths (0–10 and 10–20 cm) and for both aggregate diameter groups (3–5 and 5–7 mm). These data are in agreement with received wisdom about the effect of lime on the mechanical strength of acid soils (Khaidapova and Pestonova 2007). As for the NPK + manure variant, an increase in the physical strength of aggregates may be explained by the addition of organic substances that favour the development of coagulation bonds. When the soil dries, such bonds may be transformed into stronger mixed and cementing bond which increase the dry stability of soil aggregates (Bouajila and Gallali 2008). Thus, the addition of manure not only increased the soil organic matter content but, also, increased the strength of soil aggregates.

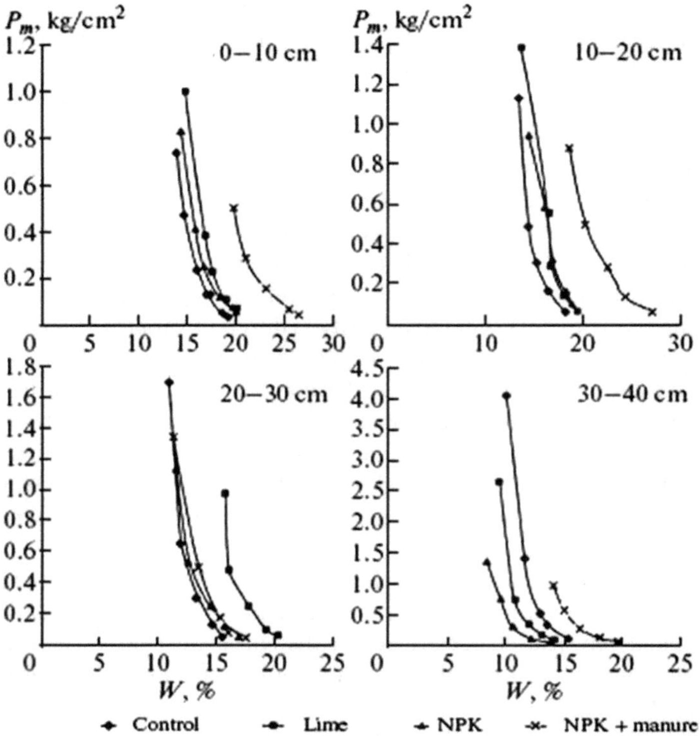

Fig. 12.1 Penetration resistance (P_m, kg/cm^2) dependent on water content (W, % of dry soil mass) in different experimental variants

Interesting results were also obtained from penetration resistance (P_m) tests at different soil water contents (W) from the liquid limit to the plastic limit. The soil is subjected to compression stress and shear stress, and dilatant properties characterizing interaction between the soil particles are clearly manifest (Fig. 12.1). The P_m–W curves for the variant NPK + manure are shifted to the right; i.e. at a given soil water content the soil resistance to penetration in this variant is higher than in other variants due to the formation of coagulation bonds between the particles; the limed variant is also characterized by an increased penetration resistance, which is seen from the steep slope of the P_m–W curves. In all variants, the strength of soil structure (soil resistance to penetration) increases sharply within a relatively narrow range of the soil water contents, which is typical of the soils with a substantial content of coarse particles. As seen from Fig. 12.1, with a decrease in the relative degree of soil moistening, the resistance to penetration increases most significantly in the limed variant; the least increase is observed in the variant with NPK and manure application which, in this case, may be explained by the lubricant action of hydrophilic organic matter. The highest penetration resistance, in the lime-treated soil, is in agreement with earlier published data (Khaidapova and Pestonova 2007); the control and the variant with NPK occupy an intermediate position.

Qualitative analysis does not enable assessment of the reliability of the observed differences between the soils of different variants. To evaluate this, we approximated the obtained curves by the power equation:

$$P_{\mathrm{m}} = \left(\frac{W}{b2}\right)^{-b1} \tag{12.3}$$

where P_{m} is the value of soil resistance to penetration , W is the soil water content, and $b1$ and $b2$ are approximation parameters.

The reliability of differences between the approximation parameters obtained for different variants of the experiment was specially evaluated. In this model, parameter $b2$ characterizes the position of the curve relative to the abscissa axis: the higher the $b2$, the higher the soil resistance to penetration at the given soil water content (i.e. the stronger the interparticle bonds). Parameter $b1$ points to the slope of the curve: the higher the $b1$, the steeper the curve, i.e. the more significant are the changes in penetration resistance with the change in the soil water content. This means that the soil particles come into close contact and display strong internal friction with a decrease in the soil water content.

The results of corresponding calculations and the assessment of reliability of the differences between different variants of the experiment show that, at all the depths, parameter $b2$ for the variant with NPK and manure is higher than for other variants; i.e. the corresponding curve lies higher, which is well seen from Fig. 12.1. Parameter $b1$ is reliably lower for the deep soil layers in this variant, which is seen from the lesser steepness of corresponding curves. Significantly lesser values of parameter $b1$ in the variant with NPK and manure point to stronger interparticle bonds in this variant and to their relatively small changes within the studied range of soil water contents. This parameter is reliably higher in the control and in the limed variant, particularly, in the 30–40 cm layer, which attests to the growing strength of interparticle bonds with a decrease in the soil water content. As already noted, this is typical for the soils with the high content of coarse particles and with the strong aggregation of the particles under the influence of lime.

The analysis of the curves showing the dependence of soil physical properties (penetration resistance) on the soil water content reveals information about the reliability of differences in the parameters of the curves and, hence, in the rheological behaviour of the soils from different variants of the experiment, whereas routine determinations of the major physical properties of the soils did not allow judgements about their differences in different variants of the experiment.

12.4 Conclusions

1. Determination of standard physical properties (particle-size distribution, bulk density, specific surface, aggregate-size distribution, wetting heat) of soddy-podzolic soils in a long-term field experiment with applications of lime, NPK, and NPK + manure revealed that the different treatments produced only minor

effects on these properties. Only the soil of the plot with NPK + manure application was characterized by a somewhat higher content of the finest particles and a higher water retention capacity.
2. Physico-mechanical properties (the strength of soil aggregates and the dependence of penetration resistance on the soil water content) were more sensitive and revealed statistically reliable changes under the impact of different treatments. These soil properties characterize interparticle bonds and their changes according to the degree of soil moistening, i.e. the rheological behaviour of the soils.
3. The parameters of approximation of the curves showing the dependence of penetration resistance on the soil water content for the NPK + manure variant differed significantly from the analogous parameters for other variants, pointing to stronger interparticle bonds and their low dependence on the soil water content (within the studied range) in the soil of this variant. Soils of the control and the limed variant also differed significantly from other soils in the parameter characterizing the increase in the strength of interparticle bonds with a decrease in the soil water content. This may be explained by a somewhat coarser texture of the soil of the control and by a stronger aggregation of the soil particles in the limed variant.

Acknowledgments This study was supported by the Russian Foundation for Basic Research, projects Nos. 10-04-00993-a and 11-04-00284-a.

References

Bouajila A, Gallali T (2008) Soil organic carbon fractions and aggregate stability in carbonate and non-carbonate soils in Tunisia. J Agron 7(2):35–39

Dutartre P, Andreux F et al. (1993) Influence of content and nature of organic matter on the structure of some sandy soils from West Africa. Geoderma 56:156–165

Khaidapova DD, Pestonova EA (2007) Strength of inter-particle bonds in soil pastes and aggregates. Eurasian Soil Sci 11(3):1330–1335

Kiryushin BD, Safonov AF (2002) Stages of the long-term experiment. In: Ninety years of the experiment at the Timiryazev Agricultural Academy: major results. TimiryazevAgricultural Academy, Moscow, pp 23–32 (Russian)

Munkholm LJ, Schjonning P, Debosz K, Christensen BT (2002) Aggregate strength and mechanical behaviour of a sandy loam soil under long-term fertilization treatment. Eur J Soil Sci 53:129–137

Rachman A, Anderson SH, Gantzer CJ, Thompson AL (2003) Influence of long-term cropping system on soil properties related to soil erodibility. Soil Sci Soc Am J 67:1328–1335

Retsch (2005) Operating instructions for sieving machine type AS200 control. Retsch GmbH, Haan

Shein EV (2005) A course in soil physics. Soil Science Faculty, Lomonosov Moscow State University, Moscow (Russian)

Shein EV, Karpachevskii LO (2007) Theory and methods of soil physics. Grif & Co, Moscow (Russian)

Shein EV, Milanovskii EY, Molov AZ (2006) The effect of organic matter on the difference between particle-size distribution data obtained by the sedimentometric and laser-diffraction methods. Eurasian Soil Sci 39(Suppl 1):84–90

Chapter 13
Evolution of Chernozem in the Complex Section at Storozheve, Ukraine

Y. Dmytruk, Z. Matviyishyna, and A. Kushnir

Abstract The Storozheve earth rampart displays three soils: the topmost on the surface of the bank, the *Cossack soil* formed in loess deposited about 18,000 years ago and buried by the building of the bank 350 years ago, and the initial soil in loess. The initial soil is depleted in heavy metals and exhibits silt redistribution, reflecting cool, wet conditions. The Cossack soil is Typical chernozem with strong accumulation of organic matter and structural development within the root zone, with a peak of heavy metals in the section but with a minimum in the soil genetic horizons, formed under cooler, wetter conditions than the present. The new soil on the bank shows weak profile development reflecting degradation of the ecosystem under a warming and drying climate.

13.1 Introduction

The study of soil genesis enables the assessment of diagenetic changes in soils which, nowadays, are exposed to intensive man-induced pressures. It requires detailed analysis of the widest possible range of attributes, especially from chronosequences. The magisterial *curriculum vitae* of the chernozem by Krupenikov et al. (2011) draws upon chronosequences in sediments deposited and subsequently exposed by early Holocene advance and recession of the Black Sea and, also, soils buried beneath Trajan's bank about 2,500 years ago. Here we add evidence from palaeosols buried 350 years ago beneath the ramparts at Storozheve, near Poltava in the Ukraine.

In 2008–2011 Z. Matviyishyna, O. Parhomenko and A. Kushnir of the Institute of Geography of NAS of Ukraine participated in studies of the Storozheve-kurgan complex, at the invitation of archaeologists O. Kovalenko and R. Lugovi

Y. Dmytruk (✉)
Chernivtsi National University, Chernivtsi, Ukraine
e-mail: yuri.dmy@gmail.com

Z. Matviyishyna • A. Kushnir
Institute of Geography NAS of Ukraine, Kiev, Ukraine

D. Dent (ed.), *Soil as World Heritage*, DOI 10.1007/978-94-007-6187-2_13,

(Matviyishyna et al. 2011). Y. Dmytruk analysed the particle-size distribution and geochemical composition of the soils and sediments which lend insight into the changing environment.

13.2 Site and Methods

The Storozheve rampart was built between 1630 and 1680 with spoil from the production of saltpetre. Saltpetre (an important military product at the time) was extracted by boiling up humus-rich topsoil. The section exhibits three soils: at the surface, the modern soil formed during the last 350 years; the middle one buried by the rampart, which we have called the *Cossack soil*; and the underlying Bug loess dating from about 18,000 years ago.

The section was described using the Ukrainian index of genetic horizons. Humus content was analysed by Tyurin's method modified by Simakova, reaction by potentiometry, particle size by Katchinsky's procedure and heavy metals dissolved in 1 M nitric acid by atomic-absorption spectroscopy.

We calculated mean and standard deviation; coefficient of radial differentiation (Kr, the ratio of indicators in each genetic horizon to the mean for their profile) elaborated because, in the absence of the soil parent material for comparison, eluvial accumulative coefficients do not completely reflect the operating processes; coefficient of silt differentiation (Kd, the ratio of maximum silt content to its minimum in a given profile), which highlights any accumulation of silt; and, finally, index of saturation of soil with heavy metals (Dmytruk 2006), the geometric mean of the coefficients of the concentration of heavy metals calculated as the ratio of their content in the sample to the background content of the same element (See also Fatyeyev and Paschenko 2003).

13.3 Results and Discussion

For Kr, values greater than 1 indicate an increase and values less than 1 decrease compared with the mean, similarly with Kd. For the index of saturation of soil with heavy metals, less than 0.9 indicates scattering of heavy metals, values between 0.9 and 1.1 indicate a balance between scattering and accumulation, and values greater than 1.10 suggest accumulation.

13.3.1 The Soil on the Embankment

This is described as:

H(k) 0–5 cm	Brown-cinnamon grey, poorly compacted sod; many roots; reacts with HCl

Hk 5–32 cm	Humus layer, light greyish brown speckled; loose, weak crumb structure and horizontal banding; many burrows 6–8 cm diameter infilled with light grey soil; weak reaction with HCl, minor carbonates (1–3 mm); clear wavy boundary
H(e)k 32–53 cm	Humus eluvial, grey brown to brown, depleted of free iron oxides; sandy silty to heavy silty; strong medium nutty structure; some burrows 7–20 cm in diameter, many worm burrows; small carbonate nodules; clear irregular boundary
Ih(k) 35–135 cm	Lumpy mixed materials making up the bank; heavy loam; many worm burrows filled with material darker than the modern soil but few large burrows; no carbonate nodules; clear boundary corresponding with many tree stumps

Low levels of heavy metals, especially cadmium, characterize the soil on the bank (Table 13.1). The mixture of materials thrown into the bank bequeaths a confused picture compared with undisturbed ecosystems, but the low content of biogenic elements (zinc, manganese, copper) indicates little biological accumulation within the modern ecosystem. At the same time, the low content of heavy metals suggests minimal human impact so this area may be taken as a background against which chemical contamination may be monitored.

Consideration of the variability of heavy-metal content, just for the upper soil (Table 13.2), confirms the diverse conditions in the surface layer of natural soils. The actual amount of heavy metals decreased with depth in each soil and down the section as a whole – apparently according to the lithology of the deeper layers. Based on the index of saturation with heavy metals, we observe that biological accumulation of metals, except for manganese, is restricted to the upper humus horizon (Table 13.3); high levels of organic matter contribute to fixation of metals. Beneath the humus horizon, silt eluviation dominates the processes of the removal of heavy metals, probably along with adsorptive complexes. The deep illuvial horizon (which has low humus content but the greatest amount of fine particles) has about the average content of heavy metals – migration and accumulation are in balance.

The content of silt in the modern soil is the least in the whole section and reflects the slowness of profile differentiation in current conditions. The surface layers collect wind-blown dust, and there is some accumulation of fine particles in the lowest layer, which we may consider an illuvial horizon. The humus content is low but its distribution even; likewise soil reaction with the surface horizon is close to neutral and there is a slight increase in pH with depth.

13.3.2 The Cossack Soil

This soil, buried 350 years ago, is fully represented in the section with clearly differentiated horizons. It is Typical chernozem with a thick humus horizon,

Table 13.1 Indicators of genetic horizons in the chronosequence

	Particle-size fractions (%)							Heavy metals (mg/kg dry soil)								
Layer	1.0–0.25	0.25–0.05	0.05–0.01	0.01–0.005	0.005–0.001	<0.001	∑ < 0.01	Pb	Cd	Cu	Ni	Cr	Zn	Mn	Humus, %	pH
H(k)	5.50	1.34	45.8	15.4	1.6	30.4	47.4	16.9	0.66	12.0	20.7	17.7	32.1	355	2.20	6.46
H(e)k	7.0	0.84	41.5	9.20	13.2	28.4	50.7	6.9	0.47	5.2	6.9	8.3	11.7	152	1.80	6.32
Ih	7.40	0.80	27.5	20.5	10.1	33.8	64.3	12.7	0.69	9.90	16.0	15.0	24.5	293	0.85	6.53
II[Hk]	5.50	0.42	34.7	17.5	8.90	33.0	59.3	14.4	7.20	13.5	19.6	14.2	26.1	466	2.37	7.01
II[Hpk]	6.50	1.10	39.1	24.5	7.90	20.9	53.3	11.9	0.5	11.4	18.0	16.5	26.9	342	1.50	7.09
II[Phk]	4.70	0.91	25.0	27.6	13.5	28.4	69.4	15.4	0.58	14.6	24.0	21.8	35.8	469	1.43	7.25
II[Pk]	4.0	0.96	35.2	18.3	10.8	30.8	59.8	15.5	0.32	10.4	20.4	12.2	28.4	325	1.23	7.24
III[Hk]	2.80	1.0	29.1	6.80	20.1	40.2	67.1	19.6	0.63	11.2	23.8	14.6	32.5	316	0.67	7.45
III[Phik]	3.30	0.50	42.8	3.90	5.40	44.2	53.4	12.7	0.32	7.2	13.8	8.4	19.8	181	0.41	7.41
III[Pk]	0.80	0.56	46.2	19.3	8.40	24.8	52.5	14.8	0.75	7.3	15.4	12.2	23.4	190	0.45	7.48

Table 13.2 Variability of soil attributes

	Soils (M ± m)		
Indices	On the bank ($n = 3$)	Cossack soil	Deep buried ($n = 3$)
Humus (%)	1.20 ± 0.69	1.65 ± 0.50	0.51 ± 0.14
pH_{water}	6.44 ± 0.18	7.15 ± 0.12	7.45 ± 0.04
Particle-size fractions (%)			
1.0–0.25	6.63 ± 1.0	5.18 ± 1.08	2.30 ± 1.32
0.25–0.05	0.99 ± 0.30	0.85 ± 0.30	0.69 ± 0.27
0.05–0.01	38.2 ± 9.57	33.5 ± 6,02	39.3 ± 9.05
0.01–0.005	15.0 ± 5.69	22.0 ± 4.88	9.98 ± 8.18
0.005–0.001	8.31 ± 6.0	10.3 ± 2.47	11.3 ± 7.79
<0.001	30.8 ± 2.73	28.2 ± 5.25	36.4 ± 10.2
$\sum < 0.01$	54.1 ± 8.98	60.5 ± 6.67	57.7 ± 8.20
Heavy metals (mg/kg dry soil)			
Pb	12.2 ± 5.02	14.3 ± 1.68	15.7 ± 3.54
Cd	0.61 ± 0.12	0.57 ± 0.17	0.57 ± 0.22
Cu	9.02 ± 3.51	12.4 ± 1.93	8.55 ± 2.25
Ni	14.6 ± 7.02	20.5 ± 2.56	17.7 ± 5.34
Cr	13.6 ± 4.83	16.2 ± 4.14	11.7 ± 3.3
Zn	22.8 ± 10.3	29.3 ± 4.44	25.2 ± 6.45
Mn	267 ± 104	400 ± 78	229 ± 76

crotovinas and clear distinction from the underlying, less loamy parent material, described in the field as:

II[Hk]1.35–1.75 m Humus horizon – uniformly dark grey to black, clay loam, granular to lumpy structure, friable, occasional crotovinas, abundant roots, gradual boundary

II[Hpk]1.75–1.90 m Upper transitional humus horizon – pale grey becoming lighter with depth, clay loam, no visible carbonates but reacts with HCl, crotovina filled with dark grey soil, many worm burrows and roots, gradual boundary

II[Phk]1.90–2.03 m Lower transitional horizon – uniform pale grey, clay loam, abundant carbonate nodules, many worm burrows and crotovinas, clear irregular boundary

II[Pk] 2.03–2.17 m Parent material – pale-brown loess, clay loam, lumpy powdery, abundant nodules and powdery carbonate efflorescences

The most notable geochemical characteristic is the concentration of heavy metals, the maximum for the whole section except for lead (Table 13.1). Despite the increased concentration, their variability is minimal – which may indicate constancy of conditions of development without much human interference. The increased content of the biogenic elements zinc, manganese and copper, compared with the profiles above and below (Table 13.2), is evidence of active biological accumulation, but the index of saturation highlights that accumulation of metals is characteristic only for the lower transitional horizon (Table 13.3). At this point

Table 13.3 Coefficients of radial differentiation for particle size and heavy metals

	Particle-size fractions							Heavy metals							
Soil layer	1.0–0.25	0.25–0.05	0.05–0.01	0.01–0.005	0.005–0.001	<0.001	$\sum < 0.01$	Pb	Cd	Cu	Ni	Cr	Zn	Mn	Is[a]
H(k)	0.95	1.47	1.29	0.81	0.17	1.03	0.82	1.26	1.14	1.09	1.15	1.17	1.21	1.03	1.15
H(e)k	1.21	0.92	1.17	0.48	1.40	0.97	0.88	0.52	0.81	0.47	0.38	0.55	0.44	0.44	0.50
Ih	1.28	0.88	0.77	1.08	1.07	1.15	1.11	0.95	1.19	0.90	0.89	0.99	0.92	0.85	0.95
II[Hk]	0.95	0.46	0.98	0.92	0.94	1.12	1.03	1.08	1.24	1.23	1.09	0.94	0.98	1.36	1.12
II[Hpk]	1.12	1.21	1.10	1.29	0.84	0.71	0.92	0.89	1.12	1.03	1.00	1.09	1.02	1.00	1.02
II[Phk]	0.81	1.00	0.70	1.45	1.43	0.97	1.20	1.15	1.00	1.32	1.34	1.44	1.35	1.37	1.28
II[Pk]	0.69	1.05	0.99	0.96	1.14	1.05	1.04	1.16	0.55	0.94	1.14	0.81	1.07	0.95	0.92
III[Hk]	1.22	1.45	0.74	0.68	1.78	1.10	1.16	1.25	1.11	1.30	1.34	1.24	1.29	1.38	1.27
III[Phik]	1.43	0.72	1.09	0.39	0.47	1.21	0.93	0.81	0.56	0.84	0.78	0.71	0.78	0.79	0.75
III[Pk]	0.35	0.81	1.17	1.93	0.74	0.68	0.91	0.94	1.32	0.85	0.87	1.04	0.93	0.83	0.96

[a]Index of saturation with heavy metals

there was a geochemical barrier associated with changing redox conditions and precipitation of carbonates. The parent material is characterized by low content of heavy metals. Redistribution of metals, including their accumulation by biological activity in soils and vegetation, takes place only with pedogenesis.

This is confirmed by the coefficients of radial differentiation: for all genetic horizons, heavy-metal content is greater than that of the parent material (Table 13.1). The lack of accumulation of metals (except cadmium) in the upper, humus horizons, where saturation indices are close to unity, may be a function of biological activity and their accumulation at the soil-bedrock boundary. Cadmium, copper and manganese *are* concentrated in the humus horizon, while the parent material contains most nickel and lead. Horizons of unambiguous scattering are not seen, which also confirms the dependence of heavy-metal concentration on soil formation.

The differentiation of silt (0.01–0.005 mm and 0.005–0.001 mm) indicates wetter conditions in the early stages of soil genesis and radical change later. Compared with the profiles above and below, the content of silt is the least but clay the most. The high silt content of the upper humus horizon may be a result of its burial by the bank which is built of siltier material; the middle part of the profile shows depletion of silt and a gradual increase towards the parent material.

The humus content remains high in spite of the mineralization that has occurred since its burial; it is still higher than the modern soil (Table 13.2). The Cossack soil remains recognizably a chernozem: its morphology and micromorphology, as well as its thickness, are just as they were when it was buried.

13.3.3 The Initial Soil in Loess

Under the Cossack soil, at a depth of 2.17–2.67 m, there is a brown soil with poorly developed horizons but still clearly differentiated from the underlying loess. It was described in the field as:

III[Hk]2.17–2.34 m	Weakly humose – uniform light brown; clay loam; poorly compacted, lumpy; carbonates; crotovina filled with grey soil; worm burrows; chips and cracks coated with carbonates; gradual boundary
III[Phik]2.34–2.46 m	Transitional – pale-brown sandy clay loam; poorly compacted; white carbonate efflorescences, especially in the lower part; many worm burrows; some crotovinas filled with whitish loess
III[Pk] 2.46–2.70 m	Parent material, loess – very pale-brown sandy silt loam; loose, lumpy; many dark-grey brown and whitish crotovinas

The initial soil developed in loess has concentrations of heavy metals close to those of the modern soil on the top of the bank and with similar variability. Here we find the highest level of lead, which confirms the absence of human influence. The index of soil saturation with heavy metals shows a clear accumulation in the upper horizon with the same specific scattering from transitional and background content in the parent material (Table 13.3). The parent loess also contains more cadmium than the overlying soil horizons. Pronounced accumulation of manganese, copper, nickel and zinc is restricted to the upper humus horizon – perhaps the result of biological activity. The scattering of values for heavy metals is characteristic of the transitional horizon; the lowest values for several metals may be explained by the thinness of the layer and the low content of organic matter (Table 13.2).

The parent material is loamy with a larger sand fraction, especially medium and coarse sand, which is characteristic of loess deposits. Compared with the parent material, the silt content is greater in the overlying soil horizons. Differentiation of the profile and its coefficient of differentiation are close to modern Grey forest soils (Tables 13.1 and 13.2) but less distinct because of the calcareous nature of the parent material and the shallow profile. The poor humus content is the result of thousands of years of the mineralization.

13.3.4 Cluster Analysis

We used cluster analysis to compare features of different age and genetic sequence. Affinity of various horizons indicates comparable conditions of formation. Clustering according to the content of heavy metals was informative. Genetic horizons of the buried Cossack soil and its parent material have similar geochemical characteristics and formed one cluster (Fig. 13.1). The other grouping combines the upper horizons of the modern soil with horizons of soil buried under the bank: grouping eluvial humus horizon of modern soil with the upper humus horizon of the buried soil.

Figure 13.2, cluster analysis on all parameters that were determined, showed that the strongest grouping is of the lower horizons of buried soil and the same humus horizon soil on the rampart. The second cluster consists of two subgroups: in the one are horizons of the buried Cossack soil and the humus horizon soil on the modern surface and in the second, horizons of different ages – parent material of the Cossack soil, the humus horizon of the initial soil on loess and, unexpectedly, the deepest layer of the modern soil.

We see that the soil buried about 350 years ago has remained close to the archetypal chernozem, with strong evidence of biological activity and humus accumulation (its humus content is the greatest of the three strata despite diagenesis). These features are inherited from a long period of favourable warmth to rainfall ratio, wetter than today's, under grassy vegetation without evident human intervention.

The lower buried soil formed in loess is not depleted of chemical elements but reflects initial hydromorphic soil formation. It has the lowest content of organic matter and most differentiation of silt – comparable to forest soils under cool conditions.

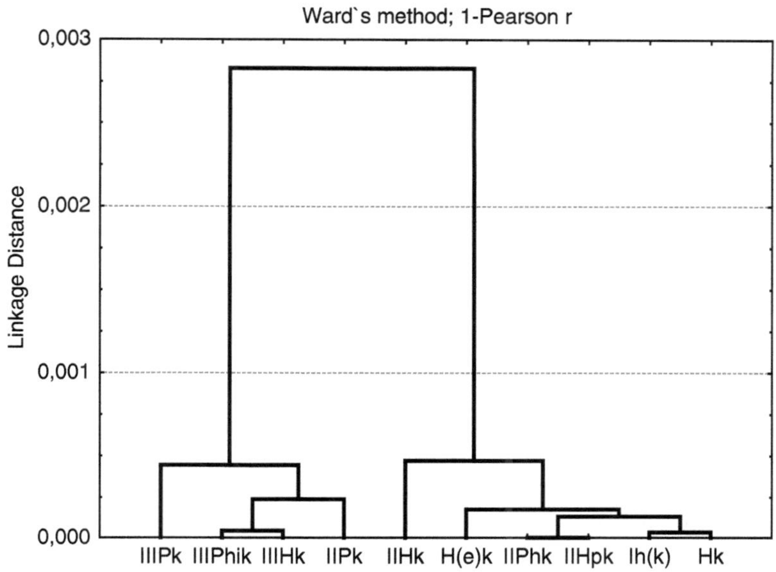

Fig. 13.1 Dendrograms of the genetic horizons of soils of different ages based on the total content of heavy metals

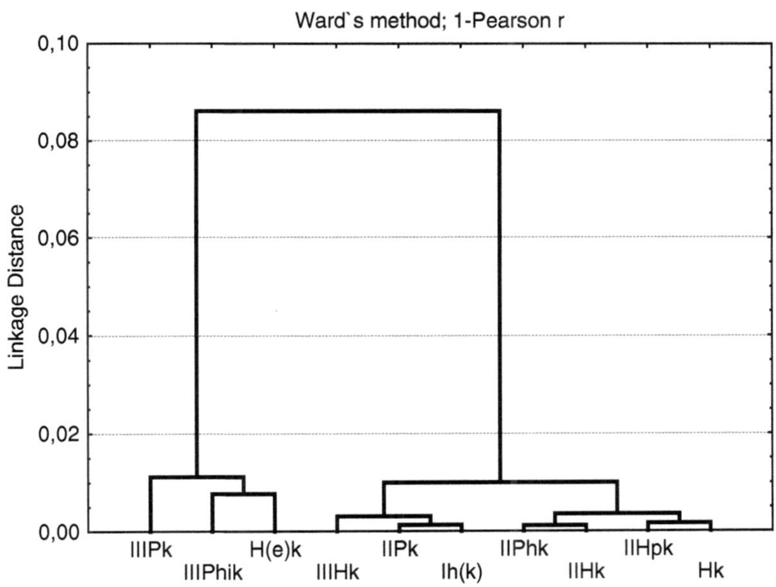

Fig. 13.2 Dendrograms of genetic horizons of soils of different ages according to humus content, soil reaction, grain size and total content of heavy metals

Over some 350 years, the modern soil on the embankment has developed some specific features: first, homogenization of the fine fraction without much redistribution, some bioaccumulation in the topmost horizon while human influence somewhat modified its heavy-metal content, and weak accumulation of humus – still not attaining the levels of the buried soil which suggests a slowing down of sod and humus formation.

13.4 Conclusions

The Storozheve chronosequence comprises an initial soil with significant downwashing of silt, suggesting a cool humid climate; a chernozem of Cossack times buried beneath the embankment 350 years ago; and the modern soil formed on the bank. Comparison between the three profiles in terms of humus content, particle-size distribution, gross content of heavy metals, coefficients of saturation of soils by heavy metals and coefficients of their radial redistribution and, finally, cluster analysis revealed:

1. The initial soil in loess retains hardly any humus, is depleted in heavy metals and exhibits clear downwashing of silt – features of strong leaching under a cool climate.
2. The soil of Cossack times, buried by the bank 350 years ago, is a Typical chernozem, exhibiting the highest content of heavy metals but the least variability within the genetic horizons, active accumulation of biogenic elements by active, humus-accumulative sod formation. A biological-chemical barrier at the interface with the parent material is evident. We deduce that this soil was formed under cooler, wetter conditions than the present.
3. The modern surface soil of the embankment has some features of chernozem formed under sod. Heavy-metal content, except for cadmium, is relatively low – suggesting insignificant biological accumulation; the process of profile differentiation is slowing down in the face of increasing aridity.

References

Dmytruk Y (2006) Ecological and geochemical analysis of soil agroecosystems. Ruta, Chernivtsi (Ukrainian)

Fatyeyev AI, Paschenko YV (2003) Background content of trace elements in soils of Ukraine. NSC of Soil Science and Agrochemical Institute, Kharkov (Ukrainian)

Krupenicov IA, Boincean BP, Dent DL (2011) The Black Earth. Ecological principles for sustainable agriculture on chernozem soils. Springer, Dordrecht

Matviyishyna Z, Parhomenko O, Kovalenko O et al (2011) Soil archeological investigation on the territory of the central steppe. In: Kovalenko O (ed) Integration of archeological and palegeographical investigation. Poltava PNPV (Ukrainian), pp 4–19

Chapter 14
Climate Change and Its Impact on Soil Productivity in Moldova

M.D. Vronskih

Abstract Analysis of long-term data (1945–2007) for Moldova reveals trends in agro-climate parameters and related harvest fluctuations of field crops, vines and fruits. Over the period 1945–1970, mean annual temperatures decreased by 0.72 °C; since then they have increased by 1.41 °C. Seasonal temperatures followed similar trends. Mean annual and seasonal temperatures in Moldova are already higher than the optimum for all the main crops; summer temperatures have the greatest effect on yields, which were reduced by higher-than-average temperatures. Mean annual rainfall during the same period recorded variable trends: from 1945 to 1965 it increased by 142 mm; from 1965 to 1977, decreased by 50 mm; and from 1976 to 2010, increased again by 52 mm, overall an upward trend, especially in summer and autumn. Analysis of long-term meteorological and statistical data shows a clear relationship between mean rainfall and crop yields: below-average rainfall (−17 %) reduced wheat yields by 11 % and maize by 15 %; very dry years (−31 %) reduced wheat yields by 16 % and maize by 23 %. Less-than-average rainfall during the growing period had the greatest impact. As we might expect, the volume of summer rain, ranging from 134 to 302 mm, had a big effect of on crop yields: decrease in summer rainfall by 33 % from the mean of 203 mm depressed yields of winter wheat by almost 10 %, maize 16.8 %, sugar beet 14.5 %, grapes 5.2 % and fruit crops by 9.2 %.

14.1 Introduction

The probable and possible consequences of global climate change have been a topic of debate over the last 20 years in scientific circles, politics, business and the media. Analysis of much published material and statistical data leads us to conclude that

M.D. Vronskih (✉)
Selectia Research Institute for Field Crops, Balti, Republic of Moldova
e-mail: selectia3@gmail.com

D. Dent (ed.), *Soil as World Heritage*, DOI 10.1007/978-94-007-6187-2_14,

the observed changes in meteorological parameters result from complex, often multidirectional interaction of many man-made and natural factors. Differences in the positions of various authors rather emphasize the leading role of some of them in the system of *factor and consequence*. I shall not discuss the causes and mechanisms of climate change but rather its impact – seeking answers to these questions:

- To what extent and in what direction will changing meteorological parameters affect the productivity of the soil and, in particular, the yields of major agricultural crops?
- What are the optimal values of temperature and rainfall (annual, seasonal and monthly) that generate the highest productivity in the soils of Moldova?
- How similar to, or different from, the optimal values are the actual meteorological indices in Moldova?
- In this context, what is the safety margin provided by the compensatory capacities of plants?
- Finally, how can (or can't) management compensate for the negative effects of climate change over such a sensitive sector as agriculture?

14.2 Data and Methodology

Statistical data on the harvests of six major crops were analysed to determine their dependence on the dynamics of air temperature and precipitation (mean annual, seasonal and monthly) for different agro-climatic zones of Moldova. Data were calculated from eight meteorological stations across the country that there have maintained observations since 1945 (about 65 years), and some data for the last 156 years were used to establish the long-term trends. From the initial meteorological parameters, GPC attributes and coefficients of temperature stability, moisture stability and continentality were calculated.

14.3 Results and Discussion

Moldova lies within the moderate-continental climatic zone characterized by long, hot summers and relatively mild winters. Mean annual temperatures range from 7.5 °C in the north to 10.5 °C in the south, accumulated temperature above 10 °C ranges from 2,750 to 3,300 °C, and the frost-free period varies between 165–196 and 260–270 days. Extreme values are a characteristic: summer maxima reach 39–40 °C in the north and 40–41 °C in the south, winter minima fall as low as −27 to −31 °C in the south and −33 to −36 °C in the north. Rainfall ranges from 380–400 mm in the south to 450–550 mm in the north.

The predominant soils are the various subtypes of chernozem that extend across 70 % of the country. One of the risk factors is hilly terrain susceptible to erosion; more than 40 % of the country has suffered some degree of soil erosion, and almost 25 % is strongly eroded. Due to the high population density (115 people per km^2), the area per person is only 0.9 ha of which 0.44 ha is arable.

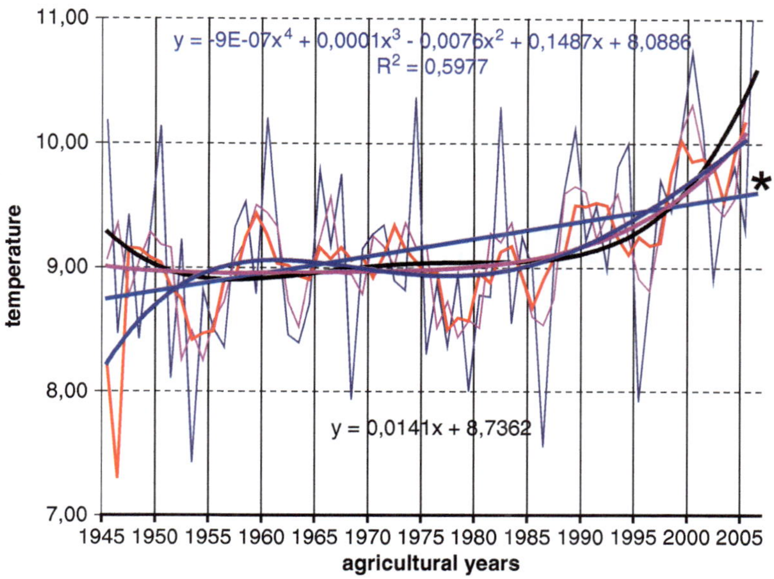

Fig. 14.1 Mean annual temperature. Key: Bold blue, starred – Multiyear linear trend; Bold blue – Polynomial 5-yearly; Black – Polynomial mean; Bold violet – Polynomial 3-yearly; Red – Multiyear 5-year means; Fine violet – Multiyear 3-year means; Fine blue – Annual means

Analysis of long-term weather data (1945–2007) reveals:

- *Zonal differences in mean temperatures* between north and south: mean annual temperatures from 8.62 to 9.53 °C, 8.91–10.47 °C in autumn, −2.84 °C to −1.35 °C in winter, 8.94–9.75 °C in spring and 19.45–20.95 °C in summer.
- *Zonal differences in rainfall*: mean annual rainfall ranges from 559.9 in the north to 520.1 mm (8 % less) in the south, 121.0–113.4 mm (7 % less in the south) in autumn. 95.5–101.9 mm (7 % more in the south) in winter, 126.3–119.8 mm (5 % less in the south) in spring and 217.1–185.0 mm (17 % less in the south) in summer.

Analysis of temperature changes over the past 65 years shows that during the period 1945–1970, the mean annual temperature decreased by 0.72 °C (−0.028 °C per year) and then for 40 years (from 1971 to 2010) consistently increased by 1.41 °C (+0.035 °C per year), an overall increase of 0.69 °C (Fig. 14.1). Seasonal temperatures follow similar trends: +0.71 °C (+0.0263 °C/year) in autumn, +2.05 °C (+0.0315 °C/year) in winter, +1.12 °C (+0.017 °C/year) in spring and +1.65 °C (+0.025 °C/year) in summer.

Mean annual rainfall recorded more variable trends (Fig. 14.2): from 1945 to 1965 it increased by 142 mm (7.1 mm/year); then from 1965 to 1977 decreased by 50 mm (−1.9 mm/year); and increased again from 1976 to 2010 by 52 mm (+3.6 mm per year). Thus, the overall trend over the past 65 years was upwards by 144 mm (+2.2 mm/year). Autumn precipitation increased by an average of 45 mm, winter precipitation decreased by 21 mm (−0.32 mm/year), spring's increased by 37 mm (+0.49 mm/year) and summer's by 46 mm (+0.71 mm/year).

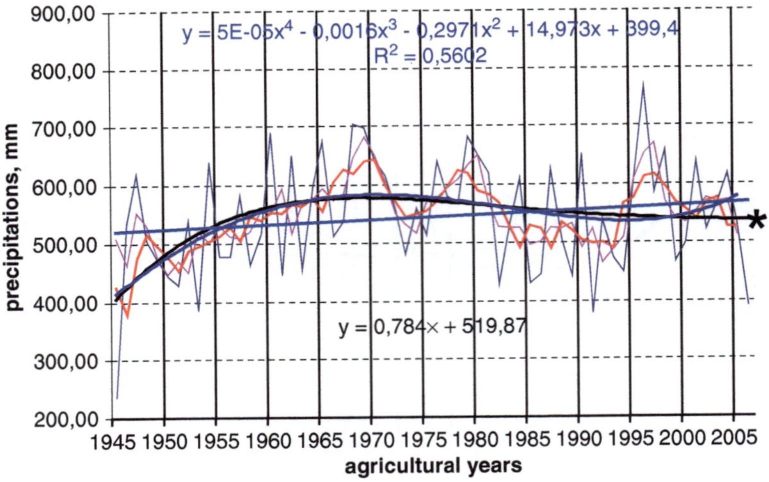

Fig. 14.2 Mean annual precipitation. Key: Bold blue, starred – Multiyear linear trend; Bold blue – Polynomial 5-yearly; Black – Polynomial mean; Bold violet – Polynomial 3-yearly; Red – Multiyear 5-year means; Fine violet – Multiyear 3-year means; Fine blue – Annual means

14.3.1 Effects on Soil Productivity

Temperature

In general, changes in temperature and rainfall were accompanied by predictable changes in crop yields. Table 14.1 summarizes relationships between harvests and mean air temperature. The average increase in mean annual temperature of 1.81 °C was accompanied by a decrease in rainfall of 70.2 mm (−14.1 %). At the same time, the yield of winter wheat decreased by 2.8 centners/ha (−12.4 %), maize by 2.8 c/ha (−10.4 %), sunflower by 1.45 c/ha (−10.6 %), sugar beet by 7.5 c/ha (−3.4 %), grapes by 8.5 c/ha (−20.8 %) and fruits by 9.9 c/ha (−23.2 %). Significantly, a further increase in mean annual temperature in the group of extremely hot summers, accompanied by a diminution of rainfall (by 127 mm or −30.7 %), was accompanied by an even greater reduction of crop yields: winter wheat by 6.6 c/ha (−31 %), maize by 3.2 c/ha (−13 %), sunflower by 1.5 c/ha (−13 %), sugar beet by 97.3 c/ha (−14 %) and grapes by 8.6 c/ha (−23 %); the reduction of fruit yield was within the limits of experimental error.

Analysis of the influence of mean annual temperatures on crop yields shows that there is a narrow range of parameter values (up to 8.7 °C) when an increase in temperature was accompanied by a proportional increase in crop yield. Any further increase in temperature was accompanied by a decrease of yield. The formulaic calculations show that, on the segment of the curve up to the optimum temperature, productivity increased along with temperature: winter wheat by 1.96 centners per hectare per degree C (9 %), maize by 2.30 c/ha (8 %), sunflower by +1.08 c/ha (8 %), sugar beet by 14.2 c/ha, grapes by 8.75 c/ha (19 %) and fruits by 5.7 c/ha (11 %) for

Table 14.1 Effects of mean annual temperature on crop yields in Moldova, 1945–2007

	No. of years	t (°C)	Mean annual rainfall, (mm)	Crop yields (centners per hectare)					
				Winter wheat	Maize	Sunflower	Sugar beet	Grapes	Fruits
Warm years	23	+9.99	493.8	22.5	26.8	13.65	221.9	40.8	42.6
Including hot years	5	+10.92	413.8	21.1	24.1	11.9	202.7	38.0	51.7
Cold years	39	+8.67	569.0	25.3	29.6	15.1	229.5	49.3	52.5
Including extremely cold years	12	+8.08	540.8	27.7	27.3	13.4	230.0	46.6	50.8
All years	62	+9.10	543.0	24.3	28.5	14.6	226.7	46.2	48.8
Warm compared with cold years	–	+1.81	−70.2	−2.8	−2.8	−1.45	−7.5	−8.5	−9.9
	–		−14 %	−12 %	−10 %	−11 %	−3 %	−21 %	−23 %
Hot compared with extremely cold years	–	+2.84	−127.0	−6.6	−3.2	−1.5	−27.3	−8.6 %	+0.9
	–		−31 %	−31 %	−13 %	−13 %	−14 %	−23 %	+2 %

each 1 °C. Beyond the optimum, further increase in temperature was accompanied by a decrease in yields: winter wheat by 1.7 c/ha (−7 %), maize by 2.36 c/ha (−8 %), sunflower by −1.69 c/ha (−11 %), sugar beet by 36.03 c/ha (−15 %), grapes by 2.28 c/ha (−4 %) and fruit by 3.73 c/ha (−7 %) for each 1 °C. There is also a zonal aspect to the response of crop yield to temperatures above the optimum – in the south of the country, yields decreased by more than in the north: winter wheat by 5 %, maize by 9 %, sunflower by 3 %, grapes by 1 % and fruits by 8 %.

Seasonal temperatures also had distinctive effects (Tables 14.2, 14.3, 14.4, and 14.5).

From Table 14.2, we see that an increase in mid-autumn temperature of 1.26 °C suppressed the yield of winter wheat by 4.09 c/ha (18 %). There were lesser decreases for the other main crops: maize by1.5 c/ha (5 %), sunflower by 0.9 c/ha (6 %), sugar beet by 7.6 c/ha (0.3 %), grapes by 1.1 c/ha (2 %) and fruit by 3.9 c/ha (8 %). Further increase in autumn temperatures (from 10.17 to 10.65 °C) further suppressed crop yields; winter wheat was again the most sensitive, declining by 8.5 c/ha (48 %); fruits declined by 6.0 c/ha (15 %), and for the other crops the loss of yield was between 2 and 6 %.

Table 14.3 shows that the increase in winter temperatures by +3.18 °C in the years with warm winters was accompanied by a significant increase in crop yields: for wheat by 4.2 c/ha (+16 %), maize 2.5 c/ha (+8 %), sunflower 0.75 c/ha (+5.0 %) and grapes 3.8 c/ha (+8 %). The yield of sugar beet was within experimental error, but fruits decreased by 3.5 c/ha (8 %). Further increase in winter temperatures (from −0.7 °C to +0.32 °C) was accompanied by a greater increase in yields: winter wheat by 9.85 c/ha (39 %), maize 5.35 c/ha (19 %), sunflower 1.0 c/ha (7 %), sugar beet 25.8 c/ha (12 %), and grapes 5.15 c/ha (12 %). Fruit yields remained suppressed, by 3.0 c/ha (−7 %).

Table 14.2 Effects of autumn temperature regime on crop yields in Moldova, 1945–2007

	No. of years	t (°C)	Mean annual rainfall	Crop yields (centners per hectare)					
				Winter wheat	Maize	Sunflower	Sugar beet	Grapes	Fruits
Years with warm autumn	32	+10.17	540.5	22.2	27.8	14.1	223.0	45.6	46.9
Including hot autumns	11	+10.65	546.1	17.7	26.5	14.15	226.8	43.3	41.2
Years with cool autumn	30	+8.91	545.5	26.4	29.3	15.0	230.6	46.7	50.8
Including cold autumns	11	+8.41	518.0	26.2	28.0	14.2	221.0	41.9	47.2
All years	62	+9.56	551.0	24.3	28.5	14.6	226.1	46.2	48.8
Warm compared with cool autumn	c/ha	+1.26	−5.0	−4.09	−1.5 %	−0.9	−7.6	−1.1	−3.9
	%		−1 %	−18 %	−5 %	−6 %	−0.3 %	−2 %	−8 %
Hot compared with cold autumn	c/ha	+2.24	+28.1	−8.5	−1.5	+0.3	+5.8	+1.4	−6.0
	%		+5 %	−48 %	−6 %	+2 %	−2 %	+3 %	−15 %

During the winter there is especial danger from periods, usually short, of extremely low temperature (−30 to 33 °C); this occurred in six of the last 65 years. Optimal winter temperatures are −1.0 °C for winter wheat and grapes and −3.5 °C for fruit crops. Other crops are not in the ground during this period.

We might expect spring temperatures to have a critical impact on crop productivity (Table 14.4). It turns out that the increase in average spring temperatures from 8.45 to 10.43 °C was accompanied by a decrease in yield of winter wheat by 3.74 c/ha (17 %), maize 2.4 c/ha (9 %), sunflower 1.1 c/ha (−8 %), sugar beet 0.4 c/ha (−0.2 %), grapes 5.6 c/ha (−13 %) and fruit by 4.2 c/ha (−9 %). Further increase of spring temperatures up to 11.02 °C was accompanied by a greater reduction in crop yields: wheat by 5.0 c/ha (23 %), maize 3.4 c/ha (13 %), sunflower 1.7 c/ha (12 %), sugar beet 8.8 c/ha (4 %), grapes 1.9 c/ha (4 %) and fruit by 1.5 c/ha (3 %). In short, the mean spring temperature, at 8.4 °C, was optimal for all six crops.

Summer temperatures had the strongest influence on crop yield (Table 14.5). The increase in summer temperatures from 20.2 to 20.9 °C was accompanied by a reduction in crop yields: wheat by 4.7 c/ha (20 %), maize 5.2 c/ha (18 %), sunflower 1.9 c/ha (13 %), sugar beet 31.3 c/ha (14 %), grapes 8.9 c/ha (19 %) and fruits by 12.2 c/ha (−25 %). Further increase of seasonal temperatures up to 21.7 °C was accompanied by a disproportionately greater reduction in crop yields:

Table 14.3 Effect of winter temperatures on crop yield

	No. of years	t (°C)		Precipitation		Yield (centners/ha)					
		Winter	Annual	Winter	Annual	Winter wheat	Maize	Sunflower	Sugar beet	Grapes	Fruits
Years with warm, including very warm, winters	32	−0.7	+9.52	94.3	537.9	26.3	29.7	14.9	231.3	48.0	47.1
	13	+0.32	+9.89	90.5	519.1	25.2	28.8	14.4	224.0	43.3	46.1
Years with cold, including extremely cold,	30	−3.83	+8.77	106.6	548.3	22.1	27.2	14.15	221.8	44.2	50.6
winters	8	−6.15	+8.14	118.2	558.0	15.3	23.4	13.4	198.2	38.15	49.1
All years, means	62	−2.21	9.16	100.2	543.0	24.3	28.5	14.6	226.7	46.2	48.8
Warm compared with cold winters %	c/ha	+3.18	+0.75	−12.3	−10.4	+4.2	+2.5	+0.75	+9.5	+3.8	−3.5
	%	one year in 4/5	+8	−13	−2	+16	+8	+5	+0.4	+7	−8
Extremely warm compared with extremely	c/ha	+6.7	+1.75	−27.7	−38.9	+9.85	+5.35	+1.0	+25.8	+5.15	−3.0
cold winters %	%	one year in 20	+18	−31	−7	+39	+19	+7	+12	+12	−7

Table 14.4 Effects of spring temperatures on crop yields

		t (°C)		Precipitation (mm)		Crop yields (centners per hectare)					
	No. of years	Spring	Annual	Spring	Annual	Winter wheat	Maize	Sunflower	Sugar beet	Grapes	Fruits
Years with warm, including very warm, springs	27	10.43	9.71	111.8	529.4	22.2	27.2	13.9	227.5	42.8	46.2
	12	11.02	10.03	99.8	488.8	20.9	26.2	13.3	219.1	45.1	48.9
Years with cool, including cold, springs	36	8.45	8.76	132.9	555.2	25.9	29.6	15.0	227.9	48.4	50.4
	9	7.35	8.31	120.9	546.7	18.5	25.0	13.8	204.2	42.8	46.1
All years, means	62	9.3	9.16	122.6	551.0	24.3	28.5	14.6	226.1	46.2	48.8
Warm compared with cool springs	c/ha	1.98	+0.95	−21.1	−25.8	−3.7	−2.6	−1.1	−0.04	−5.6	−4.2
	%			−11	−5	−17	−10	−4	+0.2	−13	−9
Warm compared with average springs	c/ha	2.72	+0.87	−22.8	−11.2	−3.4	−2.3	−1.3	−27.6	−1.1	+0.9
	%			−19	−2	−14	−8	−9	−12	−2	+2

Table 14.5 Effect of summer temperatures on crop yields

		t (°C)		Precipitation		Productivity (centners per hectare)					
	No. of years	Summer	Annual	Summer	Annual	Winter wheat	Maize	Sunflower	Sugar beet	Grapes	Fruits
Years with warm, including hot, summers	30	20.9	9.38	179.4	510.8	19.6	23.3	12.7	195.4	37.3	36.6
	11	21.7	9.71	168.8	487.9	17.0	17.75	10.4	163.6	32.0	29.7
Years with cool, including cold, summers	32	19.27	8.95	224.5	573.1	28.6	33.4	16.3	256.0	54.5	60.3
	12	18.73	8.67	238.4	566.2	31.6	34.1	16.7	256.0	60.5	65.4
All years, means	62	20.2	9.16	201.7	551.0	24.3	28.5	14.6	226.7	46.2	48.8
Warm compared with cool summers	c/ha	+0.82	+0.43	−45.1	−62.3	−9.0	−10.1	−3.6	−60.6	−17.2	−23.7
	%			−25	−12	−46	−43	−28	−31	−46	−65
Hot compared with cold summers	c/ha	+2.97	+1.04	−69.6	−78.3	−14.6	−16.4	−6.3	−101.4	−27.5	−35.7
	%			−41	−16	−86	−92	−61	−62	−86	
Warm compared with average summers	c/ha	+1.6	+0.55	−33.9	−63.1	−7.3	−10.7	−4.2	−63.1	−14.2	−19.1
	%			−17	−13	−30	−38	−29	−28	−31	−39

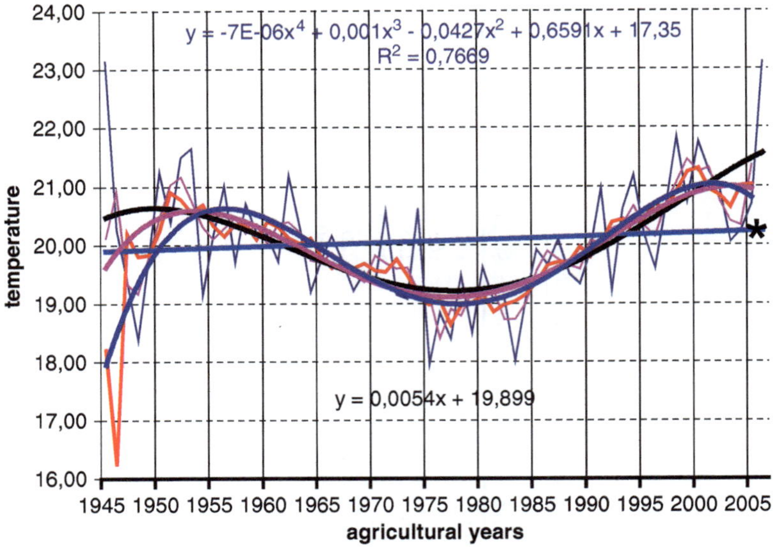

Fig. 14.3 Mean summer temperature. Key: Bold blue, starred – Multiyear linear trend; Bold blue – Polynomial 5-yearly; Black – Polynomial mean; Bold violet – Polynomial 3-yearly; Red – Multiyear 5-year means; Fine violet – Multiyear 3-year means; Fine blue – Annual means

wheat by 30 %, maize 38 %, sunflower 29 %, sugar beet 28 %, grapes 31 % and fruits by 39 %. The decrease in yields of all 6 crops can be seen across the whole range of studied temperatures from 18.8 °C up to 21.7 °C (Fig. 14.3).

Soil Moisture

Soil moisture is a critical for all crops. Analysis of long-term meteorological and statistical data shows a clear relationship between mean rainfall and crop yields (Table 14.6).

In years of below-average rainfall (less by 81 mm or 17 %), yields were suppressed: winter wheat by 2.7 c/ha (11 %), maize 3.7 c/ha (15 %), sunflower 0.8 c/ha (6 %), sugar beet 5.1 c/ha (2 %), grapes 2.4 c/ha (5 %) and fruit crops by 2.2 c/ha (5 %). Very dry years (less by 146.1 mm or 36 %), which were also warmer than average (mean annual temperature greater by 0.46 °C), further reduced crop yields: winter wheat by 3.45 c/ha (17 %), maize 5.8 c/ha (26 %), sunflower 1.8 c/ha (14 %), sugar beet 32. 9 c/ha (17 %) and grapes by 2.6 c/ha (6 %). The increase in fruit yield (1.6 c/ha or +3.2 %) was not significant.

The optimal annual rainfall for sunflower, sugar beet and grapes was 580 mm and for sunflower and fruit crops 610 mm. Any higher rainfall was accompanied by lower crop yields, except for maize which never received its optimum rainfall.

Seasonal trends were, again, distinctive. Years with dry autumns (−40 mm or −34 %) saw no significant deviation of crop yields, except for sugar beet

Table 14.6 Effect of mean annual rainfall on crop yields, 1945–2007

Indices	No. of years	Annual rainfall (mm)	Mean annual temp. (°C)	Productivity (centners per hectare)					
				Winter wheat	Maize	Sunflower	Sugar beet	Grapes	Fruits
All years, means	62	551.0	9.16	24.3	28.5	14.6	226.1	46.2	48.8
Wet years	28	631.6	8.93	27.5	33.1	15.5	233.6	49.1	51.5
Including very wet years	10	682.5	8.85	25.7	34.9	14.6	217.8	45.8	48.7
Dry years	34	470.0	9.34	21.6	24.8	13.8	221.0	43.8	46.6
Very dry years	11	404.9	9.62	20.9	22.7	12.8	193.2	43.6	50.4
Dry compared with average years	–	−81.0	+0.18	−2.7	−3.7	−0.8	−5.1	−2.4	−2.2
%	–	−17		−11	−15	−6	−2	−6	−5
Very dry compared with average years	–	−146.1	+0.46	−3.45	−5.8	−1.8	−31.9	−2.6	+1.6
%	–	−36.1		−17	−26	−14	−17	−6	+3
Very wet compared with average years	–	−131.5	−0.31	+1.4	+6.4	+−0	−8.3	−0.3	−0.15
%	–	+19.3	−3.38	+6	+18	–	−4	−1	−0.3

which increased by15.5 c/ha (7 %). Extremely low autumn rainfall (−70.9 mm, −60 %) reduced yields of winter wheat by 2.4 c/ha, but the yields of grapes and fruits, on the contrary, increased by 2.6 c/ha and 2.0 c/ha, respectively. However much-lower-than-average and much-higher-than-average winter precipitation provoked substantial yield decreases (Table 14.7).

Much-lower-than-average winter precipitation, less by 50.5 mm (50 %), combined with higher seasonal temperatures (+1.21 °C), depressed yields: for winter wheat by 0.3 c/ha (−1 %), maize 4.2 c/ha (15 %), sunflower 1.2 c/ha (8 %), sugar beet 19.9 c/ha (9 %), grapes 4.9 c/ha (11 %) and fruits by 1.9 c/ha (4 %). Very wet winters (76.0 mm or 76 % wetter than average), in combination with reduced temperatures (−0.94 °C), also depressed yields of winter wheat (by 3.9 c/ha or 16 %) and sugar beet (by 11.5 c/ha or 5 %). However, very wet winters were associated with higher yield of grapes (+1.8 c/ha or 4 %), fruit (+3.6 c/ha or 7 %) and sunflower (+0.6 c/ha or 4 %). Maize did not react to winter precipitation. The optimal winter precipitation was 113 mm for wheat, maize and sugar beet, 131 mm for sunflower and grapes and 153 mm for fruit crops.

Data for spring rainfall and associated crop yields are presented in Table 14.8. Lower-than-average spring rainfall (by 26 mm or 21 %) significantly depressed most crop yields: wheat by 2.1 c/ha (9 %), maize 3.05 c/ha (11), sunflower 1.1 c/ha (8 %), sugar beet 8.7 c/ha (4 %) and grapes by 4.6 c/ha (10 %). The yield of fruits was hardly affected, declining by 0.4 c/ha (−0.8 %). Further reduction in rainfall,

Table 14.7 Effects of winter precipitation on crop yields in Moldova, 1945–2007

Indicators	No. of years	Precipitation (mm)		t (°C)		Productivity (centners per hectare)					
		Winter	Annual	Winter	Mean annual	Winter wheat	Maize	Sunflower	Sugar beet	Grapes	Fruits
All years, means	62	100.2	551.0	−2.21	9.16	24.3	28.5	14.6	226.1	46.2	48.8
Years with wet winters	30	131.6	586.4	−2.61	8.93	24.4	30.5	15.4	226.6	48.8	49.7
Years with very wet winters	7	176.2	637.3	−3.15	9.02	20.4	28.7	15.2	215.6	48.0	52.4
Years with dry winters	32	70.8	502.3	−1.84	9.37	24.2	26.7	13.8	226.7	43.7	48.0
Including very dry winters	9	49.7	449.9	−1.00	9.72	24.0	24.3	13.4	206.2	41.3	46.9
Dry compared with average	–	−29.4	−48.7	+0.37	+0.21	−0.1	−1.85	−0.8	+0.1	−2.5	−0.8
%	–	−29.3	−8.8	–	–	−0.4	−7	−6	+0.4	−5	−2
Very dry compared with average	–	−50.5	−101.1	+1.21	+0.6	−0.3	−4.2	−1.2	−19.9	−4.9	−1.9
%	–	−50	−18	–	–	−1	−15	−8	−9	−11	−4
Very wet compared with average	–	+76.0	+86.3	−0.94	−0.14	−3.9	+0.2	+0.6	−11.5	+1.8	+3.6
%	–	+76	+16			−16	+1	+4	−5	+4	+7

Table 14.8 Effect of the spring precipitation on crop yields in Moldova, 1945–2007

Indicators	No. of years	Precipitation (mm)		t (°C)		Productivity (centners per hectare)					
		Spring	Annual	Spring	Annual	Winter wheat	Maize	Sunflower	Sugar beet	Grapes	Fruits
All years, means	62	122.6	551.0	9.29	9.16	24.3	28.5	14.6	226.1	46.2	48.8
Years with wet springs	27	157.2	578.2	9.08	9.12	28.2	32.5	15.9	238.7	52.1	49.3
Years with very wet springs	5	193.5	629.0	8.57	8.69	32.0	40.6	15.9	248.2	46.9	46.2
Years with dry springs	35	96.6	515.8	9.46	9.18	21.2	25.5	13.5	217.4	41.6	48.4
Years with very dry springs	6	59.2	444.1	10.45	9.5	15.9	17.5	11.8	159.2	36.4	40.2
Dry springs v. average	–	−26.0	−35.2	+0.17	+0.02	−2.1	−3.1	−1.1	−8.7	−4.6	−0.4
%	–	−21	−6	–	–	−9	−11	−8	−4	−10	−1
Very dry v. average	–	−63.4	−106.9	+1.16	+0.34	−8.4	−11.0	−2.8	−66.5	−9.8	−8.6
%	–	−52	−19.	–	–	−35	−39	−19	−30	−21	−18
Very wet v. average	–	+70.5	+78.0	−0.72	−0.47	+7.7	+12.1	+1.3	+22.1	+10.7	−2.6
%	–	+58	+14	–	–	+32	+42	+9	+10	+23	−5

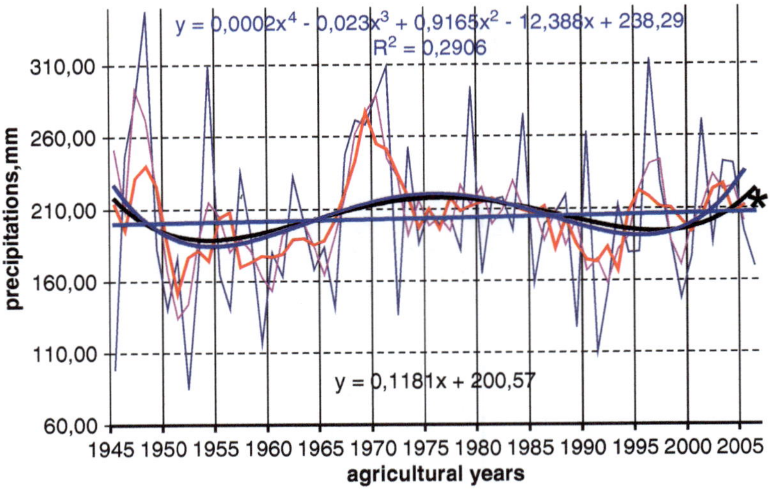

Fig. 14.4 Summer precipitation. Key: Bold blue, starred – Multiyear linear trend; Bold blue – Polynomial 5-yearly; Black – Polynomial mean; Bold violet – Polynomial 3-yearly; Red – Multiyear 5-year means; Fine violet – Multiyear 3-year means; Fine blue – Annual means

by 63.4 mm (52 %), combined with 1.16 °C higher seasonal temperatures, depressed yields across the board: winter wheat by 8.4 c/ha (35 %), maize 11.0 c/ha (39 %), sunflower 2.8 c/ha (19 %), sugar beet 66.5 c/ha (30 %), grapes 9.8 c/ha (21 %) and fruit crops by 8.6 c/ha (18 %). The analysis of data in Fig. 14.4 shows that, over the period 1945–2007, the optimum spring precipitation for all six crops is 170 mm, 39 % less than the mean.

As we might expect, the volume of summer rain, ranging from 134 to 302 mm, had a big effect on crop yields (Table 14.9). Decrease in summer rainfall from the mean of 203.1 to 165.4 mm (−19 %) depressed wheat yields by 1.0 c/ha (4 %), maize 2.6 c/ha (9 %), sunflower 0.25 c/ha (2 %), sugar beet 17.8 c/ha (8 %) and fruit crops by 3.5 c/ha (7 %). Grapes exhibited an insignificant increase. Greater reduction in summer rainfall, by 68.3 mm (34 %), provoked a much greater yield reduction: wheat by 2.4 c/ha (10 %), maize 4.8 c/ ha (17 %), sunflower 0.5 c/ha (3 %), sugar beet 32.7 c\ha (15 %), grapes 2.4 c/ha (5 %) and fruit crops by 4.5 c/ha (9 %). The optimal summer rainfall was 186 mm for sunflower and fruit crops, 225 mm for grapes and winter wheat and 250 mm for maize and sugar beet.

14.3.2 Possibilities of Adaptation Through Technology

Compensation for the negative effects of climate change (increase in temperature and water deficit) can be made by changing the structure of cropped areas by introduction of new areas (hardly possible in Moldova) and expansion of the areas sown with heat- and drought-resistant crops, introduction of varieties and

Table 14.9 Effects of summer rainfall on crop yields in Moldova, 1945–2007

		Precipitation (mm)		*t* (°C)		Productivity (centners per hectare)					
Indicators	No. of years	Summer	Annual	Summer	Annual	Winter wheat	Maize	Sunflower	Sugar beet	Grapes	Fruits
All years, means	62	203.1	551.0	20.1	9.16	24.3	28.5	14.6	226.1	46.2	48.8
Years with wet summers	26	254.3	592.6	19.67	9.10	25.6	32.5	14.8	252.1	45.7	53.5
Years with very wet summers	7	302.2	638.2	19.07	8.59	22.5	28.5	14.0	237.7	34.1	51.5
Years with dry summers	36	165.4	507.1	20.36	9.20	23.2	25.9	14.4	208.3	46.5	45.4
Years with very dry summers	14	134.8	462.0	20.47	9.36	21.9	23.7	14.1	193.4	43.8	44.3
Dry v. average	–	−37.7	−43.9	+0.26	+0.04	−1.0	−2.6	−0.25	−17.8	+0.3	+3.4
%	–	−19	−22	–	–	−4	−9	−2	−8	+0.7	−7
Very dry v. average	–	−68.3	−89.0	+0.37	+0.2	−2.4	−4.8	−0.5	−32.7	−2.4	−4,5
%	–	−34	−1	–	–	−10	−17	−3	−15	−5	−9
Very wet v. average	–	+99.1	+87.2	−1.03	−0.57	−1.8	+– 0	−0.6	+11.6	−12.1	+2.7
%	–	+488	+16	–	–	−7	–	−4	+5	−26	+6

Table 14.10 STP and the level of crop productivity in Moldova, 1945–2007

	Annual precipitation (mm)	Productivity (centners/ha)			
		Winter wheat	Maize	Sunflower	Sugar beet
STP level[a] including reduced	538.2	17.3	22.9	12.3	196.5
Wet years	622.8	19.7	27.1	13.2	203.5
Dry years	453.6	14.9	18.6	11.5	189.5
+– c\ha	–	−4.8	−8.5	−1.7	−14.0
%	–	−28	−37	−14	−7
STP level[b] including reduced	537.2	33.5	3.9	17.5	266.3
Wet years	607.1	33.1	37.4	17.5	264.0
Dry years	467.4	33.8	34.3	17.5	268.4
+– Ts\ha	–	+0.7	−3.1	+– 0	+4.7
%	–	+2.1	−8.6	–	+1.7

STP[a] Level of mechanization – 55 hp per 100 ha
Use of mineral fertilizers – 10.3 kg/ha
Use of pesticides – 1.95 kg/ha
[b] Level of mechanization −158.5 hp per 100 ha
Use of mineral fertilizers −118 kg/ha
Use of pesticides – 9.25 kg/ha

hybrids with an increased adaptive capacity, expansion of irrigation and improving the technology of cultivation. Table 14.10 summarizes the effects of the level of scientific and technical progress (STP) on the level of crop productivity.

14.4 Conclusions

1. Mean annual and seasonal temperatures in Moldova are already higher than optimum for all six studied crops. The sensitivity of crops to high temperatures experienced 11–13 years out of 65 was highest in summer:
   - In autumn, 7 % of yield (0.3–18 %)
   - In winter, 8 % of yield (0.4–17 %)
   - In spring, 9 % of yield (0.2–17 %)
   - In summer, 43 % of yield (28–65 %)
2. Depression of crop yields by rainfall deficit is greatest in spring and summer.
3. A degree of compensation for the negative effects of climate change (an increase in temperature and moisture deficit) may be provided by changing the structure of crop areas by the introduction of new areas and expansion of areas sown with heat- and drought-resistant crops, the introduction of varieties and hybrids with an increased adaptive capacity, expansion of irrigation and improving the technology of cultivation.

Bibliography

Babii L (2010) Methodological aspects of the analysis of marginal agricultural profitability. In: Materials of the international conference, Balti (Romanian)

Karnaukhov A (2010) Blame for everything – the Gulf Stream. www.aif.ru (Russian). Accessed June 2013

Lal R (2004) Soil carbon sequestration impacts on global climate change and food security. Science 304:1623–1627

Vronskih MD (2011) Climate change and the risks of agricultural production in Moldova. Grafema Libris, Chisinau (Russian)

Chapter 15
Multi-scalar Indices of Drought in Intensively Farmed Regions of the Czech Republic

V. Potop and M. Možný

Abstract Indices related to specific timescales are useful for monitoring and management of droughts. We compare the multi-scalar standardized precipitation index (SPI) and standardized precipitation-evapotranspiration index (SPEI) across the Czech Republic. The different timescales provided by the two indices can be related to different kinds of drought and their agricultural, hydrological and socio-economic impacts. Short timescales pick out variations of soil water that affect crops; water resources held in reservoirs are related to longer timescales. Data recorded since 1901 were examined to identify differences between the effects of precipitation and evapotranspiration on drought frequency. SPEI and SPI gave different indications during decades with cool dry summers (the first two decades of the twentieth century), the hottest summers (the end of the twentieth century), warm, wet springs (at the beginning of the twentieth century) and years with a big water deficit (1947, 2003, 1994, 1983 and 1933). There was similarity between the two indices for decades with warm, dry springs (1950s, 1990s and 2000s) and very long sunshine hours. The role of temperature is very evident in summer droughts that are driven by temperature anomalies, contributing to a higher potential evapotranspiration at the end of the century.

15.1 Introduction

Drought is a recurring extreme climatic event over land – characterized by below-normal precipitation over a period of months to years; severe drought profoundly affects agriculture, water resources, ecosystems and human welfare. Drought unfolds

V. Potop (✉)
Czech University of Life Sciences, Prague, Czech Republic
e-mail: potop@af.czu.cz

M. Možný
Czech Hydrometeorological Institute, Doksany, Czech Republic

Faculty of Agronomy, Mendel University in Brno, Brno, Czech Republic

D. Dent (ed.), *Soil as World Heritage*, DOI 10.1007/978-94-007-6187-2_15,

gradually so the most relevant timescale for drought predictions is seasonal (or longer) but, for agriculture, short-period meteorological information is also very relevant. Hydrological and agricultural cycles have different response times to accumulating water deficit so drought indices associated with a specific timescale are useful for the monitoring and management of drought. The multi-scalar standardized precipitation index (SPI) has been adopted by the World Meteorological Organization as the reference drought index. Recently, Vincente-Serrano et al. (2010) have proposed a new index, the standardized precipitation-evapotranspiration index (SPEI), for identifying drought periods. The main objective of this study is to compare the two indices over the lowland parts of the Czech Republic. Drought indices for lowland regions since 1901 were examined in to identify differences between the effects of precipitation variability (by SPI) and evapotranspiration (by SPEI) on drought frequency.

Studying the evolution of temperature and precipitation anomalies during the twentieth century allows us to assess their impact on the frequency of drought. Although precipitation is the main driver, rapidly increasing temperature also plays a role by increasing water loss by evapotranspiration (Dai 2011; Potop et al. 2012a, b, c), for instance, the warmest summers on record combined with unusually low rainfall in central Europe in 2003 and 2006, south and southeastern Europe in summer of 2007 and eastern Europe in summer of 2010. The hot summer of 2003 in central Europe is well known; 2007 was among the hottest summers in Romania over the entire observational period since 1901 (Bogdan et al. 2008) and the hottest in the history of instrumental observations in Moldova – when nearly all temperature records were broken – in winter, spring and, especially, in summer with Tmax ≥40 °C – 5 days for the first time (Corobov et al. 2010; Overenco and Potop 2011). The 2010 summer droughts in Europe are excellent examples of how, independently of soil moisture status, increased evapotranspiration brought on by very high temperatures had a marked impact on vegetation, in this case leading to tree deaths and wildfires (Barriopedro et al. 2011). The 'mega heatwave' during July–August 2010 affected large parts of eastern Europe and western Russia (daily maximum summer temperatures exceeded 35 °C in Moscow, the hottest since instrumental data have been available and +5.1 °C with reference to the 1961–1990 period) and further north (e.g. Helsinki 26.1 °C, Tallinn 27.0 °C, Saint Petersburg 29.3 °C).

15.2 Materials and Methods

SPEI and SPI were calculated from monthly mean temperature and precipitation totals for 1901–2010 recorded at Doksany (lat. 50°27′N, long. 14°10′E, 158 m a.s.l.), Čáslav (49°54′N,15°23′E, 251 m), České Budějovice (48°57′N, 14°28′E, 394 m), Brno (49°09′N,16°42′E, 245 m) and Olomouc (49°34′N, 17°17′E, 210 m) – representing the most productive farmland of the Czech Republic and, receiving the most irregular precipitation, most often affected by drought. The datasets are available from Czech Hydrometeorological Institute. SPEI is based on a monthly water balance (precipitation minus potential evapotranspiration [PET]), calculated using Thornthwaite's method (1948); for calculations we used the algorithm and software developed by Vincente-Serrano and others, freely available at http://digital.csic.es/handle/10261/10002.

SPI and SPEI were obtained using the same log-log probability distribution, which showed a very close fit to both the series of monthly water balance and monthly precipitation records. Use of the same probability distribution enabled accurate comparisons between the two indicators, ensuring that any differences between the series were related only to the impact of the temperature on drought conditions, not to the method used for calculation. The original SPI algorithm (McKee et al. 1993) adjusted a gamma distribution function to the precipitation series; later authors tested several distributions based on different timescales and concluded that, in central Europe, the gamma distribution is sufficiently flexible function to calculate the SPI on various timescales (Potop and Soukup 2009; Potop and Možný 2011a, b). However, for climatic areas with wide precipitation variability like southern and southeastern Europe, the Pearson III distribution is suitable (Potop and Soukup 2009; Potop 2011).

Both indices can be obtained over different timescales and, as standardized variables, they may be compared over time and space. SPEI and SPI were calculated with lags from 1 to 24 months, relevant for agricultural, hydrological and socio-economic impacts, and a drought episode was defined as a continuous period of index values less than −1. For both indices, values of −1.0 to −1.49 correspond to moderate droughts, −1.50 to −1.99 to severe droughts and below −2.0 to extreme droughts. Similarly, values 1.0–1.49 correspond to moderate wet, 1.50–1.99 to severe wet and values above 2.0 to extreme wet conditions. Values from −0.99 to +0.99 describe normal conditions.

15.3 Results and Discussion

Long-term changes in secular series of temperature and precipitation data are represented by a smoothed Gaussian filter 10-year seasonal air temperature deviation and percentage of precipitation, respectively, from the long-term mean. Mean daily temperatures and total precipitation were calculated on a seasonal basis for the secular climatological stations. As an example, the data for Čáslav are presented in Fig. 15.1 which also shows the average seasonal mean air temperature (*T*mean), mean total precipitation (*P*mean) and standard deviation patterns (StDev) from 1901 to 2010.

Summer (JJA): According to the time series of summer air temperature deviations and percentages of precipitation values at the secular stations, the lowest temperature and precipitation anomalies were recorded in the 1900s and 1910s, with the lowest negative temperature deviations in 1902 (−2.4 °C at Doksany) and 1913 (−1.7 °C at Brno, −2.5 °C at Olomouc and −1.9 °C at České Budějovice). For the Čáslav series, the 1970s had the coolest summer in 1978 ($\sigma = -1.7$ °C). The lowest percentages of

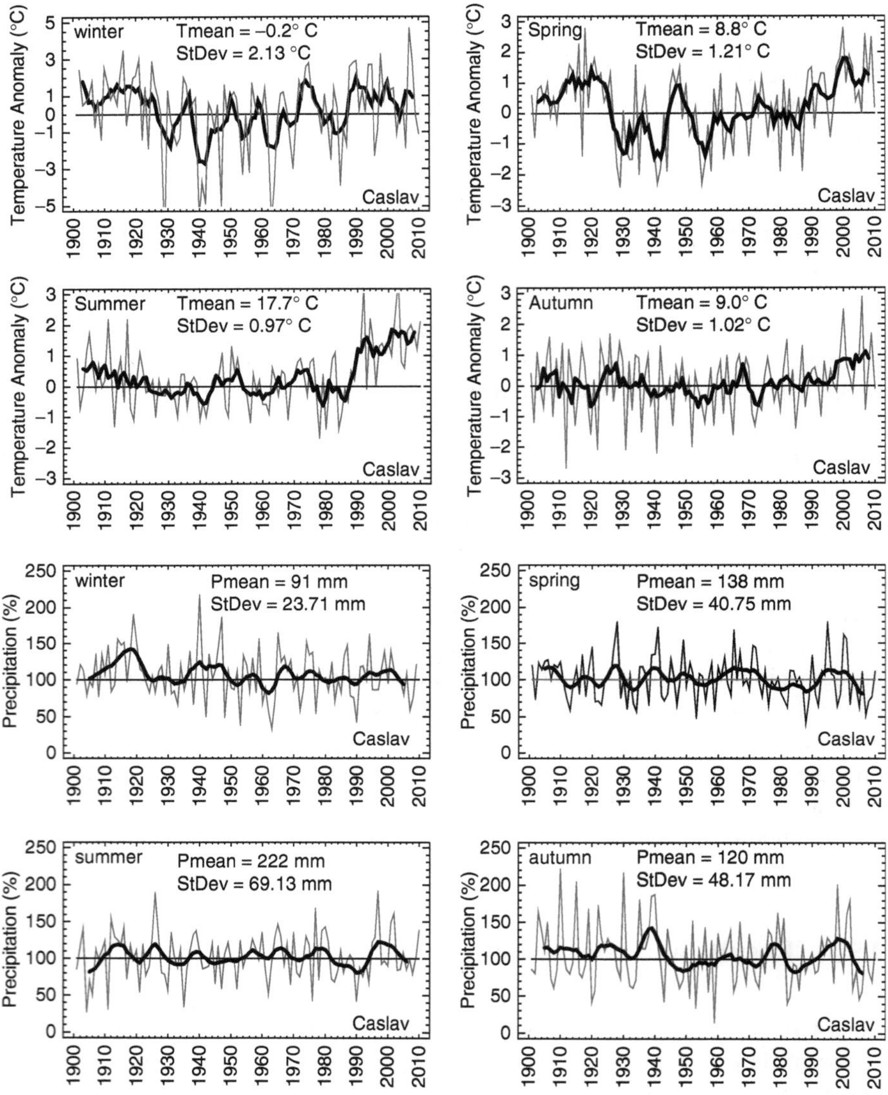

Fig. 15.1 Secular temporal evolution (1901–2010) of seasonal mean air temperature (*upper panel*) and seasonal precipitation totals (*bottom panel*). Averages are shown as deviation from the baseline 1961–1990 climate (zero line) smoothed by Gaussian filter over 10 years. Average seasonal mean air temperature and precipitation totals (*T*mean, *P*mean) and standard deviation patterns (*StDev*)

precipitation were also during this period, including 1904, 1911, 1917 and 1921 (43–27 % below normal). Overall, the lowest amounts of precipitation, expressed in terms of the per cent change from the reference period (1961–1990), were recorded in the first two decades of the twentieth century. The highest positive summer deviation

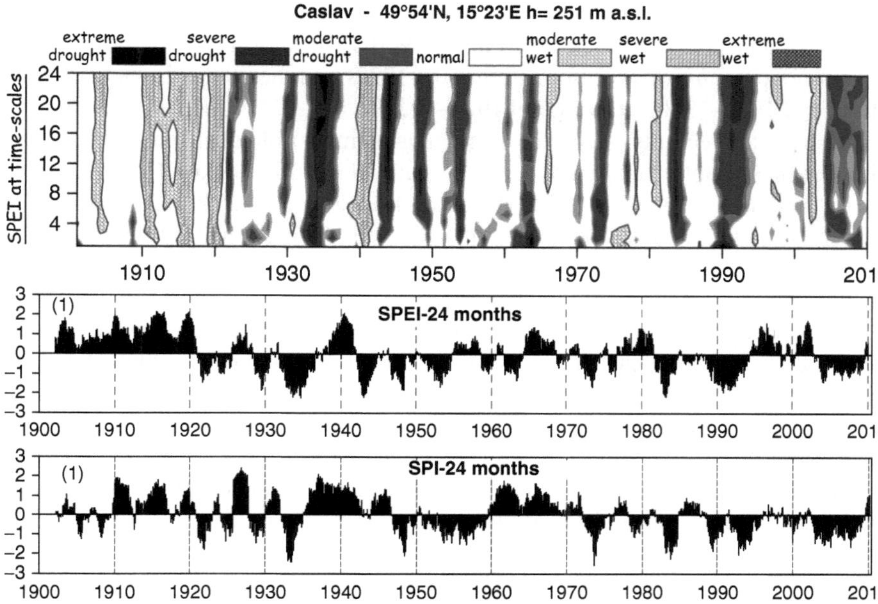

Fig. 15.2 Secular temporal evolution (1901–2010) of SPEI and SPI at timescales from 1 to 24 months (*upper panel*); SPEI and SPI at 24 months associated with hydrological drought (*bottom panel*)

of air temperature values (more than 3.0 °C) was recorded in the 1990s and 2000s for all stations. At Doksany, Brno, Čáslav and České Budějovice, the warmest summer was in 2003 with temperatures 3.2–3.7 °C higher than the norm; at Olomouc, the highest temperatures were in 1992 ($\sigma = 4.0$ °C). Therefore, the decadal summer averages of the mean daily temperature for the lowland sites over the entire 110 years revealed cooling during 1901–1920 and rapidly rising temperatures since the mid-1980s (as much as +0.6 °C per decade).

The lowest negative temperature anomalies along with the lowest precipitation occurred in the summers of the first two decades of the twentieth century: in contrast, the end of the twentieth century recorded the highest positive temperature anomalies. During these decades, SPEI and SPI showed large differences in the identification and evolution of drought conditions (upper and bottom panels of Fig. 15.2). Summer droughts tended to be longer and more severe with a greater frequency in the 1920s, 1950s and 2000s. SPEI identified extreme summer drought in summers with long, severe heat waves: the summers of 1994, 2003 and 2006 were characterized by the highest numbers of hot, dry days with intense sunshine; the longest periods of sunshine (155% of the normal amount) were recorded in the very dry June of 2006 and August 2003. SPEI also detected the maximal duration of drought episodes during the period with the greatest water deficit: in 1947, 2003, 1994, 1983 and 1933 (Potop et al. 2012a, b, c).

Winter (DJF): Conversely, during winters, the warmest periods were recorded in the 1910s and 1920s and the coldest in the 1940s. The largest positive deviation of winter temperature was recorded during the 1920s and the 2000s. However, in the decade of 1921–1930, the winter of 1928–1929 ($\sigma = -6.8$ °C, $\delta T/\sigma = 3.5$) was the coldest winter recorded since the beginning of meteorological measurements in the Czech Republic (extremely cold February in 1929, −30.6 °C at Doksany, −26.8 °C at Čáslav, −27.8 °C at České Budějovice, −20.8 °C at Brno and −19.5 °C at Olomouc). The winter of 1964 was distinguished by very negative temperature anomalies (from −2.5 to −3.0 °C) and the lowest precipitation (less than 30 %). Moreover, high fluctuations of the positive and negative anomalies of air temperature occurred during the 2001–2010 period, with two severe winters (2005–2006 and 2009–2010) and two exceptionally warm winters (2006–2007 and 2008–2009). The exceptionality was confirmed by normalized standard deviations ($\delta T/\sigma$) in both the cold and warm winters, as the temperature anomalies were greater than 3σ. The lowest negative temperature deviation ($\sigma = -6.6$ °C; $\delta T/\sigma = 3.0$) was recorded in January 2006, when the daily minimum air temperature fell to −27.1°C (Olomouc), while the highest positive temperature deviation ($\sigma = +6.3$ °C, $\delta T/\sigma = 3.0$) was recorded in January 2007. In this respect, the entire month of January and parts of December and February of the 2006–2007 winter in the Czech Republic were dominated by western and northwestern cyclones; at the same time, an extremely warm and dry winter occurred in southeastern Europe as a result of high pressure that settled over central and eastern Europe and made contact with the Azores and East European Anticyclones. This usually facilitates an eastern circulation with severe winter weather, but it was positioned further south, enabling the west winds to carry warm oceanic (Atlantic) air currents through advection (Bogdan et al. 2008). Consequently, the winters with the warmest periods and big fluctuations of precipitation occurred at the beginning of the twentieth century. In this period, SPEI detected many droughts: the most prolonged agricultural drought (3 months) starting in winter and ending in spring was recorded in 1943 (severe drought began in December and ended in May, except for at České Budějovice, where it ended in June).

Spring (MAM): For springs, the warmest decade was the 2000s, with the highest positive seasonal deviations (from 2.3 °C to 2.9 °C) recorded for all stations in 2000, 2007 and 2009; there was also more sunshine during spring droughts in April 2007 and 2009 (up to twice the norm for April). SPEI and SPI values for these months were less than −1.5 (severe drought). The next spring severe drought recorded in April of 2009 was characterized by extraordinarily long sunshine durations and longest dry spells (19–25 days) with an SPEI ≤ -2.3 (SPI ≤ -1.4). The Čáslav station clearly showed two warm periods (Fig. 15.1): the first at the beginning of the twentieth century with high fluctuations of positive seasonal temperature deviations ($\sigma = 2.8$ °C; 1918) and the second at the end of the twentieth century ($\sigma = 2.8$ °C; 2000). The coolest decade for springs was the 1930s, with the lowest negative deviation in 1929 (−2.4 °C) at Čáslav, Brno and České Budějovice; the lowest at Doksany in 1930 and 1942 (−2.4 °C); and at

Olomouc in 1955 (−2.8 °C). The lowest percentage of precipitation occurred during the1940s and 1990s.

Autumn (SON): The warmest autumns were in the 2000s (highest temperature deviation at 2.1–3.0 °C; 2006). The coldest were the 1910s and 1920s (the lowest temperature deviation at Doksany and Olomouc were found at −2.4 and −4.2 °C, respectively). The majority of autumns with precipitation less than 40 % below normal simultaneously had normal or below-normal temperatures; 2006 was an exception. This suggests that lack of rain was the primary cause of autumn droughts. The lowest percentage of precipitation during autumn occurred during the 1940s, 1950s, 1990s and 2000s, as can be seen in SPI time series at timescale of 24 months (bottom panel of Fig. 15.2). The warmest decade with the lowest autumn precipitation was the 2000s. The periods 1941–1950 and 2001–2010 were ranked as having a large number of long, severe drought events. According to SPEI, the spring-summer-autumn drought of 1947 was the most severe and longest in duration since observations began (upper panel of Fig. 15.2).

Detailed new results about the temporal evolution of SPEI at different timescales and its impact on vegetable crops are discussed and presented in broader climatological and European contexts in, e.g. Potop and Možný (2011a, b) and Potop and Soukup (2011). However, in-depth analysis is required to explore the vulnerability to drought in the context of climate change, followed by calculation of SPEI for a dense network of stations to represent the different climatological conditions across the Czech Republic (Potop et al. 2012b, c) and beyond.

15.4 Conclusions

Comparison of the results of SPI and SPEI suggests that SPEI might be used in future assessment and projections of drought events in the Czech Republic. In summary, the comparison revealed:

1. In the spring, at Čáslav secular climatological station, two warm periods were detected – the first at the beginning of the twentieth century with high positive temperature anomalies and precipitation (warm and wet) and the second at the end of the twentieth century with warm, dry conditions. In contrast, at Brno, České Budějovice and Olomouc, the periods with high positive temperature anomalies and low precipitation were recorded in the 1950s, 1990s and 2000s. There were no spring droughts in the decades 1961–1970 and 1981–1990, and the 1910s and 1930s were marked by low drought incidence. Spring droughts have been more persistent during the last 20 years.
2. The summer drought during the 2000s corresponded to higher temperatures (>2.5 °C above the norm), and SPEI has the capacity to detect the intensification of drought severity due to increasing temperature conditions independent of the analysis timescale in the 1990s and 2000s. Therefore, the role of temperature was evident in summer drought episodes that depend on temperature anomalies,

contributing to a higher water demand by potential evapotranspiration at the end of the century.

3. With respect to the spring and autumn temperature series, the slopes of the linear trends were similar to the decadal tendencies of the summer temperatures. This indicated that positive significant trends were recorded from 1991 to 2010 with values as high as 0.7 °C per decade. Our results suggested that it was a feature of unusually high temperatures and low precipitation, which were then transferred to the SPI and SPEI indices.
4. SPEI and SPI showed a large difference in the evolution and severity of drought during decades with big summer negative temperature anomalies combined with the lowest precipitation (cold and dry; the first two decades of the twentieth century), the highest summer positive temperature anomalies (the end of the twentieth century), both high spring positive temperature and precipitation anomalies (warm and wet; at the beginning of the twentieth century) and the lowest deficit of water balance (1947, 2003, 1994, 1983 and 1933).
5. On the other hand, similarities between two indices were apparent during the decades with high fluctuations of positive spring temperature and lower precipitation (warm and dry; 1950s, 1990s and 2000s); extremely long sunshine durations (155 % of the normal amount in extremely dry June of 2006 and August 2003, up to twice the norm for April of 2007 and 2009) and consecutive dry days. Therefore, the role of temperature was evident in summer drought episodes driven by the intensity and duration of temperature anomalies, generating a higher water demand by potential evapotranspiration, at the end of the last century.

Acknowledgements The research on drought conditions in the Czech Republic was supported by 6046070901 and S grants of the MSMT CR and projects CZ.1.07/2.3.00/20.0248 of the National Agency for Agriculture Research Q191C054.

References

Barriopedro D, Fischer EM, Luterbacher J et al (2011) The hot summer of 2010: redrawing the temperature record map of Europe. Science 8:220–224

Bogdan O, Marinic I, Mic LE (2008) Warm-winter risk in Romania. In: Proceedings of Balwois 2008 conference, Ohrid, Republic of Macedonia, 27–31 May, pp 1–7

Corobov R, Sheridan S, Overenco A, Terinte N (2010) Air temperature trends and extremes in Chisinau (Moldova) as evidence of climate change. Clim Res 42:247–256

Dai A (2011) Characteristics and trends in various forms of the Palmer Drought Severity Index during 1900–2008. J Geophys Res 116:D12115. doi:10.1029/2010JD015541

McKee TBN, Doeksen J, Kleist J (1993) The relationship of drought frequency and duration to time scales. In: Eighth conference on applied climatology. American Meteorological Society, Anaheim, pp 179–184

Overenco A, Potop V (2011) Summer heat episodes in Czech Republic and Republic of Moldova: a comparative analysis. J Acad Sci Moldova Life Sci 1(313):167–178

Potop V (2011) Evolution of drought severity and its impact on corn in the Republic of Moldova. Theor Appl Climatol 105:468–483

Potop V, Možný M (2011a) The application a new drought index – standardized precipitation evapotranspiration index in the Czech Republic. In: Středová H, Rožnovský J, Litschmann T (eds) Microclimate and mesococlimate of landscapes and anthropogenic environments. Skalni mlyn, 2–4 Feb 2011. Mendel University, Brno

Potop V, Možný M (2011b) Examination of the effect of evapotranspiration as an output parameter in SPEI drought index in Central Bohemian region. In: Šiška B, Hauptvogl M, Eliašová M. (eds) Bioclimate: source and limit of social development, international scientific conference, Topolcianky, 6–9 Sept 2011

Potop V, Soukup J (2009) Spatiotemporal characteristics of dryness and drought in the Republic of Moldova. Theor Appl Climatol 96(3):305–318

Potop V, Možný M, Soukup J (2012a) Drought at various time scales in the lowland regions and their impact on vegetable crops in the Czech Republic. Agric For Meteorol 156:121–133

Potop V, Boroneanţ C, Možný M, et al (2012b) Observed evolution of drought episodes assessed with the Standardized Precipitation Evapotranspiration Index (SPEI) over the Czech Republic. In: EGU General Assembly 2012. Geophysical research abstracts vol. 14. EGU2012-7681, Vienna, 22–27 April 2012

Potop V, Boroneanţ C, Štěpánek P et al (2012c). Projected changes in the evolution of drought assessed with the SPEI over the Czech Republic. In: EGU General Assembly 2012. Geophysical research abstracts vol. 14. EGU2012-7681, Vienna, 22–27 April 2012

Thornthwaite CW (1948) An approach toward a rational classification of climate. Geogr Rev 38:55–94

Vincente-Serrano SM, Beguería S, López-Moreno JI (2010) A Multi-scalar drought index sensitive to global warming: the Standardized Precipitation Evapotranspiration Index – SPEI. J Clim 23(7):1696–1718

Part II
Soil Fertility: Lessons from Long-Term Field Experiments

Chapter 16
The Continuing Value of Long-Term Field Experiments: Insights for Achieving Food Security and Environmental Integrity

D.S. Powlson, A.J. MacDonald, and P.R. Poulton

Abstract Long-term experiments are one vital tool for studying the impacts of agricultural management practices on soil properties and crop production; however, they have several limitations that must be recognized. Such experiments are resources for research – not museum exhibits that can never be altered. For example, in the Broadbalk Wheat Experiment (started 1843), the original large plots have been split so that additional cropping systems can be studied, in particular wheat grown in a crop rotation in addition to the original monoculture.

A key finding from all long-term experiments is that soil organic carbon (SOC) content tends to move from one quasi-equilibrium value to another under the influence of changes in management; it does not increase or decrease indefinitely. Compared to unfertilized plots, application of manure caused a large increase in SOC, whereas N fertilizer caused a small increase. Measurements made on the Broadbalk Experiment show that a small change in SOC content can have a disproportionately large impact on a range of soil physical properties.

Long-term experiments are valuable for investigating the range of chemical structures within SOC and their distribution between physically separated fractions. Other recent uses include studies on the microbial oxidation of methane and reduction of nitrate in soil, soil biodiversity, and losses of nitrate and phosphate to streams and groundwater.

16.1 Benefits and Limitations of Long-Term Experiments

Long-term experiments (LTEs) are one vital tool for investigating the sustainability of agricultural systems. They are the only way of measuring the crop yields that can be obtained with different levels of inputs and/or management practices in a given

D.S. Powlson (✉) • A.J. MacDonald • P.R. Poulton
Rothamsted Research, Harpenden, UK
e-mail: david.powlson@rothamsted.ac.uk

D. Dent (ed.), *Soil as World Heritage*, DOI 10.1007/978-94-007-6187-2_16,

environment and determining whether such yields can be sustained over long periods. They provide the opportunity to measure the impacts of agricultural management practices on soil properties that change slowly but influence crop production and sustainability in the long term; these include the development of acidity, declines in the content of nutrients and soil organic matter, and a range of physical properties such as compaction, aggregate stability, water infiltration and likelihood of erosion. Similarly, LTEs are the only means of examining the potential for build-up over long periods of weeds, pests and diseases as a result of different management practices. They also provide a means of assessing environmental impacts of different cropping systems and management practices, including greenhouse gas emissions and water pollution.

LTEs are valuable resources for scientific research; many topics that can be investigated have a bearing on agricultural sustainability, even if the precise links are as yet unclear – one example might be changes in the diversity of soil biological populations in response to management practices. Others may be fundamental science with no clear links to agricultural or environmental issues. Others relate to environmental science such as the use of LTEs to assess accumulation in soil or vegetation of pollutants derived from industrial or other human activities, and their impacts. In all cases, the value of an experiment is greatly enhanced if samples are archived so that they can be analysed using techniques that were unavailable at the time the samples were taken. For example, the Rothamsted Sample Archive now contains some 300,000 soil and plant samples that have been used for studying many and various topics ranging from radioactive fallout from nuclear weapons testing to atmospheric deposition of metals and organic pollutants (Poulton 1996a, 2006; Powlson and Johnston 1994; de Richter et al. 2007).

Despite the many unique benefits of LTEs, some limitations must be acknowledged. One concerns the original design and whether the existing treatments or management practices are still relevant. For example, older experiments often include no replication; fortunately many, including those at Rothamsted, comprise large plots which partially overcome issues of spatial heterogeneity within the field. There are also statistical techniques that can be applied to long runs of data in which years can substitute for spatial replication (Payne 2006).

Another type of limitation arises from the constraints of plot size. For example, *edge effects* are a serious issue with long-established plots due to soil movement resulting from tillage, water or nutrients moving between plots, or because plants near the edge receive more light than those within a large area of continuous plants. It has been found that approximately 1.0–1.5 m on each side of the 6-m-wide Broadbalk plots is affected by soil movement (Poulton 1996b) even though plots are separated by a 1.45-m path. In research on the fate of metals applied to soil as contaminants in sewage sludge (biosolids) to a sandy soil at the Woburn Market Garden Experiment, soil movement was found to be a significant factor in interpreting the results. Sewage sludge applications containing different concentrations of metals were made to 6-m-wide plots over a 20-year period; the plots were sampled after 36 years and metal concentrations were measured. The data indicated that only 40 % of the applied metals remained in the soil, which was surprising as less than 0.5 % had been removed in crops and leaching was thought to be small. Analysis of soil sampled from transects across the plots showed that a significant amount of lateral soil movement had occurred during tillage operations and this could be quantitatively estimated using a dispersion model. This model was then used to correct the measured retention of metals in the soil: taking account of the lateral movement of metals beyond the plot boundaries doubled recovery from 40 to 80 % – leading to a major reinterpretation of the original results (McGrath and Lane 1989). At the Askov Experiment in Denmark, started in 1894, it was concluded, after 100 years, that soil movement between plots could destroy the integrity of the experiment. At one site (Lermarken), it was decided to introduce semi-permanent grass strips[1] between plots to limit tillage-mediated movement of soil between plots (Christensen et al. 2006); the usable plot area within the grass strips is now 10.0 × 7.5 m. At another site (Sandmarken), where the plots are smaller, it was decided in 1997 to convert the plots to permanent grass because the soil movement problem was too severe.

Some processes occur at the landscape scale – examples include soil erosion and the spread of diseases or insect pests – so studies based on small plots, whether long term or not, are of limited value. The difficulty of quantifying landscape-scale processes is another limitation to the use of LTEs for assessing all aspects of the sustainability of agricultural systems. However, within these limitations, LTEs offer an excellent approach (and probably the only one possible) for studying a wide range of biophysical factors that determine the sustainability of agricultural systems and practices from the viewpoints of production and environmental impacts. LTEs cannot in themselves provide direct information on the economic or social factors that contribute to sustainability – e.g. the profitability of different cropping systems or management practices – or the labour constraints that may make a specific practice impracticable in certain situations. However, they do provide basic information that can be combined with knowledge from other disciplines to make some deductions on these issues.

16.2 Continuity Versus Change

Continuity of treatments and management practices are, obviously, vital if trends over periods of many years in crop yields, soil properties or other factors are to be meaningfully observed and interpreted. But this does not mean that *all* treatments have to remain static or that certain management changes can *never* be made. The Broadbalk Wheat Experiment at Rothamsted is now in its 170th year. If a totally static philosophy had been followed, we would still be growing a wheat variety almost 2 m tall, ploughing using horses and applying rates and forms of N fertilizer that are of little significance for current agriculture and its related environmental issues.

[1] The grass strips are maintained for the duration of a 4-year rotation.

Table 16.1 Major changes to treatments and management practices on the Broadbalk Experiment

Year	Change
1844	First harvest of winter wheat[a]
1849	Tile drains installed. Analysis of drainage water started in 1866
1881	Western end of all plots abandoned and long-term set-aside established (now Broadbalk Wilderness)
1885	Second farmyard manure (FYM) treatment started
1926	To control weeds, all treatments divided to create five sections. Fallowing introduced, mostly 1 year in 5
1954	Liming started to correct soil acidification occurring mainly in treatments with highest rates of N fertilizer
1964	Started routine application of herbicides (one section of all fertilizer and manure treatments excluded from herbicide treatment to facilitate studies of weed biology)
1968	All plots divided into 10 sections. Introduced crop rotation to certain sections of all treatments
	Introduced short-straw wheat varieties[b]
	Extended range of N fertilizer treatments up to 192 kgN/ha. Form of N changed from ammonium sulphate to calcium ammonium nitrate
	Third FYM treatment started; stopped in 2000
1979	Started routine application of fungicides (one section of all fertilizer and manure treatments excluded from fungicide treatment)
1985	Extended range of N fertilizer treatments up to 288 kgN/ha
1986	Started straw incorporation on Section 0
1993	New drain in Section 9
2001	Started comparison of with/without S fertilizer
	Started split application times for higher N fertilizer rates (144–288 kgN/ha) to compare with these N rates applied at a single time
	P fertilizer withheld on some treatments to allow levels of plant-available P to decline to a sensible agronomic level in plots with low crop yield resulting from constraints other than P (e.g. low N rates)
2005	FYM + 96 kgN/ha changed to FYM + 144 kgN/ha because of evidence of increasing crop yield at this input (though high nitrate leaching)

[a] Originally, crops were harvested by hand with scythes. First cut by reaper-binder in 1902. Sheaves stooked on each plot, carted and later threshed. Harvested by combine harvester since 1957
[b] Subsequently, wheat variety changed every 5–15 years

Table 16.1 summarizes some of the major changes in management practices and treatments in the Broadbalk Experiment. A key change is the variety of wheat grown. Although the cultivar had been changed several times in the first 125 years of the experiment, a major decision was taken in 1968 to change to a short-straw, *green-revolution* variety; the result of this is discussed in the next section. Since then, the variety has been changed at intervals of 5–15 years. These changes, in response to the availability of new varieties with favourable traits, are less frequent than made by commercial farmers but sufficiently frequent that the variety grown at any time is reasonably relevant to current commercial practice. Decisions regarding changes to treatments (e.g. fertilizers, manures, pesticides) are made less frequently, sometimes following a routine review and sometimes in response to a particular issue, for instance, the inclusion of a test of sulphur (S) fertilizer, taken when emerging research results indicated that many crops in the UK and Europe were affected by S deficiency as a result of a greatly decreased S deposition from the atmosphere after a maximum around 1970 (Zhao et al. 2002). Another example is the decision to increase the range of applied fertilizer N rates as it became clear that grain yields of newer varieties responded to high rates; N rates changed in the first 8 years of the experiment but then remained unchanged on most plots. From 1852, the rates were 0, 48, 96 and 144 kgN/ha; the highest rate was considered extremely high at the time but was included in order to bracket the range considered normal in the mid-19th century. When a short-straw variety was introduced in 1968, a higher rate (196kgN/ha) was included. Subsequently, in the light of additional information from other experiments for responses to higher rates, additional treatments of 240 and 288 kgN/ha have been added.

An important principle in making changes to long-term experiments is that the decision should be taken by a multidisciplinary group. This is the practice followed at Rothamsted; as far as possible, the group includes members with expertise in soil science, plant nutrition, plant physiology, plant pathology, pesticide science, entomology and agronomy. The group also includes a member with responsibilities for farm operations – to ensure that decisions made by academic members do not conflict with practical necessities. Constraints of space mean that, whenever a new treatment is added, another must be discontinued. The multidisciplinary group has to assess the likelihood of the various existing treatments being valuable in the future. Weighing this against the potential value of the proposed new or changed treatment is a matter of judgement rather than an exact science (Johnston and Powlson 1994; Leigh et al. 1994; Poulton 1996b). Among the guiding principles are:

1. *Whether, on balance, a change in treatment or management practice will enhance either the scientific value or practical relevance of the experiment.* In general, scientific value would have the greater weight because numerous short-term experiments are conducted with a focus on practical relevance. The aim is to ensure that the experiment is enhanced as a scientific resource whilst not becoming divorced from current agricultural practices. It is also recognized that some extreme treatments are scientifically necessary (e.g. no fertilizer and unusually high rates of fertilizer or manure), even though they are abnormal from a practical viewpoint.
2. *Changes necessary to maintain the integrity and future value of the experiment.* An example is the need to control soil acidification. By the 1940s, some Broadbalk plots had become acid. It was decided that, unless this trend was halted, yields on many plots would decline to such an extent that the results would be uninformative. Applications of chalk (calcium carbonate), as and when needed, began in the 1950s.

16.3 Crop Yield Trends and Influence of Management Practices

"Food security is one of this century's key global challenges" (Royal Society 2009). LTEs are the ideal means of obtaining information on the crop production attainable from different cropping systems with a given set of management practices and inputs. Importantly, they are the only way of assessing the extent to which yields can be sustained in the long term and the environmental implications of the practices necessary to obtain them. It is well recognized that food *production* is only one part of food security; *access* and *affordability* are also essential. But the production stage is fundamental, and it is this part for which LTEs are an essential and unique source of data.

Figure 16.1 shows grain yields of winter wheat in selected treatments of the Broadbalk Experiment between 1844 and 2011, mainly expressed as means over 5 or 10 years. With no inputs of nutrients as manure or fertilizer (the control treatment), yields were mostly in the range 1–1.5 t/ha. This result prompted investigations on the sources of N in this treatment – in which the crop removes about 30 kgN/ha/year. Mineralization of soil organic matter is obviously one source, but the total N content of the soil in this treatment has remained constant at around 3,000 kgN/ha within the plough layer (0–23 cm) for over 160 years (Jenkinson 1990). If mineralization were the only source of N for the crop, it would have declined to zero within 100 years – so we must assume annual inputs of at least 30 kgN/ha from other sources. Based on results from experiments conducted in the 1980s with ^{15}N-labelled fertilizer (Powlson et al. 1986), annual inputs of 48 kgN/ha were calculated. Non-symbiotic biological nitrogen fixation was shown to be very small, and there are very few leguminous weeds present. Subsequent direct measurements showed that the major input of N was from a combination of wet and dry deposition from the atmosphere (wet deposition comprises ammonium and nitrate dissolved in rain; dry deposition is a combination of ammonia, oxides of nitrogen, nitric acid and particulate N-containing materials). Measurements made in the 1990s led to an estimate of 43 kgN/ha/year total N deposition to winter wheat at the Broadbalk site, of which 79 % was dry deposition (Goulding et al. 1998). These authors used modelling to determine the fate of deposited N; this indicated that about 53 % was directly taken up by the crop within the year of deposition with 30 % being immobilized into soil organic matter (thus replenishing soil organic N and contributing to N mineralization in future years), 5 % was leached as nitrate, and 12 % returned to the atmosphere as N_2O and N_2. The recent report of the European Nitrogen Assessment (Sutton et al. 2011) draws attention to the impacts of N deposition on plant biodiversity and impacts on human health of the various N species in the atmosphere. Deposition may have now decreased in southeast England, but it is increasing in rapidly developing regions – recent measurements in the Beijing region showed dry deposition alone to be 55 kgN/ha/year (Shen et al. 2009). Thus, even an observation as simple as measurements of crop yield in an unfertilized plot of an LTE pointed the way for research on aspects of the N cycle that were previously unforeseen.

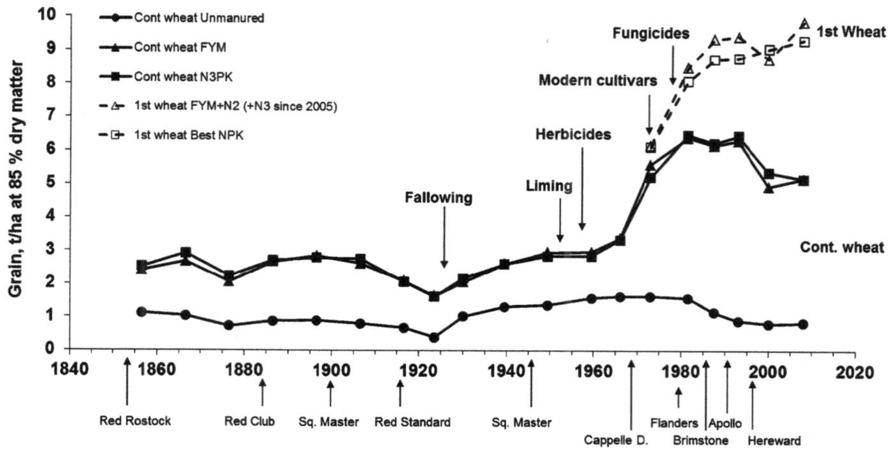

Fig. 16.1 Wheat grain yields in selected treatments of the Broadbalk Experiment 1844–2011 and some key management changes

The Broadbalk wheat yields show a marked increase in both the farmyard manure (FYM) and inorganic fertilizer (NPK) treatments following the change to short-straw varieties in 1968. The different harvest index of modern varieties means a larger proportion of photosynthate goes to grain. When old and new varieties were grown side by side on Broadbalk plots, it was found that total above-ground biomass was approximately equal for both varieties (Austin et al. 1993), so the efficiency of conversion of applied N into grain increased. Another clear result shown in Fig. 16.1 is that the FYM and inorganic fertilizer treatments (NPK, with N applied at 144 kgN/ha) gave almost the same yields for continuous (monoculture) winter wheat for the entire duration of the experiment. With the tall-straw varieties (pre-1968), yields in both treatments were about double that of the control, averaging about 2.5 t/ha; with the short-straw varieties (post-1968), yields increased to 5–6 t/ha.

A crop rotation was introduced in certain sections of the experiment in 1968 in order to compare wheat yields after a break crop with those in monoculture. For the NPK treatment, yields of first wheat after a break were about 2 t/ha greater than for continuous wheat. This is not surprising in view of the well-known build-up of pests and diseases when a crop is grown continually in the same soil. With wheat and other small-grain crops, the root disease take-all (caused by the fungus *Gaeumannomyces graminis* var. *tritici*) is one of the most damaging diseases in most parts of the world. It can cause severe loss of yield and decreases the efficiency with which wheat plants take up N. It typically increases to become very severe if susceptible crops are grown consecutively for about 2–5 years but, thereafter, typically becomes less severe as a consequence of a natural biological control known as take-all decline (TAD). The disease and TAD have been studied on Broadbalk and in other experiments at Rothamsted over many years (Gutteridge et al. 1996; Hornby et al. 1998; Macdonald and Gutteridge 2012). Although much remains to be learnt about the specific cause or causes of TAD, it provides modest but useful decreases in the severity of the disease and the damage it causes. For example, on Broadbalk, continuous wheat typically yields less than the first wheat after a break, but more than the second and third wheat crops after a break. Although 100-year experiments are not essential for such research, sequences of relevant crops running for periods of 5–20 years are necessary. A recent advance in research on take-all disease was the observation that wheat cultivars differ in their build-up of inoculum during the first year after a break crop, indicating that crop breeding may offer a solution to this damaging disease (McMillan et al. 2011).

Figure 16.1 shows lower yields of continuous wheat since about 2000 compared to those in the previous 20 years, though yields for first wheat after a break have generally been sustained. There are indications that the current wheat variety grown on Broadbalk (Hereward) is particularly susceptible to take-all, which would explain the sustained yields of first wheat after a break. There have also been several years with adverse weather conditions in the last 10 years, in some cases leading to very late sowing. As yet, it is not known whether these factors entirely explain the decline or whether other factors may be affecting the growth of wheat in the monoculture sections of the experiment. This is obviously a crucial issue for the sustainability of the continuous wheat system.

16.4 Nutritional Quality of Wheat Grain

There is evidence from the Broadbalk Experiment that an unintended consequence of the global change to short-straw wheat varieties has been a decreased concentration of minerals in grain – with potential impacts on human nutrition and health (Fan et al. 2008). Analysis of archived grain samples showed that concentrations of zinc, iron, copper and magnesium decreased following the change to short-straw varieties. There has been discussion on whether intensive farming practices have led to a depletion of these minerals in soil, but evidence from Broadbalk does not support this; concentrations of these minerals in soil (both total and EDTA-soluble) either increased or remained constant during the duration of the experiment. EDTA solubility is often taken as an indication of the availability of minerals for crop uptake, so the change in mineral composition in grain appears to be related to changed wheat varieties rather than to long-term impacts of management practices on soil. This research is an example of the value of archiving both crop and soil samples from long-term experiments for uses not foreseen earlier in the life of the experiment.

16.5 Management Impacts on Soil Organic Matter

Figure 16.2 shows the long-term changes in soil organic carbon (SOC) content in several treatments of the Broadbalk Experiment. Measured data (a combination of analyses on modern samples and those from soils stored in the archive) are shown

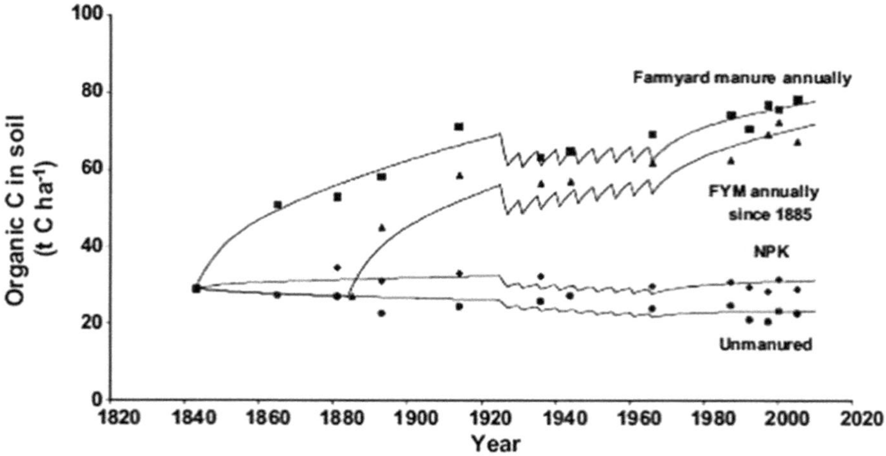

Fig. 16.2 Changes in soil organic carbon content in the 0–23-cm soil layer of selected treatments of the Broadbalk Experiment. Symbols are measured data and lines are simulations using the Rothamsted Carbon Model (After Powlson et al. 2012)

as points and the line is a simulation using the RothC model (Coleman and Jenkinson 1996). By 2005, the SOC content in the treatment given a high rate of FYM every year since the start of the experiment in 1843 was almost three times that in the control. But the increase occurred slowly; the tendency of SOC to move asymptotically towards a new equilibrium value is clear. About half of the final observed increase occurred within the first 30–40 years of manure application. During the period 1926–1966, most plots on Broadbalk were bare fallowed 1 year in five, to control weeds, and no manure was applied in the fallow years. Thus, organic C inputs were reduced in these years, giving rise to the sawtooth shape in the simulations. The overall effect during this 40-year period was a small decline in SOC content in all treatments but is most clearly seen in the FYM treatment. After fallowing ceased (because weeds were then controlled by herbicides), the increasing trend in the FYM treatment resumed but at a slower rate as the soil tended towards a new equilibrium value. If the soil had not been fallow years during this 40-year period, the new equilibrium value would presumably have been attained earlier and the annual rate of SOC increase in the later years would have been even less. The same trend can be seen in the later FYM treatment that started in 1885.

The SOC content of the control (with no manure or fertilizer inputs) remained stable for the duration of the experiment. This is because the site had been arable for many centuries before the start of the experiment. It is thought that it was initially cleared from native forest at least 1,500 years previously, so the SOC content inherited from forest presumably declined during this long period and reached a new, low equilibrium value that is maintained by the small crop inputs in the control treatment.

In the treatment receiving inorganic fertilizers, crop yields were much higher than in the control as were the organic inputs to soil from crop roots and stubble. This small increase in the annual input of organic C led to a small increase in SOC

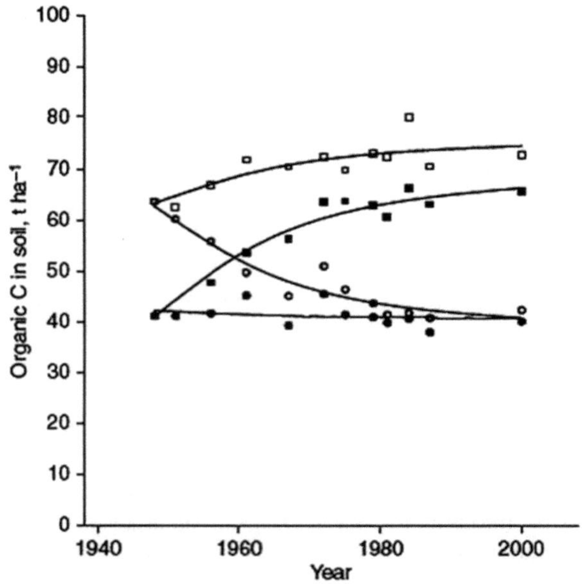

Fig. 16.3 Changes in SOC content (0–23 cm) in a silty clay loam in two ley-arable experiments at Rothamsted 1949–2002. Highfield Experiment: old grassland soil remaining in grass □ or ploughed and kept in arable cropping ○. Fosters Experiment: old arable soil remaining in arable ● or sown to grass and kept in grass ■ (After Johnston et al. 2009)

compared to the control. This is a general trend, seen in many long-term studies; in 135 studies at 114 long-term sites worldwide, application of inorganic N (often with P and K) led to the SOC and soil organic nitrogen (SON) contents of soils being 8 and 12 %, respectively, greater than in the corresponding controls (Ladha et al. 2011). However, at many sites (in contrast to Broadbalk), there was a general decline of SOC and SON over time in all treatments which was not reversed by inorganic fertilizer applications; rather the decrease was slightly less where inorganic N fertilizers were applied. These observations at so many sites contradict the claim that is sometimes made that inorganic fertilizers, especially N, cause a loss of soil organic matter; there has been recent discussion of this erroneous claim and the misinterpretation of results from one LTE from which it arose (Powlson et al. 2010).

Figure 16.3 illustrates the importance of land management changes in determining SOC content, both examples coming from sites at Rothamsted on a silty clay loam soil. In the Highfield Experiment, an area that had been under grass for over 100 years, and contained >60 t SOC/ha to a depth of 23 cm, was ploughed and converted to arable. Over the next 20–30 years, its SOC content declined rapidly; within 50 years, it had lost one third of its original SOC and had reached the same content as that of a soil that had been arable for well over 100 years. A section of the grassland that continued under grass increased slightly in SOC because the intensity of management increased, including increased fertilizer N applications. This led to increased grass yields and increased inputs of root material into soil. At the arable site (the Fosters Experiment), the area that continued under arable cropping remained at an almost constant SOC content throughout the 50 years shown in Fig. 16.3, though at a higher level than that in the Broadbalk Experiment (Fig. 16.2). The section of the arable field sown to grass, and managed in the same way as the grass area of Highfield, increased in SOC from 40 to 65 t/ha within 50 years but was still below the value in the continuing grass section of Highfield.

16.6 Influence of Soil Organic Matter Content on Crop Yields

In the Broadbalk Experiment, yields of winter wheat in the NPK treatment (with N applied at 144 kg/ha) and the FYM treatment were approximately equal for the duration of the experiment, with both the old and new (post-1968) varieties (shown for continuous wheat in Fig. 16.1). This is surprising in view of the greatly improved soil physical conditions in the FYM treatment compared to NPK as a result of the far higher SOC content (Fig. 16.2). This led to the conclusion that, for winter wheat on this soil type, the additional organic matter in soil derived from FYM did not increase yields through mechanisms other than supply of nutrients. The improved soil physical conditions did not translate into increased crop yield: yields attained with FYM could be equalled by additions of inorganic fertilizers. Results obtained in recent years may alter this interpretation somewhat, though this is not certain. Since 1968, one of the FYM-treated plots was given additional N fertilizer; FYM is incorporated into soil in autumn, prior to sowing winter wheat, and the additional N applied in spring at the same time as to the NPK treatments. In most years, but not all, the FYM + N treatment has given the largest yield of all treatments in the experiment, greater than FYM alone and greater than any of the NPK treatments including the current highest N rate of 288 kgN/ha. This is shown for first wheat after a break crop in Fig. 16.1. One possible explanation is that improved soil conditions and/or nutrient availability in the early growth stages in the FYM-treated soil (compared to NPK treatments) permits increased early growth and a larger yield potential, this potential being realized when additional fertilizer N is applied in spring. Another possibility is that the result reflects the poor synchrony between nitrate produced from mineralization of organic N in manure and the rapid uptake of N by the crop during spring. The lack of synchrony leads to major N losses, illustrated by the large loss of nitrate through over-winter leaching from the FYM treatments shown in Fig. 16.5. If this is the explanation for the larger yield in the FYM + N treatment, the additional fertilizer N is simply meeting the N shortfall from mineralization of manure during rapid spring growth. This may be especially important for modern wheat varieties in which grain yield responds to high rates of N. It is possible that both factors are contributing, so the situation is complex: in the context of sustainability, understanding and better defining the role of organic inputs in determining crop yields requires further detailed research.

Adjacent to the Broadbalk Wheat Experiment is the Hoosfield Barley Experiment where spring barley has been grown continuously with a range of manure and fertilizer treatments since 1852 (Poulton 2006). In this experiment, there is clear evidence of a crop yield benefit from increased SOC (in addition to benefits from nutrients supplied in manure), in contrast to the general situation with winter wheat on Broadbalk. Figure 16.4 shows grain yields of spring barley in two main treatments of the experiment in four periods during which different crop varieties were grown. With both main treatments (FYM or inorganic fertilizers, termed the PK treatment), four rates of fertilizer N (including zero N) are superimposed. In the earliest period shown (1976–1979), the grain yield in the FYM treatment could be matched in the

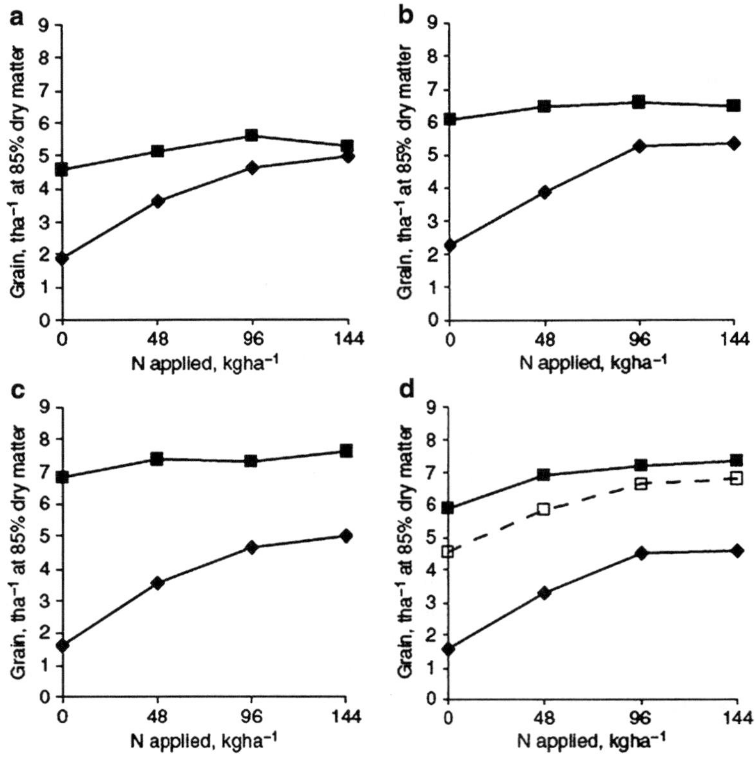

Fig. 16.4 Grain yields of spring barley (t/ha) in the Hoosfield Continuous Barley Experiment. Annual treatments 1852–2006: PK fertilizers ♦; 35 t/ha farmyard manure (*FYM*) ■. Annual treatment only from 2001 to 2006: 35 t/ha FYM □. (**a**) *cv*. Julia, 1976–1979; (**b**) *cv*. Triumph, 1988–1991; (**c**) *cv*. Cooper, 1996–1999; (**d**) *cv*. Optic, 2005–2007 (After Johnston et al. 2009)

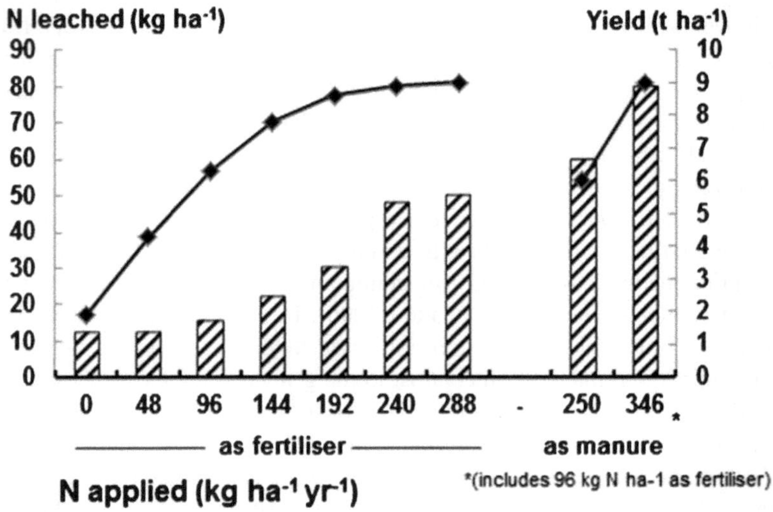

Fig. 16.5 Winter wheat grain yields (*points and lines*) and quantities of nitrate-N leached through drains (*bars*) in selected treatments of the Broadbalk Experiment (Data refer to *cv*. Hereward grown as first wheat after a break crop (After Goulding et al. 2000))

PK treatment provided sufficient N fertilizer was applied (panel A). In later periods, when more modern and N-responsive spring barley varieties with greater yield potential were grown, this was no longer the case, especially where yield potential was protected by the use of fungicides. For example, in the two latest periods shown (panels C and D), yields with FYM were about 2 t/ha greater than in 1976–1979 and could not be matched in the PK treatment, even with the highest rate of N. Probably, the improved physical structure in the topsoil in the treatments receiving FYM permits more rapid and extensive rooting, greater uptake of water and nutrients and greater resilience to any stresses such as water shortage or disease occurring later in the growing season compared to plants in the less-well-structured soil of the PK treatment. Other factors may also contribute, including the time course of N mineralization and the depth distribution of nitrate and the larger and different microbial community in the FYM treatment compared to PK; for example, the composition of the rhizosphere population may give roots greater protection from soilborne pathogens.

Of course, the same arguments can be applied to the situation of winter wheat in the Broadbalk Experiment – but here (with the possible exception discussed earlier) the results show no unique benefit to crop yield from the higher SOM content in the FYM treatment. The most likely explanation of the difference between the two crops is that spring barley, having a growing season of only 5–6 months, is more sensitive to soil structure than winter wheat which is growing for about 10 months; in particular, winter wheat has a much longer time for roots to become well established.

16.7 Influence of Soil Organic Matter Content on Soil Physical Properties

Soil physical properties are determined to a great extent by soil texture, but it has been known for centuries that they are strongly modified by organic matter. For example, an increase in soil organic matter (SOM) content increases water-holding capacity and facilitates the formation of stable aggregates and stable pore spaces that improve aeration and make root penetration easier (Johnston and Dawson 2010). More recently, it has been found that small changes in SOM content can have a disproportionately large impact on certain soil physical properties, almost certainly through the role of specific fractions within the totality of SOM.

A specially designed apparatus was used to measure the energy required to pull a plough through soil on all the plots of the Broadbalk Experiment (Watts et al. 2006). The measurements were used to calculate specific plough draught (S), defined as the force in kPa per cross-sectional area of worked soil. A larger value of S indicates that greater force is required to pull the implement through the soil; a low value of S makes for ease of working and decreased fuel requirement. Clay content was known to have a major influence on the value of S, so this was also measured with greater

Table 16.2 Effect of soil organic C concentration on specific draught in the Broadbalk Experiment

Long-term treatment	Organic C concentration in plough layer (0–23 cm) soil %	Specific plough draught, *S* kPa
Nil	0.84	88
Farmyard manure	2.80	75
Inorganic fertilizers	1.08	77

After Watts et al. (2006)

spatial detail than had previously been done. It was found that clay content in the topsoil varied between 19 and 39 % between the different plots (a greater range than had previously been realized) and this factor had the greatest impact on *S*. But organic matter content had a strong modifying effect. Table 16.2 shows data for three treatments after correcting for variation in clay content. The treatment receiving farmyard manure (FYM) contained over three times the amount of SOC compared with the Nil plot (receiving no manure or fertilizer). This very considerable increase in SOC led to a 15 % decrease in *S* compared with the Nil treatment. The treatment receiving inorganic fertilizers contained only 29 % more SOC than the Nil treatment, yet the force required to work the soil was decreased almost as much as in the FYM treatment – a relatively small change in total SOC concentration had a disproportionately large impact on the energy required to pull an implement through the soil. This is significant in practical terms because it is difficult to increase SOC by a large amount in normal agricultural practice: this finding shows that even small changes can be beneficial and, therefore, worthwhile.

It is well known that increased SOM content contributes to increased stability of soil macroaggregates. Aggregate stability is commonly assessed by measuring aggregate sizes after wet sieving, with results expressed as mean weight diameter (MWD); a larger value of MWD indicates more stable aggregates. This was measured in different treatments within LTEs in the UK (Broadbalk), Germany and Australia; as expected, aggregate stability was greater in soils with greater SOC content. But in this work, a specific fraction within total SOC termed *labile C* was also measured. This fraction represents organic C in forms that are readily oxidizable under defined conditions; it often accounts for about 10 % of total SOC and is thought to comprise microbial biomass (about 2 % of total SOC) plus microbial metabolites. It was consistently found that aggregate stability was more closely correlated with the labile C fraction than with total SOC (Blair et al. 2006); there was a similar trend for the relationships between total or labile C and water infiltration rate.

In the debate on whether or not it is acceptable to remove straw for use in bioenergy generation, Powlson et al. (2011a) reviewed the effects of removal or incorporation of cereal straw on SOC using results from 25 experiments of between 6- and 56-year duration. Although there was a trend for total SOC content to be slightly greater where straw was incorporated, the effect was small: SOC increases with straw were only statistically significant in 6 out of the 25 experiments, and in the majority of cases increases were <10 %. But, in several of the studies, authors noted improvements in the occurrence and stability of water-stable aggregates even when there was no measurable effect on total SOC content. In at least one case, in Canada (Malhi and Lemke 2007), it was concluded that this improvement in soil physical structure was beneficial for seedling emergence, root growth and decreased soil erosion.

In recent years, there has been renewed interest in physical fractionation methods as a means of studying the composition and significance of different forms of organic C within the total. Sohi et al. (2001) identify two *light fractions* by dispersion of soil in a heavy liquid (sodium iodide with a density of 1.8 g/cm^3). The first fraction termed *free light fraction* (FR-SOM) floats to the surface when the soil is gently dispersed. The second, only released when soil aggregates were broken using ultrasonic dispersion and termed *intra-aggregate light fraction* (IA-SOM), was assumed to comprise material partially stabilized through its physical location. Using several types of spectroscopy, it was concluded that the FR-SOM was predominantly organic matter from recent plant inputs whereas IA-SOM was a more processed fraction, having undergone a greater degree of decomposition by the soil microbial population. A third fraction was the organic C strongly associated with the mineral components of the soil. Conclusions regarding the significance of these identifiable fractions, such as their possible use in modelling C dynamics, were initially based on soil from experiments in the UK. In order to test whether the conclusions were of general applicability, the fractionation technique was subsequently applied to soils from a common set of treatments (control, inorganic fertilizers, FYM) within eight long-term experiments in different countries having diverse environments and cropping systems (Sohi et al. 2005). They included a semiarid site (Syria) that had run for 14 years at the time of sampling and one in a wet monsoonal region (Philippines) that had run for 34 years. The initial conclusions regarding the significance of the different SOM fractions were confirmed. This is a good example of (a) networking between LTEs and (b) using them for purposes that could not have been predicted when they were started.

16.8 Nutrient Cycling Studies

16.8.1 Nitrogen

N fertilizer labelled with ^{15}N was applied to wheat within micro-plots of the Broadbalk Experiment in the 1980s in order to gain more detailed information on the fate of N within the different treatments (Powlson et al. 1986). Of course, valuable information on the fate of fertilizer-derived N can be obtained within short-term experiments. Embedding such a study within Broadbalk, where it was known that soil N and C content had reached a steady state, enables additional calculations – especially on the inputs of N from atmospheric deposition. Recovery of ^{15}N in above-ground crop (averaged over separate applications in 4 years at 144 kgN/ha) was 62 %; it was 68 % in two of the years and 57 % in the others. Recovery was lower in years with high spring rainfall (in the few weeks following N fertilizer application); this was consistent with greater N losses through denitrification in years with wetter conditions in spring. The trend was observed on other soil types and with other crops in southeast England (Macdonald et al. 1997). On average, an additional 17 % of applied N was retained in organic matter in the plough layer soil. Total recovery of applied labelled N in crop plus topsoil ranged from 70 to >90 %. The labelled N retained in soil was rapidly converted into relatively stable forms and only remineralized slowly during subsequent years. After four subsequent crops, only 16 % of the labelled N initially retained in soil (0–70 cm) plus stubble had been taken up by crops, 29 % had been mineralized and lost, and 55 % remained in soil (Hart et al. 1993). This pattern of rapid immobilization of applied inorganic N into organic forms of N that are remineralized slowly over several years has been observed in other situations where ^{15}N labelling has been superimposed within ongoing long-term experiments, for example, with spring barley using the Hoosfield Experiment and with herbage using the Park Grass Experiment (Glendining et al. 2001; Jenkinson et al. 2004).

Under tropical conditions, recovery of applied fertilizer N in crop tends to be lower and losses greater. For example, in an extensive coordinated set of experiments with ^{15}N-labelled fertilizer (six different crops at 13 field sites in nine countries, mainly tropical and subtropical), the mean recovery of applied N in crop in the first growing season was only 33 % but with a wide range of 7–63 % (Dourado-Neto et al. 2010). In these experiments, which continued for 3 years so that the fate of ^{15}N immobilized into soil organic matter could be quantified, the ^{15}N micro-plots were set within the main plots of ongoing field experiments which had already been established for some years.

Lawes et al. (1881/1882) published studies on nutrient loss via drains over 130 years ago. Nitrate leaching from agricultural land continues to be a significant environmental issue that has prompted regulations to restrict agricultural practices in an attempt to improve water quality (in the EU, regions covered by such regulations are termed Nitrate Vulnerable Zones) but achieving significant reductions in nitrate leaching whilst maintaining large crop yields remains a challenge. The new drains installed beneath one section of the Broadbalk Experiment in 1993 have been used for measuring nitrate leaching under different fertilizer and manure regimes (Goulding et al. 2000).

Under the maritime climate of the UK and northwest Europe, most nitrate leaching occurs during winter rather than immediately after spring application of N fertilizer. The quantity of nitrate leached represents a combination of any unused fertilizer still present in soil as nitrate, plus nitrate derived from mineralization of soil organic matter. Experiments with ^{15}N-labelled fertilizer have shown that in the UK, at least 80–90 % of the leached N is from mineralization of soil organic matter plus nitrate derived from any manure applied (Macdonald et al. 1989). These findings rest on a combination of studies on long-term field sites and shorter-term experiments. In other environments, especially where excessive rates of N fertilizer are used such as is often the case in China, the direct contribution from unused fertilizer can be dominant

(Ju et al. 2009; Zhou et al. 2010). Measurements of nitrate in water from the drains beneath plots of the Broadbalk Wheat Experiment (Fig. 16.5) show that some nitrate entered the drains even when no N fertilizer or manure was applied; this must be derived from mineralization plus a contribution from N deposited from the atmosphere. There was a marked increase in nitrate leaching when the quantity of N fertilizer applied in the previous spring approached or exceeded that required to obtain maximum crop yield. The largest amounts leached were from the FYM treatment, with or without additional inorganic N fertilizer.

16.8.2 Phosphorus

LTEs are essential for studying the supply of phosphate (P) to crops. Soil contains P in a wide range of inorganic and organic forms but, in general, that which is immediately available for plant uptake is in soil solution. The quantity in this form is usually very small so a continuing supply depends on equilibration between this form and others. The reactions involved occur over periods ranging from seconds to decades, or more, but it has been found that changes occurring over the years-to-decade timescale are particularly important for the sustainability of P supply and, hence, crop production. Rates of change of P reserves in soil have been studied in one of the lesser known LTEs at Rothamsted, the Exhaustion Land Experiment (Poulton 2006). For 83 years (1902–1985), no P was applied and the rate of P exhaustion was studied. By 1985, crop-available P as assessed by extraction with sodium bicarbonate solution[2] had declined to about 2–7 mg/kg (depending on the treatment before 1901). From 1986, applications of P fertilizer at three rates were started in order to measure the rate of increase in crop-available P. Within 3 years, it had increased to values up to 27 mg/kg. Figure 16.6 shows grain yields of spring barley at this time in treatments where availability of N or K was not limiting yield. The results show a yield plateau with yield increases up to a *critical value*, in this case about 10 mg/kg, but with no further increase after this (Poulton et al. 2013).

The critical-value approach is the basis of P fertilizer advice in the UK and several other countries (Johnston 2001; Syers et al. 2008); farmers are advised to maintain soil significantly above the relevant critical value, to avoid yield loss and to have their soils analysed periodically. The critical value varies according to soil type and crop; in order to determine the value for different situations, it is ideal to have a long-term site with treatments spanning a range of Olsen P levels including low values, so that curves such as that in Fig. 16.6 can be constructed. Worldwide there are far too few such sites, and a priority for agricultural development is to establish additional sites in diverse environments, especially in regions where agriculture is developing rapidly such as South America and China. In a network of experiments in China (at eight sites with cropping systems representing some

[2] Olsen's reagent, which is commonly used in advisory work in the UK and many other countries.

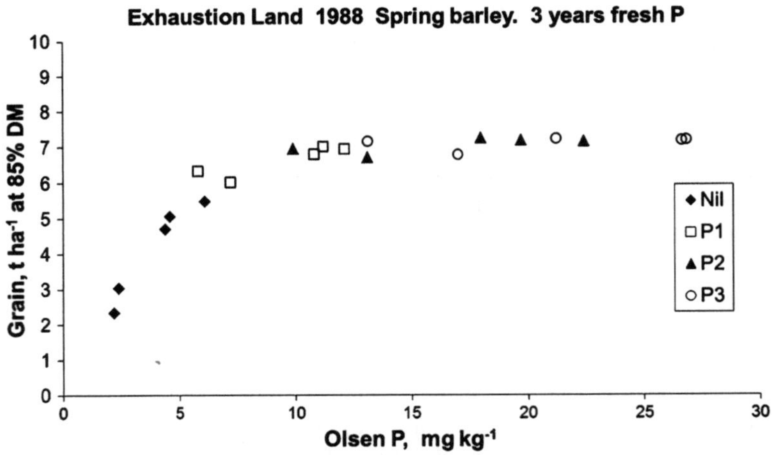

Fig. 16.6 Grain yield of spring barley in the Exhaustion Land Experiment in relation to concentration of sodium bicarbonate-extractable phosphate (Olsen P) in situations where other nutrients were not limiting (After Poulton 2006)

70 % of the cropped area), trends in the first 15 years of different fertilizer treatments clearly showed that Olsen P declined within a few years to levels causing severe loss of yield if P was not applied as fertilizer or manure; in contrast, yield loss due to decline of plant-available K was slower to appear. In several cases with wheat, maize or rice, grain yields in the absence of applied P were as much as 4 t/ha less than in the treatment where P was supplied (Zhao et al. 2010).

Ensuring an adequate supply of P to crops is essential for global food production and has major implications for farmers' access to P fertilizer, the need for novel approaches for resource-poor smallholder farmers such as precision placement close to individual plants and for efficient recycling of P from manures and other organic residues. But ensuring that the concentration of P in soil is not too high is also important from the viewpoint of minimizing water pollution. P can move from soil to waters through surface runoff (usually of soil particles carrying adsorbed P) and also by P leaching. Very small concentrations of P in surface waters are enough to trigger algal growth, so P losses that are insignificant agronomically can have major adverse environmental impacts. The phenomenon of P leaching, mainly through macro-pore flow such that water avoids contact with a large surface area of soil, was not recognized until measurements were made on drainage water from the Broadbalk plots (Heckrath et al. 1995). This work demonstrated that P losses were small until Olsen P reached a certain value (the *change point*) after which losses increased greatly. The value of this change point varies greatly between soil types, and the factors influencing its value are still not fully understood. The practical result of the finding is that crop-available P concentrations in soil should be maintained above the critical value for crop yield but below the change point value to avoid P loss to water. Thus, P management, at least in situations where rainfall is sufficient to cause P leaching, is more complex and knowledge-intensive than previously thought, which would not have been known without experimentation using long-term sites.

16.9 Microbiological Studies Using Rothamsted Long-Term Experiments

16.9.1 Crop Disease Populations

Every year, two septoria blotch diseases of wheat cause major losses of grain worldwide. Plant disease surveys in the UK show that the relative importance of the two diseases, caused by either *Phaeosphaeria nodorum* or *Mycosphaerella graminicola*, has changed. *P. nodorum* was prevalent in the 1970s but *M. graminicola* became dominant since the 1980s; similar changes have occurred elsewhere. In research to identify the relative importance of the two diseases over a longer timescale, DNA was extracted from archived straw samples from the Broadbalk Experiment (Bearchell et al. 2005). It was found that the relative abundance of *P. nodorum* had increased fairly consistently from the 1840s until about 1970 but, thereafter, decreased sharply. Unexpectedly, this change correlated strongly with changes in SO_2 concentration in the atmosphere (Fig. 16.7); because data on these concentrations were not available for the early period, annual emissions of S (that can be estimated quite accurately from coal burning statistics) were used as a proxy.

It was concluded that the action of SO_2 (perhaps in combination with atmospheric ozone) was responsible for the change in disease abundance. The mechanisms of penetration of wheat leaves by the two fungi are different: *M. graminicola* penetrates through stomata rather than directly through the cuticle and has a longer latent period

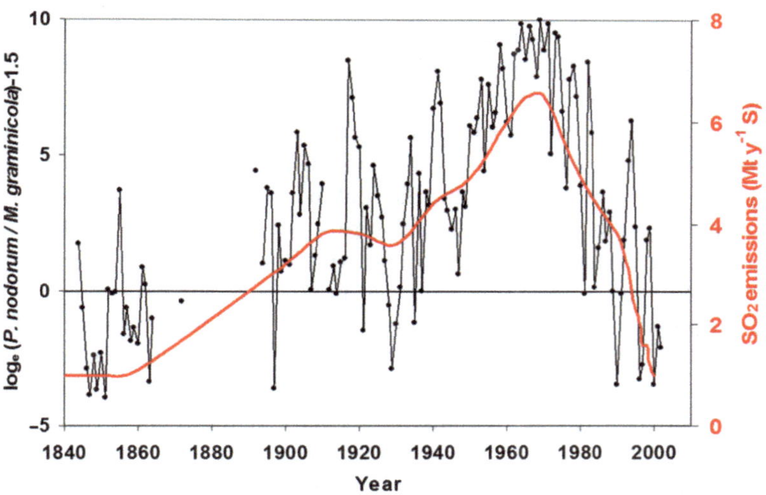

Fig. 16.7 Changes in relative abundance of two fungi causing septoria blotch disease and relationship with changes in sulphur dioxide (SO_2) in the atmosphere, using historical SO_2 emissions from coal burning as an indicator. Abundance of the two fungi based on analysis of DNA extracted from straw in the Rothamsted Sample Archive (Bearchell et al. 2005)

than *P. nodorum*, so may be more exposed to air pollutants and hence more susceptible to SO_2. Its recovery of dominance when the atmospheric concentration of SO_2 diminished is consistent with this.

16.9.2 Soil Biodiversity

It has long been recognized that a wide range of soil organisms is crucial for numerous processes essential for both crop production and ecosystem services. A new Global Soil Biodiversity Initiative has recently been announced: http://www.globalsoilbiodiversity.org. Molecular techniques developed in recent decades have added greatly to the range of approaches available for studying soil biodiversity and the impacts of management or environmental changes. A recent example, using a Rothamsted long-term site, quantified biodiversity in an extreme situation (Hirsch et al. 2009); measurements were made on soil from a site adjacent to the Highfield Experiment that has been kept under bare fallow for 50 years (the experiment was described earlier, and changes in SOC content in the arable and grass treatments are shown in Fig. 16.3). The microbial biomass C content of the soil kept bare, with virtually no plant inputs over this period, was only 7 % of that in the soil that continued under permanent grass and 33 % of that in the arable treatment. The free light fraction revealed by density separation (Sohi et al. 2001), mainly representing relatively fresh organic inputs, was only 3 % of that in the grassland treatment. Intra-aggregate light fraction, a somewhat more processed and stabilized fraction, was 13 % of that in the grassland soil. As expected, the abundance and diversity of mesofauna (mites and collembola), which feed mainly on fresh plant inputs, were much lower in the bare fallow than either the grassland or arable soils. In the case of the bacterial population, *abundance* (as assessed by culturing on a low-nutrient agar) was much lower in the bare fallow soil than in soil with fresh plant inputs – consistent with the greatly decreased microbial biomass C content. By contrast, bacterial *diversity* was similar in the bare fallow soil to that in the soils receiving fresh plant inputs from grass or arable crops – whether assessed by the chemical fingerprint of their fatty-acid composition (PLFA), Biolog cell phenotyping, or extracted DNA and RNA. DNA examined in this way is considered to represent the species present, though the definition of species in bacteria is a matter of debate (Prosser et al. 2007), so distinct 16S rRNA gene PCR products revealed by DNA-based techniques were referred to as *operational taxonomic units* (OTUs). The OTUs derived from rRNA comprise a small fraction of those identified from genomic DNA and are commonly assumed to represent cells that are, or recently have been, metabolically active. So the finding of considerable bacterial diversity and activity in the 50-year bare-fallowed soil is remarkable – and indicates the resilience of the soil bacterial community.

These findings have implications for the survival strategies of soil bacteria and may have relevance for the restoration of soils that have been damaged (e.g. through engineering or mining activities or inappropriate tillage or management). This scientific discovery was only possible because of the existence of the unusual and extreme long-term bare fallow treatment that had no obvious practical value and the use described here could not have been foreseen when it was established. New treatments have now been established within the long-term bare fallow area in which grass and wheat are being grown for the first time in 50 years, so in future it will be possible to study the recovery of biological activity in this soil in response to new plant inputs.

16.9.3 Aerobic Soil as a Sink for Methane

Methane (CH_4) is a greenhouse gas that has approximately 30 times the greenhouse-warming potential of CO_2. Its concentration in the atmosphere has increased considerably during the past century with emissions from various human activities including the recovery of fossil fuels (coal mining, oil exploration, leakage of pipelines) but also from production in anaerobic soils, whether natural wetlands or soil flooded for growing paddy rice. It is destroyed in the atmosphere through photochemical reactions, but its only known terrestrial sink is oxidation by bacteria in aerobic soils. In soil, CH_4 is oxidized to CO_2, a less powerful greenhouse gas, so the process is environmentally favourable. It is therefore important to determine the rate of CH_4 oxidation in different environments and factors affecting the process. Research at several LTEs, including Rothamsted (UK) and Bad Lauchstädt (Germany), has addressed the issue, but the microbial processes involved are complex and have proved difficult to study. Comparisons of soils from different land uses at Rothamsted show that the highest rate of CH_4 oxidation occurs in undisturbed soils under forest, followed by grassland. Soil cultivated annually for arable cropping showed the lowest rate, about 15 % of the rate in a forest soil (Powlson et al. 1997). Addition of inorganic N fertilizer had a further negative effect in the Rothamsted experiments but not always at Bad Lauchstädt. Understanding of the influences of different management practices is complicated by the fact that the organisms responsible for CH_4 oxidation, the methanotrophs, are not well characterized and CH_4 can also be oxidized by ammonium oxidizers (nitrifiers).

A fairly recent approach has been to incubate soil cores in the presence of ^{13}C-labelled CH_4 and then extract phospholipid fatty acids (PLFAs) from soil. The specific combination of PLFAs present can be used to characterize different bacterial groups and methanotrophs that are active in CH_4 oxidation and retain some ^{13}C in their cells – providing an excellent means of identifying the group(s) that are operating in a given situation. The technique is termed PLFA-stable isotope probing (PLFA-SIP). Results from woodland and grassland sites at Rothamsted showed that the estimated methanotrophic biomass in soil, based on ^{13}C incorporation, was considerably smaller than at grassland and forest sites previously studied in the UK (Maxfield et al. 2011) – yet rates of CH_4 oxidation were of the same order. Tillage appeared to decrease the methanotrophic biomass as this tended to be small in arable (but unfertilized) soil. Long-continued application of inorganic N fertilizer increased the size of the methanotrophic population, but decreased the rate of CH_4 oxidation. The results are consistent with the hypothesis that, in the Rothamsted soils, much of the CH_4 oxidation is mediated by nitrifying bacteria; these can perform the oxidation due to the similarity between the methane monooxygenase and ammonium monooxygenase enzymes, but do not assimilate C from CH_4 for growth. Much still remains to be understood about the role of different bacterial groups in the oxidation of CH_4 in soils and the sometimes confusing and contradictory impacts of various management practices. But LTEs are one of the key resources for such research.

In a study on CH_4 oxidation using the Park Grass Experiment, soil cores were incubated with radioactive ^{14}C-labelled CH_4, again on the assumption that methanotrophs responsible for oxidation would retain some of the CH_4-derived C and thus provide a label (Stiehl-Braun et al. 2011). The soil cores were subsequently impregnated with resin and thin sections prepared that were used for autoradiography to identify the physical location of active methanotrophs with respect to depth and soil pores in different treatments of this grassland soil. It had previously been found that low soil pH inhibited CH_4 oxidation; in one of the Park Grass plots studied, soil pH had decreased to 4.0 due to nitrification of ammonium N added over many years with no lime addition to counteract the effect. In this soil, the new visualization method showed that CH_4 oxidation activity was confined to soil below a depth of 12 cm where soil was at a slightly higher pH.

16.9.4 Nitrous Oxide Emissions from Soil

Agriculture globally is responsible for about 14 % of all human-induced greenhouse gas emissions, of which 70 % is associated with N fertilizer (Pachauri and Reisinger 2007). This is partly attributable to emission of CO_2 during N fertilizer manufacture and partly to emission of nitrous oxide (N_2O) through the processes of bacterial nitrification and denitrification in soil (Smith 2010). N_2O has a greenhouse-warming potential almost 300 times that of CO_2, so even emissions that are small in terms of agronomic N loss can be highly damaging environmentally. In some low-rainfall regions, such as north China, nitrification can be the main source of N_2O from soil (Ju et al. 2011), but in regions with higher rainfall, such as northwest Europe, denitrification is dominant (Morales et al. 2010). Figure 16.8 shows N_2O emissions from several plots in the Broadbalk Experiment and in two adjacent woodland areas as measured in the field. In the arable plots, N_2O emission increased with increasing rates of N fertilizer application and was especially high in the treatment given a rate of N fertilizer that usually exceeded crop requirement (288 kgN/ha). Emission was low in the treatment receiving FYM, probably because the peak concentration of nitrate in soil in this treatment was lower than in the inorganic fertilizer treatments because it was produced gradually through mineralization of organic N. However, nitrate leaching was higher in the FYM treatment

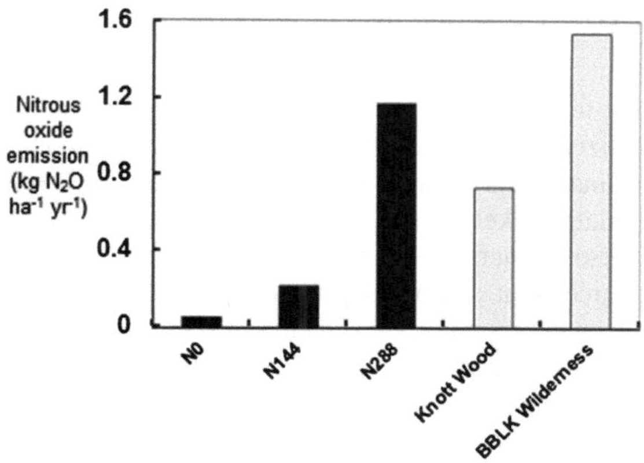

Fig. 16.8 Nitrous oxide emissions from arable plots in the Broadbalk Experiment and in two forest sites. Measurements made monthly in the field using cover box methodology and annual emission rates calculated (Based on data from Goulding et al. 1998)

than from plots receiving inorganic N fertilizer (Fig. 16.5), so indirect N_2O emission from FYM would be greater. Emissions from the woodland areas were large and similar to those from arable plots given a high rate of N fertilizer, possibly due to N deposition from the atmosphere, as discussed earlier.

Denitrification, the microbial reduction of nitrate, has two gaseous products: N_2O and N_2, their ratio depending on the extent of reduction. Some populations of denitrifying bacteria lack the *nosZ* gene that controls conversion of N_2O to N_2, and there are preliminary indications that the proportions of the different populations vary between soils and can be quantified using molecular techniques (Morales et al. 2010). If confirmed, this would appear to be a significant development, potentially providing a basis for decreasing N_2O emissions by designing management practices specific for different situations (Powlson et al. 2011b). In recent research using soils from the Broadbalk Experiment, molecular methods were used to characterize the denitrifying populations in different treatments and compare their abundance with denitrification potential as assessed by N_2O emission in laboratory incubations with added nitrate under conditions conducive to denitrification (Clark et al. 2012). It was found that abundance of the *nosZ* gene was less in soil from the woodland regeneration area than from the arable plots and this was consistent with the very high emission of N_2O from this soil under the conditions tested (Fig. 16.8). It was found that two genes controlling an earlier step in the denitrification pathway (reduction of nitrite) differed in their response to long-term N inputs. The *nirK* gene was more abundant in soils with increasing N inputs from fertilizer or manure. By contrast, the *nirS* gene was most abundant in the soil never receiving N fertilizer and its abundance was inversely related to N inputs. The full implications of these findings are still to be understood, but they do show that soils with large inputs of N, or those having a large organic C content, show a greater likelihood of producing N_2O when conditions are suitable for denitrification; this provides a further incentive to avoid excessive N inputs. However, maintaining organic C at a higher level (e.g. through manure applications) leads to improved soil structure and better aeration so, probably, decreasing the chance of reaching conditions conducive to denitrification at intermediate soil moisture contents.

16.10 Conclusions

Long-term experiments with a range of different cropping systems, tillage methods, fertilizer or manure treatments are one vital component of research to develop more sustainable agricultural systems. Sustainable systems are needed to feed a global population likely to reach about nine billion by 2050 but, simultaneously, minimize adverse impacts on the environment. Beyond this crucial role, LTEs are scientific resources that can be used for many and various purposes in the agricultural and environmental sciences (and other disciplines) – purposes that are often unrelated to the original aims of the experiment. Recognition of these benefits has led to some LTEs in the UK, including the Classical Experiments at Rothamsted (some started almost 170 years ago) and the more recently established North Wyke Farm Platform (based on grazed grassland farming) being classified as National Capabilities by the main funding body, the Biotechnology and Biological Sciences Research Council. In addition, the Rothamsted LTEs and Sample Archive are included within the new EU-funded ExpeER project (Distributed Infrastructure for EXPErimentation in Ecosystem Research) (www.expeer.fr) which aims to improve access to state-of-the-art research facilities in Europe through its coordinated program of Transnational Access.

Factors that should be considered if any LTE is to be of lasting scientific value and continue to attract the necessary funding include:

1. The ability to make carefully considered modifications to ensure the integrity of the experiment and to increase its value as a scientific resource, whilst maintaining continuity of core elements
2. Rigorous attention to detail in the management of the site and recording of all aspects of the treatments, field operations and sampling
3. Archiving soil and crop samples for future use
4. A culture of scientific openness that encourages collaboration with researchers, beyond the group directly responsible for running the experiments, who can bring complementary skills and insights
5. Networking with other LTEs to gain added value through comparing and contrasting with findings from different environments and cropping systems

Acknowledgements Rothamsted Research receives grant aid from the UK Biotechnology and Biological Sciences Research Council. Some of the work reported was funded by the UK Department of Environment, Food and Rural Affairs.

References

Austin RB, Ford MA, Morgan CL, Yeoman D (1993) Old and modern wheat cultivars compared on the Broadbalk wheat experiment. Eur J Agron 2:141–147

Bearchell SJ, Fraaije BA, Shaw MW, Fitt BDL (2005) Wheat archive links long-term fungal pathogen population dynamics to air pollution. Proc Natl Acad Sci USA 102:5438–5442

Blair N, Faulkner RD, Till AR, Poulton PR (2006) Long-term management impacts on soil C, N and physical fertility. Part 1: Broadbalk experiment. Soil Tillage Res 91:30–38
Christensen BT, Petersen J, Trentemøller UM (2006) The Askov long-term experiments on animal manure and mineral fertilizers: the Lermarken site. Plant Production Report 121, Danish Institute of Agricultural Sciences, Foulum
Clark IM, Buchkina N, Jhurreea D et al (2012) Impacts of nitrogen application rates on the activity and diversity of denitrifying bacteria in the Broadbalk wheat experiment. Philos Trans R Soc Lond Ser B 367:1235–1244
Coleman K, Jenkinson DS (1996) RothC-26.3 – a model for the turnover of carbon in soil. In: Powlson DS, Smith P, Smith JU (eds) Evaluation of soil organic matter models using existing long-term datasets. Springer, Heidelberg, pp 237–246
de Richter DB, Hofmockel M, Callaham MA et al (2007) Long-term soil experiments: keys to managing Earth's rapidly changing ecosystems. Soil Sci Soc Am J 71:266–279
Dourado-Neto D, Powlson DS, Abu Bakar R et al (2010) Multi-season recoveries of organic and inorganic nitrogen-15 in tropical cropping systems. Soil Sci Soc Am J 74:139–152
Fan MS, Zhao FJ, Fairweather-Tait SJ et al (2008) Evidence of decreasing mineral density in wheat grain over the last 160 years. J Trace Elem Med Biol 22:315–324
Glendining MJ, Poulton PR, Powlson DS et al (2001) Availability of the residual nitrogen from a single application of ^{15}N-labelled fertilizer to subsequent crops in a long-term continuous barley experiment. Plant Soil 233:231–239
Goulding KWT, Bailey NJ, Bradbury NJ et al (1998) Nitrogen deposition and its contribution to nitrogen cycling and associated soil processes. New Phytol 139:49–58
Goulding KWT, Poulton PR, Webster CP, Howe MT (2000) Nitrate leaching from the Broadbalk wheat experiment, Rothamsted, UK, as controlled by fertiliser and manure inputs and the weather. Soil Use Manage 16:244–250
Gutteridge RJ, Jenkyn JF, Poulton PR (1996) Occurrence of severe take-all in winter wheat after many years of growing spring barley, and effects of soil phosphate. Aspects Appl Biol 47:453–458
Hart PBS, Powlson DS, Poulton PR et al (1993) The availability of the nitrogen in the crop residues of winter wheat to subsequent crops. J Agric Sci Camb 121:355–362
Heckrath G, Brookes PC, Poulton PR, Goulding KWT (1995) Phosphorus leaching from soils containing different phosphorus concentrations in the Broadbalk experiment. J Environ Qual 24:904–910
Hirsch PR, Gilliam LM, Sohi SP et al (2009) Starving the soil of plant inputs for 50 years reduces abundance but not diversity of soil bacterial communities. Soil Biol Biochem 41:2021–2024
Hornby D, Bateman GL, Gutteridge RJ et al (1998) Take-all disease of cereals. A regional perspective. CABI, Wallingford
Jenkinson DS (1990) The turnover of organic carbon and nitrogen in soil. Philos Trans R Soc Lond B 329:361–368
Jenkinson DS, Poulton PR, Johnston AE, Powlson DS (2004) Turnover of nitrogen-15-labeled fertilizer in old grassland. Soil Sci Soc Am J 68:865–875
Johnston AE (2001) Principles of crop nutrition for sustainable food production. Proc Int Fertil Soc 459:1–38
Johnston AE, Dawson CJ (2010) Physical, chemical and biological attributes of agricultural soils. In: Proceedings of the International Fertilizer Society, 675. International Fertilizer Society, York
Johnston AE, Powlson DS (1994) The setting-up, conduct and applicability of long-term continuing field experiments in agricultural research. In: Greenland DJ, Szabolcs I (eds) Soil resilience and land use. CABI, Wallingford, pp 395–421
Johnston AE, Poulton PR, Coleman K (2009) Soil organic matter: its importance in sustainable agriculture and carbon dioxide fluxes. Adv Agron 101:1–57
Ju XT, Xing GX, Chen XP et al (2009) Reducing environmental risk by improving N management in intensive Chinese agricultural systems. Proc Natl Acad Sci 106:3041–3046

Ju XT, Lu X, Gao ZL et al (2011) Processes and factors controlling N_2O production in an intensively managed low carbon calcareous soil under sub-humid monsoon conditions. Environ Pollut 159:1007–1016

Ladha JK, Reddy CK, Padre AT, van Kessel C (2011) Role of nitrogen fertilizer in sustaining organic matter in cultivated soils. J Environ Qual 40:1756–1766

Lawes JB, Gilbert JH, Warrington R (1881/1882) On the amount and composition of the rain and drainage-waters collected at Rothamsted. Parts I and III. J Royal Agric Soc Engl 17:241–268, & 18, 1–71

Leigh RA, Prew RD, Johnston AE (1994) The management of long-term agricultural experiments: procedures and policies evolved in the Rothamsted Classical Experiments. In: Leigh RA, Johnston AE (eds) Long-term experiments in agricultural and ecological sciences. CABI, Wallingford, pp 253–268

Macdonald AJ, Gutteridge RJ (2012) Effects of take-all (*Gaeumannomyces graminis var. tritici*) on crop N uptake and residual mineral N in soil at harvest of winter wheat. Plant Soil 350:253–260

Macdonald AJ, Powlson DS, Poulton PR, Jenkinson DS (1989) Unused fertiliser nitrogen in arable soils – its contribution to nitrate leaching. J Sci Food Agric 46:407–419

Macdonald AJ, Poulton PR, Powlson DS, Jenkinson DS (1997) Effects of season, soil type and cropping on recoveries, residues and losses of ^{15}N-labelled fertilizer applied to arable crops in spring. J Agric Sci Camb 129:125–154

Malhi SS, Lemke R (2007) Tillage, crop residue and N fertilizer effects on crop yield, nutrient uptake, soil quality and nitrous oxide gas emissions in a second 4-yr rotation cycle. Soil Tillage Res 9:6269–6283

Maxfield PJ, Brennand EL, Powlson DS, Evershed RP (2011) Impact of land management practices on high-affinity methanotrophic bacterial populations: evidence from long-term sites at Rothamsted. Eur J Soil Sci 62:56–68

McGrath SP, Lane PW (1989) An explanation for the apparent loss of metals in a long-term field experiment with sewage sludge. Environ Pollut 60:235–256

McMillan VE, Hammond-Kosack KE, Gutteridge RJ (2011) Evidence that wheat cultivars differ in their ability to build up inoculum of the take-all fungus, *Gaeumannomyces graminis var. tritici*, under a first wheat crop. Plant Pathol 60:200–206

Morales SE, Cosart T, Holben WE (2010) Bacterial gene abundances as indicators of greenhouse gas emissions in soils. ISME J 4:1–10

Pachauri RK, Reisinger A (eds) (2007) Climate change 2007 synthesis report. Contribution of Working Groups I, II and III to the fourth assessment report of the Intergovernmental Panel on Climate Change. IPCC, Geneva

Payne RW (2006) New and traditional methods for the analysis of unreplicated experiments. Crop Sci 46:2476–2481

Poulton PR (1996a) The Rothamsted long-term experiments: are they still relevant? Can J Plant Sci 76:559–571

Poulton PR (1996b) Management and modification of procedures for long-term field experiments. Can J Plant Sc 76:587–594

Poulton PR (2006) Rothamsted Research: Guide to the classical and other long-term experiments datasets and sample archive. Rothamsted Research, Harpenden. http://www.rothamsted.ac.uk/Content.php?Section=Resources&Page=ExperimentsGuide

Poulton PR, Johnston AE, White RP (2013) Plant-available soil phosphorus. Part I: the response of winter wheat and spring barley to Olsen P on a silty clay loam. Soil Use Manage 29:4–11

Powlson DS, Johnston AE (1994) Long-term field experiments: their importance in understanding sustainable land use. In: Greenland DJ, Szabolcs I (eds) Soil resilience and land use. CABI, Wallingford, pp 367–394

Powlson DS, Pruden G, Johnston AE, Jenkinson DS (1986) The nitrogen cycle in the Broadbalk Wheat Experiment: recovery and losses of 15N-labelled fertilizer applied in spring and inputs of nitrogen from the atmosphere. J Agric Sci Camb 107:591–609

Powlson DS, Goulding KWT, Willison TW et al (1997) The effect of agriculture on methane oxidation in soil. Nutr Cycl Agroecosyst 49:59–70

Powlson DS, Jenkinson DS, Johnston AE et al (2010) Comments on 'Synthetic nitrogen fertilizers deplete soil nitrogen: a global dilemma for sustainable cereal production', by Mulvaney RL, Khan SA, Ellsworth TR in the J Environ Qual 2009 38 2295–2314. J Environ Qual 39 749–752

Powlson DS, Glendining MJ, Coleman K, Whitmore AP (2011a) Implications for soil properties of removing cereal straw: results from long-term studies. Agron J 103:279–287

Powlson DS, Gregory PJ, Whalley WR et al (2011b) Soil management in relation to sustainable agriculture and ecosystem services. Food Policy 36:S72–S87

Powlson DS, Bhogal A, Chambers BJ et al (2012) The potential to increase soil carbon stocks through reduced tillage or organic material additions in England and Wales: a case study. Agric Ecosyst Environ 146:23–33

Prosser JI, Bohannan BJM, Curtis TP et al (2007) The role of ecological theory in microbial ecology. Nat Rev Microbiol 5:384–392

Royal Society (2009) Reaping the benefits: science and the sustainable intensification of global agriculture. Royal Society Policy document 11/09. Royal Society, London

Shen JL, Tang AH, Liu XJ et al (2009) High concentrations and dry deposition of reactive nitrogen species at two sites in the North China Plain. Environ Pollut 157:3106–3113

Smith KA (2010) Nitrous oxide and climate change. Earthscan, London

Sohi SP, Mahieu N, Arah JRM et al (2001) A procedure for isolating soil organic matter fractions suitable for modeling. Soil Sci Soc Am J 65:1121–1128

Sohi SP, Mahieu N, Powlson DS et al (2005) Investigating the chemical characteristics of soil organic matter fractions suitable for modeling. Soil Sci Soc Am J 69:1248–1255

Stiehl-Braun PA, Powlson DS, Poulton PR, Niklaus PA (2011) Effects of fertilizers and liming on the micro-scale distribution of soil methane assimilation in the long-term Park Grass experiment at Rothamsted. Soil Biol Biochem 43:1034–1041

Sutton MA, Howard CM, Erisman JW et al (eds) (2011) The European nitrogen assessment. Cambridge University Press, Cambridge

Syers JK, Johnston AE, Curtin D (2008) Efficiency of soil and fertilizer phosphorus use. Fertilizer and Plant Nutrition Bulletin 18. FAO, Rome

Watts CW, Clark LJ, Poulton PR et al (2006) The role of clay, organic carbon and long-term management on mouldboard plough draught measured on the Broadbalk wheat experiment at Rothamsted. Soil Use Manage 22:334–341

Zhao FJ, McGrath SP, Blake-Kalff MMA et al (2002) Crop responses to sulphur fertilization in Europe. In: Proceedings of the International Fertiliser Society 504. International Fertiliser Society, York

Zhao B-Q, Li XY, Li XP et al (2010) Long-term fertilizer experiment network in China: crop yields and soil nutrient trends. Agron J 102:216–230

Zhou JB, Chen ZJ, Liu XJ et al (2010) Nitrate accumulation in soil profiles under seasonally open 'sunlight greenhouses' in northwest China and potential for leaching loss during summer fallow. Soil Use Manage 26:332–339

Chapter 17
Long-Term Field Experiments with Fertilizers in Romania: Their Relevance to Sustainable Agriculture

C. Hera

Abstract Long-term field experiments are indispensible for understanding, proving and monitoring changes in soil fertility resulting from many years of husbandry, especially from fertilization. Because they are long-term, they are costly to maintain but their scientific and practical information cannot be replaced by other means – and their value keeps on growing. The longest-running continuous experiments, now 169 years old, were established at Rothamsted, in England, by Lawes and Gilbert in 1843 to investigate issues of nutrient cycling that could only be resolved by experiment. Experimentation in Romania has also focused on long-term field experiments. Systematic trials were established 46 years ago in 15 locations with different soil and climatic conditions, and long-term experiments of more than 46 years are continuing at eleven sites. These experiments comprise a geographic network spanning different soil and climatic conditions with a unique design to track the evolution of differential soil fertility and the impact of fertilizers on the environment. First of all, the results characterize their particular sites. More than this, they focus attention on universally valid, surprisingly actual, cause-and-effect relations, and they help greatly to reconcile ecological and economic interests and to clear up real issues of environmentally friendly nutrient supply and sustainable husbandry.

17.1 Introduction

Long-term field experiments are critical to understanding the complex interactions between plants, soil, climate and management practices and their combined effects on the productivity of farming systems. It is well-recognized that they hold valuable information about the dynamics and sustainability of intensive farming systems. Interest has redoubled as a result of a growing perception that certain soil processes

C. Hera (✉)
Romanian Academy, Bucharest, Romania
e-mail: hers@acad.ro

D. Dent (ed.), *Soil as World Heritage*, DOI 10.1007/978-94-007-6187-2_17,

Table 17.1 Important long-term field experiments with duration >100 years

Location	Country	Start
Rothamsted (Broadbalk, Hoosfield, Park Grass etc.)	UK	1843
Grignon	France	1875
Illinois (Morrow Plots)	USA	1876
Halle/Saale (Eternal Rye)	Germany	1878
Missouri (Sanborn Field)	USA	1888
Dakota	USA	1892
Askov (Sandmarken and Lermarken)	Denmark	1894
Auburn	USA	1896
Bad Lauchstadt (Static Fertilization Experiment)	Germany	1902
Dikopshof	Germany	1904

are long term in nature and must be studied as such. Agroecosystems are, in many ways, unique. Their long-term changes merit particular attention because they are intensively managed and what might be considered best management practice, today, may disrupt the complex array of nutrient and energy cycles and impose unanticipated changes on soil resources.

Fertilizers are key to enhancing agricultural production and their consumption is still increasing rapidly. The need to know their impact on crop yield and quality and, also, on soil and the environment calls for long-term studies at representative sites to monitor changes in nutrient behaviour and to develop strategies for intervention. The sustainability of an agricultural practice depends on its long-term effects on soil productivity and health, but many modern inputs like fertilizers, plant-protection chemicals and herbicides leave behind adverse effects, resulting in declining yields and pollution. Again, long-term experiments are essential to understand and counter these effects.

The world's oldest long-term agricultural experiments were established between 1843 and 1856 by Lawes and Gilbert at Rothamsted in south-east England. They began as agronomic experiments to determine the nutrient requirements of farmland, drawing on Liebig's theory of soil fertility which, in those days, prompted heated arguments about nutrient cycling that could only be resolved by experiment. Although the original questions have been answered, the experiments continue to provide invaluable agronomic, ecological, environmental and scientific information (see Powlson and others in this symposium); the same applies to many other long-established field experiments (Table 17.1).

This chapter highlights some management problems associated with long-term field experiments and suggests ways to mitigate them. Management tasks in establishing new long-term studies include selecting a suitable site, designing the experiment to facilitate statistical analyses and future expansion, collecting data commensurate with the objectives, and incorporating non-destructive changes in the experiments as technology and objectives change. All the long-established field experiments have undergone changes. Technological advances in plant populations, varieties and fertilizers must be incorporated to maintain topicality and generate useful data; incidental changes such as soil erosion, soil and fertility creep between plots, build-up of soil acidity and weeds, and weather effects are inevitable features of long-term experiments but, usually, do not violate the integrity or invalidate the results of the experiment. If properly managed, long-term experiments generate unique, interesting and valuable information that provides both a link to the past and, also, insights into agronomic practices for sustainable future crop production.

17.2 Objectives

Long-term field experiments are undertaken to (1) test the sustainability of a particular agroecosystem and determine what changes, if any, are needed to maintain sustainability; (2) provide data of immediate value to farmers, ecologists and environmental scientists and to improve best practice; (3) provide a resource of soil and plant materials for further research into processes that control soil fertility, plant productivity and the quality of water and habitats; (4) assess the effects of non-farm activities on soil fertility and crop quality; and (5) provide long-term data for developing predictive models of the likely effects of management practices and climate change on soil properties, land capability and the environment. Within this broad remit, the agricultural value of long-term experiments includes testing of new practices, evaluation of the effects of soil type and farming systems on soil organic matter and the assessment of sustainability.

17.3 Long-Term Field Experiments in Romania

Since 1966, long-term field experiments with lime and fertilizers have been established in all the Agricultural Research Stations belonging to the Agricultural Research and Development Institute of Fundulea, and trials in all 14 research stations (Fig. 17.1) were reorganized as long-term experiments under rain-fed and irrigated conditions. Unlike long-term experiments set up elsewhere in just a few locations, those in Romania comprise a geographic network of different soil and climatic conditions with a unique design to track the evolution of differential soil fertility and the impact of fertilizers on the environment. First of all, the results characterize their particular sites, but they also focus attention on universally valid, surprisingly actual, cause-and-effect relations, and they help greatly to reconcile ecological and economic interests and to clear up real issues of environmentally friendly nutrient supply and sustainable husbandry.

17.3.1 Experiment with $N \times P$ Fertilization

The experimental device:

Factor A = N: $a_1 = N_0$; $a_2 = N_{40}$; $a_3 = N_{80}$; $a_4 = N_{120}$; $a_5 = N_{160}$
Factor B = P: $b_1 = P_0$; $b_2 = P_{40}$; $b_3 = P_{80}$; $b_4 = P_{120}$; $b_5 = P_{160}$ (Fig. 17.2)

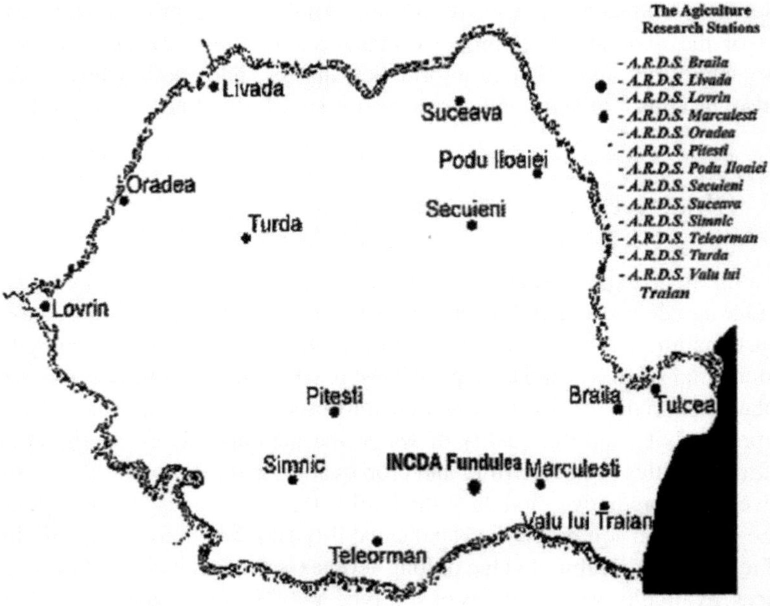

Fig. 17.1 Location of Fundulea National Research and Development Institute and agricultural experimental stations in Romania with long-term fertilizer experiments

10	8	6	9	7	25	23	21	22	24	18	20	16	19	17	12	15	11	13	14	3	1	5	4	2
17	18	16	20	19	5	3	2	1	4	10	9	7	6	8	23	25	21	24	22	11	13	14	12	15
13	11	16	12	14	18	16	2	17	19	23	25	24	21	22	2	5	4	1	3	6	8	7	10	9
1	2	3	4	5	6	7	8	9	10	11	12	13	14	15	16	17	18	19	20	21	22	23	24	25

Fig. 17.2 The influence of NP on yield in long-term field experiments in the research network from Romania

1. N_0P_0	6. N_0P_{40}	11. N_0P_{80}	16. N_0P_{120}	21. N_0P_{160}
2. $N_{40}P_0$	7. $N_{40}P_{40}$	12. $N_{40}P_{80}$	17. $N_{40}P_{120}$	22. $N_{40}P_{160}$
3. $N_{80}P_0$	8. $N_{80}P_{40}$	13. $N_{80}P_{80}$	18. $N_{80}P_{120}$	23. $N_{80}P_{160}$
4. $N_{120}P_0$	9. $N_{120}P_{40}$	14. $N_{120}P_{80}$	19. $N_{120}P_{120}$	24. $N_{120}P_{160}$
5. $N_{160}P_0$	10. $N_{160}P_{40}$	15. $N_{160}P_{80}$	20. $N_{160}P_{120}$	25. $N_{160}P_{160}$

NPK rates are different depending the crop: for wheat and sunflower: N_0, N_{40}, N_{80}, N_{120}, N_{160}; for maize: N_0, N_{50}, N_{100}, N_{150}, N_{200}

Phosphate fertilizer is applied in a single dose in the autumn. Nitrogen fertilizer is split equally between autumn and spring. The crop rotation is sunflower – soybean – wheat – maize.

17.3.2 *Experiment with N P K Fertilization*

The factors researched are the applied rates of fertilizer:

A. Potassium rate: K_0, K_{40}, K_{80}, K_{120}
B. NP rates: N_0 P_0, N_{80} P_{40}, N_{80} P_{80}, N_{160} P_{80} (Fig. 17.3)

Ammonium nitrate is applied in spring and superphosphate and potash in autumn. The crop rotation is sunflower – wheat – maize – wheat. The influence of rates of fertilizers on the yield of wheat was determined and a statistical interpretation made of the yield differences between different treatments.

17.3.3 *Experiment with Lime and Fertilizers*

Factor A = lime amendment dose: a_1 control; a_2 3t $CaCO_3$/ha; a_3 6 t $CaCO_3$/ha; a_4 9 t $CaCO_3$/ha
Factor B = chemical fertilization: b_1 N_0P_0; b_2 $N_{80}P_{80}$; b_3 $N_{120}P_{80}$; b_4 $N_{240}P_{80}$; b_5 $N_{240}P_{80}K_{80}$ (Fig. 17.4)

11	10	12	9	13	15	16	14	2	3	1	4	6	7	5	8	R4
14	16	13	15	3	2	1	4	8	7	5	6	12	11	9	10	R3
6	5	7	8	10	11	12	9	16	13	15	14	3	4	2	1	R2
1	2	3	4	5	6	7	8	9	10	11	12	13	14	15	16	R1

Fig. 17.3 The influence of NPK chemical fertilizers on yield in long-term field experiments in the research network from Romania

1. $N_0P_0K_0$	5. $N_{80}P_{40}K_0$	9. $N_{80}P_{80}K_0$	13. $N_{80}P_{80}K_0$
2. $N_0P_0K_{40}$	6. $N_{80}P_{40}K_{40}$	10. $N_{80}P_{80}K_{40}$	14. $N_{80}P_{80}K_{40}$
3. $N_0P_0K_{80}$	7. $N_{80}P_{40}K_{80}$	11. $N_{80}P_{80}K_{80}$	15. $N_{80}P_{80}K_{80}$
4. $N_0P_0K_{120}$	8. $N_{80}P_{40}K_{120}$	12. $N_{80}P_{80}K_{120}$	16. $N_{80}P_{80}K_{120}$

NPK rates are different depending for wheat and sunflower: N_0, N_{80}, N_{160}; for maize: N_0, N_{100}, N_{200} P and K rates are the same at all the crops: P_0, P_{40}, P_{80}; K_0, K_{40}, K_{80}, K_{120}

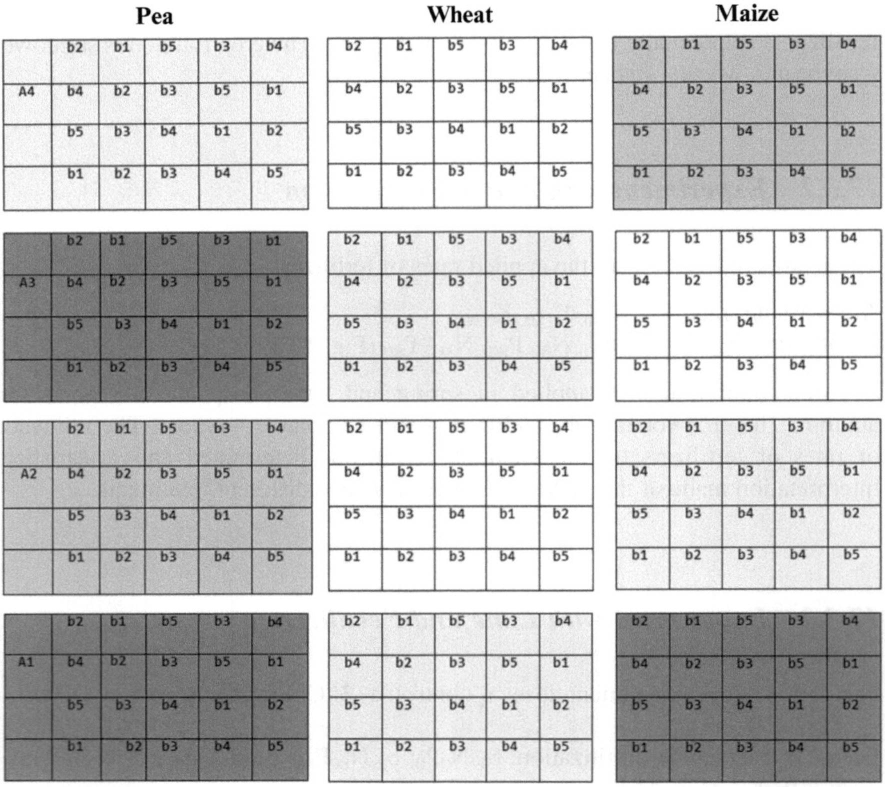

Fig. 17.4 Scheme of long-term experiments on the influence of lime and chemical fertilizers on crop yield

Legend: A-rates $CaCO_3$: Al nil; A2 3t/ha; A3 6t/ha; A4 9t/ha
B-rates NPK(kg/ha):

Rates of fertilizer are different for different crops: for wheat the rates are: N_0P_{80}, $N_{80}P_{80}$, $N_{120}P_{80}$, $N_{160}P_{80}$, $N_{160}P_{80}K_{80}$; for maize the rates are: N_0P_{80}, N_0P_{80}, $N_{160}P_{80}$, $N_{240}P_{80}$, $N_{240}P_{80}K_{80}$; and for peas the rates are: N_0P_{80}, $N_{30}P_{80}$, $N_{60}P_{80}$, $N_{90}P_{80}$, $N_{90}P_{80}K_{80}$.

Phosphate and potassium fertilizers are applied as a single dose in autumn; nitrogen is split equally between autumn and spring applications. The crop rotation is sunflower – soybean – wheat – maize.

17.3.4 Experiment with Organo-Mineral Fertilization

Factor A = chemical fertilization: a_1 N_0P_0; a_2 $N_{50}P_0$; a_3 $N_{50}P_{50}$; a_4 $N_{100}P_{100}$
Factor B = manuring: b_1 unmanured; b_2 20 t/ha; b_3 40 t/ha; b_4 60 t/ha (Fig. 17.5)

14	16	13	15	10	11	12	9	7	5	6	8	3	4	2	1	R4
7	5	6	8	2	4	1	3	15	14	16	13	11	10	12	9	R3
10	12	9	11	13	15	14	16	3	4	1	2	5	7	6	8	R2
1	2	3	4	5	6	7	8	9	10	11	12	13	14	15	16	R1

Fig. 17.5 The influence of NP chemical fertilizers + manure on yield in long-term field experiments in the research network from Romania

1. $G_0N_0P_0$	5. $G_0N_{50}P_0$	9. $G_0N_{50}P_{50}$	13. $G_0N_{100}P_{100}$
2. $G_{20}N_0P_0$	6. $G_{20}N_0P_0$	10. $G_{20}N_{50}P_{50}$	14. $G_{20}N_{100}P_{100}$
3. $G_{40}N_0P_0$	7. $G_{40}N_0P_0$	11. $G_{40}N_{50}P_{50}$	15. $G_{40}N_{100}P_{100}$
4. $G_{80}N_0P_0$	8. $G_{60}N_0P_0$	12. $G_{60}N_{50}P_{50}$	16. $G_{60}N_{100}P_{100}$

Rates of N and P + manure are the same in all crops (rotation wheat-maize)

Phosphate and potash are applied as a single dose in autumn; nitrogen is split equally between autumn and spring applications; farmyard manure is applied every 4 years. The rotation is alternating wheat and maize.

17.4 Results and Discussion

Long-term fertilizer experiments are usually undertaken to monitor both changes in yield and product quality and, also, changes in soil properties and the environment. Knowledge of these, often gradual, changes helps in the development of strategies and policies for rational use of fertilizer to improve productivity, sustainability and environmental integrity. In Romania, the long-term trials with lime, fertilizers and manure conducted over the last 40 years (Table 17.2) have yielded valuable information that has been applied by researchers and extension workers. In the first place, the experiments have established optimum rates of lime, mineral fertilizers and manure according to the crop, soil and climatic conditions. Optimum doses of N vary between 80–120 kg/ha under rain-fed conditions and 140–170 kg/ha under irrigation, depending on soil and climate. Because the content of soluble phosphorous in soil is increasing in proportion to the dose of phosphate fertilizer applied, the dose needed to obtain the maximum production can be gradually decreased.

Periodically, at the end of every fourth year, the status of the main soil chemical indices has been determined. These data have been used to establish different fertilization strategies:

Table 17.2 Soil-climatic characteristics of long-term field experiments in Romania

Research station	Type of soil	pH	Humus %	P mobile ppm	K mobile ppm	*V* %	*T* °C	Precipitation mm
Fundulea	Phaeozem	6.0–6.7	2.5–3.0	23–49	190–260	77–83	12.5	476
Teleorman	Chernozem	6.2	3.5–3.9	50–60	300–350	81–83	12.6	434
Valul lui Traian	Chernozem	7.4	3.5–3.8	52–64	260–270	80–81	9.8	457
Brăila	Chernozem	8.6–8.9	1.9–2.3	13–18	372–440	78–82	9.3	450
Perieni	Chernozem		2.6–4.1	36–41	101–142	76–78	9.2	512
Podu Iloaiei	Chernozem	6.8–7	4.0–4.5	40–46	12S–132	72–74	9.1	542
Suceava	Chernozem	5.6–5.8	3.1.3.3	32–38	150–170	75–78	7.8	587
Livada Satu Mare	Luvisol	5.8	1.21	34–36	63–65	63–68	9.7	742
Oradea	Pre-luvisol	5.5	2.3–2.5	23–30	83.0–91.0	68–70	10.2	615
Lovrin	Chernozem	6.4–6.6	2.7–3.2	51–54	94–102	72–75	10.4	590
Dăbuleni	Arensol		0.3–1.6	–	–	–	11.1	548
Caracal	Phaeozem	5.55–5.65	2.2–2.5	45–52	244–280	73–76	10.6	537
Turda	Chernozem	6.3–6.8	3.1–4.2	45–47	249	75–78	8.4	545
Mărculeşti	Chernozem	8.0–8.3	3.0–3.2	35–38	99–126	63–68	11.2	457
Albota Piteşti	Luvisol	5.16	2.0	28–36	39–41	76.84	10.6	460
Secuieni	Chernozem		2.1–2.8	42–48	93–101	68–71	8.7	547

- The optimum economic dose of potash decreases in proportion to the content of exchangeable potassium or the soluble-activity index that measures easily soluble soil potassium.
- N fertilizers, while increasing crop yields, bring about increased consumption of bases and, consequently, soil acidification.
- Soluble phosphatic fertilizers, on the other hand, buffer soil acidity. When the rates of phosphate applied as manure and fertilizer are higher than the P consumed by the crops, the content of mobile phosphate in the topsoil generally increases.
- Periodic application of farmyard manure maintains a weakly-acid-to-neutral reaction and can neutralize harmful acidity in strongly acidic soils. Manure increases the soil's humus content; the rate of humus accumulation per unit of applied manure tends to increase over time and with the frequency of manuring.
- N fertilizer applied at recommended economic doses does not cause accumulation of soil nitrogen. The toxic symptoms of excess N in plants are manifested alongside deficiencies of molybdenum, phosphorus, sulphur and potassium. In strongly acid soils, excess N can manifest toxic symptoms combined with symptoms of aluminium toxicity and deficiency of phosphorus and potassium. The main measures to prevent excess nitrogen in soils, crops and groundwater are correct nitrogen fertilizer rates (depending on the actual needs of the crop and the amount of mineralizable nitrogen in the soil) and judicious seasonal fertilizer application.

- Nitrogen deficiency is associated with soil acidity, low humus content, fine texture and leaching – all of which reduce the activity of soil microorganisms.
- Excess phosphate (>80 ppm PAL[1]) can induce secondary deficiencies of micronutrients that can be very damaging for both quantity and quality of the harvest (secondary zinc deficiency in corn, sorghum, beans, etc.). In acid soils, mobile P does not exceed 60 ppm PAL. Liming, which reduces soil acidity, reduces the ratio of these elements in ecosystems and disturbs tissue multiplication. The negative consequences of excess phosphate are exacerbated by a weakly-acid-to-neutral reaction, light texture and waterlogging. Secondary Zn deficiency is enhanced by neutral-to-weakly-alkaline reaction, light texture and low humus content (less than 2 %).
- When P fertilizers are not applied, or applied at rates lower than consumption by the crops, levels of mobile P in the topsoil diminish (at a rate depending on the initial mobile P values). Phosphate deficiency is manifest depending on soil reaction and the provision of other nutrients; deficiency is highlighted early on by poor development of the root system. On acid soils fertilized only with nitrogen, symptoms of phosphate deficiency may be observed in association with nitrate and aluminium toxicity and potassium deficiency.
- Potash fertilizer and manure gradually increase mobile K in the topsoil according to the annual dose but levelling off at doses greater than 150 kgK_2O/ha. Excess potassium leads to passive accumulation in tissues.
- Potassium deficiency occurs on soils containing less than 80 ppm mobile K, typically acid soils; soils rich in carbonates and soluble salts of Ca and Mg; and soils of high cation exchange capacity, fine texture and poor aeration. Potassium deficiency may be corrected by correcting the soil reaction and application of potash fertilizer.
- Early-season deficiency of calcium in crops is observed in cold, rainy springs and fine-textured soils (>35 % clay). The problem can be rectified by liming and economic application of manure or potash fertilizer.
- Deficiency of magnesium can occur because of low values of mobile Mg, acid reaction (pH <5.8) and abundance of mobile potassium (>200 ppm) in the topsoil. Luvisols in north-west Romania and coarse-textured soils registering Mg deficiency have less than 75 ppm mobile Mg.
- Deficiency of sulphur disturbs protein synthesis so amides, amino acids and sulphur-free mineral compounds of nitrogen (nitrates and ammonium compounds) accumulate in the tissues.
- Manganese deficiency, associated with <1 ppm exchangeable Mn, is common in annual crops fertilized with potash, especially during drought. A high concentration of manganese in waterlogged soils impairs the uptake of iron; this is caused by mobilization of Mn^{2+} through reduction of manganese oxides, but the concentration of Mn in soil solution is reduced 100-fold for every unit increase of pH.
- On luvisols, liming exacerbates boron deficiency in flax, sunflower and other crops.

[1] Phosphate by the acetate-lactate-ammonia method.

17.5 Proposals for Future Research

The Romanian long-term fertilizer experiments are evaluating the long-term effects of manure and fertilizers on crop production and soil health – with special emphasis on sustainability. Systematic studies over 40 years have demonstrated that unbalanced and non-integrated fertilizer use can create a lot of problems. The foremost challenge for the coming years is to make a critical interpretation of the results obtained and adapt fertilizer schedules and the strategy for future experimentation on soils and plant nutrition.

In the light of widespread deficiencies of secondary and micronutrients caused by unbalanced nutrient management, the urgent need to improve soil health and crop quality and the increasing interest in organic farming, it is imperative to determine the long-term effects of organic farming on soil health (especially on the availability of micronutrients) and on crop productivity and quality.

Continuous monocultures of wheat, maize, sunflower and sugar beet are susceptible to micronutrient deficiencies, especially of Zn and Mo. Since Zn and Mo are giving good responses in maize and wheat, assessment of the effects of these nutrients, alone and alongside organic manures, and the efficiency of different methods of application of these nutrients should be investigated.

Another imperative is periodic evaluation of the quality parameters of production under long-term manuring and fertilization. This means building up a database for all soil chemical properties during the period of experimentation. Such a database can be a ready source for understanding nutrient-use efficiency and, also, the changes in soil nutrient status brought about over many years.

Acknowledgement I am grateful to Dr. Ana Popescu for her contribution to the data selection.

Bibliography

Borlan Z, Hera C, Dornescu D et al (1994) Fertilizers and soil fertility. Compendium of agrochemistry. Editura Ceres, Bucharest (Romanian)

Ciobanu G, Ciobanu C, Domuta C, et al (2009) Influence of K × NP fertilizers in long-term field experiments on agrochemical soil indexes and on winter wheat yield and quality in the pre-luvosoil conditions from northwest Romania. J Environ Prot Ecol/Workshop UAB – BENA, pp 6–7

Ciobanu G, Domuta C, Ciobanu C et al (2012) Effect of potassium on grain quality and yield of winter wheat in pre-luvosoil conditions in western Romania. J Balkan Environ Assoc/Workshop UAB – BENA

Hera C (2010) 77 years dedicated to enhancing soil fertility for the benefit of sustainable and productive agriculture. In: Hera C, Dumitru M, Schnug E et al (eds) Proceedings of the 15th World Fertilizer Congress of the International Scientific Centre for Fertilizers (CIEC). Publishing House of the Romanian Academy, Bucharest, pp 17–30 (Romanian)

Hera C, Mihaila V (1981) Changes in soil agrochemical soil indices by fertilization. Ann RICIC Fundulea 47:319–327 (Romanian)

Hera C, Burlacu G, Mihaila V et al (1976) Interaction between nitrogen and phosphorus in long-term experiments. Ann RICIC Fundulea 41:83–91 (Romanian)

Hera C, Idriceanu A, Dobrescu E et al (1986) Nutrient interactions due to mineral fertilization and their influence on maize yield quality. Probleme de Agrofitotehnie Teoretica si Aplicata VII (2):171–181 (Romanian)

Idriceanu A, Hera C (1985) Nutrient interactions in long-term experiments with chemical fertilizer applications to maize. Probleme de Agrofitotehnie Teoretica si Aplicata VII(2):175 (Romanian)

Jackson ML (1973) Soil chemical analysis. Prentice Hall of India, New Delhi

John SK, Ravindran CS, George J (2005) Long-term fertilizer experiments: three decades experience in cassava. Sreekariyam Central Tuber Crops Research Institute, Thiruvananthapuram

Korschens M (2006) The importance of long-term field experiments for soil science and environmental research - a review. Plant Soil Environ 52(Special Issue):1–8

Leigh RA, Johnston AE (eds) (1994) Long-term experiments in agricultural and ecological sciences. CABI, Wallingford

Mihaila V, Hera C (1994) Evolution of the chemical proprieties of the chernozem at Fundulea as a result of application of fertilizers. Probleme de Agrofitotehnie Teoretica si Aplicata XVI(1):29 (Romanian)

Powlson DS, Poulton PR (2003) Long-term experiments in the 21st century: continuity or change?. In: 22nd Congress of the Nordic Association of Agricultural Scientists, Turku, 1–4 July

Rathke GW, Korschens M, Diepenbrock W (2002) Substance and energy balances in the static fertilization experiment Bad Lauchstadt. Arch Agric Soil Sci 48:423–433

Chapter 18
The Beginnings of Long-Term Field Experiments on Crop Rotations at Balti

J. Libershteyn

Abstract The Balti field experiments were set up in the 1950s in order to meet a pressing need for agronomic knowledge on field crops and crop rotation following collectivization of farms, a big increase in the area of industrial crops like sunflower and sugar beet, accelerated erosion, and plagues of weeds, pests and diseases. The new Selectia Institute for Field Crops at Balti built upon the experience of the Moldovan Experimental Plant Breeding Station and other experimental stations in-country and across the Soviet Union. What were, at the time, novel combinations and treatments were also introduced. Dedicated efforts over the years and the accumulating data provided well-grounded, practical recommendations for improving the structure and management of field crops in the Republic. They have also laid a foundation for modern, advanced cropping systems that combine agrochemical, environmental and economic feasibility.

18.1 In the Beginning: Needs and Objectives

I have written before about the founding of long-term experiments on crop rotations and continuous cropping in the north of Moldova. Now, on the 50th anniversary of this foundation, I come back to the beginnings of our scientific institution.

In the late 1950s, there was a pressing need for reliable information on the optimal structure of sown field crops, their management and alternation in crop rotation:

- The tradition of peasant farms in our region was to maintain an equal ratio of row crops and continuously sown crops (like wheat). Collectivization in the mid-1950s led to a big increase of row crops in the north of the country, particularly industrial crops like sunflower and sugar beet. During this period,

J. Libershteyn (✉)
Formerly Selectia Research Institute for Field Crops, Balti, Republic of Moldova,
Present Address: Sydney, NSW, Australia
e-mail: peterlene@gmail.com

D. Dent (ed.), *Soil as World Heritage*, DOI 10.1007/978-94-007-6187-2_18,

the limited capacities of the farms to complete harvesting and tillage meant that, when the time came for sowing winter crops, only the areas vacated by row crops were free. This was the reason for widespread cultivation of winter wheat and a catastrophic spread of grain ground beetles that caused mass destruction; as much as 20 % of the crop had to be reseeded.

- Continuous cultivation of row crops, including incompatible species like sugar beet and sunflower, also opened up an array of problems, especially the accelerated loss of soil fertility and even the topsoil itself, and uncontrolled spread of diseases and weeds – all leading to poor yields.
- Alongside the unresolved issue of the optimal balance between the field crops, there were less-studied questions regarding the components of crop rotation and the consequences of their actions and interactions. Mistakes were often made, which caused significant harvest losses.

We should not forget that, in the early post-war years, the Moldovan Experimental Breeding Station carried out field experiments on some 2–3-year links between alternating field crops (winter wheat, maize, perennial grasses). This helped to clarify some of the recommendations for crop management but such short-term trials could not unearth fundamental solutions. Transformation of the experimental breeding station into a scientific research institute with increased human and financial resources, in the late 1950s and early 1960s, made it possible to build on the early work. Preparations were made for long-term investigations and development of scientifically justified crop rotations: firstly, to determine the optimal structure of crops in field in accord with the fundamental principles of agricultural science, environmental safety (as understood at the time) and cost-effectiveness; and secondly, to establish reasonable requirements of the specified crop rotations in space and time, taking into account their direct effects and after-effects. Simultaneous operation of each of the investigated schemes in full was essential, so that yield data and other parameters were received annually for each plot.

18.2 Establishing the Experiments

After long and detailed deliberations and on-site reviews of current investigations in the USSR, it was decided to lay out eight crop rotations. In each scheme, winter wheat was allotted 30 %, which corresponded to its proportional area in the cropland of the Republic. The proportion of row crops in the original rotations ranged from 10 to 70 %, including 10 to 30 % sugar beet, 10–20 % sunflower and 20–40 % maize. Some of the crop rotations included perennial grasses, but, at that time and in spite of the established facts, Premier Khrushchev declared perennial grass crops to be low-yielding and they were to be replaced by corn. Under these circumstances, research on perennial grasses in the crop rotations was not included in any accounting documents for 3 years, and they were formally re-entered into the records only in 1964 – after the resignation of Mr Khrushchev.

Within the scheme of experiments, there was one unit of continuous winter wheat, one of continuous maize, one of perennial grasses and one of black fallow. Tobacco was not included in the original crop rotations because at the time that the research scheme was drawn up, it occupied only about 7,000–8,000 ha in the whole country, mainly on Calcareous chernozem in the central district of Orhei and in the northern part of Floresti. In the following years, when areas of this culture 'of questionable value' (the words of Academician D Pryanishnikov) soared to more than 50,000 ha, the question arose about its place in crop rotations. Actually, research at Balti by K Cebotari had already demonstrated the value (unexpected for many people) of planting tobacco after sugar beet, a fact reflected in recommendations for production.

At the time, my proposal to include one crop rotation with three fields of sugar beet was blasphemy against the canons of agronomy. Today, many years later, such a scheme has confirmed its right to life – on condition that sunflower is excluded from the same rotation. Including two successive row crops within all schemes was also innovative and, according to the then-authoritative views of adherents of V Williams, inadmissible in terms of preservation of soil structure. And there were several germane issues about the placement of different cultures in crop rotations. For example, the feasibility of planting sugar beet after winter wheat was never in doubt, but the consequences of different predecessors of the winter wheat on the productivity of sugar beet was unexplored and required special experimental testing, including comparison with fallow.

Speaking of fallow, in those days, when conditions were favourable, *i.e.* sufficient soil moisture and high nutrient status, the varieties of winter wheat then cultivated would lodge badly when grown after bare fallow because of their tall straw. Losses at harvest amounted to as much as 3 t/ha so that harvested yields were no better than those of winter wheat following other predecessors. This circumstance encouraged a steady decrease in bare fallow and, by the time that high-yielding but lodging-resistant varieties of winter wheat arrived on the scene – varieties that could successfully follow bare fallow – the area under bare fallow had been reduced to nothing. When, in the mid-1960s, 200 ha of Bezostaja-1 winter wheat on the Pobeda collective farm in Vulkaneshti District recorded an unprecedented 72 centners/ha for 2 years in succession following bare fallow, many of us began to doubt whether fallow should be eliminated from Moldova, especially in the south.

18.3 The Selectia Long-Term Experiments in Operation

Preparations culminated in the autumn of 1961 and the following spring. Following the traditions of the experimental breeding station, the experiments on crop rotations were established and conducted in full accordance with the rules of procedure of experimental work and have always been maintained in excellent condition. The research centre and the field experiments have received visits from numerous participants of republic, regional and international meetings held at the Institute, as well as many specialists: they have become a living, working museum of agricultural science.

Our research centre has been fortunate to enjoy longevity. This may be attributed to the fact that it has attracted highly professional researchers – people who have well understood the importance of the work and have continued and developed the experiments with great skill. Nowadays, however, the attitude towards long-established field experiments is not always favourable: I shall give just one example from personal experience. In 1948, I founded a state strain-testing station at Marandeni, where I worked for 7 years before being transferred to Balti. An 80 ha area was divided into a nine-field crop rotation and, over the next 2 years, a framework of shelter belts was planted around it. The intensive work of the experimental station became a shining example to many farms in the surrounding area; the agronomic activities not only were informative but also provided practical assistance in their cropping. But in 1993, without warning and after accomplishing five full crop rotations, the long-term experiments were terminated by order of the head of the strain-testing network of the Republic. The motive for this hasty decision was that the head of the collective farm and head of the strain-testing station could never agree!

Coming back to our research at Balti, I should emphasize that, from the outset, it has provided both scientific output and well-grounded, effective, practical recommendations for improving the structure and the alternation of crop rotations in the Republic. As new facts have accumulated, this information has provided the theoretical foundation for the construction of advanced cropping systems that combine agrochemical, environmental and economic feasibility. This is well illustrated by the fundamental studies of the current head of the Research Centre, Professor Boris Boincean.

In conclusion, on my own behalf and on behalf of the staff, both living and deceased, who launched the research on crop rotations (N Lebedev, P Kibasov, I Libershteyn, V Kazanzhi, Y Bondarenko, G Shontsu and S Zhurat), let me thank all those who have followed us. They not only have preserved but have increased the worth of the research for science and society.

Chapter 19
Fifty Years of Field Experiments with Crop Rotations and Continuous Cultures at the Selectia Research Institute for Field Crops

B.P. Boincean

Abstract Experimental data are presented from long-term field experiments on Typical chernozem in the north of Moldova. Yields of crops in rotations and continuous monocultures are evaluated and compared with national averages. The effects of fertilization and crop rotation have been determined, including links within rotations – especially the different predecessors for winter wheat; generally, the better the predecessors, the less are the benefits of fertilization. From both economic and ecological perspectives, it is better to respect crop rotation than to compensate for its absence with an excess of agrochemicals for crop nutrition and protection. Changes in the stocks of soil organic matter have been established under crop rotations, black fallow, continuous monocultures and meadow; losses are much higher under black fallow and continuous crops than under crop rotations. Under meadow, it takes 25 years to restore the initial stock of soil organic matter.

19.1 Introduction

Field experiments are the main method of agronomic research. Results from long-term field experiments in various countries have been described by, amongst others, Johnstan (1989), Jenkinson (1991), Boincean (1999), Safonov (2002), Powlson and others (this symposium) and, recently, reviewed by Krupenikov et al. (2011). Only by such experiments can we establish real changes and trends of crop yields and soil fertility; they allow us to better understand the complex interactions between climate, soils and crops under management, and they are the basis for practical recommendations to farmers. The longer the duration of the field experiment, the greater is its scientific and practical value.

B.P. Boincean (✉)
Selectia Research Institute for Field Crops, Balti, Republic of Moldova
e-mail: bboincean@gmail.com

D. Dent (ed.), *Soil as World Heritage*, DOI 10.1007/978-94-007-6187-2_19,

We are very grateful to our predecessors for the establishment and continuation of the Balti long-term experiments; the role of honour is the frontispiece of this publication. Special appreciation is due to Nicolae Lebedev, a pupil of Pacosski who compiled the first flora of Bessarabia – not neglecting the weeds! Pacosski, himself, was a pupil of Dokuchaev who first described the *Typical chernozem* in this very locality.

19.2 Experimental Site and Methods

The Balti long-term field experiments on crop rotations were established in 1962; crops grown as continuous monocultures on fertilized and unfertilized plots were added in 1965. The soil is Typical chernozem heavy clay. Laboratory analysis of the 0–20 cm layer in 1993 revealed 4.8–5.0 % organic matter (by Tiurin's method), pH_{water} 7.3 and pH_{CaCl2} 6.2 and total nitrogen, phosphorus and potassium contents of 0.20–0.25, 0.09–0.11 and 1.22–1.28 %, respectively.

The field experiment with crop rotations includes eight rotations with different proportions of row crops – from 40 up to 70 % including 10–30 % sugar beet, 10–20 % sunflower and 20–40 % maize. The proportion of winter wheat is 30 % in all rotations but it is sown after different predecessors: in one field after early-harvested predecessors, in the second after maize silage, and in the third after maize for grain. Details of the fertilizer regimes were reported earlier (Boincean1999). Each plot in the crop rotations is 283 m^2, with three replicates, and in the continuous monocultures 450 m^2 without replicates.

After three full rotations (1962–1991), the experiment was interrupted for 2 years to estimate differences in inherent soil fertility through the yields of two crops sown on all plots – winter rye for green mass in 1992 and oats for grain in 1993. The fourth rotation resumed in 1994 but without applying fertilizer to crop rotation no. 7, where the sequence of crops is similar to crop rotation no. 3, to establish the influence of fertilization in both continuous monocultures and in the crop rotations.

19.3 Results and Discussion

The yields for different crops in the long-term experiments and the average yields for the Republic of Moldova are shown in Tables 19.1, 19.2, 19.3, and 19.4. For winter wheat, the greatest yield increase for both the long-term experiment and under production conditions coincided with the second rotation (1972–1981), amounting to 1.86 and 1.75 t/ha, respectively (Table 19.1). Absolute yields were higher in the experimental fields than in the country as a whole by 1.43 and 1.54 t/ha for the first and second rotations, respectively. In the long-term experiments, yields continued to increase during the third rotation (by 0.29 t/ha), but the country average declined by 0.05 t/ha during the same period. Since then, yields have levelled off in both the experimental plots and across the country as a whole.

Table 19.1 Average yields of winter wheat in Moldova and in the Selectia long-term experiments 1962–2011

	Average for Moldova		Long-term field experiment (after vetch and green oats, fertilized plots)		
Years	t/ha	± t/ha relative to previous period	t/ha	± t/ha relative to previous period	Yields in long-term experiment relative to national average
1962–1971	1.85	–	3.28	–	+1.43
1972–1981	3.60	+1.75	5.14	+1.86	+1.54
1982–1991	3.55	−0.05	5.43	+0.29	+1.88
1994–2003	2.35	−1.2	5.15	−0.28	+2.80
2004–2011[a]	2.44	+0.09	5.03	−0.12	+2.59

[a] 8-year means

Table 19.2 Average yields of sugar beet in Moldova and in the Selectia long-term experiments 1962–2011

	Average for Moldova		Long-term field experiment following vetch and green oats, and winter wheat, fertilized plots		
Years	t/ha	± t/ha relative to the previous period	t/ha	± t/ha relative to the previous period	Yields in long-term field experiment relative to the national average
1962–1971	23.65	–	40.06	–	+16.41
1972–1981	26.85	+3.20	45.79	+5.73	+18.94
1982–1991	27.58	+0.73	44.96	−0.83	+17.38
1994–2003	19.81	−7.77	44.04	−0.92	+24.23
2004–2011[a]	26.97	+7.16	40.90	−3.14	+13.93

[a] 8-year means

Table 19.3 Average yields of sunflower in Moldova and in the Selectia long-term experiments 1962–2011

	Average for Moldova		Long-term experiment, fertilized plots		
Years	t/ha	± t/ha relative to the previous period	t/ha	± t/ha relative to the previous period	Long-term experiment relative to national average
1962–1971	1.60	–	2.59	–	+0.99
1972–1981	1.70	+0.10	2.24	−0.35	+0.54
1982–1991	1.86	+0.16	2.78	+0.54	+0.92
1994–2003	1.29	−0.57	2.08	−0.70	+0.79
2004–2011[a]	1.32	+0.03	2.27	+0.19	+0.95

[a] 8-year means

Table 19.4 Average grain yields of maize in Moldova and in the Selectia long-term experiments 1962–2011

Years	Average for Moldova		Long-term experiment, fertilized plots		Long-term experiment relative to national average
	t/ha	± t/ha relative to previous period	t/ha	± t/ha relative to previous period	
1962–1971	3.17	–	5.60	–	+2.43
1972–1981	3.55	+0.38	6.78	+1.18	+3.23
1982–1991	3.96	+0.41	6.86	+0.08	+2.90
1994–2003	2.79	−1.17	5.84	−1.02	+3.05
2004–2011[a]	2.94	+0.15	5.87	+0.03	+2.93

[a] 8-year means

Overall, winter wheat yields in the long-term field experiments were greater than the national average by 2.80 and 2.59 t/ha during 1994–2003 and 2004–2011, respectively.

For sugar beet (Table 19.2), as with wheat, yields increased during the second rotation: 5.73 t/ha in the long-term experiment, coinciding with an increase of 3.20 t/ha across the country as a whole. During the first and second rotations, the absolute yields in the field experiments were 16.4 and 18.9 t/ha above the national average. During the third and following rotations, yields in the field experiments decreased by 0.83 and 3.14 t/ha relative to the second rotation while yields across the country fluctuated; nevertheless, the yields in the long-term experiment remained substantially above the national average.

Sunflower is sensitive to a complex of diseases but introduction of hybrids brought about an increase in yield of 0.54 t/ha from the second to the third rotation of the long-term experiment (Table 19.3). Average yields for the country as a whole increased throughout the period 1962–1991. During fourth rotation, yields decreased both nationally and in the long-term experiments by 0.57 and 0.70 t/ha, respectively, and have only recovered slightly since then. As with other crops investigated, yields of sunflower in the long-term experiments exceeded the national average, in this case by 0.54–0.99 t/ha over 48 years.

The introduction of hybrids also increased maize yields during the second rotation of the long-term experiments (Table 19.4). Grain yields increased during the second and third rotations but declined in the fourth rotation and have hardly recovered since. Nevertheless, the yields of maize grain in the long-term experiment exceeded the national average by 2.43–3.23 t/ha for the whole period.

For all crops, both in the long-term field experiments and across the country, the coefficients of variation (CoV) of yields are high (Table 19.5) and so was the CoV for precipitation over the 48-year period (Table 19.6).

In the long-term experiments, yields of most crops increased over the first two or three rotations (1962–1991) and then levelled off or decreased. This levelling off has occurred even under similar conditions of precipitation and accumulated temperature: by calculation of the hydrothermic coefficient, the number of wet and temperate years was the same in both periods (1972–1981 and 1982–1991).

Table 19.5 Coefficients of variation (%) for crop yields in Moldova and in the Selectia experiments 1962–2011

	Crops			
	Winter wheat	Sugar beet	Sunflower	Maize for grain
Average for Moldova	32.3	25.1	20.4	21.8
In the long-term experiments	30.4	18.6	32.4	27.6

Table 19.6 Precipitation and accumulated temperature at Selectia RIFC 1962–2011

Years	Total precipitation mm	Accumulated temperature day/degrees above 10 °C	Hydrothermic coefficient	No. of years: Wet (>2.0) Temperate (1–2) Dry (<1.0)
1962–1971	545	3,194	1.7	3 wet 7 temperate
1972–1981	608	2,970	2.1	4 wet 6 temperate
1982–1991	542	3,037	1.8	4 wet 6 temperate
1994–2003	552	3,334	1.7	2 wet 8 temperate
2004–2011	458[a]	2,927[a]	1.6	7 temperate 1 dry
Mean	544	3,099	1.8	
CoV	20.9	7.7		

[a] 8-year mean

Further decrease in yields countrywide during the fourth and fifth rotations has been caused mainly by social and economic change, but we should note that accumulated temperatures were higher for 1994–2003 and precipitation lower for 2004–2011, with higher extremes during the growing period.

Winter wheat: The experimental data for 48 years (Table 19.7) demonstrate the advantage of early-harvested crops and black fallow as predecessors for winter wheat. Mean yield after winter vetch and rye was 4.89 t/ha; after spring vetch and oats, 5.20 t/ha; after lucerne in the third year after first cut, 4.90 t/ha; after peas, 4.76 t/ha; and after maize silage, 4.33 t/ha. We have also calculated 38-year mean wheat yields so as to cover the same period under crop rotations and continuous wheat: the 38-year means after early-harvested predecessors (winter vetch and rye, spring vetch and oats, lucerne in the third year after first cut) and after black fallow were 5.2–5.29 t/ha; the same yield was achieved after peas for grain – 5.14 t/ha. Black fallow is not recommended. It offers no advantage to the following crop compared with other early-harvested predecessors, even in dry years, and it accelerates the decomposition of soil organic matter.

Table 19.7 Yields of winter wheat in crop rotations and continuous wheat 1962–2011, t/ha, fertilized plots

	Predecessors							
Years	Winter vetch and rye	Spring vetch and oats	Lucerne 3rd year after first cut	Peas for grain	Black fallow	Maize silage	Maize for grain	Continuous winter wheat
1962–1971	3.44	3.28	3.43	3.32	–	3.27	2.40	2.95[b]
1972–1981	5.10	5.14	5.22	4.98	5.10	4.84	–	3.45
1982–1991	5.47	5.43	5.58	5.51	5.55	4.45	3.51	3.69
1994–2003	5.22	5.15	5.10	4.94	5.12	4.67	3.31	2.69
2004–2011[a]	5.32	5.03	5.26	5.12	5.42	4.42	3.94	2.90
48-year mean	4.89	4.80	4.90	4.76	–	4.33	–	–
38-year mean	5.27	5.20	5.29	5.17	5.29	4.60	3.56	3.20

[a] 8-year means
[b] Mean for 1966–1971

Table 19.8 Stocks of soil moisture (mm) after different predecessors to the optimal time for sowing winter wheat, soil layers 0–20 and 0–100 cm, average for 1982–1991including the drought year 1986

Stocks of available soil moisture, soil layers	Predecessors							
	Winter vetch and rye	Spring vetch and oats	Alfalfa, 3rd year after first cut	Peas for grain	Black fallow	Maize silage	Maize for grain	Continuous winter wheat
0–20 cm	29	32	31	31	31	24	20	33
0–100 cm	136	154	145	144	153	118	91	140
Drought year 1986								
0–20 cm	19	23	28	26	36	9	10	27
0–100 cm	78	80	133	92	165	57	57	80

Compared with early-harvested predecessors, the yield of winter wheat after maize silage was less by 0.60–0.69 t/ha and after maize for grain by 1.64–1.73 t/ha; yields of continuous wheat on fertilized plots are less by 2.0–2.09 t/ha. The importance of early-harvested predecessors is evident. On the Balti steppe, soil moisture is a limiting factor for yield formation and stocks of soil moisture are significantly higher after early-harvested than after late-harvested predecessors, especially in drought years (Table 19.8). Tillering of the crop in autumn determines the capacity to over-winter and subsequently produce a relatively high yield. The yield depression of continuous wheat, compared with wheat grown in rotation, is not related to soil moisture but to weeds, pests and diseases.

Early-harvested predecessors enable higher and more reliable yields (Table 19.9). The coefficients of variation for both traditional and more intensively managed varieties of winter wheat are lower when the crop is sown after early-harvested

Table 19.9 Coefficient of variation (CoV%) for yields of fertilized and unfertilized varieties of winter wheat in crop rotation and continuous wheat in the Selectia long-term experiment 1994–2011

	Predecessors							
	Vetch and oats for green mass		Maize silage		Maize for grain		Winter wheat	
	Fertilization							
Variety	Unfert.	Fert.	Unfert.	Fert.	Unfert.	Fert.	Unfert.	Fert.
Odessa 51	30.1	34.8	44.1	36.9	55.8	43.7	57.0	40.9
New, more intensive	30.1	29.0	37.8	30.4	–	–	58.0	45.0

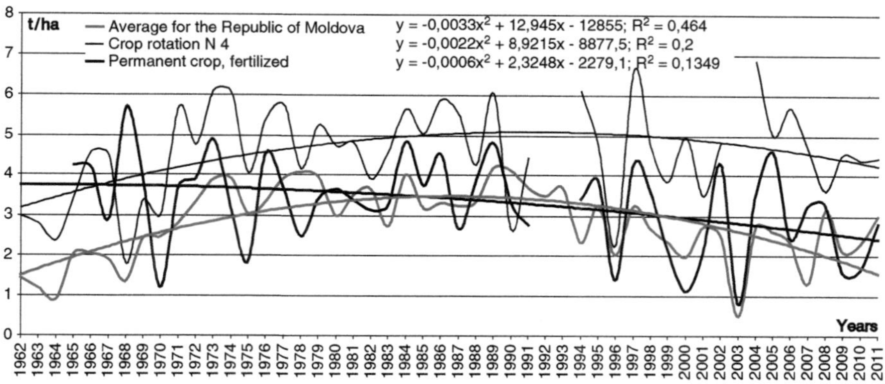

Fig. 19.1 Yields of winter wheat in the long-term field experiment of RIFC Selectia (crop rotation and permanent crop) and on average for the Republic of Moldova, 1962–2011, t/ha

predecessors and higher when it is sown after late-harvested predecessors, especially the continuous wheat.

Figure 19.1 depicts the trends in yield of winter wheat in crop rotation, continuous wheat and the Moldovan national average; the trends for different varieties of winter wheat after different predecessors are shown in Figs. 19.2 and 19.3.

To some extent, fertilization evens out variation in yields but the above-mentioned tendencies remain. Considering the lower level of soil fertility in rotation no. 2 (black fallow and the lowest rates fertilizers) relative to rotation no. 4 (the highest rates of fertilizers with 12 t/ha farmyard manure), we might expect a higher yield in rotation no. 4. Actually, the yields of winter wheat have been the same for the whole period (Table 19.10 and Fig. 19.4); the crop yield is not equivalent to the fertility of the soil. This was the reason why we interrupted the crop rotations for 2 years (1992–1993) to sow the same crop in all experimental plots. By sowing winter rye for green mass during the first year of the break, we established that the highest yields were in fact achieved following crop rotations four and five – which were receiving the most crop residues and manure (Boincean 1999).

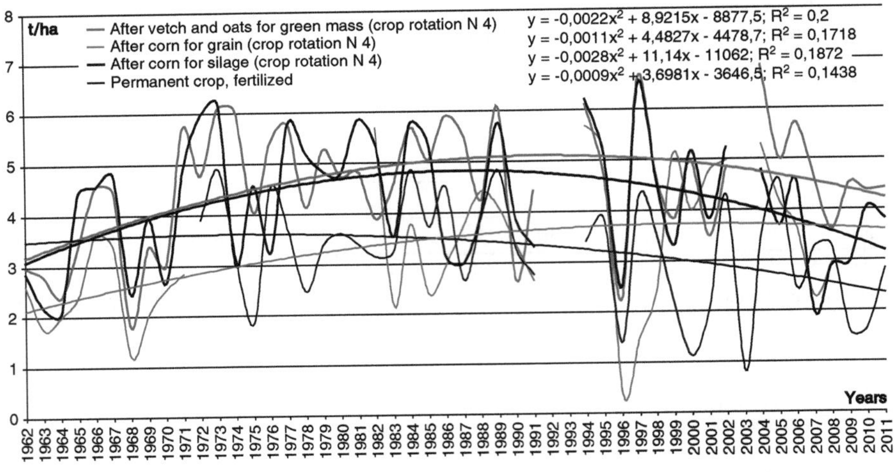

Fig. 19.2 Yields of winter wheat (variety Odessa 51) in different links of crop rotations and in continuous monoculture in the long-term field experiments of RIFC Selectia, 1962–2011, t/ha

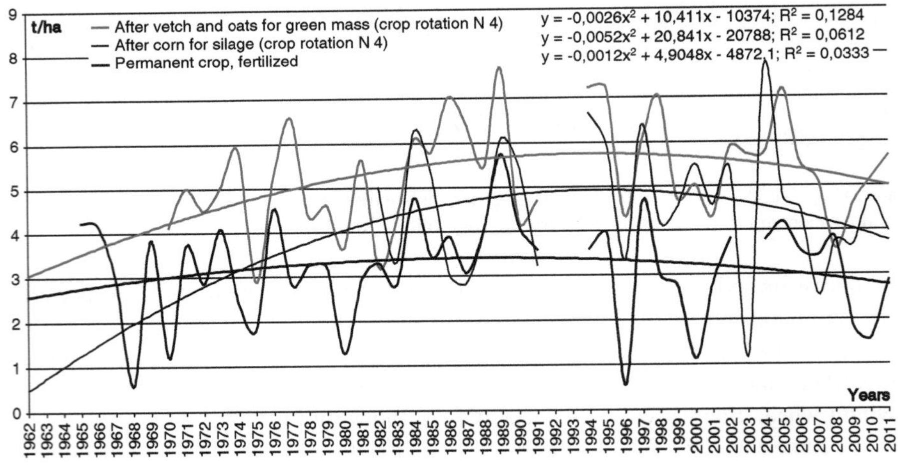

Fig. 19.3 Yields of winter wheat (intensive varieties) in different links of crop rotations and in continuous monoculture in the long-term field experiments of RIFC Selectia, 1962–2011, t/ha

We need to know the influence of different factors on yield formation. Therefore, after three full crop rotations, we introduced an unfertilized rotation (no. 7) which is otherwise the same as fertilized rotation no. 3. In this way, over the last 18 years, it has been possible to establish the influence of rotations, fertilization and varieties on wheat yield (Table 19.11).

The highest wheat yields have been achieved following vetch and oats for green mass; the lowest have been from continuous wheat on both fertilized and unfertilized plots. The yield benefit from fertilization after an early-harvested predecessor

Table 19.10 Average yields of winter wheat Odessa 51 (t/ha) in rotations with different levels of soil fertility 1962–2011

Years	Crop rotation no. 2	Crop rotation no. 4	Difference, t/ha
1962–1971	–	3.45	–
1972–1981	5.10	5.10	–
1982–1991	4.80	4.79	−0.01
1994–2003	4.12	4.15	+0.03
2004–2011	4.89	4.89	0
48-year mean	–	4.46	–
38-year mean	4.72	4.72	0

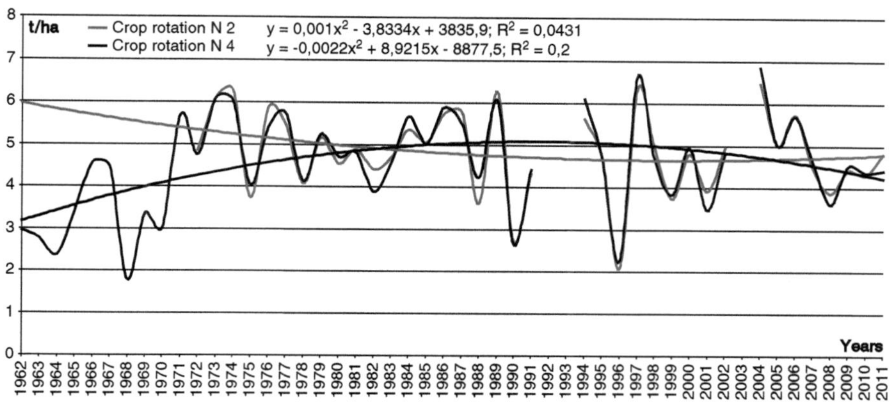

Fig. 19.4 Yields of winter wheat in the long-term field experiment of RIFC Selectia (crop rotation N 4 and N 2), 1962–2011, t/ha

was 0.34 t/ha (8 %) for the traditional variety and 0.37 t/ha (almost 8 %) for newer, high-yielding varieties. The benefit from fertilizer was higher after late-harvested predecessors and, especially, under continuous wheat: for Odessa 51, compared with the benefit of 0.34 t/ha after early-harvested predecessors, the benefit after maize silage was 0.71 t/ha (22 %), after corn for grain 1.02 t/ha (40 %) and for continuous wheat 0.96 t/ha (53 %).

The yield benefit from crop rotation depends on the predecessor; the better the predecessor, the higher is the yield increase. The yield difference between varieties isn't significant; the effect of rotation for Odessa 51 on unfertilized plots is 2.38 t/ha (131 %) but on fertilized plots 1.48 t/ha (81 %) after maize silage and 0.75 t/ha (41 %) after maize for grain. Fertilization halves the effect of rotation but the regularities remain. Again for Odessa 51 on unfertilized plots, yield decrease from late-harvested relative to early-harvested predecessors was 0.90 t/ha (21 %) after maize silage, 1.63 t/ha (39 %) after maize for grain and 2.38 t/ha (57 %) for continuous wheat. Fertilizer mitigated the negative influence of late-harvested predecessors.

Table 19.11 The influence of crop rotation, predecessors, fertilization and varieties on yields of winter wheat in the Selectia long-term experiment, mean yields 1994–2011

			Yield of winter wheat		± from fertilization		± from crop rotation				± from predecessors				± from varieties			
							Unfertilized		Fertilized		Unfertilized		Fertilized		Unfertilized		Fertilized	
Rotational sequence	Predecessors	Varieties	t/ha	%	t/ha	%	t/ha	%	t/ha	%	t/ha	%	t/ha	%	t/ha	%	t/ha	%
Crop rotation	Vetch and oats	Odessa 51	4.20	4.5	+0.34	8	2.38	131	1.76	63	–	–	–	–	–	–	–	
		New high-yielding	4.73	5	+0.37	8	2.78	143	2.26	80	–	–	–	–	+0.53	13	+0.56	12
	Maize silage	Odessa 51	3.30	4	+0.71	21.5	1.48	81	1.23	44	−0.90	21	−0.53	12	–	–	–	–
		New high-yielding	3.53	5	+1.03	29	1.58	81	1.72	61	−1.2	25	−0.54	11	+0.23	7	+0.55	14
	Maize for grain	Odessa 51	2.57	4	+1.02	40	0.75	41	0.81	23	−1.63	39	−0.95	21	–	–	–	–
Continuous crop	Winter wheat	Odessa 51	1.82	3	+0.96	53	–	–	–	–	−2.38	57	−1.76	39	–	–	–	–
		New high-yielding	1.95	3	+0.89	46	–	–	–	–	−2.78	59	−2.26	44	+0.13	7	+0.06	2

Table 19.12 Effect of rotational sequence and continuous cropping on sugar beet yield (t/ha) 1962–2011, fertilized plots

Years	Links of crop rotations			
	Winter vetch and rye > winter wheat > sugar beet	Maize silage > winter wheat > sugar beet	Maize for grain > winter wheat > sugar beet	Continuous sugar beet, fertilized
1962–1971	41.34	39.86	38.52	–
1972–1981	46.20	44.53	45.02	40.40[b]
1982–1991	45.50	46.27	42.15	38.93[b]
1994–2003	44.28	42.02	39.43	22.63
2004–2011[a]	37.91	34.97	34.16	9.57
Average	43.26	41.81	40.09	

[a] 8-year mean
[b] 7-year mean

The efficiency of mineral fertilizer is determined by the soil's nitrogen-supplying capacity; the higher the nitrogen-supplying capacity, the lower is the fertilizer efficiency. This should be taken in consideration to avoid over-fertilization – with the economic and environmental consequences of nitrogen leaching and volatilization.

New, high-yielding varieties have contributed to a yield increase of 7 % (0.13 t/ha) for continuous wheat on unfertilized plots up to 13 % (0.53 t/ha) in rotations (Table 19.11). There was no response to fertilizer from the new varieties sown after early-harvested predecessors, but fertilizer increased the yield of new-variety wheat sown after maize silage by 0.55 t/ha (14 %). The reason for this response, or lack of response, is an open question. We suppose that new varieties respond more to fertilizer under less favourable growing conditions because they have a less-robust root system – but this has yet to be proven.

Sugar beet is sown after winter wheat which, itself, follows crops with different times of harvesting. From the links of crop rotations with sugar beet, the most favourable conditions are seen to be when it follows wheat with early-harvested predecessors; over 48 years, sugar beet in this link of the rotation yielded an average of 43.26 t/ha (Table 19.12). Sugar beet in the links of rotation with maize silage and maize for grain yielded 41.81 and 40.09 t/ha, respectively. The experiment was halted after a collapse in yield of continuous sugar beet during 1972–1981; it was re-established in 1984, but yields on fertilized plots again declined dramatically from 38.93 t/ha in 1984–1991 down to 9.57 t/ha today. Figures 19.5 and 19.6 depict the trends in yield of sugar beet in various links of crop rotations, under continuous cropping and as a national average.

Over the last 18 years, we can separate the influence of fertilization and crop rotation with different links of crop rotations (Table 19.13). Fertilizer is most effective on continuous sugar beet – increasing yields by 8.29 t/ha (97 %) compared with 8.69–10.1 t/ha (27–31 %) in rotations. The benefit of crop rotation is higher on unfertilized plots (22.7–24.0 t/ha or 266–281 %) compared with fertilized plots (23.1–25.8 t/ha or 137–153 %). The lesser benefit of crop rotation under the

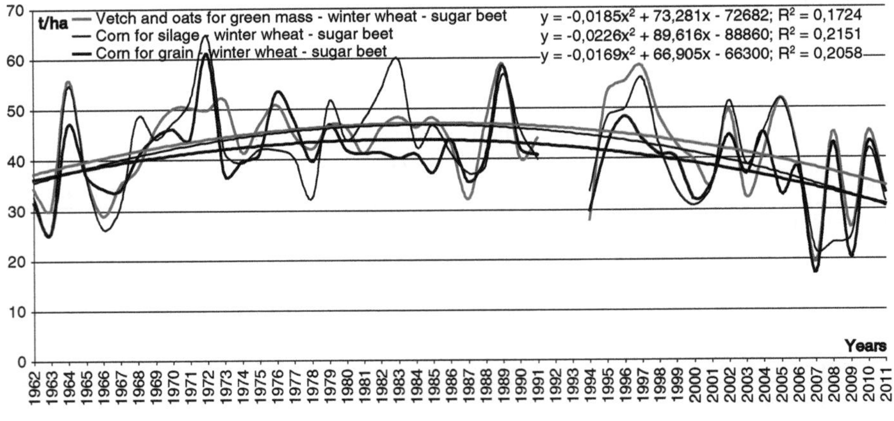

Fig. 19.5 Yields of sugar beet in different links of crop rotations and in permanent crop, long-termfield experiments of RIFC Selectia, 1962–2011, t/ha

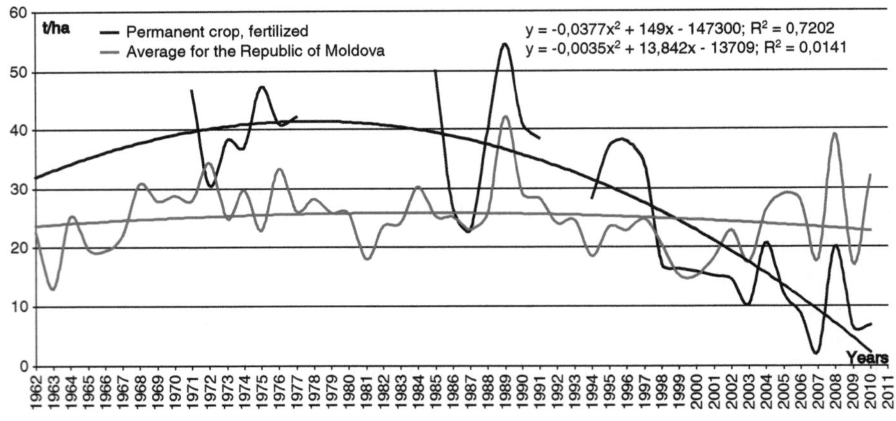

Fig. 19.6 Yields of sugar beet on average for the Republic of Moldova and in continuous monoculture long-term field experiments of RIFC Selectia, 1962–2011

influence of fertilizer was one of the reasons for the neglect of crop rotation during the green revolution era. At that time, it was easy to compensate for the lack of crop rotation with higher rates of mineral fertilizer; fertilizer was cheap and its negative consequences were discounted. Now, the situation has changed: nonrenewable sources of energy and their derivates, including nitrogen fertilizer, are no longer cheap and the aggravation of ecological problems is better appreciated.

The position of sugar beet a crop rotation has less effect on yields than growing continuous sugar beet as opposed to crop rotation; the yield under continuous cropping is much less than in any of the rotations. Similarly, the influence of crop rotation on beet yield is significantly higher than the influence of fertilizer

Table 19.13 The influence of fertilization on crop rotations and continuous sugar beet in the Selectia long-term field experiment, averages for 1994–2011

		Yield of sugar beet, t/ha		± from fertilization		± from crop rotation				± from different links of crop rotation			
						Unfertilized		Fertilized		Unfertilized		Fertilized	
Rotational sequence	Links of crop rotation	Unfert.	Fertil.	t/ha	%	t/ha	%	t/ha	%	t/ha	%	t/ha	%
Crop rotation	Spring vetch and oats- Winter wheat- Sugar beet	32.55	42.65	+10.1	31	+24.01	281	+25.82	153	–	–	–	–
	Maize silage- Winter wheat- Sugar beet	31.24	39.93	+8.69	28	+22.7	2668	+23.1	137	−1.31	4	−2.72	6
Continuous sugar beet	Sugar beet	8.54	16.83	+8.29	97	–	–	–	–	−24.01	74	−25.82	60

Table 19.14 Influence of the proportion of sugar beet in rotations in the Selectia long-term experiment, means 1962–2011

Proportion of sugar beet %	Links of crop rotations	Yield, t/ha
10	Lucerne on 3rd year after first cut – winter wheat – sugar beet	44.25
20	Vetch and oats – winter wheat – sugar beet	43.24
	Maize silage – winter wheat – sugar beet	41.88
	Mean	42.56
30	Winter vetch and rye – winter wheat – sugar beet	43.26
	Maize silage – winter wheat – sugar beet	41.81
	Maize for grain – winter wheat – sugar beet	40.09
	Mean	41.72
100[a]	Continuous sugar beet	23.01

[a] 25-year mean

Table 19.15 Sunflower yield (t/ha) in rotation for different return periods to the same field, 1962–2011 means, fertilized plots

	Return period, years			
Years	9	5	3	Continuous sunflowers
1962–1971	2.59	–	–	–
1972–1981	2.24	1.93	2.09	–
1982–1991	2.79	2.55	2.35	2.14[a]
1994–2003	2.08	1.92	2.03	1.45
2004–2011	2.28	2.07	2.09	1.71
38-year means	2.35	2.12	2.14	–
25-year means	2.11	1.91	2.02	1.73

[a] 7-year mean

(Table 19.13). However, increasing the proportion of sugar beet in the rotation from 10 to 30 % did not influence yield significantly (Table 19.14).

Sunflower is grown in crop rotations in proportions of 10 and 20 % with return to the same field in 9, 5 and 3 years. On average over 38 years, the highest yield (2.35 t/ha) was achieved in the rotation with 10 % sunflower, where the crop returned to the same field after 9 years (Table 19.15). Yields achieved by returning sunflower to the same field after 3 and 5 years were 2.12 and 2.14 t/ha, respectively. Yields were significantly higher in crop rotations than from continuous sunflowers.

Sunflower doesn't respond to fertilizer like wheat and sugar beet. The yield response was similar in both crop rotation and continuous sunflowers: +0.13–0.15 t/ha or 7.5–9 % (Table 19.16).

The yield increase from crop rotation relative to continuous sunflowers was 0.59 t/ha (38 %) on fertilized plots and 0.57 t/ha (40 %) on unfertilized plots. Figures 19.7 and 19.8 depict the trends of yields in crop rotations, continuous sunflowers and average yields for the Republic of Moldova.

Maize for grain was not affected by different predecessors in rotations. Over 48 years, the average yields were the same in both links of crop rotation. The

Table 19.16 Influence of fertilization on yields of sunflower in rotation and continuous cropping 1994–2011

	Yields of sunflower		± from fertilization		± from crop rotation			
					Unfertilized		Fertilized	
Rotational sequence	Unfert.	Fertilized	t/ha	%	t/ha	%	t/ha	%
Crop rotation	2.01	2.16	+0.15	7.5	+0.57	40	+0.59	38
Continuous sunflowers	1.44	1.57	+0.13	9	–	–	–	–

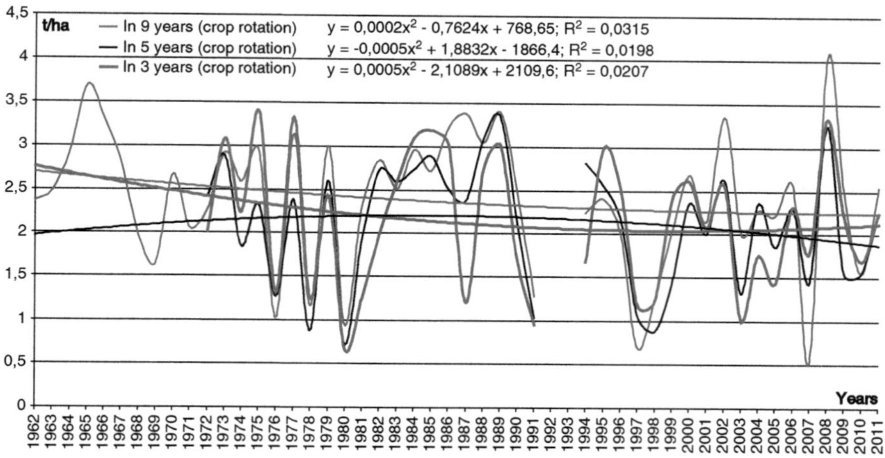

Fig. 19.7 Yield of sunflower in crop rotation with different terms of crop returning on the same field and in continuous monoculture, 1962–2011, RIFC Selectia, t/ha

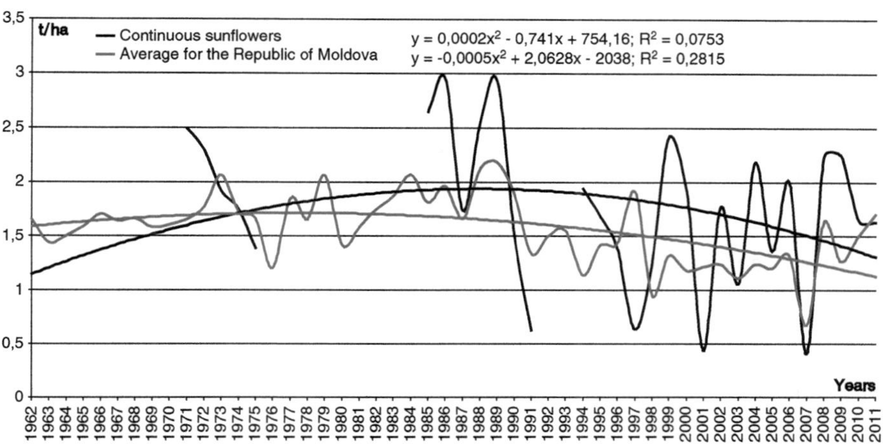

Fig. 19.8 Yield of sunflower in average for the Republic of Moldova and in continuous monoculture, 1962–2011, RIFC Selectia, t/ha

Table 19.17 Influence of rotational sequence on yields of maize for grain in the Selectia long-term experiment, means for fertilized plots 1962–2011, t/ha

Years	Links of crop rotation		
	Spring vetch and oats > winter wheat > sugar beet > maize for grain	Maize silage > winter wheat > sugar beet > maize for grain	Continuous maize
1962–1971	5.60	5.66	4.80[a]
1972–1981	6.78	6.81	5.83
1982–1991	6.86	6.40	5.30
1994–2003	5.84	5.76	5.66
2004–2011	5.87	5.82	5.23
48-year means	6.20	6.10	5.66
38-year means	6.36	6.22	5.52

[a] 1965–1971

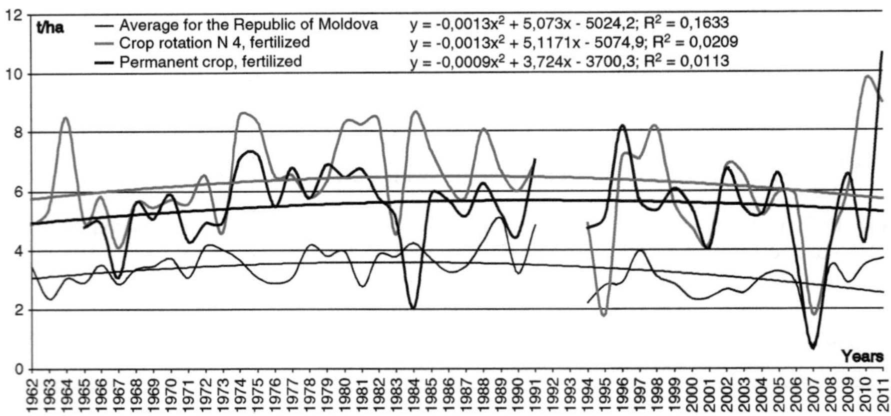

Fig. 19.9 Yields of maize for grain in the long-term field experiment of RIFC Selectia (crop rotation and continuous monoculture) and on average for the Republic of Moldova, 1962–2011, t/ha

38-year means were 6.22–6.36 t/ha in rotation but 5.52 t/ha for continuous maize (Table 19.17).

The trends of grain yields for maize in the long-term experiment with crop rotations, continuous maize and the Moldovan average are shown in Fig. 19.9. Maize has a less pronounced response to fertilizer in crop rotation (+0.42 t/ha or 8 %), but for continuous maize the benefit is very high, at +1.66 t/ha or 44 % (Table 19.18). The beneficial effect of crop rotation is greatest on unfertilized plots. Compared with continuous maize, the increase in grain yield from crop rotation was 1.62 t/ha (42.5 %) without fertilizer but only 0.38 t/ha (7 %) on fertilized plots.

Table 19.18 Influence of fertilization on yields of maize in rotation and continuous maize in the Selectia long-term experiment, 1994–2011 means

Crop sequence	Grain yield		± from fertilization		± from crop rotation			
					Unfertilized		Fertilized	
	Un-fertilized	Fertilized	t/ha	%	t/ha	%	t/ha	%
Crop rotation nos. 3 and 7	5.43	5.85	+0.42	8	+1.62	42.5	+0.38	7
Continuous maize	3.81	5.47	+1.66	44	–	–	–	–

Table 19.19 Yields of maize for grain (t/ha) in crop rotations N2 and N4, 1962–2011

Years	Crop rotation no. 2	Crop rotation no. 4	Difference t/ha
1962–1971	5.56	5.59	+0.03
1972–1981	6.78	6.96	+0.18
1982–1991	6.56	6.84	+0.28
1994–2003	6.02	5.69	−0.33
2004–2011	6.62	5.97	−0.65
Mean	6.29	6.22	–

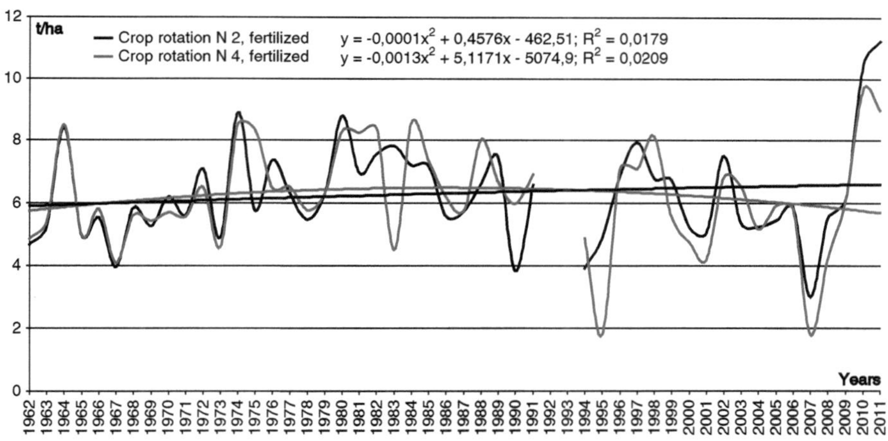

Fig. 19.10 Yields of maize for grain in the long-term field experiment of RIFC Selectia (crop rotation N4 and N2), 1962–2011, t/ha

As with winter wheat, we observe that the yields of maize do not reflect differences in soil fertility. Grain yields are the same in crop rotations nos. 4 and 2 despite substantial differences in soil fertility (Table 19.19 and Fig. 19.10).

Winter barley has much the same reaction to crop rotation and fertilization as maize for grain (Table 19.20). In crop rotation, the response to fertilizer is 0.94 t/ha (30 %), but under continuous barley, it is 1.66 t/ha (86 %). Compared with continuous barley, the increment from crop rotation was 1.21 t/ha (62 %) on unfertilized plots but only 0.49 t/ha (14 %) on fertilized plots. It is clear that there are grounds for reducing the rates of fertilizer application in crop rotations relative to continuous cultures.

Table 19.20 Influence of fertilizer on yields of winter barley in crop rotation and continuous barley, 1994–2011 means

	Yields, t/ha		± from fertilization		± from crop rotation			
					Unfertilized		Fertilized	
Rotational sequence	Unfertilized	Fertilized	t/ha	%	t/ha	%	t/ha	%
Crop rotation	3.15	4.09	+0.94	30	+1.21	62	+0.49	14
Continuous barley	1.94	3.60	+1.66	86	–	–	–	–

19.3.1 Soil Organic Matter Stocks

Soil organic matter is an integral index of soil fertility. Changes in the stocks and annual losses of soil organic matter during the first 22 years and the following 25 years are shown in Table 19.20 for the soil layer 0–20 cm and in Table 19.21 for the 20–40 cm layer. During the first 22 years, the biggest losses of soil organic matter occurred in black fallow: 21.1 t/ha from unfertilized plots and 19.7 t/ha from fertilized plots, which is more than one quarter of the initial stocks. Under continuous maize, losses were 15.8 and 14.4 t/ha from unfertilized and fertilized plots, respectively, and under continuous winter wheat 12.9 and 12.5 t/ha from unfertilized and fertilized plots, respectively.

In crop rotations, stocks of soil organic matter vary according to the proportion of row crops in the rotation. With 40, 50, 60 and 70 % of row crops, annual losses during the first 22 years of the experiment amounted to 0.20, 0.49, 0.55 and 0.66 t/ha, respectively. During the first 22 years of the experiment, annual losses of soil organic matter under rotations with 70 % of row crops on fertilized plots were the same as losses under continuous maize on fertilized plots – 0.65–0.66 t/ha. The annual loss of soil organic matter on black fallow was 0.89 and 0.96 t/ha on fertilized and unfertilized plots, respectively. Under meadow, in contrast, annual losses of soil organic matter over the same period were 0.19 t/ha from the fertilized plot and 0.39 t/ha from the unfertilized plot. During the following 25 years, the annual losses of soil organic matter have been significantly lower but stocks of soil organic matter have continued to decrease under both continuous monocultures and crop rotations, with the exception of fertilized plots under continuous maize and winter wheat – the latter began to accumulate soil organic matter although total losses during 47 years have been much higher than the accumulation.

The loss of soil organic matter under crop rotations may be prevented by applying higher rates of farmyard manure. In a rotation with 40 % of row crops that includes three fields of lucerne and with 4 t/ha of manure, annual losses of soil organic matter increased from 0.2 t/ha during the first 22 years up to 0.3 t/ha over the last 25 years; so three courses of lucerne in rotation together with 4 t/ha manure are not compensating for the losses of soil organic matter from the rotation as a whole. Annual losses of soil organic matter are higher in the rotation with 50 % row crops and 10 % black fallow (0.41 t/ha) but have remained the same during both periods. By applying more manure (8–12 t/ha), the annual losses of soil organic

Table 19.21 Change in soil organic carbon stocks (t/ha) since 1962 in the long-term field experiments with crop rotations and continuous crops, 0–20 cm soil layer, field no. 6

		Continuous crops								Crop rotations					
										Proportion of row crops %					
										40	50	60			70
		Meadow		Black fallow		Winter wheat		Maize for grain		NPK+ 4 t/ha	NPK.+ 4 t/ha	Un-	Fertilized NPK + manure, t/ha		
Year	Indices	Unfertilized	Fertilized	Unfertilized	Fertilized	Unfertilized	Fertilized	Unfertilized	Fertilized	manure	manure	fertilized	8.0	12.0	8.0
1962 carbon stock: 78.7 t/ha															
1984	Carbon stocks	70.1[a]	74.6[a]	57.6	59.0	65.8	66.2	62.9	64.3	74.2	67.9	–	66.7	66.5	64.1
	Stock reduction	−8.6	−4.1	−21.1	−19.7	−12.9	−12.5	−15.8	−14.4	−4.5	−10.8	–	−12.0	−12.2	−14.6
	% of initial stocks	11	5	27	25	16	16	20	18	6	14	–	15	16	19
	Annual losses, t/ha	0.39	0.19	0.96	0.89	0.59	0.57	0.72	0.65	0.20	0.49	–	0.54	0.55	0.66
2009	Carbon stocks	76.6	82.3	55.2	55.9	60.0	73.4	60.5	67.4	64.8	59.5	60.7	62.2	63.8	63.1
	Stock reduction	−2.1	+3.6	−23.5	−22.8	−18.7	−5.3	−18.2	−11.3	−13.9	−19.2	−18.0	−16.5	−14.9	−15.6
	% of initial stocks	3	+5	309	29	24	7	23	14	18	24	23	21	19	20
	Annual losses, t/ha	−0.04	+0.08	0.5	0.48	0.40	0.11	0.39	0.24	0.30	0.41	0.38	0.35	0.32	0.33
Losses in 2009 relative to 1984	t/ha	+6.5	+7.7	−2.4	−3.1	−5.8	+7.2	−2.4	+3.1	−9.4	−8.4	–	−4.5	−2.7	−1.0
	% of initial stocks	+8	+10	3	4	7	+9	3	+4	12	11	–	6	3	1

[a] 1993

matter were reduced from 0.54–0.66 t/ha during the first 22 years of the experiment to 0.32–0.35 t/ha during the last 25 years.

The best conditions for the accumulation of soil organic matter have been in the meadow where the stocks of soil organic matter on fertilized plots have been restored and, even, surpassed the initial level over the last 25 years; even the unfertilized meadow is close to the initial level (Table 19.21).

Less soil organic matter was lost from the 20–40 cm soil layer than from 0 to 20 cm (Table 19.22). Under continuous monocultures and black fallow, on both fertilized and unfertilized plots, losses of soil organic matter from the 20–40 cm layer during the last 25 years have been significantly less than in the first 22 years. However, losses have remained the same in crop rotations with 60 and 70 % of row crops and higher rates of manure (8–12 t/ha), and they have increased in crop rotation no. 5 (with lucerne) and no. 2 (with black fallow and low rates of manure). It seems that low rates of manuring are leading to decreasing stocks of organic matter in the deeper soil layers, notwithstanding the perennial legumes. To test this hypothesis, we sampled the soil profile down to 120 cm (Table 19.23). The content of soil organic matter determined in 1992 was higher in crop rotation no. 5 than in rotation no. 2 for the whole soil profile. It remained so in 2011 – but only in the upper 30 cm. The content of soil organic matter in the deeper layers under rotation five has become significantly lower than in rotation two. This means that, to avoid depletion in soil organic matter from deeper soil layers under crop rotations, perennial legumes should be augmented by higher rates (>4 t/ha) of manure . The same is true for the content of total nitrogen (Table 19.24).[1] And we still need to take account of the destructive influence of black fallow.

19.4 Conclusions

1. During the initial period of agricultural intensification from 1962 to 1981, yields of winter wheat, sugar beet, sunflower and maize increased both in the long-term field experiments and, on average, across the Republic of Moldova. Between 1982 and1991, yields levelled off and, more recently, even decreased. Crop yields in the Selectia long-term field experiments have been significantly above the national average, but, over 48 years, the coefficients of variation for yields of arable crops have ranged between 18.6 and 32.3 %, both in the long-term experiments and countrywide.
2. Over 48 years, the average yields of winter wheat in the Selectia long-term experiment have been 5.20–5.28 t/ha after black fallow and early-harvested predecessors, 5.14 t/ha after peas, 4.60 t/ha after maize silage, 3.56 t/ha after maize for grain and 3.20 t/ha from continuous wheat on fertilized plots. The

[1] Changes in the quality of soil organic matter are analysed in a separate paper.

Table 19.22 Change in soil organic carbon stocks (t/ha) since 1962 in the long-term field experiments with crop rotations and continuous crops, 20–40 cm soil layer, field no. 6

										Crop rotations						
										Proportion of row crops %						
						Permanent crops				40	50	60			70	
		Meadow		Black fallow		Winter wheat		Maize for grain		NPK+ 4 t/ha	NPK.+ 4 t/ha	Un-	Fertilized + manure, t/ha			
Year	Indices	Unfertilized	Fertilized	Unfertilized	Fertilized	Unfertilized	Fertilized	Unfertilized	Fertilized	manure	manure	fertilized	8.0	12.0	8.0	
1962 carbon stock: 76.7 t/ha																
1984	Stocks of carbon	74.4[a]	80.1[a]	60.8	62.7	69.7	68.4	65.8	67.9	75.7	71.2	–	71.2	70.7	68.4	
	Stock reduction	−2.3	+3.4	−15.9	−14.0	−7.0	−8.3	−10.9	−8.8	−1.0	−5.5	–	−5.5	−6.0	−8.3	
	% of initial stocks	3	+4	21	18	9	11	14	12	1	7	–	7	8	11	
	Annual losses, t/ha	0.10	+0.15	0.72	0.64	0.32	0.38	0.50	0.40	0.04	0.25	–	0.25	0.27	0.38	
2009	Stocks of carbon	74.9	80.1	58.0	58.8	64.2	76.2	64.2	68.1	67.9	62.1	59.8	65.5	65.5	66.3	
	Stock reduction	−1.8	+3.4	−18.7	−17.9	−12.5	−0.5	−12.5	−8.6	−8.8	−14.6	−16.9	−11.2	−11.2	−10.4	
	% of initial stocks	2	+4	24	23	16	1	16	11	12	19	22	15	15	14	
	Annual losses, t/ha	0.04	+0.07	0.40	0.38	0.27	0.01	0.27	0.18	0.19	0.32	0.36	0.24	0.24	0.22	
Losses in 2009 relative to 1984	t/ha	+0.5	0	−2.8	−3.9	−5.5	+7.8	−1.6	+0.2	−7.8	−9.1	–	5.7	−5.2	−2.1	
	% to the initial stocks	+0.7	0	4	5	7	+10	−2	+0.3	10	12	–	7	7	3	

[a]1993

Table 19.23 Content of organic carbon in soil profile in crop rotations nos. 2 and 5, field no. 5, t/ha

Year	Soil layer, cm											
	0–10	10–20	20–30	30–40	40–50	50–60	60–70	70–80	80–90	90–100	100–110	110–120
Crop rotation no.2												
1992	3.07	2.98	2.79	2.45	2.15	1.81	1.44	0.77	0.70	0.67	0.64	0.61
2011	2.44	2.34	2.31	2.27	2.01	1.52	1.33	0.87	0.77	0.69	0.50	0.36
±	−0.63	−0.64	−0.48	−0.18	−0.14	−0.29	−0.11	+0.10	+0.07	+0.02	−0.14	−0.25
Crop rotation no.5												
1992	3.22	3.26	3.03	2.76	2.33	1.99	1.44	1.22	–	0.92	0.61	0.46
2011	2.92	2.79	2.56	2.10	1.36	1.27	1.17	0.82	0.49	0.46	0.41	0.36
±	−0.30	−0.47	−0.47	−0.66	−0.97	−0.72	−0.27	−0.40	–	−0.46	−0.20	−0.10

Table 19.24 Content of total nitrogen in soil profile in crop rotations nos. 2 and 5, field no. 5, year 2011, t/ha

Year	Soil layer, cm											
	0–10	10–20	20–30	30–40	40–50	50–60	60–70	70–80	80–90	90–100	100–110	110–120
Rotation no. 2	0.252	0.252	0.229	0.235	0.224	0.168	0.168	0.157	0.117	0.117	0.095	0.095
Rotation no. 5	0.268	0.257	0.232	0.190	0.154	0.140	0.140	0.112	0.095	0.084	0.084	0.084

limiting factor on yields of winter wheat in steppe conditions is the stock of soil moisture, especially at the optimal time of sowing.

3. The benefit of fertilization on yields of winter wheat is much less after early-harvested predecessors than after late-harvested predecessors and, especially, continuous wheat. The same applies to the benefits of crop rotation. The reduction in yields of winter wheat after late-harvested predecessors relative to early-harvested predecessors is equivalent to or greater than the benefit of fertilization.
4. New, high-yielding varieties of winter wheat achieve the highest increase in yield relative to traditional varieties (greater than 12 %) only after early-harvested predecessors.
5. Average yields of sugar beet grown in rotation after winter wheat depend on the wheat's predecessor. Yields of 43.26, 41.81 and 40.09 t/ha were achieved on fertilized plots where the wheat followed vetch and oats, maize silage and maize for grain, respectively. Only 8.7–10.1 t/ha of this yield is attributed to fertilization. Yields of continuous sugar beet have decreased fourfold relative to yields in crop rotation.
6. In the case of sunflowers, the longer the period for return to the same field in crop rotation, the higher is the yield. The benefit from fertilization under continuous sunflowers and in crop rotation is 0.13 and 0.15 t/ha (7 and 9 %), respectively. The benefit of crop rotation relative to continuous sunflowers is 0.57–0.59 t/ha (38–40 %).
7. Maize grain yield does not respond to different links of crop rotations. The yield increase due to fertilization is lower in crop rotations than in continuous maize – 0.42 t/ha (8 %) compared with 1.66 t/ha (44 %). The benefit of crop rotation is significantly higher on unfertilized than on fertilized plots – 1.62 t/ha (43 %) compared with 0.38 t/ha (7 %).
8. The stock of soil organic matter may be an integral index of soil fertility but, in chernozem, it is not mirrored by actual crop yields. Annual losses of soil organic matter were higher during the first 22 years of the experiment compared with the following 25 years, especially for black fallow and continuous monocultures. In rotations with lucerne and black fallow with meagre applications of farmyard manure, annual losses increased or remained the same during the last 25 years, but increasing the manure application to 8–12 t/ha mitigated the loss of soil organic matter during the latter period.
9. Over the 47-year period, loss of soil organic matter from the topmost 20 cm of chernozem was:

   - From black fallow, unfertilized and fertilized – 23.5 and 22.8 t/ha, respectively (30 and 29 % of the initial stock).
   - From winter wheat, unfertilized and fertilized – 18.7 and 5.3 t/ha (24 and 7 % of the initial stock).
   - From maize for grain, unfertilized and fertilized – 18.2 and 11.3 t/ha (23 and 14 % of the initial stock).

- From crop rotations with 40, 50, 60 and 70 % row crops – 13.9, 19.2, 14.9 and 15.6 t/ha (18, 24, 19 and 20 % of the initial stock).
- Under fertilized meadow, the initial content of soil organic matter was restored in 25 years.

10. Future research should establish the causes of differences in soil fertility and crop yields for crop rotations with different components and rates of application of farmyard manure.

Acknowledgements This work was possible thanks to the time and effort of many researchers, technicians and staff over the whole period of 50 years.

References

Boincean BP (1999) Ecological agriculture in the Republic of Moldova (Crop rotation and soil organic matter). Stiinta, Chisinau (Russian)

Jenkinson DS (1991) The Rothamsted long-term experiments: are they still of use? Agron J 83(1):2–10

Johnstan AE (1989) The value of long-term experiments in agricultural research. In: Proceedings of the Sanborn field centennial: a celebration of 100 years of agricultural research. University of Missouri, Columbia, pp 12–20

Krupenikov IA, Boincean BP, Dent DL (2011) The black earth. Ecological principles for sustainable agriculture on chernozem soils. Springer, Dordrecht

Safonov AF (ed) (2002) Long-term field experiment of the KA Timiriazev Agricultural Academy. 90 years: the summary of scientific researches. KA Timiriazev Agricultural Academy, Moscow (Russian)

Chapter 20
Long-Term Field Experiments as a Foundation for Conserving and Enhancing Soil Fertility

S. Andrieş, V. Lungu, and N. Leah

Abstract Long-term field experiments have been maintained by the *Nicolae Dimo* Institute of Pedology, Agrochemistry and Soil Protection for some 50 years. The long-term experiments on Grey forest soil (*Phaeozem*), Leached chernozem and Carbonate chernozem (*Haplic and Calcic chernozem*) that are now included in the EuroSOMNET Information Research System 2000 lend insights to the quality of arable soils under systematic application of fertilizers and support practical recommendations for soil management. Keys to this information are presented. Long-term field experiments should be continued in order to obtain new information on the optimization of soil properties, management regimes and increasing fertility by means of agro-technical and agrochemical measures.

20.1 Introduction

In the Republic of Moldova, the main natural factors that determine high and stable crop yields are timely rainfall and soil fertility. But to obtain and maintain high yields, it is necessary to have scientific support. According to locally established norms, optimum application of mineral and organic fertilizer increases yields by 1.2 tonne/ha in the case of winter wheat, 1.4 t/ha for maize (for grain), 13.8 t/ha for sugar beet and 0.5 t/ha for sunflower.

Long-term field experiments constitute the scientific basis for check-up, prognosis and measures for the conservation and increase of fertility. Such experiments have been carried by the Nicolae Dimo Institute of Pedology, Agrochemistry and Soil Protection (N Dimo Institute) for more than 50 years. We argue that these research objectives must be maintained in order to obtain new information on soil quality and as the foundation for the rational use of land resources.

S. Andrieş (✉) • V. Lungu • N. Leah
Nicolae Dimo Institute of Pedology, Agrochemistry and Soil Protection,
Chisinau, Republic of Moldova
e-mail: ipaps_dimi@mtc.md

D. Dent (ed.), *Soil as World Heritage*, DOI 10.1007/978-94-007-6187-2_20,

20.2 Experimental Sites and Methods

Three long-term field experiments have been established on grey forest soils and different subtypes of chernozem that occupy 85 % of the arable land of the Republic of Moldova (Krupenikov and Ursu 1985).

20.2.1 Grigorievca: Experiments on Carbonate Chernozem

In 1961, Academician I Dicusar and Dr B Tulcinschi founded a long-term field experiment at Grigorievca in Căuşeni District (lat. 49°41′ N, long. 29°21′E). The soil is *Carbonate chernozem,*[1] which is widely distributed, especially in the southern region, occupying more than 654,000 ha or 19.4 % of the country. The landform, characteristically, is a plain with slopes less than 1° at an elevation of 121 m. The station lies within the warm, drought-prone southern agro-climatic zone: cumulative temperature above 10 °C is 3,200–3,400 day degrees, there are 310–320 sunny days annually, mean annual temperature is 9.8 °C with the hottest month (July) 21.7 °C, and growing period is 180–190 days. Annual precipitation is 380–500 mm with 235–275 mm during the growing period; potential evaporation is 850–900 mm giving an Ivanov-Vysotsky moisture coefficient of 0.5–0.6; meteorological drought occurs 3 years in ten (Cerbari 2010).

On the basis of this experiment, the Experimental Station of Pedology and Agrochemistry of the N Dimo Institute, encompassing 22 ha, was founded by Government Decision No. 469 on July 11, 1994. The objective is to monitor soil quality under agricultural use and to elaborate a system of fertilization appropriate for crop rotations on Carbonate chernozem in southern Moldova, specifically to:

- Assess the efficiency of agricultural exploitation of soil fertility
- Determine soil chemical, physical and biological properties depending on the system of fertilization
- Establish optimal agrochemical parameters for high yields
- Study the balance of humus and nutrients in the soil
- Improve soil-plant diagnosis
- Evaluate the moisture regime and determine the water requirements for given levels of crop yield
- Determine the quality of production
- Assess the economic efficiency of fertilizers.

The experiment occupies four fields. At different periods, the number of variants has ranged from 8 to 14, with four replicates; individual plots occupy 100–200 m^2. The crop rotation includes winter wheat, maize for grain and silage, sunflower, barley, peas and mixture of vetch and oats for forage. Mineral, organic and

[1] Calcic chernozem in the *World reference base for soil resources 2006.*

combined organic-mineral systems of fertilization are tested using applications of superphosphate, ammophos, ammonium nitrate, potassium chloride and farmyard manure. The analyses of the soil and plant samples have been performed according to the standards. In 2002, the experiment was included in the EuroSOMNET International Research System.

20.2.2 Ivancea: Experiments on Leached Chernozem

In 1964, I Dicusar and N Turtureanu founded the long-term field experiment on *Leached chernozem*[2] at Ivancea in Orhei District (lat. 47 °18′N, long. 28 °54′E, elevation 170–180 m). Leached chernozem occupies more than 395,000 ha or 11.7 % of the country in the north and central regions. The climate of the central zone, in which Ivancea is located, is warm sub-humid; accumulated temperature above 10 °C is 3,000–3,250 day degrees, sunny days 390–310, mean annual temperature 8.5–9.0 °C, and growing season 178–182 days. Mean annual precipitation is 550–600 mm, potential evaporation 800–820 mm, and Ivanov-Vysotsky moisture coefficient 0.7–0.8; meteorological drought occurs 1 or 2 years in ten.

The experiment employs five fields with the number of variants increasing from four in 1965–1970 to 14 in 1985–2011, each with four replicates; each individual plot is 200 m^2. Since 1986, different levels of mobile phosphorus (from 1.0 to 1.5 mg/100 g soil to 4.5 mg/100 g) have been maintained by compensating for the nutrients exported at harvest by application of fertilizers. Different levels of mineral nitrogen have been created by application of nitrogen fertilizers (from N_0 to N_{300}) against a constant-PK background, and different levels of exchangeable potassium have been created by applying K_0, K_{60} and K_{120} fertilizer against a constant-NP background. The systems of mineral and organic-mineral fertilization under test include the use of vegetable wastes as organic fertilizers.

The objective is to develop a system of fertilization appropriate to crop rotations on Leached chernozem, aiming for high crop yields, maximum profit from fertilization and protection of the environment against pollution by excess nutrients, specifically to:

- Study the soil property response to cropping systems and fertilization
- Establish optimal levels of soil nutrients to obtain the desired yields and avoid pollution of the groundwater
- Study water regimes for the rational use of moisture in the period of yield formation
- Assess the balances of humus and nutrients in the soil, depending on cropping system and fertilization
- Assess the economic and energy efficiency of the fertilizer treatments.

[2] Haplic chernozem in the *World reference base for soil resources 2006*.

20.2.3 Ivancea: Experiments on Grey Soil

In 1964, Professor V Grati founded the long-term experiment on *Grey forest soil*[3] (Grati 1977), also at Ivancea, lat. 47°19′N, long. 28°53′E at an elevation of 182 m, on a broad ridge cleared from forest about 100 years ago. Grey soils are widespread in the north and central regions of Moldova and make up about 10 % of the arable. The agro-climatic conditions are the same as for the experiments on Leached chernozem.

Initially, the experiment comprised five fields, now four. Depending on the purpose and specific tasks of the research, the number of variants was changed in 1971, 1981 and 1986. Individual plots are 100 m^2 with four replicates. The objective is to develop a system of fertilization appropriate to crop rotations on Grey soil, aiming for high yields and protection of the environment against pollution, specifically:

- To monitor the modification of regime and chemical, physical and biological properties of grey soil depending on the cropping system and fertilization
- To establish optimal doses of fertilizers for field crops
- To assess the quality of the produce
- To estimate crop water requirements
- To evaluate the balances of humus and nutrients in the soil
- To estimate the economic efficiency of the fertilizers.

On the basis of the experiments at Ivancea, including the experiment founded in 1976 by Professor M Ţurcanu on the efficacy of various organic wastes as fertilizers, the Experimental Pedology, Agrochemistry and Ecology Station of the N Dimo Institute was established by Government Decision No. 469 of July 11, 1994, with an area of 176 ha. In 2000, both experiments at Ivancea were included in the EuroSOMNET system. The results of all these experiments are presented in the annual reports of the Soil Agrochemistry and Plant Nutrition Laboratory and archived by the N Dimo Institute and since 2006 at the State Agency on Intellectual Property of the Republic of Moldova.

20.3 Results and Discussion

The Carbonate chernozem at Grigorievca is fine textured with about 57 % clay (35 % fine clay) and 17 % calcium carbonate in the topsoil, humus content 3.4 and total nitrogen 0.17 %, total phosphorus 0.15 %, pH_{water} 7.3, exchangeable cations 40.2 me/100 g soil, mobile phosphorus about 1.0 mg/100 g soil, and exchangeable potassium 30.0 mg/100 g. Carbonate chernozem in the southeast of Moldova has satisfactory chemical and physicochemical properties for field crops (although low

[3] Phaeozem in *World reference base for soil resources 2006.*

in mobile phosphorus), but they are susceptible to loss of humus and soil structure and, consequently, to compaction under cultivation (Cerbari 2010). Their *bonitet* rating is somewhat lower than other chernozem: according to soil properties 71 points and according to crop yield 66–68 points (Ursu 1986, 2011).

The Leached chernozem is a clay loam with 35–36 % fine clay. The humus content declines from 3.3 % in the topsoil to 1.5 % in the Bhw2 horizon. For the Ah horizon, total nitrogen is 0.15–0.19 %, total phosphorus 0.12 %, pH_{water} 6.8 and hydrolytic acidity 2.5me/100 g soil, total exchangeable cations 37.4 mg/100 g, mobile phosphorus about 1.0 mg/100 g soil, and exchangeable potassium 23 mg/100 g – all very satisfactory for cropping but, as result of agricultural exploitation, Leached chernozem have suffered significant loss of humus and, also, secondary compaction.

The characteristics of virgin and arable grey soil are presented by Cerbari (2010): clay comprises 60 % and fine clay 40 % and, under cultivation, humus 2.3 and total nitrogen 0.14 %, total phosphorus 0.12 and total potassium 1.2 %. pH_{water} is 6.3, base saturation 94 %, hydrolytic acidity 3.6 me/100 g soil, exchangeable calcium 4.8 me/100 g, mobile phosphorus (Machigin method) 2.2 mg/100 g soil, and exchangeable potassium 22.2 mg/100 g soil.

The scientific and practical results of the experiments have been presented in journal articles and collections, including thematic collections (N Dimo Institute 1965, 1970, 1971, 1978; Krupenikov 1977a, b; Turtureanu et al. 1976, 1979; Ungureanu et al. 1982; Ursu 1985, 1986). Results have also been generalized in monographs (Andrieş 1993, 2007, 2011; Andrieş et al. 2000; Cerbari 2000, 2010; Corduneanu 1978; Corduneanu et al. 1984; Donos 2008; Ganenko 1991; Grati 1977; Marinescu 1991; Sinkevici 1989; Turtureanu 1976). The main issues are:

- The economic efficiency of fertilizers applied to field crops on different soils (Turtureanu 1976)
- Soil organic matter and nutrient balances in the soil-plant system (Corduneanu 1978)
- Fertilization systems and soil fertilization (Corduneanu et al. 1984; Ursu 1986)
- Soil-forming processes under agricultural use (Grati 1977; Sinkevici 1989)
- Soil properties and regimes, especially the status of humus and soil biodiversity according to fertilization (Ganenko 1991; Marinescu 1991)

On the basis of field experiments (Andrieş 1993, 2007, 2011; Donos 2008), we have established:

- The regularities of effects of fertilization on crop yields in the different pedo-climatic regions
- Optimal nitrogen, phosphorus and potassium regimes for the main soil types and subtypes in order to obtain high yields and quality
- Soil nutrient requirements for different levels of crop yield
- Soil water requirements for the main field crops, depending on the fertilization system
- Values of diagnostic indicators of plant mineral nutrition
- The optimal systems of fertilization, depending on soil type and subtype and crop rotation.

The field experiments are a scientific foundation for development of recommendations on the rational use of fertilizers. All research and educational institutions in the Republic of Moldova participate in the elaboration of these recommendations, which have been updated periodically (Andrieş et al. 2001b, 2012; Toma 2008; Ţurcanu et al. 1994; Ungureanu et al. 1981; Ursu et al. 1987). They deal with soil fertility assessment methods, crop nutrient requirements, fertilization of field crops and systems of fertilization in crop rotations, the safe storage of mineral fertilizers, health and safety, and environmental protection. The scientific-practical results have been and are used to elaborate State programs for conservation and increase of soil fertility that have been adopted by government decisions (Andrieş et al. 2001a, 2004; Government of Moldova 2011).

20.4 Conclusions

The long-term field experiments of the *Nicolae Dimo* Institute of Pedology, Agrochemistry and Soil Protection, the *Selectia* Institute of Field Crops, the *Porumbeni* Phytotechnical Institute and the Agrarian State University of Moldova are located on different soil types and subtypes in all the pedo-climatic zones of Moldova. They are of incontestable scientific and practical worth for monitoring soil quality and the development of prognosis and measures for the increase of the fertility of agricultural land. To date, they have been protected and financed by the State and, in the absence of a better alternative, this arrangement should continue.

Primary data on the long-term field experiments (47–51 years) carried out by the N Dimo Institute on Grey soil, Leached and Carbonate chernozem are recorded in the annual reports of the N Dimo Institute, published as scientific papers and monographs, and incorporated in State programs on soil conservation and fertility.

References

Andrieş SV (1993) Regulation of nutrient regimes of soils according to planned yields of winter wheat and corn. Ştiinta, Kishinev (Russian)

Andrieş SV (2007) Optimization of soil nutrient regimes and crop productivity. Pontos, Chisinau (Romanian)

Andrieş SV (2011) Agrochemistry of nutrient elements. Soil fertility and ecology. Pontos, Chisinau (Romanian)

Andrieş S, Ţîganoc V et al (2000) Bulletin of ecopedological (agro-chemical) monitoring. Issue VII. Pontos, Chisinau (Romanian)

Andrieş SV et al (2001a) National complex program for the increase of soil fertility. Pontos, Chisinau (Romanian)

Andrieş SV, Donos A, Ţîganoc V et al (2001b) Recommendations on the application of fertilizers for field crop rotations on different soils in the post-privatization period. Pontos, Chisinau (Romanian)

Andrieş SV et al (2004) Complex program for the enhancement of deteriorated fields and the increase of soil fertility. Part II. Pontos, Chisinau (Romanian)
Andries SV et al (2012) Recommendations on the application of fertilizers for field crops on different soils. Pontos, Chisinau (Romanian)
Cerbari V (2000) Information system on the quality of soil cover of the Republic of Moldova (data base). Pontos, Chisinau (Romanian)
Cerbari V (2010) Monitoring of soil quality of the Republic of Moldova: data base, conclusions, prognosis, recommendations. Pontos, Chisinau (Romanian)
Corduneanu PN (1978) Cycling of the main nutrient elements of crops. Ştiinta, Kishinev (Russian)
Corduneanu PN, Turtureanu NA, Moldovan AI (1984) The system of fertilizer application and soil fertility. Cartea Moldoveneasca, Kishinev (Russian)
Dimo N Institute (1965) Effectiveness of fertilizers under the conditions of Moldova. Issue 4. Cartea Moldoveneasca, Kishinev (Russian)
Dimo N Institute (1970) Effectiveness of fertilizers under the conditions of Moldova. Issue 5. Cartea Moldoveneasca, Kishinev (Russian)
Dimo N Institute (1971) Effectiveness of fertilizers under the conditions of Moldova. Issue 6. Cartea Moldoveneasca, Kishinev (Russian)
Dimo N Institute (1978) The soils of Moldova and their use in the conditions of intensive tillage. Ştiinta, Kishinev (Russian)
Donos A (2008) Nitrogen accumulation and modification in soil. Pontos, Chisinau (Romanian)
Ganenko VP (1991) The humus of Moldovan soils and its modification under the influence of fertilizers. Ştiinta, Kishinev (Russian)
Government of Moldova (2011) Program for the conservation and increase of soil fertility. Official Monitor of the Republic of Moldova No. 139-145 of 26.08.2011, art. 696. Chisinau (Romanian)
Grati VP (1977) Forest soils of Moldova and their rational use. Ştiinta, Kishinev (Russian)
Krupenikov IA (ed) (1977a) Genesis and rational use of the soils of Moldova. Ştiinta, Kishinev (Russian)
Krupenikov IA (ed) (1977b) Research and use of the soils of Moldova. Ştiinta, Kishinev (Russian)
Krupenikov IA, Ursu AT (1985) The soils of Moldova, volume 2. Soil geography, description of soil provinces, districts and sub-districts. Ştiinta, Kishinev (Russian)
Marinescu KM (1991) Microbial association of ameliorative soils. Ştiinta, Kishinev (Russian)
Sinkevici ZA (1989) Modern processes in Moldovan chernozems. Ştiinta, Kishinev (Russian)
Toma SI (2008) Application of fertilizers in sustainable agriculture. AES Publishing House, Chisinau (Romanian)
Ţurcanu M, Andrieş S et al (1994) Recommendations on the application of fertilizers. Agroinformreclama, Chisinau (Romanian)
Turtureanu NA (1976) Effectiveness of fertilizer use in the soils of Moldova. Cartea Moldoveneasca, Kishinev (Russian)
Turtureanu NA, Corduneanu PN, Ţurcanu MA et al (1976) Accumulation and export of nutrient elements by agricultural crops. Ştiinta, Kishinev (Russian)
Turtureanu NA, Corduneanu PN, Moldovan AI et al (1979) The system of fertilizers under conditions of intensive tillage. Ştiinta, Kishinev (Russian)
Ungureanu VG, Ţurcanu MA, Vainberg NL et al (1981) Recommendations on the use of fertilizers in the Moldavian SSR in the period 1981-1985. Timpul, Chisinau (Romanian)
Ungureanu VG, Ursu AT et al (1982) Yield and quality of crops under systematic use of fertilizers. Ştiinta, Kishinev (Russian)
Ursu AT (ed) (1985) Fertilizers, soil fertility and productivity of crops in Moldova. Ştiinta, Kishinev (Russian)
Ursu AT (1986) The soils of Moldova, volume 3. Soil management, protection and improvement. Ştiinta, Kishinev (Russian)
Ursu AT (2011) The soils of Moldova. Ştiinta, Chisinau (Romanian)
Ursu AT, Andries S et al (1987) Recommendations on the use of fertilizers in the Moldavian SSR in the period 1986–1990. KSAI, Kishinev (Russian)

Chapter 21
Productivity and Fertility of the Balti Chernozem Under Crop Rotation with Different Systems of Fertilization

B.P. Boincean, L.T. Nica, and S.S. Stadnic

Abstract Forty-year yields are presented for various crops in rotation under different systems of fertilization on Typical chernozem. Winter wheat and the rotation as a whole showed an initial increase in yields, followed by stagnation or decline during the last 15–20 years. Over the period, stocks of soil organic carbon and total nitrogen declined under mineral fertilization, whereas farmyard manure and combined organic and mineral fertilizers contributed to a modest accumulation of carbon and nitrogen – mainly in soil layers deeper than 40 cm. However, taking account of harvested crops and mineralization of humus, the carbon budget was profoundly negative: the highest carbon deficit was incurred with mineral fertilizers (3,527–4,233 kgC/ha/year); the least deficit was incurred under combined application of farmyard manure and mineral fertilizers (1,861–2,444 kgC/ha/year); unfertilized plots held an intermediate position. The nitrogen balance was also negative, especially on the unfertilized plots and those receiving only mineral fertilizers.

21.1 Introduction

Agricultural intensification in the era of green revolution depended on cheap energy and mineral fertilizers, especially nitrogen, applied to responsive crop varieties. This model is now problematic because the inputs are no longer cheap and because of their negative influence on the environment – earlier we demonstrated the risk of nitrate leaching, even from unfertilized plots (Boincean and Nica 2007; Boincean et al. 2010). Soil organic matter management should be a priority for farmers and researchers, but the complexity of this issue demands a holistic approach to farm

B.P. Boincean (✉) • L.T. Nica • S.S. Stadnic
Selectia Research Institute for Field Crops, Balti, Republic of Moldova

Alecu Russo Balti State University, Balti, Republic of Moldova
e-mail: bboincean@gmail.com

D. Dent (ed.), *Soil as World Heritage*, DOI 10.1007/978-94-007-6187-2_21,

management in which the system of fertilization is one component but not the dominant one (Boincean 1999; Gliessman 2000).

Organic and mineral fertilizers have different effects on both crop yields and soil fertility. Recently, research on the Morrow Plots in Illinois appeared to demonstrate degradation of soil under mineral fertilization; Mulvaney and Khan have generalized data from long-term field experiments worldwide that purport to show the same trend (Khan et al. 2007; Mulveney et al. 2009) but this is disputed by Powlson et al. (2010) and Powlson et al. in this symposium. Enriching soil with fresh organic matter creates good conditions for a transition to more sustainable agriculture, less dependent upon external inputs (Seiter and Horwath 2004; Drinkwater et al. 2008).

Very often, changes in the topsoil are assessed without considering changes in the whole soil profile relative to unfertilized plots and the initial stocks and quality of soil organic matter. Here we include evidence of such changes. Nitrogen-use efficiency is calculated for different systems of fertilization. This is important from scientific and practical points of view but, to optimize the rates of mineral fertilizers, farmers also need their own trials with fertilized and unfertilized strips of different crops in each field.

21.2 Experimental Site and Methods

A long-term field experiment with different systems of fertilization was established at Balti in 1973 with a crop rotation of vetch-with-oats for green mass – winter wheat – sugar beet – maize for grain – spring barley – sunflower. The sown varieties and hybrids are registered in the Republic of Moldova, and the practices for growing field crops are those recommended for the country. The soil is Typical chernozem[1] heavy loam with initial contents of 4.35–5.08 % soil organic matter, 0.24–0.26 % total nitrogen, 0.12–0.13 % phosphate, 1.20–1.40 % potassium and pH_{water} 6.6–7.1. Figure 21.1 summarizes the precipitation and accumulated

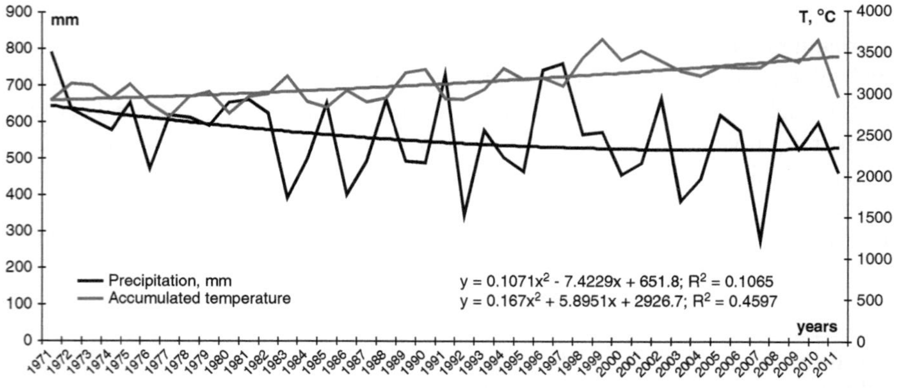

Fig. 21.1 Precipitation and temperature 1971–2011

[1] Haplic chernozem in *World reference base for soil resources 2006*.

temperature over 40 agricultural years, during which there has been a tendency for less precipitation and increasing temperature.

Four systems of fertilization are compared (Tables 21.1 and 21.2):

- Unfertilized control (I)
- Mineral fertilizers (II – NPK, III – NPK_2 and IV – NPK_3)
- Combined manure and mineral fertilizers (V – 10 t/ha manure+NPK_1, VI – 10 t/ha manure+NPK_2, VII – 10 t/ha manure+NPK_3, VIII – 15 t/ha+NPK_1, IX – 15 t/ha+NPK_2, X – 15 t/ha+NPK_3)
- Manure (XI – 15 t/ha manure, XII – residual action of previous fertilization)

Composted farmyard manure is ploughed in prior to sugar beet (60 t/ha) and sunflower (30 t/ha). Mineral fertilizers are applied annually for winter wheat, sugar beet, maize and sunflower before autumn tillage, except for winter wheat where half of the nitrogen is applied in the autumn and the other half in spring. Spring barley and vetch-with-oats benefit from the residual activity of fertilizers applied to the previous crops in the rotation. The experiment has four replicates, two located in the bottom half of the field and two in the upper half of the same field; in the first replicate the layout of plots is systematic but, for the others, plots are randomized. Each experimental plot is 242 m^2 (5.6 m 43.2 m).

21.3 Results and Discussion

21.3.1 Yield Response to Fertilization

Forty-year trends of yields are depicted in Figs. 21.2, 21.3, 21.4, 21.5, 21.6, and 21.7. Initially, yields of winter wheat increased under all systems of fertilization but stagnated or, even, declined over the last 15–20 years (Fig. 21.2). The increase in yield attributable to fertilization also decreased during the last 15–20 years. Sugar beet shows the best response to fertilization relative to the unfertilized plots (Fig. 21.3); yields declined in the early years but then stabilized. Maize (Fig. 21.4) and sunflower (Fig. 21.5) hardly responded to different systems of fertilization although yields from fertilized plots were higher than those from unfertilized plots. Spring barley and vetch-with-oats (Figs. 21.6 and 21.7) did not receive fertilizer but benefitted from the fertilizer applied to previous crops in the rotation.

Weather has a tremendous influence on yields (Table 21.3). The yield increase attributable to fertilization and yield variation caused by weather conditions were, respectively, for winter wheat – 22–30 and 26–30 %; for sugar beet – 38–48 and 22–26 %; maize – 15 and 29–33 %; sunflower – 9 and 30–32 %; spring barley – 33–54 and 33–39 %; for vetch-and-oats – 27–40 and 43–46 %; and for the rotation as a whole 24–31 and 27–29 %. Yield variation due to weather was greater than the yield increase attributable to fertilization in the cases of maize, sunflower and vetch-with-oats but, for the rotation as a whole, yield increase due to fertilization and yield variation due to weather were much the same.

Table 21.1 Fertilizer regimes for field crops in the six-field crop rotations I–VI, 1973–2008

			Without farmyard manure									15 t/ha of farmyard manure									
			NPK_1			NPK_2			NPK_3			NPK_1			NPK_2			NPK_3			Only
Rotations	Crops	Control, unfertilized	N	P	K	N	P	K	N	P	K	N	P	K	N	P	K	N	P	K	manure
Systems of fertilization																					
1973–1980																					
0 1973–1975	Winter wheat	–	–	–	–	–	20	–	–	20	–	90	90	90	–	40	–	–	40	–	–
	Sugar beet	–	–	–	–	120	120	120	120	120	120	120	120	120	120	120	120	120	120	120	–
	Maize for grain	–	–	–	–	90	90	90	90	90	90	–	20	–	90	90	90	90	90	90	–
I 1976–1980	Peas for grain	–	–	–	–	–	–	–	–	–	–	–	–	–	–	–	–	–	–	–	–
	Winter wheat	–	–	–	–	90	90	90	90	90	90	90	90	90	90	90	90	90	90	90	–
	Sugar beet	–	–	–	–	120	120	120	120	120	120	120	120	120	120	120	120	120	120	120	–
	Maize for grain	–	–	–	–	90	90	90	90	90	90	90	90	90	90	90	90	90	90	90	–
	Sunflower	–	–	–	–	–	–	–	–	–	–	–	–	–	–	–	–	–	–	–	–
1981–1990																					
II 1981–1985	Vetch + oats	–	–	–	–	–	–	–	–	–	–	–	–	–	–	–	–	–	–	–	–
III 1986–1990	Winter wheat	–	–	–	–	–	60	60	–	60	60	90	60	60	120	60	60	120	60	60	–
	Sugar beet	–	60	60	60	90	90	90	120	160	120	60	60	60	90	90	90	120	180	120	75
	Maize for grain	–	60	30	30	90	60	30	150	60	60	60	30	30	90	60	60	150	60	60	–
	Sunflower	–	30	60	30	60	90	60	60	120	60	30	60	30	60	90	60	60	120	60	–
1991–2008																					
IV 1991–1996	Vetch + oats	–	–	–	–	–	–	–	–	–	–	–	–	–	–	–	–	–	–	–	–
V 1997–2002	Winter wheat	–	60	30	30	90	60	60	120	60	60	60	30	30	90	60	60	120	60	60	–
VI 2003–2008	Sugar beet	–	30	30	30	60	60	60	90	120	90	30	30	30	60	60	60	90	120	90	60
	Maize for grain	–	60	30	30	90	45	45	150	60	60	60	30	30	90	45	45	150	60	60	–
	Spring barley	–	–	–	–	–	–	–	–	–	–	–	–	–	–	–	–	–	–	–	–
	Sunflower	–	60	30	60	60	90	60	60	120	60	30	60	30	60	90	60	60	120	60	30

The productivity of the crop rotation shows the general trend of an initial increase, followed by stagnation or decline during the last 15–20 years (Fig. 21.8). The averages for the rotation under different systems of fertilization are represented in Fig. 21.9: the lowest yields have been from unfertilized plots and those receiving the lowest rates of mineral fertilizers. But these data do not indicate real changes in soil fertility; the crucial role of inherent soil fertility is illustrated by its contribution to the yields in the *fertilized* plots, amounting to 70–87 % as well as 100 % of the yields from the unfertilized control (Fig. 21.10). It is noteworthy that application of manure increased the recovery of mineral fertilizers, especially when they are used at low rates (Fig. 21.11).

21.3.2 Changes in the Stocks of Soil Organic Matter

Tables 21.4 and 21.5 present the changes in soil organic carbon compared with the unfertilized control. It appears that mineral fertilizers are contributing to losses throughout the soil profile of 19.0–26.3 tC/ha (11–15 % of the initial value). On the other hand, manure and combined organic and mineral fertilizers increased stocks of soil organic matter by 23.9–40.5 t/ha (14–23 %). These changes are not uniform down profile. In plots receiving mineral fertilizer, losses from 0 to 40 cm were 7–16.7 tC/ha (28–34 % of the initial stocks), while 16.4–33 tC/ha (66–72 %) was lost from the 40 to 100 cm layer. In plots with combined mineral and organic fertilization, any organic matter lost from the 0 to 40 cm layer was compensated by gains below.

Annual losses of soil organic matter from 0 to 100 cm were 600 kgC/ha on unfertilized plots and 1,087.2–1,274.4 kgC/ha under mineral fertilization. Farmyard manure alone and in combination with mineral fertilizer contributed to accumulation of 46.1–438.5 kgC/ha for the 0–100 cm soil layer (Table 21.6). We suppose that, in the absence of a supplementary source of carbon, mineral fertilizers intensify decomposition of humus and crop residues. Organic fertilizer also increases decomposition in the 0–40 cm layer, but the products of decomposition then accumulate deeper in the soil. Dokuchaev, in his famous *Russian chernozem* (1887/1952), postulated development of the characteristic thick profile through two processes: decomposition of plant residues and leaching of humic substances from upper to lower layers. This was disputed by Kosticev (1949) who recognized only the decomposition of plant residues as the main and unique process of chernozem formation. However, the translocation of humic substances down profile was subsequently demonstrated experimentally by Ponomareva and Plotnicova (1980).

21.3.3 Changes in Total Nitrogen

Over 39 years, there has been a substantial loss of total nitrogen from the unfertilized control and plots fertilized only with NPK; the greater part of the loss was

Table 21.2 Total and annual amounts of mineral (N:P:K kg active ingredient/ha) and organic (t/ha) fertilizers used in different systems of fertilization, 1971–2011, Field No.6

Variant	1971–1980	1981–1990	1991–2011	1971–2011	Annual mean/ha
I	–	–	–	–	–
II	–	480:420:360	630:450:420	1,110:870:780	27:21:19
III	510:570:570 + 40 t manure/ha	720:600:480	990:870:780	2,220:2,040:1,830 + 40 t manure/ha	54:50:45 + 1 t manure/ha
IV	510:570:570 + 40 t manure/ha	960:840:600	1,500:1,260:960	2,970:2,670:2,130 + 40 t manure/ha	72:65:52 + 1 manure/ha
V	570:570:570 + 40 t manure/ha	480:420:360 + 100 t manure/ha	630:450:420 + 230 t manure/ha	1,680:1,440:1,350 + 370 t manure/ha	41:35:33 + 9 t manure/ha
VI	570:570:570 + 40 t manure/ha	720:600:480 + 100 t manure/ha	990:870:870 + 230 t manure/ha	2,280:2,040:1,830 + 370 t manure/ha e	56:50:45 + 9 t manure/ha
VII	600:570:570 + 40 t manure/ha	960:840:600 + 100 t manure/ha	1,500:1,260:960 + 230 t manure/ha	3,060:2,670:2,130 + 370 t manure/ha	75:65:52 + 9 t manure/ha
VIII	600:570:570 + 40 t manure/ha	480:420:360 + 150 t manure/ha	630:450:420 + 315 t manure/ha	1,710:1,440:1,350 + 505 t manure/ha	42:35:33 +12.3 t manure/ha
IX	630:570:570 + 40 t manure/ha	720:600:480 + 150 t manure/ha	990:870:780 + 315 t manure/ha	2,340:2,040:1,830 + 505 t manure/ha	57:50:45 +12.3 t manure/ha
X	630:570:570 + 40 t manure/ha	960:840:600 + 150 t manure/ha	1,500:1,260:960 + 315 t manure/ha	3,090:2,670:2,130 + 505 t manure/ha	75:65:52 +12.3 t manure/ha
XI	740:540:590	1,460:1,040:1,200	315 t manure/ha	2,200:1,580:1,790 + 315 t manure/ha	54:39:44 + 7.7 t manure/ha
XII	740:540:590	1,700:1,140:1,500	Residual action	2,440:1,680:2,090	60:41:51

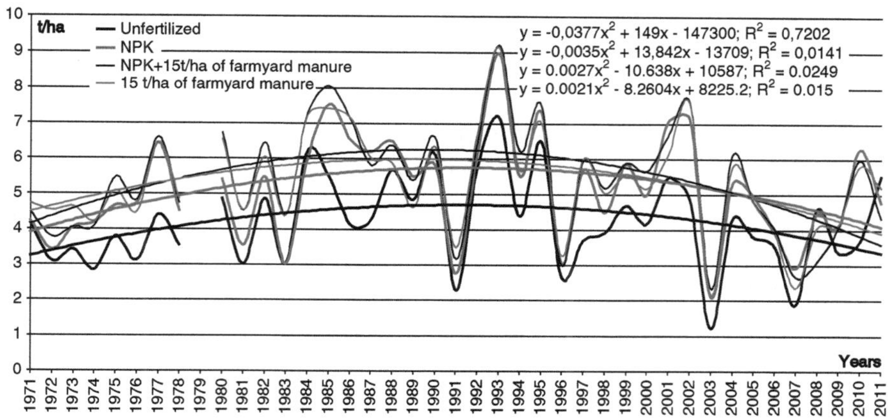

Fig. 21.2 Yields of winter wheat under different systems of fertilization, 1971–2011

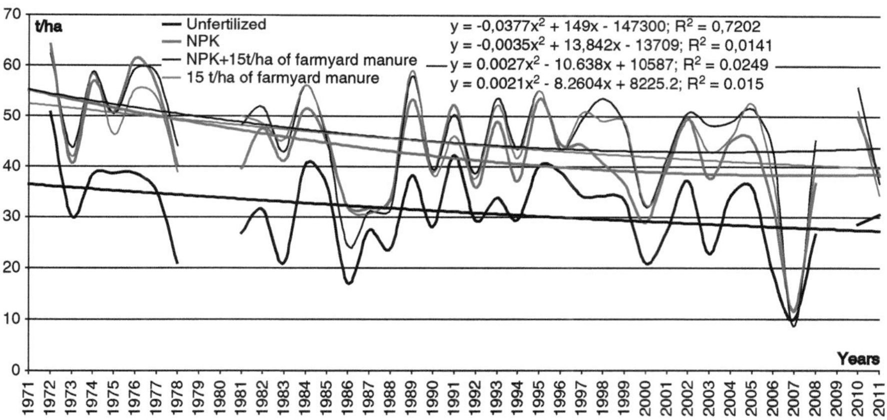

Fig. 21.3 Yields of sugar beet under different systems of fertilization, 1971–2011

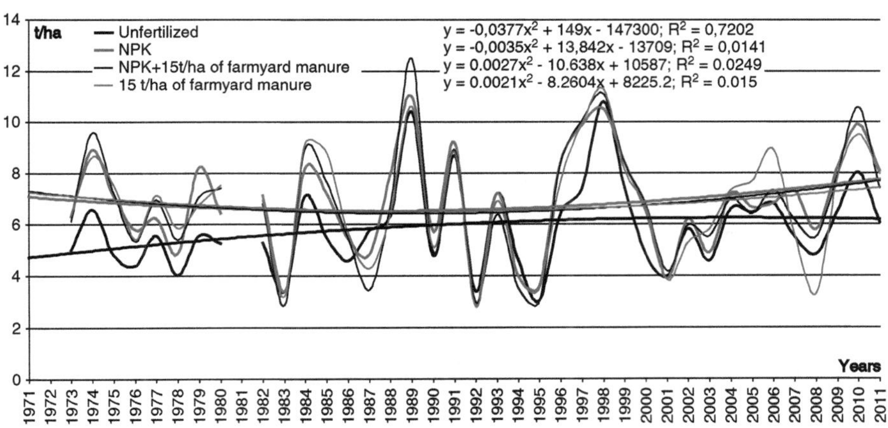

Fig. 21.4 Yields of maize grain under different systems of fertilization, 1971–2011

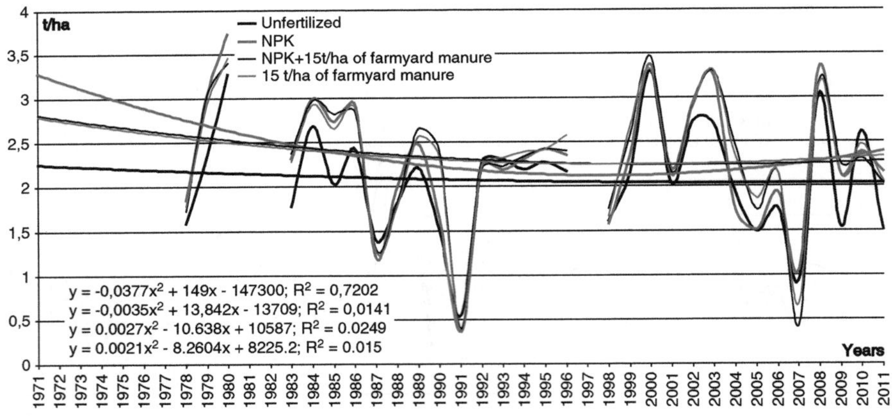

Fig. 21.5 Yields of sunflower under different systems of fertilization, 1971–2011

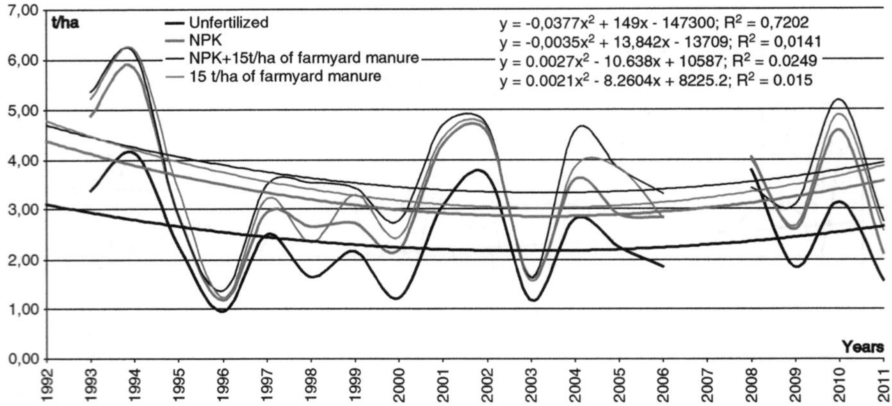

Fig. 21.6 Yields of spring barley under different systems of fertilization, 1971–2011

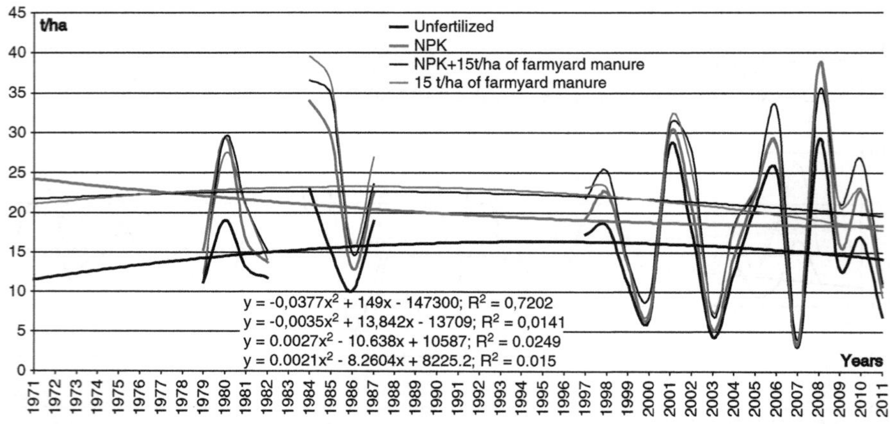

Fig. 21.7 Yields of vetch-and-oats for green mass under different systems of fertilization, 1971–2011

Table 21.3 Yield increase from fertilization (t/ha/%) and variation due to weather conditions (kv%), 1971–2011

Systems of fertilization in crop rotation	Crops																		Productivity of crop rotation		
	Winter wheat			Sugar beet			Maize for grain			Sunflower			Spring barley[a]			Vetch-and-oats green mass[b]					
	t/ha	±	kv %	t/ha	±	kv %	t/ha	±	kv %	t/ha	±	kv %	t/ha	±	kv %	t/ha	±	kv %	t c.u./ha	±	kv %
Unfertilized	4.22	–	30.1	31.1	–	26.0	5.9	–	28.9	2.1	–	29.8	2.4	–	38.6	15.4	–	45.2	4.5	–	29.0
NPK_2	5.15	0.93	28.4	43.1	12.0	23.0	6.8	0.9	28.6	2.3	0.2	31.7	3.2	+0.8	36.7	19.5	4.1	46.2	5.6	+1.1	27.4
		22.0			38.6			15.2			9.5			33.3			26.6			24.4	
10 t manure/ha + NPK_2	5.49	1.27	28.0	43.0	11.9	22.8	6.8	0.9	32.7	2.3	0.2	30.5	3.5	1.1	33.9	21.0	5.6	45.0	5.7	+1.2	27.8
		30.1			38.3			15.2			9.5			45.8			36.4			26.7	
15 t manure/ha + NPK_2	5.42	1.20	28.7	46.1	15.0	23.1	6.8	0.9	32.8	2.3	0.2	30.9	3.7	1.3	33.1	21.6	6.2	43.1	5.9	+1.4	29.1
		28.4			48.2			15.2			9.5			54.2			40.2			31.1	
15 t manure/ha	5.36	1.14	26.4	44.9	13.8	21.8	6.8	0.9	30.9	2.3	0.2	29.8	3.5	1.1	36.0	21.3	5.9	45.8	5.8	+1.3	27.9
		27.0			44.4			15.2			9.5			45.8			38.3			28.9	

kv% Coefficient of variation
[a]Average for 1993–2011
[b]Average for 1979–1987 and 1997–2011

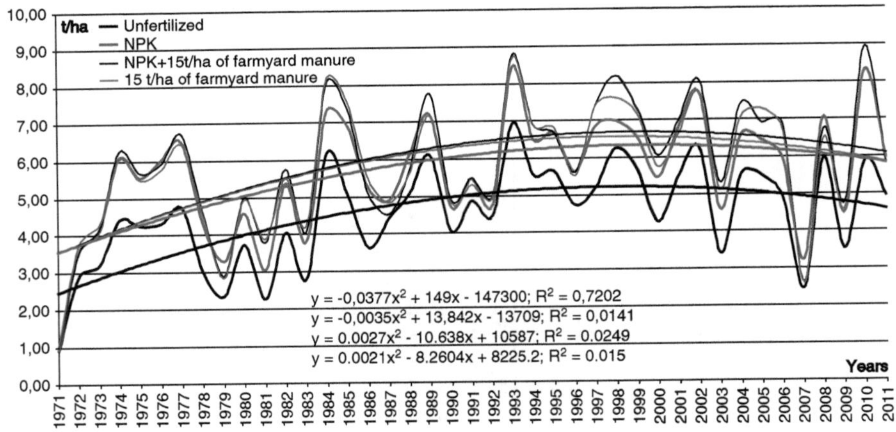

Fig. 21.8 Productivity of the crop rotation (tonne cereal units/ha) under different systems of fertilization, 1971–2011

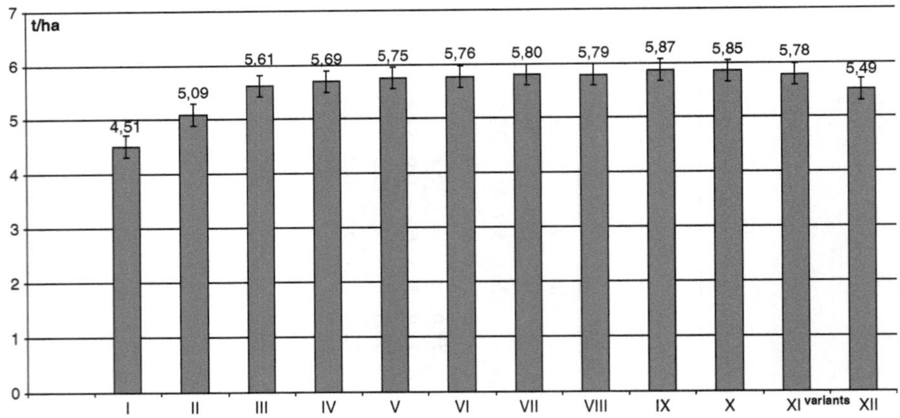

Fig. 21.9 Productivity of crop rotation (tonne cereal units/ha) under different systems of fertilization, average 1971–2011

suffered from the deeper layers. Stocks of total nitrogen have remained much the same under the influence of mineral fertilizers as on unfertilized plots, with a small gain (0.22 t/ha or 1.5 %) under the highest rates of fertilizer (Table 21.7). The data for fertilization solely with farmyard manure and for combined manure and mineral fertilizers do not show a consistent picture but, in all cases, application of manure greatly reduced nitrogen depletion. Combined organic and mineral fertilizers increased the stocks nitrogen in the soil profile relative to unfertilized plots by 3.17–5.14 t/ha (22–36 %), mainly in the 40–100 cm layer (69–100 %), less in the 0–40 cm layer (0–31 %) (Table 21.8).

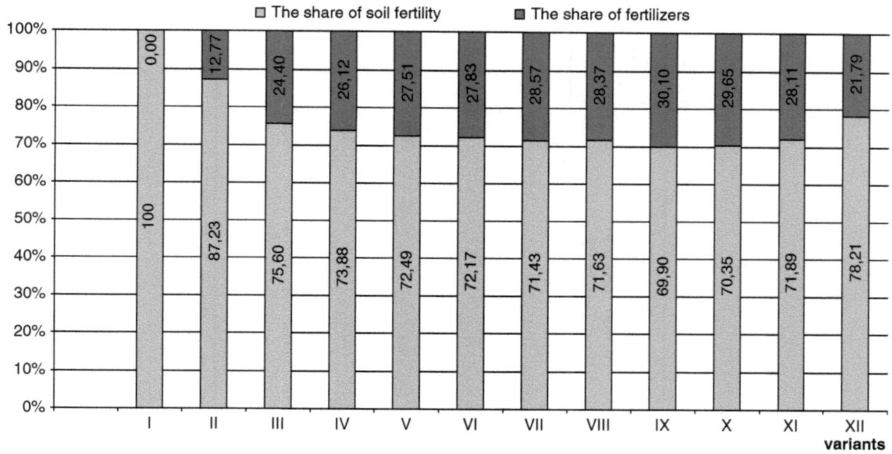

Fig. 21.10 The share of inherent soil fertility and fertilizers in the productivity formation of crop rotation, average for 1971–2011

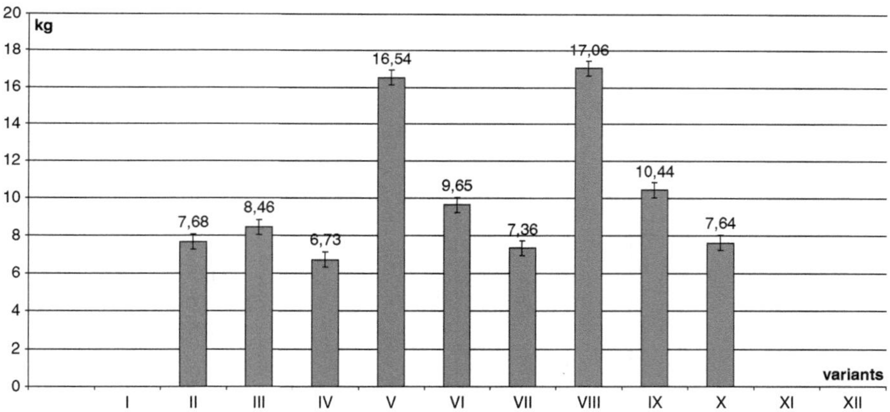

Fig. 21.11 Recovery of mineral fertilizers with the yield increase of crops in crop rotation (kg of cereal units per 1 kg of active ingredients of NPK), average for 1971–2011

For the period 1970–2009, annual losses of total nitrogen from the soil profile (0–100 cm) were from:

- Unfertilized plots – 189.5 kg/ha
- Fertilized with NPK – 188.7–195.1 kg/ha
- Fertilized with combined manure and mineral fertilizers – 57.7–108.2 kg/ha
- Fertilized only with manure – 76.1 kg/ha

The losses of nitrogen from the 0 to 40 cm layer relative to total losses from 0 to 100 cm (Table 21.9) were:

Table 21.4 Changes in soil organic matter stock (as carbon) under different systems of fertilization relative to the unfertilized control 2009, t/ha and %

Soil layers cm	Control	Increase relative to the control, ± t/ha and %[a]						
		NPK_1	NPK_2	NPK_3	15 t/ha farmyard manure + NPK_1	15 t/ha farmyard manure + NPK_2	15 t/ha farmyard manure + NPK_3	15 t/ha farmyard manure
0–20	50.6	+4.6	−2.0	−2.2	+0.7	+2.4	+3.3	+1.1
	100 %	9.1	3.9	4.3	1.4	4.7	6.5	2.2
20–40	55.2	−9.6	−7.7	−7.2	−0.5	−2.4	+1.9	−2.4
	100 %	17.4	13.9	13.0	0.9	4.3	3.4	4.3
40–60	31.7	−1.8	−7.0	−3.9	+20.8	+7.8	+14.8	+11.2
	100 %	5.7	22.1	12.3	65.6	24.6	46.7	35.3
60–80	22.4	−6.5	−6.5	−8.1	+3.3	+8.5	+14.5	+11.4
	100 %	29.1	29.0	36.2	14.7	38	64.7	50.9
80–100	16.6	−5.7	−3.1	−3.6	+9.1	+7.6	+6.0	+3.9
	100 %	34.3	18.7	21.7	54.8	45.8	36.1	23.5
0–100	176.5	−19.0	−26.3	−25.0	+33.4	+23.9	+40.5	+25.2
	100 %	10.8	14.9	14.2	18.9	13.5	23.0	14.3

[a]Increase relative to control, t/ha; increase relative to control, %

Table 21.5 Accumulation and loss of soil organic matter relative to control from 0 to 40 and 40 to 100 cm under different systems of fertilization, %, 2009

Soil layers cm	Fertilization						
	NPK_1	NPK_2	NPK_3	15 t/ha manure + NPK_1	15 t/ha manure + NPK_2	15 t/ha manure + NPK_3	15 t/ha manure
0–40	−26.3	−36.9	−37.6	+0.6	0	+12.8	0
40–100	−73.7	−63.1	−62.4	+99.4	+ 100.0	+87.2	+100.0

- Unfertilized plots – 2.69 t/ha (36.4 %)
- Fertilized with NPK – 3.23–3.51 t/ha (43.9–46.9 %)
- Fertilized with 15 t/ha manure+NPK – 1.29–2.71 t/ha (57.3–72.1 %)
- Fertilized with 15 t/ha manure 1.33 t/ha (44.8%).

21.3.4 *Carbon Budget Under Different Systems of Fertilization*

The soil organic carbon budget for the whole period of the experiment, 1971–2009 (Table 21.10), shows the lowest annual harvest of dry matter from unfertilized plots – 10.1 t/ha; the first increment (rate) of mineral fertilizers increased the yield by1.5 t/ha (14.8 %); the second and third higher rates increased the yield relative to the control by 3.5 and 3.2 t/ha (34.6 and 31.7 %), respectively. Farmyard manure applied at 15 t/ha of crop rotation together with the lowest rate of mineral fertilizer gave the same yield of dry matter as the highest rate of mineral fertilizers applied alone, an increase of 3.2 t/ha (31.7 %). Manure in combination with the second and third incremental rates of mineral fertilizer increased the yield of dry matter by 4.1 t/ha (40.6 %) and 4.8 t/ha (47.5 %), respectively, but the highest yield, an increase of 5.0 t/ha (49.5 %) was achieved on plots fertilized only with manure.

Carbon input from crop residues was calculated taking into consideration the ratio between above-ground crop production and the mass of root system using our own experimental data and generalized data from the literature (Boincean 1999). The amount of farmyard manure was calculated according the rates of manure applied at different periods. The carbon content of crop residues was taken as 40 %, and for composted farmyard manure 13.2 %. Carbon output was determined as the amount of carbon carried away as harvested crops together with the mineralization of soil organic matter from the soil profile (0–100 cm).

The balance was negative for all systems of fertilization. The greatest carbon deficit was under mineral fertilization: 3,527–4,233 kgC/ha/year. Under combined organic and mineral fertilization, the deficit was 1,861–2,444 kg C/ha/year, and for manure, alone, 2,976 kg C/ha/year. Unfertilized plots returned an intermediate position with a deficit of 2640 kgC/ha. The output of carbon was significantly higher than the input for all systems of fertilization in crop rotation and for both 0–20 and 0–100 cm soil layers. Mineral fertilizers increased the carbon deficit, especially for the deeper soil layers. However, with combined manure and NPK fertilizers, carbon accumulation was greater than mineralization in the subsoil.

Table 21.6 Changes in stocks of soil organic carbon (t/ha) under different systems of fertilization relative to the unfertilized control, 1970–2009

	Soil layers, cm													Total C losses and share between layers	
	0–20		20–40		40–60		60–80		80–100		0–100				
System of fertilization	2009	±	2009	±	2009	±	2009	±	2009	±	2009	±	Annual losses 0–100 cm kgC/ha	0–40 cm t/ha, %	40–100 cm t/ha %
Initial stocks (1970), t/ha	59.8		23.0		41.5		25.2		20.4		199.9				
Control (unfertilized)	50.6	−9.2	55.2	+2.2	31.7	−9.8	22.4	−2.8	16.6	−3.8	176.5	−23.4	−600.0	−7.0 29.9	−16.4 70.1
NPK_1	55.2	−4.6	45.6	−7.4	29.9	−11.6	15.9	−9.3	10.9	−9.5	157.5	−42.4	−1087.2	−12.0 28.3	−30.4 71.7
NPK_2	48.6	−11.2	47.5	−5.5	24.7	−16.8	15.9	−9.3	13.5	−6.9	150.2	−49.7	−1274.4	−16.7 33.6	−33.0 66.4
NPK_3	48.4	−11.4	48.0	−5.0	27.8	−13.7	14.3	−10.9	13.0	−7.4	151.5	−48.4	−1241.0	−16.4 33.9	−32.0 66.1
15 t/ha farmyard manure + NPK_1	51.3	−8.5	54.7	+1.7	52.5	+11.0	25.7	+0.5	25.7	+5.3	209.9	+10.0	+256.4	−6.8 –	+16.8 168.0
15 t/ha farmyard manure + NPK_2	53.0	−6.8	52.8	−0.2	39.5	−2.0	30.9	+5.7	24.2	+3.8	200.4	+0.5	+12.8	−7.0 –	+7.5 1500
15 t/ha farmyard manure + NPK_3	53.9	−5.9	57.1	+4.1	46.5	+5.0	36.9	+11.7	22.6	+2.2	217.0	+17.1	+438.5	−1.8 –	+18.9 110.5
15 t/ha farmyard manure	51.7	−8.1	52.8	−0.2	42.9	+1.4	33.8	+8.6	20.5	+0.1	201.7	+1.8	+46.1	8.3 –	+10.1 561.1

Table 21.7 Changes in soil profile stocks of total nitrogen under different systems of fertilization relative to the unfertilized control 2009, t/ha and %

		Increase relatively to the control (unfertilized), ± t/ha and %						
Soil layers cm	Control	NPK_1	NPK_2	NPK_3	15 t/ha manure + NPK_1	15 t/ha manure + NPK_2	15 t/ha manure + NPK_3	15 t/ha manure
0–20	3.82	+0.12	+0.25	−0.12	+1.11	+0.36	+0.87	+1.22
	100 %	3.1	6.5	3.1	29.1	9.4	22.8	31.9
20–40	4.56	−0.94	−0.79	−0.67	+0.29	−0.38	+0.14	+0.14
	100 %	20.6	17.3	14.7	6.3	8.3	3.1	3.1
40–60	2.47	+0.57	+0.29	+0.57	+2.03	+1.01	+1.46	+1.33
	100 %	23.1	11.7	23.1	82.2	40.9	59,1	53.8
60–80	1.90	0	+0.13	0	+0.86	+1.01	+1.58	+1.01
	100 %	–	6.8	–	45.3	53.2	83.2	53.2
80–100	1.46	+0.15	+0.15	0	+0.85	+1.17	+1.01	+0.72
	100 %	10.3	10.3	–	58.2	80.1	69.2	49.3
0–100	14.2	−0.10	+0.03	−0.22	+5.14	+3.17	+5.06	+4.42
	100 %	0.7	0.2	1.5	36.2	22.3	35.6	31.1

Table 21.8 Partitioning of accumulation and loss of total nitrogen relative to the control between soil layers 0–40 and 40–100 cm, 2009, tN/ha and %

Soil layers, cm		System of fertilization in crop rotation						
		NPK_1	NPK_2	NPK_3	15 t/ha + NPK_1	15 t/ha + NPK_2	15 t/ha + NPK_3	15 t/ha
0–40	t/ha	−0.82	−0.54	−0.79	+1.4	−0.02	+1.01	+1.36
	%				27.2	0	20.0	30.8
40–100	t/ha	+0.72	+0.57	+0.57	+3.74	+3.19	+4.05	+3.06
	%				72.8	100.0	80.0	69.2
0–100	t/ha	−0.1	+0.03	−0.2	+5.14	+3.17	+5.06	+4.42
	%	100	100	100	100	100	100	100

The coefficient of humification for farmyard manure was calculated as the share of accumulated carbon in soil profile to the amount of manure applied. For the whole soil profile (0–100 cm), the coefficient for manure applied with the lowest and highest rates of mineral fertilizers was 44.7 and 76.6 %, respectively. The apparent decrease at the middle rate of mineral fertilizers has yet to be explained. The coefficient of humification for farmyard manure applied alone was 14 % for the whole soil profile.

21.3.5 Nitrogen Balance and Nitrogen-Use Efficiency

Nitrogen balance and nitrogen-use efficiency were calculated for two periods of time in field No.2 (Tables 21.11 and 21.12). For inputs we have taken into consideration not only the nitrogen from mineral and organic fertilizers but also residual nitrogen from previous crops in the rotation (the difference between potential capacity of nitrogen mineralization by soil and the amount of nitrogen actually taken up by the real crop in the same year). The unused amount of nitrogen is carried over to the crop grown in the following year (although there is a risk of loss through leaching or volatilization), so we have subtracted the amount of nitrogen carried over from the amount of nitrogen taken up by the harvested crops.

Nitrogen-use efficiency was calculated as the ratio between the additional nitrogen taken up on fertilized plots relative to unfertilized plots (taking into consideration the amount of unused nitrogen carried over from the previous crop) as a percentage of the amount of nitrogen applied as mineral fertilizer (Table 21.11). Increased rates of mineral fertilizer decreased nitrogen-use efficiency from 31.4 down to 14.8 % However, the same rates of mineral fertilizer applied together with 15 t/ha of farmyard manure stabilized nitrogen-use efficiency at 20.0–25.7 % – and the highest value of nitrogen-use efficiency from mineral fertilizers (49.1 %) was achieved on the plots with, otherwise, only organic fertilizer. We should keep in mind that on plot with only manure treatment since 1990, mineral fertilizer was applied during the previous period of time (see data in Tables 21.1 and 21.2).

Table 21.9 Changes in stocks of total nitrogen under different systems of fertilization in crop rotation relative to the control 1970–2009, t/ha

System of fertilization	0–20		20–40		40–60		60–80		80–100		0–100		Annual losses 0–100 cm kgN/ha	Total losses and partitioning between layers t ha/%	
	2009	±	2009	±	2009	±	2009	±	2009	±	2009	±			
Initial stocks (1970) t/ha	5.64		5.43		4.19		3.44		2.90		21.6			0–40 cm	40–100 cm
Control (unfertilized)	3.82	−1.82	4.56	−0.87	2.47	−1.72	1.90	−1.54	1.46	−1.44	14.21	−7.39	−189.5	−2.69	−4.7
														36.4	65.6
NPK_1	3.94	−1.70	3.62	−1.81	3.04	−1.15	1.90	−1.54	1.61	−1.29	14.11	−7.49	−192.0	−3.51	−3.98
														46.9	53.1
NPK_2	4.07	−1.57	3.77	−1.66	2.76	−1.43	2.03	−1.41	1.61	−1.29	14.24	−7.36	−188.7	−3.23	−4.13
														43.9	56.1
NPK_3	3.70	−1.94	3.89	−1.54	3.04	−1.15	1.90	−1.54	1.46	−1.44	13.99	−7.61	−195.1	−3.48	−4.13
														45.7	54.3
15 t/ha farmyard manure + NPK_1	4.93	−0.71	4.85	−0.58	4.50	+0.31	2.76	−0.68	2.31	−0.59	19.35	−2.25	−57.7	−1.29	−0.96
														57.3	42.7
15 t/ha farmyard manure + NPK_2	4.18	−1.46	4.18	−1.25	3.48	−0.71	2.91	−0.53	2.63	−0.27	17.38	−4.22	−106.2	−2.71	−1.51
														64.2	35.8
15 t/ha farmyard manure + NPK_3	4.69	−0.95	4.70	−0.73	3.93	−0.26	3.48	+0.04	2.47	−0.43	19.27	−2.33	−59.7	−1.68	−0.65
														72.1	−27.9
15 t/ha farmyard manure	5.04	−0.60	4.70	−0.73	3.80	−0.39	2.91	−0.53	2.18	−0.72	18.63	−2.97	−76.1	−1.33	−1.64
														44.8	55.2

Table 21.10 Soil organic carbon budget under different systems of fertilization in crop rotation, 1971–2009, Field No.2

	Yield, dry matter/year		C input, kg/ha/year			C output, kg/ha/year			C balance, kg/ha/year		Coefficient of humification for manure, %	
						Mineralization losses						
System of fertilization	t/ha	Increase, t/ha, %	As crop residues	As manure	Total	0–20 cm	0–100 cm	Off-take at harvest	0–20 cm	0–100 cm	0–20 cm	0–100 cm
Control (unfertilized)	10.1	–	2,000		2,000	−236	−600	−4,040	−2,276	−2,640	–	–
NPK_1	11.6	+1.5 14.8	2,200	–	2,200	−118	−1,087	−4,640	−2,558	−3,527	–	–
NPK_2	13.6	+3.5 34.6	2,430	50.8	2,481	−287	−1,274	−5,440	−3,246	−4,233	0	0
NPK_3	13.3	+3.2 31.7	2,510	50.8	2,561	−292	−1,241	−5,320	−3,051	−4,000	0	0
15 t/ha manure + NPK_1	13.3	+3.2 31.7	2,630	573	3,203	−218	+256	−5,320	−2,335	−1,861	0	44.7
15 t/ha manure + NPK_2	14.2	+4.1 40.6	2,650	573	3,223	−174	+13	−5,680	−2,631	−2,444	0	2.3
15 t/ha manure + NPK_3	14.9	+4.8 47.5	2,660	573	3,233	−151	+439	−5,960	−2,878	−2,288	0	76.6
15 t/ha manure	15.1	+5.0 49.5	2,690	328	3,018	−208	+46	−6,040	−3,230	−2,976	0	14.0

Table 21.11 Nitrogen-use efficiency under different systems of fertilization in the crop rotation, average for 1973–2008, Field No.2

		N inputs, kg/ha				Annual mineralization losses, kgN/ha			N balance kg/ha/year		Nitrogen-use efficiency		
System of fertilization	Yield of rotation, t/ha dry matter	As mineral fertilizer	As manure	From previous crop	Total	0–20 cm	0–100 cm	N uptake by crops kg/ha/year	0–20 cm	0–100 cm	N off-take by harvest less N from previous crop, kg/ha	Additional N uptake, fertilized plots,kg/ha	N-use efficiency from mineral fertilizer, %
Control	10.1	–	–	7.4	7.4	46.7	189.5	94.3	−133.6	−276.4	86.9	–	–
NPK_1	11.6	25.8	–	25.4	51.2	43.6	192.1	120.4	−112.8	−261.3	95	8.1	31.4
NPK_2	13.6	58.3	5.5	33.1	96.9	40.3	188.7	135.2	−78.6	−22.7	102.1	15.2	26.1
NPK_3	13.3	71.7	5.5	37.9	115.1	49.7	195.1	135.4	−70.0	−215.4	97.5	10.6	14.8
15 t/ha manure + NPK_1	13.3	42.5	62	40.0	114.5	18.2	57.7	135.4	−9.1	−48.6	95.4	8,5	20.0
15 t/ha manure + NPK_2	14.2	59.2	62	39.1	160.3	37.4	108.2	141.2	−18.3	−89.1	102.1	15.2	25.7
15 t/ha manure + NPK_3	14.5	75.8	62	38.6	176.4	24.4	59.7	142.3	+9.7	−25.6	103.7	16.8	22.2
15 t/ha manure	15.1	53.3	35.5	31.3	120.1	15.4	76.1	144.4	−39.7	−100.4	113.1	26.2	49.1

Table 21.12 Nitrogen-use efficiency under different systems of fertilization in crop rotation, average for 1991–2008, Field No.2

System of fertilization	Yield of rotationt/ha dry matter	N inputs, kg/ha				Annual mineralization losses, kg N/ha		N uptake by crops kg/ha/year	N balance kg/ha/year		N off-take by harvest less N from previous crop, kg/ha	AdditionalN uptake, fertilized plots, kg/ha	N-use efficiency from mineral fertilizer, %
		As mineral fertilizer	As manure	From previous crop	Total	0–20 cm	0–100 cm		0–20 cm	0–100 cm			
Control	10.1	–	–	14.3	14.3	46.7	189.5	89.4	−121.8	−264.6	75.1	–	–
NPK_1	11.9	30	–	39.0	69.0	43.6	192.1	114.5	−119.1	−267.6	75.5	+0.4	1.3
NPK_2	12.3	50	–	44.6	94.6	40.3	188.7	118.9	−114.6	−263	74.3	–	0
NPK_3	13.0	70	–	46.5	116.5	49.7	195.1	123.1	−126.3	−271.7	76.6	+1.5	2.1
15 t/ha manure + NPK_1	13.0	30	14.2	52.7	96.9	18.2	57.7	123.1	−88.6	−128.1	70.4	–	0
15 t/ha manure + NPK_2	14.3	50	14.2	51.7	115.9	37.4	108.2	129.9	−115.6	−186.4	78.2	+3.1	6.2
15 t/ha manure + NPK_3	14.0	70	14.2	53.7	137.9	24.4	59.7	127.9	−98.6	−133.9	74.2	–	0
15 t/ha manure	14.4	–	14.2	42.2	56.4	15.4	76.1	132.4	−105.6	−166.3	90.2	+15.1	–

Surprisingly, nitrogen-use efficiency decreased during the second period of the crop rotation (1991–2009) when an additional field was included, making it a six-field instead of five-field rotation (Table 21.12). Increasing crop diversity allowed more nitrogen to be carried over from previous crops and, on the unfertilized plots, the amount of nitrogen carried over doubled from 7.4 to 14.3 kgN/ha. Previously, we have demonstrated that the lower the diversity of crops, the higher is the efficiency of mineral fertilizers (Boincean 1999, 2004). This is a crucial issue for sustainable farming systems, and transdisciplinary cooperation is needed to elucidate it. In view of the escalating cost of mineral fertilizer and the risks of soil degradation and pollution of groundwater, the other practical problem for farmers is to match the applied rates of mineral nitrogen with the soil's own capacity to provide nitrogen for crops. This depends on several factors that are different in different farms and for different fields. Therefore, we recommend comparative trials on fertilized and unfertilized strips in each farm.

21.4 Conclusions

1. Crop yields have shown diverging trends over 40 years: with all systems of fertilization, yields of winter wheat increased initially but, during the last 15–20 years, stagnated or even decreased. Yields of sugar beet and vetch-and-oats decreased. Sunflower, maize for grain and spring barley yields decreased initially but have increased in recent years. The productivity of the whole crop rotation under different systems of fertilization initially increased but, in recent years, has stagnated or even declined.
2. In the case of winter wheat, the gain under different systems of fertilization (22–30 %) was equivalent to the fluctuations in yield caused by weather conditions (26–30 %). This tendency held for the crop rotation as a whole although sugar beet and spring barley responded more to mineral fertilizers and maize for grain and sunflower rather less.
3. For most systems of fertilization, inherent soil fertility was responsible for 70–76 % of crop yields; the greatest share of soil fertility (87 %) was under the lowest rate of mineral fertilizers.
4. Recovery of mineral fertilizers by harvested crops decreased with increasing rates of fertilizer application. The lowest rates of mineral fertilizers providing the highest recovery when applied together with organic fertilizers.
5. Over a 38-year period, mineral fertilizers contributed to a reduction of the stocks of soil organic matter over the whole soil profile of 19.0–26.3 tC/ha (11–15 %) relative to unfertilized plots. Combined mineral and organic fertilizers contributed to an accumulation of soil organic carbon of 23.9–40.5 tC/ha (13.5–23 %) relative to unfertilized plots – but mainly in soil layers deeper than 40 cm. In absolute terms, the annual losses of soil organic carbon from 0 to 100 cm soil layer were 1,087.2–1,274.4 kg/ha under mineral fertilization and 600 kg/ha for unfertilized plots. Over the same period, manure and combined

organic and mineral fertilizers increased soil organic carbon by 12.8–438.5 kg/ha for the whole soil profile.

6. Relative to the unfertilized control, stocks of total nitrogen were unchanged under mineral fertilization, whereas manure and combined mineral and organic fertilizers increased the amount of total nitrogen for the whole profile by 3.17–5.14 t/ha (22–36 %). Losses of total nitrogen from the 0 to 40 cm layer were compensated by accumulation in the soil layer 40–100 cm under mineral fertilization, and under organic and combined mineral and organic fertilization, relative accumulation of total nitrogen of 3.06–4.05 t/ha (69–100 %) was mainly in the subsoil.
7. In absolute terms, annual losses of total nitrogen from the 0 to 100 cm profile during 39 years were 188.7–195.1 kgN/ha under mineral fertilization, much the same as for the unfertilized plots (189.5 kgN/ha). Annual losses of nitrogen from the soil profile were less with manure and combined mineral and organic fertilizers at 57.7–108.2 kgN/ha. Losses occurred mainly from the 0 to 40 cm soil layer under organo-mineral fertilization (57–72 %) and mainly from 40 to 100 cm under mineral fertilization (53–66 %).
8. For the period 1973–2008, nitrogen-use efficiency decreased with increased rates of mineral fertilizers (from 31 down to 15 %) but the combination of manure and mineral fertilizers stabilized N-use efficiency in the range 20–26 %. N-use efficiency was drastically reduced by increasing the diversity of crops in crop rotation during 1991–2008. The greater the diversity of crops in crop rotation, the greater is the capacity of soil to meet the crop's nitrogen requirements – so less is the need for nitrogen from mineral fertilizers. The potential capacity of soil to provide crops with nitrogen has to be evaluated experimentally in each field, but this will much reduce the expense of mineral fertilizers and, simultaneously, the losses of nitrogen through leaching and volatilization.

Acknowledgements We thank all the research, technical and field staff involved at different times in the long-term field experiment and supporting laboratory studies. Special thanks to Miss Lidia Bulat, who carried out many of the laboratory analyses.

References

Boincean BP (1999) Ecological agriculture in the Republic of Moldova (crop rotation and soil organic matter). Stiinta, Chisinau (Russian)

Boincean BP (2004) Crop rotation and yields of field crops on chernozem soils of Moldova. In: Losacov VG (ed) Crop rotation in modern agriculture, materials for the international conference. K Timiriazev Moscow Agricultural Academy, Moscow, pp 43–49 (Russian)

Boincean BP. Nica LT (2007) Productivity, fertilization and fertility of chernozem soil in the steppe zone of Moldova. In: de Neve S, Salomez J (eds) Mineral versus organic fertilization. Conflict or synergies? Proceeding of the 16th international symposium of the International Scientific Centre of Fertilizers, CIEC, Ghent, 16–19 September, pp 102–109

Boincean BP, Nica LT, Stadnic SS (2010) Nitrogen leaching for the field crops in the Balti steppe. Akademos 1:16–98 (Romanian)

Dokuchaev VV (1952) Russian chernozem, vol 2, Report for the Economic Society. State Publisher of Agricultural Literature, Moscow, 634 p (Russian)
Drinkwater LE, Schipanski M, Snap SS, Jackson LE (2008) Ecologically based nutrient management. In: Snapp S, Pound B (eds) Agricultural systems. Agroecology and rural innovation for development. Elsevier, Amsterdam, pp 161–210
Gliessman SR (2000) Agroecology. Ecological processes in sustainable agriculture. Lewis, Boca Raton
Khan SA, Mulveney RL, Ellsworth TR, Boast CW (2007) The myth of nitrogen fertilization for soil carbon sequestration. Environ Qual 36:1821–1832
Kosticev PA (1949) Soils in chernozem regions of Russia. Their provenance, composition and characteristics. State Publisher of Agricultural Literature, Moscow (Russian)
Mulveney RL, Khan SA, Ellsworth TR (2009) Synthetic nitrogen fertilizers deplete soil nitrogen: A global dilemma of sustainable cereal production. Environ Qual 38:2295–2314
Ponomareva VV, Plotnicova TA (1980) Humus and soil formation. Science, Leningrad
Powlson DS, Jenkinson DS, Johnstan AF et al (2010) Comments on 'Synthetic nitrogen fertilizers deplete soil nitrogen: a global dilemma for sustainable cereal production' by Mulvaney and others. J Environ Qual 39:749–752
Seiter S, Horwath W (2004) Strategies for managing soil organic matter to supply plant nutrients. In: Magdoff F, Weil R (eds) Soil organic matter in sustainable agriculture. CRC Press, Boca Raton, pp 269–293

Chapter 22
Long-Term Field Experiment with Irrigation on the Balti Chernozem

B.P. Boincean, M.P. Martea, A.I. Ungureanu, and P.M. Hropotinschi

Abstract The long-term field experiment with irrigation on Typical chernozem on the Balti steppe of Moldova shows a bigger response from winter wheat than from sugar beet. Over 42 years, wheat yields initially increased, stabilized, and, then, decreased. Short-straw wheat varieties outperformed medium-straw varieties under optimum irrigation and fertilization but yielded less under rain-fed and unfertilized conditions. Similarly, newer hybrids of sugar beet offered advantages only under optimum management.

Even with supplementary irrigation, rainfall provides 67 and 68 % of the water used by fertilized wheat and sugar beet, respectively, and 89–90 % of the water used by the same crops without fertilizer. Native soil fertility contributes 79–80 % of yield formation in irrigated and nonirrigated wheat and 61 % in the case of sugar beet.

The stock of soil organic matter decreased on all experimental plots; irrigation did not affect the rate carbon loss in unfertilized conditions but accelerated the degradation of soil organic matter when combined with fertilizer. Under rain-fed winter wheat, the combined annual losses of nitrogen from the soil profile and unused nitrogen from applied fertilizer was twice the amount of nitrogen fertilizer applied – and three times higher under irrigation. There is also serious wastage of nitrogen under sugar beet.

22.1 Introduction

Inadequate soil water curtails the yields of most field crops in Moldova (Mihalcevschi 1980). By calculation from the mean annual precipitation 1968–2011 and the water consumption per unit of production, yields of winter wheat and sugar beet should be 4.0 and 34.6 t/ha, respectively. But yields fluctuate dramatically because precipitation

---

B.P. Boincean (✉) • M.P. Martea • A.I. Ungureanu • P.M. Hropotinschi
Selectia Research Institute for Field Crops, Balti, Republic of Moldova
e-mail: bboincean@gmail.com

D. Dent (ed.), *Soil as World Heritage*, DOI 10.1007/978-94-007-6187-2_22,

is not distributed regularly; autumn rainfall over the same period ranged from 25 to 307 mm, during spring from 42 to 213 mm, and during summer from 82 to 391 mm. Therefore, supplementary irrigation can make a big difference. Sustainability requires economically efficient production without damage to the environment and the stability of the rural community. Long-term field experiments enable evaluation of the effects of irrigation on both crop yields and soil fertility – information which is critical to plans for increasing the area under irrigation in the near future.

22.2 Experimental Site and Method

The long-term field experiment at Balti, in the north of Moldova, is conducted on Typical chernozem. Mean annual precipitation recorded at Selectia RIFC for the farming years 1968–2011 was 577 mm with a seasonal distribution of 128 mm (22 %) in autumn, 96 mm (17 %) in winter, 128 mm (22 %) in spring, and 226 mm (39 %) during summer. Most of the rain falls during the warm parts of the year but drought is common, whether in spring, summer, or autumn. Mean annual temperature is 8 °C; the average for the warmest month (July) is 19.5 °C and for the coldest (January) minus 4.5 °C; the absolute maximum and minimum temperatures are 41 °C and minus 34 °C. The average length of the growing season is 260–270 days and the frost-free period 165 days, with fluctuations of 20–25 % for these periods. The accumulated temperature above 10 °C is 2,700–3,000 °C.

The six-field experimental crop rotation is 3 years of lucerne followed by winter wheat, sugar beet, and maize-for-grain; each field is 3 ha. Three systems of fertilization are compared: without fertilization, farmyard manure applied at 80 t/ha for the sugar beet crop, and farmyard manure plus mineral fertilizers – $N_{60}P_{90}K_{40}$ for wheat and $N_{70}P_{90}K_{60}$ for sugar beet. The experiment has four replicates; each experimental plot is 200–400 m^2 and 50–100 m^2 for yield measurements.

During the growing season, the soil water content is maintained at 50–80 % of field capacity. For wheat, irrigation is applied at different rates before and after sowing and during critical periods of development. For sugar beet, irrigation is scheduled for the three main periods of development, with rates of 300 up to 900 m^3/ha, different regimes maintaining soil water at 60, 70, and 80 % of field capacity. The source is a semi-artesian well, pumped from a depth of 117 m.

22.3 Results and Discussion

Here, we focus on the productivity of two main crops in the rotation and changes in soil fertility under the combined influence of irrigation and fertilization.

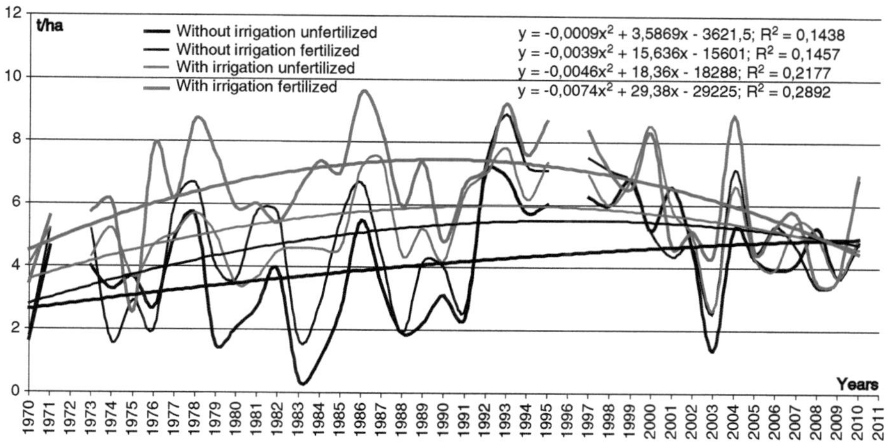

Fig. 22.1 Winter wheat yields under irrigation and fertilization, 1970–2011

22.3.1 Winter Wheat

Figure 22.1 depicts the yields of winter wheat, which increased in the early years, stabilized, then decreased. In this experiment, wheat follows a good predecessor, lucerne in its third year after the first cut; even so, the yield increase under fertilization was quite high: 1.13 t/ha (21 %) on irrigated plots and 0.81 t/ha (20 %) on rain-fed plots (Table 22.1). Irrigation increased yields on unfertilized plots by 1.31 t/ha (32 %) and by 1.6 t/ha (33 %) on fertilized plots so the contribution to yields of native soil fertility was about 80 % on both irrigated and rain-fed plots.

Supplementary irrigation contributes more to yield formation than fertilizer – 67 % on fertilized plots and 68 % on unfertilized plots – so the importance of soil organic matter, rainfall, and stocks of soil water remain crucial to arable farming. Building up humus and soil structure increases the sustainability of agriculture and lessens dependence on costly inputs of mineral fertilizer and irrigation.

Even on irrigated plots, climate drives much of the fluctuation of crop yield. The coefficient of variation (CoV) on plots without irrigation was 43.8 % on unfertilized plots and 39.3 % on fertilized plots; under irrigation, CoV on unfertilized and fertilized plots was still 25.2–26.5 %. Although there was no clear correlation between wheat yields and precipitation in the September to March and April to July periods, inadequate soil water was limiting at some periods; irrigation applied when soil water was less than 60 % of field capacity brought about a substantial yield increase. Table 22.2 shows that the yield increase under irrigation was greater than 1 t/ha in half of the years of the experiment.

In these dry years, the yield increase from fertilizer was 1.25 t/ha on rain-fed plots and 0.91 t/ha on irrigated plots, whereas irrigation increased yields by 2.41 t/ha (82 %) on unfertilized plots and 2.75 t/ha (72 %) on fertilized plots. In the other, wetter years, the response to fertilization and irrigation was less, although higher

Table 22.1 Yields of winter wheat in rotation after lucerne under fertilization and irrigation, means, 1970–2010

		Rain-fed		Irrigated		± from fertilizer		± from irrigation	
Indices		Unfertilized	Fertilized	Unfertilized	Fertilized	Rain-fed	Irrigated	Unfertilized	Fertilized
Wheat yield	t/ha	4.14	4.95	5.45	6.58	+0.81	+1.13	+1.31	+1.63
	%					20	21	32	33
Coefficient of variation, %		43.8	39.3	25.2	26.5				
Share of soil fertility and natural precipitation in yield formation, %						80	79	68	67

Table 22.2 Yields of winter wheat in rotation after lucerne in years with dry spells (when irrigation increased yield by >1 t/ha): 1970, 1974, 1976, 1979–1980, 1983–1991, 1995, 2000, 2003–2004, 2007

		Rain-fed		Irrigated		± from fertilization		± from irrigation	
Index		Unfertilized	Fertilized	Unfertilized	Fertilized	Rain-fed	Irrigated	Unfertilized	Fertilized
Yields of short-straw varieties of winter wheat	t/ha	2.93	3.84	5.34	6.59	+0.91	+1.25	+2.41	+2.75
	%					31.0	32.5	82.2	71.6

from fertilization on irrigated plots (+0.73 t/ha – 14 %) and from irrigation on fertilized plots (+0.62 t/ha – 11.5 %).

Different varieties of winter wheat respond differently to fertilizers and irrigation (Tables 22.3 and 22.4). For short-straw varieties, the yield increase attributable to fertilizer was 1.5 t/ha (19 %) under irrigation and 0.79 t/ha (21.5 %) under rain-fed conditions; for medium-straw varieties the increase was 0.89 t/ha (16 %) and 0.77 t/ha (20 %). Short-straw varieties have the advantage under irrigation, but they are more demanding of nutrients and need care in restoration of soil fertility. On average for 1986–1992 and, especially, in dry years, the short-straw varieties produced the lowest yields under rain-fed conditions: medium-straw varieties are better adapted to unfavorable soil water and nutrient regimes because of their more vigorous root systems.

22.3.2 Sugar Beet

Compared with winter wheat, sugar beet was more responsive to fertilizer but less responsive to irrigation. Figure 22.2 depicts sugar beet yields under irrigation and fertilization in the long-term field experiment: after the first few years, the trend has been steadily downwards. On average for 1968–2009, the yield increase attributable to fertilizer was 16.4 t/ha on irrigated plots and 14.7 t/ha on rain-fed plots, in both cases about 39 % (Table 22.5). Yield increase under irrigation was 5.7 t/ha on fertilized plots and 4.0 t/ha on unfertilized plots (in both cases about 11 %). Fertilization reduced the fluctuation of yields under changeable weather conditions.

Irrigation increases beet yields by more than 5 t/ha about 1 year in three – when dry spells occurred during the critical part of the growing season (Table 22.6). In these years, the benefit from fertilization on irrigated and rain-fed plots was 14.4 t/ha (27 %) and 14.1 t/ha (37 %), respectively, and the yield increase from irrigation on fertilized and unfertilized plots was 9.9 t/ha (19 %) and 9.6 t/ha (25 %). The most efficient rates of irrigation were those less than 500 m^3/ha (Table 22.7); the small gains from heavier applications did not cover the additional cost.

Table 22.8 summarizes the response to fertilizers of different varieties and hybrids under irrigation over the period 1995–2011. The hybrid Vilia outyielded the standard varieties by 1.42 and 2.34 t/ha on unfertilized and fertilized plots, respectively. Newer hybrids (Ghiocel, ICCC-25, and Scorpion) did not outperform Vilia on unfertilized plots, but Ghiocel and ICCC-25 yielded well on fertilized plots; the same applied to yields in dry years (Table 22.9). In short, hybrids can increase yields but only under good crop management.

Table 22.3 Response of different varieties of winter wheat to irrigation and fertilization, means t/ha, 1986–1992

	Rain-fed		Irrigated		± from fertilization		± from irrigation	
Variety	Unfertilized	Fertilized	Unfertilized	Fertilized	Rain-fed	Irrigated	Unfertilized	Fertilized
Short-straw varieties:	3.67	4.46	5.95	7.10	+0.79	+1.15	+2.28	+2.64
Piticul (1986–1988)					21.5 %	19 %	62 %	59 %
Dnestreanca (1989–1991)								
Belceanca 5 (1992)								
Medium-straw varieties:	3.91	4.69	5.67	6.56	+0.77	+0.89	+1.76	+1.87
MPC 82/04 (1986)					20 %	16 %	45 %	40 %
Eritrospermum 127 (1987–1991)								
Scorospelii mold. (1992)								
± from short- to medium-straw varieties	–0.24	–0.23	+0.28	+0.54	+0.02	+0.26	+0.52	+0.77

Table 22.4 Response of different varieties of winter wheat to irrigation and fertilization in dry years, t/ha

	Rain-fed		Irrigated		± from fertilization		± from irrigation	
Variety	Unfertilized	Fertilized	Unfertilized	Fertilized	Rain-fed	Irrigated	Unfertilized	Fertilized
Short-straw varieties:	2.33	3.27	5.82	7.52	+0.94	+1.70	+3.49	+4.25
Piticul (1986–1988)					40 %	29 %	150 %	130 %
Dnestreanca (1989–1991)								
Belceanca 5 (1992)								
Medium-straw varieties:	3.20	3.91	5.65	6.73	+0.71	+1,08	+2,45	+2.82
MPC 82/04 (1986)					22 %	19 %	77 %	72 %
Eritrospermum 127 (1987–1991)								
Scorospelii mold. (1992)								
± from short- to medium-straw varieties	–0.87	–0.64	+0.17	+0.79	+0.23	+0.62	+1.04	+1.43

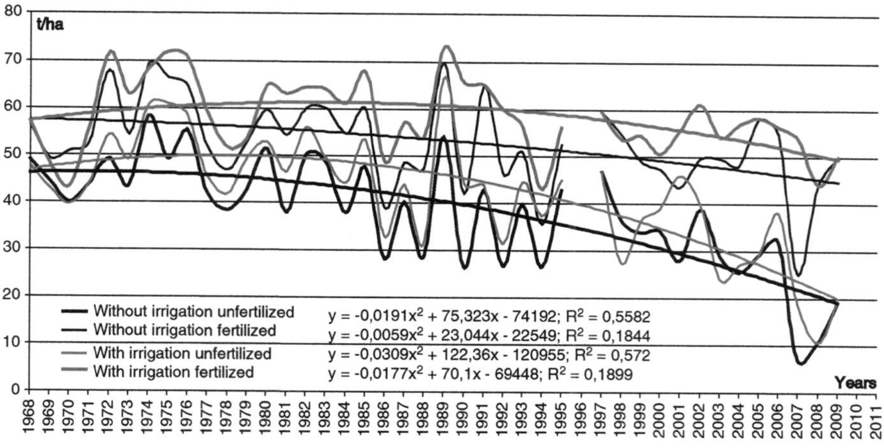

Fig. 22.2 Sugar beet yields under irrigation and fertilization, 1968–2011

22.3.3 Changes in Stocks of Soil Organic Carbon and Total Nitrogen

The stock of soil organic matter decreased on all experimental plots (Table 22.10). Under rain-fed conditions, the losses from the whole soil profile over 42 years were 24.9 tC/ha from unfertilized plots and 11.8 tC/ha from fertilized plots. Losses under irrigation were 24.2 and 18.7 tC/ha from unfertilized and fertilized plots, respectively. Irrigation made no difference to losses of soil organic matter from unfertilized plots but increased the rate of loss fertilized plots.

The content of total nitrogen was lower under irrigation than in the rain-fed plots. This was reflected in a wider C/N ratio, especially in the deeper soil layers (Table 22.11), and was most evident in plots receiving farmyard manure as well as mineral fertilizers. Moreover, irrigation increased the content of the labile fractions of soil organic matter, especially in fertilized plots (Table 22.12).

Stocks of total nitrogen decreased over the 41 years of the experiment, especially from irrigated plots with fertilization (Tables 22.13 and 22.14). Under rain-fed conditions, losses of total nitrogen from the whole soil profile amounted to 2.95 t/ha from fertilized plots and 4.41 t/ha from unfertilized plots. Under irrigation, the losses were 4.47 from unfertilized plots and 5.59 t/ha from fertilized plots.

Nitrogen use efficiency (NUE) was calculated as the ratio of fertilizer nitrogen taken up by the crop to the total amount of nitrogen applied as fertilizer (Olk et al. 1999). For winter wheat and sugar beet, the only crops to which fertilizer was applied, nitrogen was lost not only from soil but also from the applied fertilizers. NUE was 49 % for rain-fed winter wheat and 62 % for the irrigated crop; total potential losses from rain-fed plots for the period 1973–2010 were 4 220 kgN/ha (114 kgN/ha/year) under fertilization and almost as much from unfertilized plots. Under irrigation, potential annual losses amounted to 176.9kgN/ha. These losses

Table 22.5 Yields of sugar beet in rotation under the influence of fertilization and irrigation, means 1968–2009

Indices		Rain-fed		Irrigated		± from fertilization		± from irrigation	
		Unfertilized	Fertilized	Unfertilized	Fertilized	Rain-fed	Irrigated	Unfertilized	Fertilized
Yield	t/ha	38.0	52.7	42.0	58.4	+14.7	+16.4	+4.0	+5.7
	%					39	39	11	11
Coefficient of variation, %		58.9	17.7	29.3	13.0				
Contribution of soil fertility, soil water, and rainfall to yield formation, %						61	61	89.5	89

Table 22.6 Yields of sugar beet in rotation under fertilization and irrigation in dry years (1972–1973, 1975, 1979, 1981–1982, 1984–1985, 1989–1990, 1994, 2001, and 2007)

		Rain-fed		Irrigated		± from fertilization		± from irrigation	
Index		Unfertilized	Fertilized	Unfertilized	Fertilized	Rain-fed	Irrigated	Unfertilized	Fertilized
Yields of sugar beet	t/ha	38.5	52.6	48.1	62.5	+14.1	+14.4	+9.6	+9.9
	%					37	27	25	19

Table 22.7 Influence of irrigation rates on the yields of sugar beet, t/ha, means 1968–2009

Rates of irrigation, m^3/ha	Fertilization		± from fertilization on irrigated plots	± from irrigation	
	Unfertilized	Fertilized		Unfertilized	Fertilized
>500	44.8	58.1	+13.3	–	–
500–100	45.4	61.2	+15.8	+0.6	+3.1
<1,000	44.5	63.2	+18.7	–0.3	+5.1

Table 22.8 Response of irrigated varieties and hybrids to fertilization, means t/ha, 1995–2011

Variety, hybrid	Fertilization		± from fertilization	± from the hybrids relatively to the variety	
	Unfertilized	Fertilized		Unfertilized	Fertilized
Victoria (1995–2008)	35.85	56.72	+20.87	–	–
Barracuda (2009–2011)					
Vilia (1995–2011)	37.27	59.06	+21.79	+1.42	+2.34
Ghiocel (1998–2007)	35.59	63.88	+28.29	+0.08	+5.41
ICCC-25 (2008)					
Scorpion (2009–2011)	35.51	58.47	+22.96	–	–

Table 22.9 Response of irrigated varieties and hybrids in dry years (1998, 1999, 2007–2010), t/ha

Variety, hybrid	Fertilization		± from fertilization	± from the hybrids relative to varieties	
	Unfertilized	Fertilized		Unfertilized	Fertilized
Victoria (1995–2008)	27.71	52.54	+24.83	–	–
Barracuda (2009–2011)					
Vilia (1995–2011)	29.04	55.56	+26.52	+1.33	+3.02
Vilia (1998–2011)					
Ghiocel (1998–2007)	27.89	60.86	+32.97	+0.51	+5.71
ICCC-25 (2008)					
Scorpion (2009–2011)	27.38	55.15	+27.77	–	–
relative to Vilia					

are 2–3 times the amount of nitrogen applied as fertilizer. In the case of sugar beet, NUE was 38 % on rain-fed plots and 42 % under irrigation. For rain-fed crops, annual losses of nitrogen were about the same as the amount of nitrogen applied as mineral fertilizer and manure (194.5 kgN/ha); losses from irrigated plots were 1.3 times greater. Fertilizer nitrogen not used by crops, together with nitrogen losses from the soil, may be leached to the groundwater or volatilized, particularly in the late autumn and early spring when evapotranspiration is less than precipitation and snowmelt (Boincean et al. 2003) (Tables 22.15, 22.16, and 22.17).

Table 22.10 Changes in the stocks of soil organic matter under fertilization and irrigation, 1968–2010, Field no.1

Soil layer, cm	tC/ha 1968	tC/ha 2010				± tC/ha			
		Without irrigation		With irrigation		Without irrigation		With irrigation	
		Unfertilized	Fertilized	Unfertilized	Fertilized	Unfertilized	Fertilized	Unfertilized	Fertilized
0–20	70.1	62.6	64.1	62.6	66.2	−7.5	−6.0	−7.5	−3.9
20–40	58.8	59.5	61.6	58.8	58.0	+0.7	+ 2.8	0	−0.8
40–60	46.3	40.8	43.3	39.7	41.6	−5.5	−3.0	−6.6	−4.7
60–80	28.8	22.4	27.2	23.5	23.2	−6.4	−1.6	−5.3	−5.6
80–100	20.2	14.0	16.2	15.4	15.6	−6.2	−4.0	−4.8	−3.7
Total, t/ha	224.2	199.3	212.4	200.0	205.5	−24.9	−11.8	−24.2	−18.7
% change from initial stock						11.1	5.3	10.8	8.3
Annual losses, t/ha of carbon						0.59	0.28	0.58	0.45

Table 22.11 Effect of fertilization and irrigation on the content of total carbon and total nitrogen in the long-term field experiment with irrigation, 2010, Field no.1

Soil layer, cm	Without irrigation						With irrigation					
	Unfertilized			Fertilized			Unfertilized			Fertilized		
	C %	N %	C/N	C %	N %	C/N	C %	N %	C/N	C %	N %	C/N
0–10	2.79	0.235	11.9	2.85	0.274	10.4	2.62	0.224	11.7	2.86	0.235	12.2
10–20	2.43	0.229	10.6	2.50	0.235	10.6	2.60	0.218	11.9	2.67	0.213	12.5
20–30	2.41	0.218	11.0	2.60	0.224	11.6	2.54	0.218	11.6	2.44	0.218	11.2
30–40	2.18	0.201	10.8	2.14	0,201	10.6	1.98	0.201	9.8	2.02	0.168	12.0
40–50	1.65	0.168	9.8	1.75	0,179	8.9	1.68	0.185	9.1	1.71	0.179	9.5
50–60	1.33	0.157	8.5	1,41	0,151	9.3	1.23	0.134	9.2	1.33	0.134	9.9
60–70	0.88	0.140	6.3	1,06	0,145	7.3	0.93	0.117	7.9	0.88	0.129	6.8
70–80	0.72	0.095	7.6	0,89	0,129	6.9	0.75	0.112	6.7	0.78	0.084	9.3
80–90	0.59	0.089	6.6	0,60	0,089	6.7	0.57	0.101	5.6	0.61	0.078	7.8
90–100	0.41	0.078	5.3	0,56	0,089	6.3	0.54	0.089	6.1	0.58	0.078	7.4

Table 22.12 Content of total and labile carbon under irrigation and fertilization, Field no. 5, 2009

		1968			2009											
					Without irrigation						With irrigation					
					Unfertilized			Fertilized			Unfertilized			Fertilized		
Soil layer, cm		C total	Water-soluble	Soluble in 0.1 M NaOH	C total	Water-soluble	Alkali-soluble	C total	Water-soluble	Alkali-soluble	C total	Water-soluble	Alkali-soluble	C total	Water-soluble	Alkali-soluble
0–20	t/ha	66.7	0.86	1.35	59.0	0.73	1.39	66.0	0.88	1.24	59.3	0.72	0.86	62.9	1.00	1.00
	%	100	1.3	2.0	100	1.2	2.4	100	1.3	1.9	100	1.2	1.5	100	1.6	1.6
20–40	t/ha	57.2	0.69	1.23	54.3	0.55	0.87	65.8	0.97	1.11	62.6	0.78	0.94	65.0	1.09	1.72
	%	100	1.2	2.2	100	1.0	1.6	100	1.3	1.7	100	1.2	1.5	100	1.7	2.6
40–60	t/ha	47.9	0.67	1.11	35.5	0.85	1.02	50.9	0.68	1.36	44.8	0.50	0.67	44.5	1.17	1.51
	%	100	1.4	2.3	100	2.4	2.9	100	1.3	2.7	100	1.1	1.5	100	2.7	3.4
60–80	t/ha	25.2	0.50	0.76	23.2	0.34	0.34	33.0	0.68	0.85	21.5	0.50	0.67	22.1	0.50	1.00
	%	100	2.0	3.0	100	1.5	1.5	100	2.1	2.6	100	2.3	3.1	100	2.3	4.6
Mean	t/ha	49.25	0.68	1.11	43.0	0.62	0.90	53.9	0.78	1.14	47.1	0.62	0.78	48.6	0.94	1.30
	%	100	1.4	2.2	100	1.4	2.1	100	1.5	2.1	100	1.3	1.7	100	1.9	2.7

Table 22.13 Changes in the stocks of total nitrogen under irrigation and fertilization 1968–2009, Field no.1

	1968		2009							
			Without irrigation				With irrigation			
			Unfertilized		Fertilized		Unfertilized		Fertilized	
Soil layer cm	%	t/ha	%	t/ha	%	t/ha	%	t/ha	%	t/ha
0–20	0.27	6.36	0.23	5.57	0.25	6.10	0.22	5.30	0.22	5.38
20–40	0.23	6.01	0.21	5.43	0.21	5.51	0.21	5.43	0.19	5.02
40–60	0.19	5.32	0.16	4.44	0.17	4.52	0.16	4.36	0.16	4.27
60–80	0.16	4.42	0.12	3.28	0.14	3.84	0.11	3.19	0.11	2.97
80–100	0.2	3.30	0.08	2.32	0.09	2.49	0.10	2.66	0.08	2.18
Total		25.41		21.04		22.46		20.94		19.82

Table 22.14 Losses of soil total nitrogen (t/ha) under fertilization and irrigation 1968–2009, Field No. 1

	Rain-fed		Irrigated	
Soil layer cm	Unfertilized	Fertilized	Unfertilized	Fertilized
0–20	−0.79	−0.26	−1.06	−0.98
20–40	−0.58	−0.50	−0.58	−0.99
40–60	−0.92	−0.80	−0.96	−1.05
60–80	−1.14	−0.58	−1.23	−1.45
80–100	−0.98	−0.81	−0.64	−1.12
Total losses	−4.41	−2.95	−4.47	−5.59
Annual losses	−0.108	−0.072	−0.109	−0.136

Table 22.15 Nitrogen use efficiency of rain-fed and irrigated winter wheat and sugar beet 1968–2010

	Rain-fed					Irrigated				
	N applied, kg/ha					N applied, kg/ha				
Crop	Mineral fertilizer	Manure	Total	N uptake in extra yield kg/ha	NUE %	Mineral fertilizer	Manure	Total	N uptake in extra yield kg/ha	NUE %
Winter wheat	67.3	–	67.3	33.0	49.0	67.3	–	67.3	41.5	61.7
Sugar beet	84.5	110.5	195.0	73.7	37.8	84.5	110.5	195.0	81.8	42.0

Table 22.16 Losses of nitrogen from soil and fertilizers under winter wheat, kg/ha 1973–2010

	Rain-fed				Irrigated			
	Not	Fertilized			Not	Fertilized		
Indices	fertilized	Mineral	Manure	Total	fertilized	Mineral	Manure	Total
Total N input as fertilizers	250	2,490	–	2,490	250	2,490	–	2,490
Nitrogen use efficiency, %				49				61.7
Fertilizer N not used by crop				1,270				953.7
Nitrogen losses from soil profile	4,410			2,950	4,470			5,590
Total N losses from fertilizer and soil	4,410			4,220	4,470			6,543.7
Annual N losses from soil and fertilizers	119.2			114	120.8			176.9

Table 22.17 Losses of nitrogen from soil and fertilizers under sugar beet, kg/ha 1968–2009

	Rain-fed				With irrigation			
	Not	Fertilized			Not	Fertilized		
Indexes	fertilized	Mineral	Manure	Total	fertilized	Mineral	Manure	Total
Total N input as fertilizers	925	3,465	4,530	7,995	925	3,465	4,530	7,995
Nitrogen use efficiency, %				37.8				42.0
Fertilizer N not used by crop				4,973				4,637
Nitrogen losses from soil profile	4,410			2,950	4,470			5,590
Total N losses from fertilizer and soil	4,410			7,923	4,470			10,227
Annual N losses from soil and fertilizers	107.6			193.2	109.0			249.4

22.4 Conclusions

1. Winter wheat responds well to fertilizer and irrigation. On average over 40 years, fertilization increases yield by 20 % and irrigation increases yield by 32–33 %. However, in both rain-fed and irrigated plots, native soil fertility contributes about 80 % of yield formation. Even in plots with supplementary

irrigation, soil moisture supplied by natural precipitation contributes 67–68 % of crop water use. However, irrigation moderates the variability of yields; CoV was 25–26 % under irrigation compared with 39–44 % under rain-fed conditions.

2. Yields of winter wheat rose during the early years of the experiment, stabilized, and have declined over the last 15–20 years – although yields have continued to increase on unfertilized, rain-fed plots. New, short-straw varieties respond better to irrigation but do less well under rain-fed conditions, especially under drought.
3. Over the 43 years, yields of sugar beet have declined on all experimental plots. Sugar beet responds more to fertilization (38–39 %) than to irrigation (11 %); rates of irrigation less than 500 m^3/ha are the most efficient. Fertilization significantly moderates the fluctuation of rain-fed yields. New hybrids are more productive on irrigated and fertilized plots but offer no advantage under rain-fed and unfertilized conditions.
4. Fertilization decreases the loss of soil organic matter under rain-fed conditions but not under irrigation. Annual losses from the soil profile amount to 0.58–0.59 tC/ha on unfertilized plots, 0.28 tC/ha on fertilized rain-fed plots, and 0.45 tC/ha on fertilized, irrigated plots. Irrigated, fertilized plots exhibited the greatest amounts of mobile soil organic matter.
5. Fertilized, irrigated plots suffered the highest annual losses of total nitrogen – 0.136 t/ha. Nitrogen use efficiency by winter wheat was 49 % on rain-fed and 62 % under irrigation and for sugar beet, 38 and 42 %, respectively. Under winter wheat, total losses of nitrogen from soil profile together with unused nitrogen from fertilizers amounted to two-to-three times the amount applied as fertilizer. For sugar beet, the losses are equal to the amount of fertilizers applied on rain-fed plots but 1.3 times higher on irrigated plots.

Acknowledgements The authors are sincerely grateful for the care and efforts of many researchers, technicians, field, and laboratory staff over the last 43 years, on which this account depends.

References

Boincean BP, Nica L, Bugaciuc M et al (2003) Long-term productivity and fertility of arable chernozem soils of Moldova. In: Boincean BP (ed) Sustainable agriculture: results, problems and perspectives. Proceedings of the international conference, State University of Balti, Balti (Romanian), 21–22 June, pp 21–22

Mihalcevschi VD (1980) Irrigation of field crops in Moldavia. Ştiinţa, Chisinau (Russian)

Olk DC, Cessman KG, Simbahan G (1999) Interpreting fertilizer-use efficiency in relation to soil nutrient-supplying capacity, factor productivity, and agronomic efficiency. In: Nutrient cycling in agroecosystems. Kluwer, Dordrecht, pp 35–41

Chapter 23
Quality of Soil Organic Matter Under Crop Rotations and Continuous Cultures

B.P. Boincean, L.I. Bulat, M.A. Bugaciuc, M. Cebotari, and V.V. Cuzeac

Abstract Humification of organic matter in the soil plays many parts. Classical chemical fractionation of soil organic matter differentiates between different kinds of soil but throws no light on the effects of husbandry within any one soil type. Therefore, to evaluate the quality of the soil organic matter in an arable chernozem, we have applied indices of the water-soluble and alkali-soluble fractions of soil organic matter, coefficient of extinction and C:N ratio. Results indicate that, under arable, an insufficient supply of fresh organic matter is contributing to the progressive loss of humus, leaching of the water-soluble fraction and other negative consequences.

23.1 Introduction

A soil's organic matter is an integrating index of fertility but, in evaluating soil quality, we should consider qualitative changes of composition of the soil organic matter as well as its total amount. Classical chemical fractionation of soil organic matter reveals differences between soil types (Kononova 1963; Orlov 1974; Alexandrova 1980; Licov et al. 1981, 2004) but does not distinguish the effects of different agronomic practices on the quality of soil organic matter in any one type of soil.

Soil organic matter is an amalgam of fractions at different stages of decomposition, but the direction and intensity of humification is determined by the *labile fraction*. Earlier, we investigated several indicators of the quality of soil organic matter (Licov et al. 1981, 2004; Boincean 1999). In this chapter we apply a few of these indicators to evaluate soil organic matter quality in the Balti chernozem under the Selectia long-term field experiments.

B.P. Boincean (✉) • L.I. Bulat • M.A. Bugaciuc • M. Cebotari • V.V. Cuzeac
Selectia Research Institute of Field Crops, Balti, Republic of Moldova
e-mail: bboincean@gmail.com

D. Dent (ed.), *Soil as World Heritage*, DOI 10.1007/978-94-007-6187-2_23,

23.2 Materials and Methods

The design of the Selectia long-term field experiment with different crop rotations and continuous cultures has already been described. Here we use the contents of water-soluble soil organic matter and the fraction soluble in 0.1 M NaOH as indicators of soil organic matter quality under contrasting regimes in the long-term experiment. A key point is that analyses were carried out on the whole extract, without the further denaturing involved in the conventional procedure for separating humic and fulvic acids. The content of carbon in the alkaline extract was determined at different stages of the growing season using the Tiurin spectrophotometric procedure. The coefficient of extinction, which indicates the capacity of dissolved humic substances to decrease the intensity of light penetrating through a 1 cm layer of the alkaline extract, was determined for the whole visible spectrum. Following Kononova (1963), we take it that the *younger* the humic substances, the lower are the coefficients of extinction, and the *older* the humic components, the higher are the coefficients of extinction. This regularity is determined by the dominance of more condensed aromatic carbon compounds in *older* humic substances and more aliphatic components in *younger* humic substances.

23.3 Results and Discussion

Tables 23.1 and 23.2 present data on content of labile organic matter under meadow, crop rotations, continuous wheat, continuous maize and black fallow in the 0–20 cm and 20–40 cm layers, respectively. Both layers show similar trends, although these are more pronounced in the surface layer.

The lowest amount of both water-soluble and alkali-soluble organic matter was found in both fertilized and unfertilized plots under continuous maize, even compared with black fallow. The coefficient of extinction for alkali solution was highest in the unfertilized plot of continuous maize.

The labile fraction comprises aliphatic and low-phenol components that are most active in stabilizing soil structure and which are, also, readily decomposed to yield soluble plant nutrients. Loss of labile soil organic matter means loss of such favourable agronomic characteristics as tilth, permeability and capacity to supply crops with nutrients. Such soils are perceptibly more compact and we may suppose that they will need greater applications of mineral fertilizers in order to achieve the same yield as those with more labile organic matter. This is proved by the long-term field experiments which demonstrate that the efficiency of mineral fertilizers is higher for continuous monocultures, maize in particular, than in crop rotations. The same holds for continuous winter wheat, although compared with continuous maize, continuous wheat provides more labile soil organic matter and the labile fraction contains more phenolic components.

Table 23.1 Content of labile soil organic matter (Percentage of water-soluble and alkali-soluble carbon is calculated relative to the total organic carbon content) under crop rotations and continuous cultures, Field no.6 2011, soil layer 0–20 cm. Numbers in parentheses identify the rotation

	Meadow		Black fallow		Continuous crops				Crop rotations					
									Level of saturation with row crops, %					
					Winter wheat		Maize-for-grain		40	50	60			70
Indices	Unfert.	Fert.	Unfert.	Fert.	Unfert.	Fert.	Unfert.	Fert.	Fertil. (no.5)	Fertil. (2)	Unfert. (7)	Fertil. manure 8 t/ha (3)	Fertil. manure 12 t/ha (4)	Fertil. (1)
Water-soluble organic	307	430	307	368	276	246	123	184	367	214	214	520	306	397
matter (mg/kg, %)	9.6	12.5	13.3	15.8	11.0	8.0	4.9	6.5	13.6	8.6	8.4	20.0	11.5	15.1
Alkali soluble	491	799	338	430	399	461	307	368	734	642	367	551	428	612
(mg/kg, %)	15.3	23.3	14.7	18.2	15.9	15.0	12.2	13.1	27.2	25.9	14.5	21.3	16.1	23.2
Coefficient of extinction	4.4	4.1	4.5	4.3	4.1	5.0	6.8	5.9	5.7	5.0	4.9	4.9	5.9	4.6

Table 23.2 Content of labile soil organic matter under crop rotations and continuous cultures, Field no.6 2011, soil layer 20–40 cm. Numbers in parentheses identify the rotation

	Meadow		Black fallow		Continuous crops				Crop rotations					
									Level of saturation with row crops, %					
					Winter wheat		Maize grain		40	50	60			70
Indices	Unfert.	Fert.	Unfert.	Fert.	Unfert.	Fert.	Unfert.	Fert.	Fertil. (no.5)	Fertil. (2)	Unfert. (7)	Fertil. manure 8 t/ha (3)	Fertil. manure 12 t/ha (4)	Fertil. (1)
Water-soluble organic matter (mg/kg, %)	246 8.5	307 11.0	245 10.9	307 13.9	368 14.9	307 10.4	123 4.9	184 7.0	122 4.6	184 7.7	184 8.0	367 14.5	275 10.9	214 8.4
Alkali-soluble (mg/kg, %)	461 15.9	584 21.0	368 16.5	307 13.9	307 12.4	399 13.6	215 8.7	307 11.7	428 16.4	459 19.2	336 14.6	367 14.5	367 14.5	489 19.2
Coefficient of extinction	4.6	4.5	4.2	5.0	4.4	5.4	6.2	6.5	5.5	6.6	6.0	6.6	6.6	6.1

Table 23.3 Content of labile soil organic matter, mg/kg and % (Percentage water-soluble and alkali-soluble carbon is calculated relative to the total organic carbon in each soil layer) in soil layers 0–100 cm under crop rotations with black fallow (no.2) and with lucerne (no.5) in the Selectia long-term field experiments, 2011, Field no.5

	Soil layers, cm									
Indices	0–10	10–20	20–30	30–40	40–50	50–60	60–70	70–80	80–90	90–100
Crop rotation no.2										
Water-soluble soil organic matter	60	330	150	420	180	180	360	120	178	179
	2.4	14.1	6.5	18.6	8.9	11.8	27.0	13.8	23.1	25.8
Alkali-soluble soil organic matter	570	810	630	690	475	330	480	300	118	118
	23.3	34.6	27.2	30.5	23.6	21.7	36.0	34.4	15.3	17.1
Crop rotation no.5										
Water-soluble soil organic matter	178	267	178	59	178	118	59	59	59	59
	6.1	9.5	6.9	2.8	13.1	9.2	5.0	7.2	12.3	12.9
Alkali-soluble soil organic matter	297	534	356	178	475	178	119	178	59	59
	10.2	19.1	13.9	8.5	35.1	13.9	10.1	21.7	12.3	12.9

The contents of water-soluble soil organic matter on unfertilized plots in crop rotation no.7 and rotation no.2 (with black fallow and the lowest rates of farmyard manure) are similar to, though higher than, under continuous maize. We suppose that a low input of fresh organic matter leads to more intensive destruction of the existing soil organic matter, especially when higher rates of mineral nitrogen are applied; black fallow is very destructive of soil organic matter (see also Table 23.3 in respect of the whole soil profile).

Crop rotation no.5, with perennial legumes, and rotations with higher rates of farmyard manure (1, 3 and 4) all have enhanced levels of water-soluble soil organic matter. However, each has a different content of alkali-soluble organic matter; the highest amounts are found in rotation no.5, which incorporates perennial legumes, and no.4 which receives the highest application of farmyard manure (12 t/ha). The coefficients of extinction for the labile organic fraction are also highest under these rotations, which indicate greater return of fresh organic matter to the soil – accompanied by much improved soil structure.

The content of labile soil organic matter under crop rotations is closer to that under meadow than to continuous cereals and black fallow. However, meadow and black fallow exhibit a similar structure of the labile fraction as indicated by the coefficient of extinction for the alkali-soluble fraction which is, in both cases, lower than under continuous cereals and much lower than under crop rotations. However, the directions of transformation of soil organic matter are opposite: under black fallow the humus being lost by decomposition, whereas under meadow, humus is accumulating.

Table 23.3 illustrates changes in the quality of the labile fraction of soil organic matter down the soil profile to a depth of one metre under crop rotation no.2 (with

Table 23.4 C/N ratios of labile organic matter in crop rotations and continuous cereals, soil layer 0–20 cm, Field no.6, 2009

	Meadow		Black fallow		Continuous cultures				Crop rotations					
					Winter wheat		Maize-for-grain		Level of saturation with row crops, %					
												Fertilized		
Index	Unfert.	Fertil.	Unfert.	Fertil.	Unfert.	Fertil.	Unfert.	Fertil.	40 (no.5)	50 (no.2)	Unfert. (no.7)	(no.3)	(no.4)	70 (no.1)
C %	3.19	3.43	2.30	2.33	2.50	3.06	2.52	2.81	2.70	2.48	2.53	2.59	2.66	2.63
												8 t/ha	12 t/ha	
N %	0.269	0.308	0.207	0.218	0.207	0.263	0.196	0.224	0.241	0.211	0.222	0.239	0.229	0.245
C/N	11.8	11.1	11.1	10.7	12.1	11.6	12.8	12.5	11.2	11.7	11.4	10.8	11.6	10.7

Table 23.5 C/N ratios under crop rotations and continuous cultures, soil layer 20–40 cm, Field no.6, 2009

	Meadow		Black fallow		Continuous cultures				Crop rotations					
					Winter wheat		Maize-for-grain		Level of saturation with row crops, %					
											60			
												Fertilized		
Index	Unfert.	Fertil.	Unfert.	Fertil.	Unfert.	Fertil.	Unfert.	Fertil.	40 (no.5)	50 (no.2)	Unfert. (no.7)	8.0 (no.3)	12.0 (no.4)	70 (no.1)
C %	2.88	3.08	2.23	2.26	2.47	2.93	2.47	2.62	2.61	2.39	2.30	2.52	2.52	2.55
N %	0.246	0.246	0.185	0.201	0.190	0.252	0.185	0.218	0.224	0.198	0.201	0.226	0.218	0.217
C/N	11.7	12.5	12.0	11.2	13.0	11.6	13.3	12.0	11.6	12.1	11.4	11.1	11.6	11.7

black fallow and low rates of farmyard manure) and rotation no.5 (which includes 30 % of perennial legumes). Under the rotation with black fallow, soil organic matter (especially the alkali-soluble fraction) is more mobile than in crop rotation with perennial legumes. Speaking about higher mobility, we mean a higher content of *young* humic substances.[1] The water-soluble fraction of soil organic matter is also more mobile under rotation no.2, except in the upper 30 cm. We deduce that compared with black fallow, the perennial legumes in the crop rotation are arresting the leaching of labile soil organic matter. This matters to farmers because loss of labile organic matter stimulates further decomposition of all the soil organic matter. Insufficient input of fresh organic matter intensifies soil degradation, and nitrates and pesticides can be leached through the soil to the groundwater; in short, the less-favourable the conditions for soil organic matter transformation in the topsoil, the greater the danger for soil degradation and pollution.

Tables 23.4 and 23.5 present total organic carbon and total nitrogen contents and the C/N ratio in soil layers 0–20 cm and 20–40 cm under different management. The highest contents of carbon and nitrogen are found under meadow. The lowest content of carbon is found under black fallow; however, the content of nitrogen in unfertilized black fallow was the same as the unfertilized plots of winter wheat (0.21 %) and total nitrogen was even lower in unfertilized plots under maize-for-grain (0.20 %). The C/N ratio was higher under continuous cereals (12.1–12.8), except for fertilized continuous winter wheat (which had the lowest content of total nitrogen).

Because fertilized plots of continuous cereals now have the lowest contents of soil organic matter, they present better opportunities for the accumulation of soil organic matter than plots under crop rotations. The lowest content of carbon in crop rotations is in unfertilized plots (no.7) and in the rotation with black fallow (no.2). The situation is the same in respect of total nitrogen. Both are determined by the turnover of soil organic matter, which is a complex process, and we need more detailed research to understand just what is going on.

23.4 Conclusions

1. The content of labile soil organic matter is higher under crop rotations than under continuous cereals, especially on fertilized plots. Coefficients of extinction are higher under crop rotations and continuous cereals than under meadow and black fallow. The highest value (under continuous maize-for-grain) indicates accelerated decomposition of soil organic matter; decomposition is also severe under black fallow.

[1] Research is needed to further establish the agronomic significance of the ratio of soluble and mobile humic substances in the total amount of extracted organic components, e.g. in respect of the availability of nitrogen for crops.

2. The various indicators of soil organic matter quality should be interpreted with care, considering the simultaneous quantitative changes in soil organic matter. For instance, black fallow and meadow have different contents of alkali-soluble soil organic matter but similar values for the coefficient of extinction, which actually indicate opposite directions of the humification–decomposition processes.
3. Insufficient addition of fresh organic matter to the soil is contributing to more-intensive decomposition of soil organic matter throughout the whole soil profile, which has many negative ecological consequences.

References

Alexandrova LN (1980) Soil organic matter and processes of its transformation. Science, Leningrad (Russian)
Boincean BP (1999) Ecological agriculture in the Republic of Moldova (crop rotation and soil organic matter). Stiinta, Chisinau (Russian)
Kononova MM (1963) Soil organic matter. Academy of Science of USSR, Moscow (Russian)
Licov AM, Cernicov VA, Boincean BP (1981) Evaluation of soil organic matter through the characteristics of its labile fraction. Izvestia TSHA 5:65–70
Licov AM, Esikov AI, Novikov MN (2004) Soil organic matter of arable non-chernozem soils. VNIIPTIOU Academy of Agricultural Sciences of Russia, Moscow (Russian)
Orlov DS (1974) Humic acids in soils. Lomonosov Moscow State University, Moscow (Russian)

Chapter 24
Soil Organic Matter and Soil Microbial Biomass in the Balti Long-Term Experiments

S. Corcimaru, G.H. Merenuic, and B.P. Boincean

Abstract The long-term field experiments at Balti have been used to examine and to relate the effects of different agricultural management practices on soil organic matter (SOM) content, soil microbial biomass, and microbial activity. Compared with SOM, microbial biomass carbon, basal respiration, and both microbial and metabolic quotients were found to be more sensitive to changes in soil quality and management practices. Soil microbial parameters were depressed by conventional crop rotations, with and without applications of manure and fertilizer. In contrast, ecological crop rotations with organic amendments increased SOM and stimulated microbial activity. The observed equilibrium between SOM and soil microbial parameters can be disturbed by crop rotations and/or amendments that increase SOM content without comparable stimulation of soil microbial activity.

The results can be used to develop tools for soil quality monitoring and sustainability assessment in arable systems. Increase in SOM, of itself, may not indicate improvement in soil quality: a combination of positive SOM balance with favorable changes in soil microbial parameters presents better evidence of sustainability of arable cropping systems.

24.1 Introduction

Soil organic matter (SOM) has been suggested as the single most important indicator of soil quality and productivity (Gregorich et al. 1997). For any given soil and climate, the amount of SOM is determined by land use and management, but the usefulness of SOM data for soil quality monitoring is constrained by the difficulty of

S. Corcimaru (✉) • G.H. Merenuic
ASM Institute of Microbiology and Biotechnology, Chisinau, Republic of Moldova
e-mail: sergiu@cc.acad.md

B.P. Boincean
Selectia Research Institute for Field Crops, Balti, Republic of Moldova

D. Dent (ed.), *Soil as World Heritage*, DOI 10.1007/978-94-007-6187-2_24,

experimentally verifying changes over short periods. Long-term experiments would seem to offer a solution to this problem (Christensen and Johnston 1997).

Soil microbial biomass (soil organisms with a volume less than 5,000 μm^3) accounts for only 1–3 % of total soil organic carbon but it plays a key role in soil forming processes, not least in SOM transformations. It has been suggested as an integrated measure of soil quality; changes in microbial biomass could be used for early predictions of changes in SOM brought about by various management practices (Dalal 1998; Winding et al. 2005). Seeking to contribute to the development of effective tools for soil quality and sustainability assessment in arable cropping systems, we have taken advantage of the Balti long-term field experiments, in the north of the Republic of Moldova, to examine and relate the effects of different agricultural management practices on SOM content and on the soil microbial biomass and its activity.

24.2 Experimental Site and Methods

The Balti long-term field experiments are established on moderately humified Typical chernozem (CSRM 1999 and Ursu and others in this volume)[1]; under the relatively undisturbed conditions of a shelterbelt, $pH_{(water)}$ was 7.26, SOM 6.91 %, microbial biomass carbon 639.9 μg/g, and basal respiration rate 0.81μgCO_2-C/g soil/ha (Merenuic et al. 2009). The study included experimental plots under continuous black fallow, winter wheat, and maize (all started in 1965); plots with continuous winter barley, sugar beet, and sunflower (all started in 1984); as well as plots with the same crops grown in three 10-field conventional crop rotations (rotations 3, 5, and 7, started in 1962) and two 7-field ecological crop rotations (1 and 3, started in 1989). The essential difference between ecological and conventional rotations is that the ecological rotation includes perennial legumes and grasses whereas the conventional rotation comprises only annual crops, although these include one field of vetch and oats for green mass. In each case, there were subplots with and without application of farmyard manure and mineral fertilizers and an additional subplot with application of farmyard manure only (in the ecological crop rotations). The initial SOM content was 5.65, 4.74, and 4.52 % for the experiments started in 1962–1965, 1984, and 1989, respectively. For a more detailed description of the experiments, see Boincean (1999 and Chap. 19 in this volume).

Soil samples were collected in 2011, immediately before spring sowing. Soil was sampled from a depth of 0–20 cm; passed through a 2 mm sieve and plant material, stones, and visible organisms removed manually; adjusted to 40 % of field capacity; and pre-incubated for 14 days at 25 °C in the dark and in aerated plastic bags (to prevent accumulation of CO_2) with periodic adjustment of moisture (to prevent drying). Soil organic carbon (SOC) was determined by dichromate oxidation followed by back titration of the excess dichromate (Nelson and Sommers 1982). Microbial biomass carbon (MB-C) was determined by substrate-induced

[1] Haplic chernozem in the *World Reference base or soil resources 2006.*

respiration (SIR) (Corcimaru 2009): three-gram soil samples were placed in vials and 0.5 ml of a substrate solution with glucose (36.0 mg/ml), K_2HPO_4 (3.6 mg/ml), and $(NH_4)_2SO_4$ (3.6 mg/ml) added dropwise; the vials were closed airtight and the samples were incubated for 3 h at 25 °C in the dark; and CO_2 emission was measured using a Chrom-5 gas chromatograph. MB-C was derived according to the equation: MB-C (μgC/g soil) = 37.75 × SIR (μgCO_2-C/g soil/h). Microbial quotient was calculated as MB-C/SOC. Basal respiration (BR) was determined by the CO_2 emission from soil incubated for 24 h at 25 °C in the dark. Measurement of BR (μgCO_2-C/g soil/h) was done as described for SIR except that no substrate solution was added. Metabolic quotient (qCO_2) was calculated as BR expressed per mg of MB-C.

24.3 Results and Discussion

The SOM content in the experimental plots was much lower than in the undisturbed soil (on average by 1.5 times) and lowest under continuous black fallow. In general, agricultural management practices had a negative impact on SOM as well as on microbial biomass and its activity (Fig. 24.1).

Of the eight treatments, a positive SOM balance was observed in only three continuous sugar beet and sunflower and ecological crop rotations – in each case with the benefit of organo-mineral or organic fertilizers. Only in the manured ecological crop rotations did the increase in SOM coincide with statistically significant stimulation of the microbial biomass and respiration. The size and activity of the soil microbial community depends on the bioavailability of carbon (Stockdale and Brookes 2006), which is a key factor in defining its quality (Christensen and Johnston 1997). It would appear that manured ecological crop rotations are the only arable management that improves SOM quality.

The SOM content in the experimental plots depended on both the cropping system and fertilization. Crops, grown either as continuous monocultures or in rotations, brought about an average 18.5 % increase in SOM compared with black fallow. At the same time, applications of organo-mineral fertilizers increased SOM by 17.6 % compared with the parallel situations without fertilizer.

Compared with undisturbed soil, microbial biomass carbon (MB-C) was less than one third of that in the undisturbed soil and basal respiration (BR) less than a quarter, on average, *across all the experimental plots*. The size of the soil microbial biomass depended more on crop type and rotation than fertilizer applications. Compared with black fallow, continuous monocultures (with the exception of maize) increased MB-C by 58.0–77.4 %, conventional crop rotations decreased MB-C by 16.6 %, and the ecological crop rotations had MB-C on average 35.8 % higher than in the black fallow. In most cases, application of fertilizers did not produce statistically significant changes in MB-C compared with the parallel plots without fertilizer. Within the conventional rotations, the smallest MB-C values were accompanied by the smallest microbial quotients and the highest metabolic quotients – strong signs of adverse impacts of conventional rotations on the soil

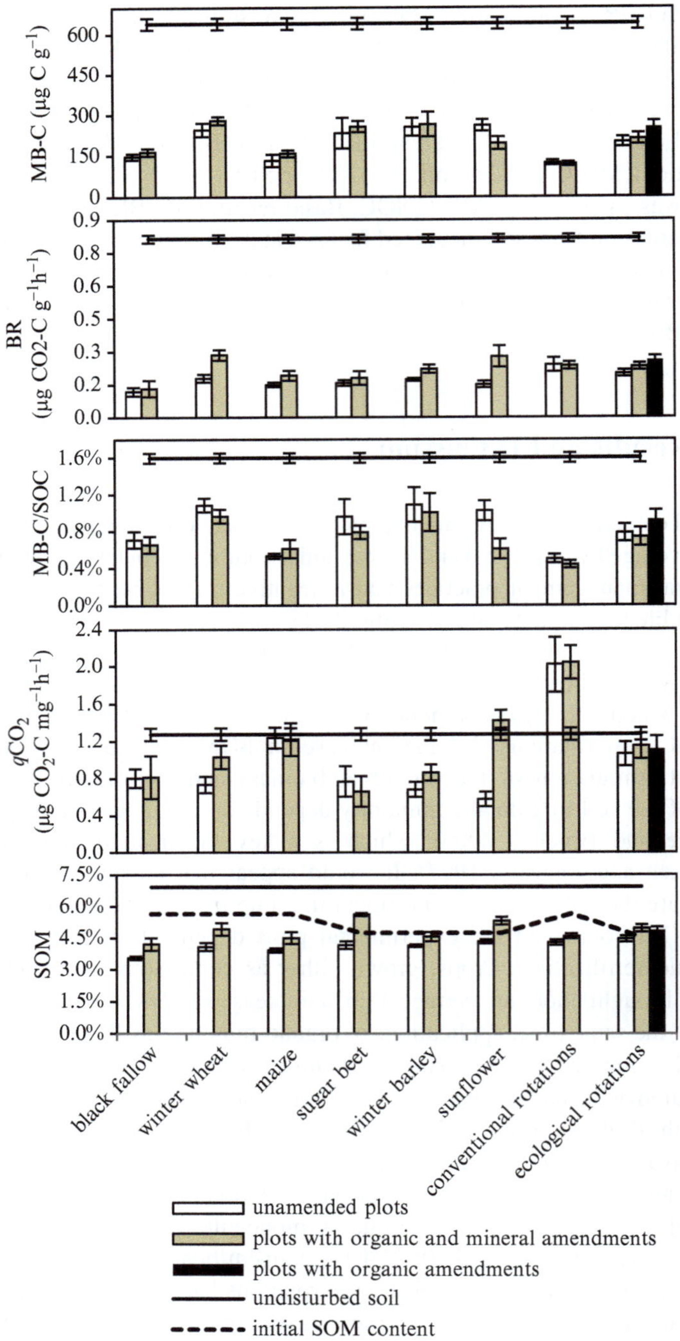

Fig. 24.1 Microbial biomass carbon (*MB-C*), basal respiration (*BR*), microbial quotient (*MB-C/SOC*), metabolic quotient (qCO$_2$), and soil organic matter content (*SOM*) in soil samples from long-term experimental plots with continuous black fallow, continuous monocultures, and conventional and ecological rotations. *Error bars* indicate the confidence interval values

Table 24.1 Regression analysis of relationship between SOM (x) and microbial parameters (y) for undisturbed soil and unamended plots under continuous black fallow and crops

Microbial parameter	Regression equation	r
Microbial biomass carbon	$y = -391.960 + 14{,}974.000\,x$	0.984
Basal respiration	$y = -0.720 + 21.826\,x$	0.983
Microbial quotient	$y = -0.002 + 0.259\,x$	0.860
Metabolic quotient	$y = 0.259 + 13.710\,x$	0.559

microbial biomass and its activity. It is alarming to find that these impacts were statistically more pronounced than in the 46-year continuous black fallow.

High correlations were observed between BM-C, BR, BM-C/SOC, and SOM for the undisturbed soil and the unamended plots under continuous black fallow and crops (Table 24.1). This implies a connection between SOM quantity and SOM quality in all the cases without fertilization and crop rotations. Fertilization and crop rotations, either separately or together, were unable to maintain this connection: SOM increases in these cases (as compared to the black fallow) did not consistently coincide with proportional increases in the microbial parameters.

24.4 Conclusions

1. In eight out of nine cases, agricultural management practices substantially depressed soil microbial biomass and its activity and, to a lesser extent, SOM content and/or SOM quality. Contrary to received wisdom, conventional crop rotations – with and without organo-mineral amendments – caused the greatest negative effect on the soil microbial biomass, surpassing even 46 years of black fallow. If confirmed by further investigations, this should be a cause for concern.
2. Ecological crop rotations with addition of farmyard manure were the only case of positive SOM balance and simultaneous improvements in soil microbial indicators of SOM quality.
3. SOM content and the SOM balance were unable to discriminate between positive and negative changes in the soil microbial biomass/SOM quality caused by different agricultural management practices. Microbial biomass carbon, basal respiration, and both microbial and metabolic quotients were more sensitive to changes in soil quality and possessed a greater discriminatory power relative to different agricultural management practices.
4. Increase in SOM quantity, of itself, may not indicate improvement in soil quality: a combination of positive SOM balance with favorable changes in soil microbial parameters presents a better evidence for sustainability of arable cropping systems.
5. Depending on the nature of agricultural management practices, there may or may not be equilibrium between SOM content and soil microbial parameters. Crop rotations and applications of organo-mineral fertilizer disturb this equilibrium by failing to adequately stimulate the soil microbial biomass.

References

Boincean B (1999) Ecological agriculture in the Republic of Moldova. Stiinta, Chisinau (Romanian)

Christensen BT, Johnston AE (1997) Soil organic matter and soil quality – lessons learned from long-term experiments at Askov and Rothamsted. In: Gregorich EG, Carter MR (eds) Soil quality for crop production and ecosystem health, vol 25, Developments in soil science. Elsevier, Amsterdam/New York, pp 399–430

Corcimaru S (2009) Assessment of the substrate-induced respiration method for estimation of microbial biomass in Moldovian soils. Buletinul Academiei de Ştiinţe a Moldovei Ştiinţe Biologice, Chimice şi Agricole 307:134–142 (Romanian)

CSRM (1999) Classification of the soil of the Republic of Moldova. SMMSS, Chisinau (Romanian)

Dalal RC (1998) Soil microbial biomass – what do the numbers really mean? Aust J Exp Agric 38:649–665

Gregorich EG, Carter MR, Doran JW, Pankhurst CE, Dwyer LM (1997) Biological attributes of soil quality. In: Gregorich EG, Carter MR (eds) Soil quality for crop production and ecosystem health, vol 25, Developments in soil science. Elsevier, Amsterdam/New York, pp 81–113

Merenuic G, Ursu A, Corcimaru S, Sasco E (2009) Microbiological description of zonal soils of Moldova. Mediul Ambiant 47:4–8

Nelson DW, Sommers LE (1982) Total carbon, organic carbon, and organic matter. In: Page AL et al (eds) Methods of soil analysis. Part 2. Agronomy 9, 2nd edn. ASA/SSSA, Madison, pp 961–1010

Stockdale EA, Brookes PC (2006) Detection and quantification of the soil microbial biomass – impacts on the management of agricultural soils. J Agric Sci 144:285–302

Winding A, Hund-Rinkeb K, Rutgers M (2005) The use of microorganisms in ecological soil classification and assessment concepts. Ecotoxicol Environ Saf 62:230–248

Chapter 25
Total and Labile Organic Matter in Typical Chernozem Under Crop Rotation, Continuous Cropping, Perennial Crops, and Fertilization

M. Nicorici

Abstract Soil organic matter is an integrated indicator of soil fertility. However, its stability within any particular soil type makes it hard to detect the effects of different farm practices on the continuous transformation of soil organic matter. Separation of soil organic matter into fractions according to their ability to resist microbiological decomposition enables assessment of the effects of crop rotations, continuous monocropping, and fertilization on the vitality of the soil.

25.1 Introduction

Soil organic matter plays a key role in soil fertility; its reserves and quality largely determine the soil's behavior and productivity under agriculture. A soil rich in organic matter is well provided with nutrients: the soil's organic fraction contains 94–96 % of the total reserves of nitrogen, 60–70 % of sulphur, and 40–45 % of phosphate. In the case of Moldovan chernozem, it has been established by experiment that an increase of 1 % in humus stocks ensures an additional yield of 1 t/ha of maize or 0.8 t/ha of winter wheat (Andrieş 1993).

Without considering the soil as a living system in its own right, it is hard to increase soil productivity and the effectiveness of inputs, let alone mitigate the negative effects of regional pollution of groundwater or global warming (Boincean 1999). For instance, soil fertility depends on physiologically balanced amounts of nutrients and many other factors in the symbiosis of plants and microorganisms (Ştefanic et al. 2006). Various agricultural practices can help to maintain or, even, increase soil organic matter: crop rotation, tillage (or, rather, no till), and application of manure and fertilizers that increase the accessibility of plant nutrients and alter the biological, chemical, and biochemical structure of the soil. On the other

M. Nicorici (✉)
Alecu Russo Balti State University, Balti, Republic of Moldova
e-mail: mnicorici@mail.ru

D. Dent (ed.), *Soil as World Heritage*, DOI 10.1007/978-94-007-6187-2_25,

hand, black fallow abruptly increases the mineralization of soil organic matter (Ursu 2000; Likov et al. 2004; Boincean et al. 2007; Corceaghin and Terentiev 2007; Andrieş et al. 2008). This research investigates the effects of permanent grass, arable crop rotation, continuous monocultures, and fertilization on the stocks of soil organic carbon and the labile and microbial carbon fractions in the Typical chernozem of the Balti steppe; in particular, it aims to identify the specific effects of management practices on the more responsive fractions of soil organic matter.

25.2 Experimental Site and Methods

Research was conducted in 2004–2007 within the long-term field experiments of the Selectia Research Institute of Field Crops (RIFC) on Typical chernozem[1] under winter wheat and sugar beet in crop rotation, continuous winter wheat and continuous sugar beet, black fallow, and meadow – both fertilized and unfertilized. The average rates of fertilizer applied to the fertilized plots were: N84.7, P97.2, K 67.1 plus 12tonne/ha farmyard manure per rotation. The long-term experiments on crop rotations began in 1962; continuous crops of wheat and sugar beet were added in 1965 (see Boincean Chapter 19 in this symposium). Soil samples were collected from the topmost 20 cm in three series: under winter wheat at straw formation, flowering and ripening; under sugar beet at the two-leaf stage, rhizocarp development and harvest. Soil organic matter was determined using Tiurin's method (1946), labile organic matter following Cambardella (1992), and soil microbial biomass following Zveaghintev (1991).

25.3 Results and Discussion

Figure 25.1 and Table 25.1 present the values of soil organic carbon, labile, and microbial carbon under 12 different cropping systems. Absolute values of total organic carbon in the topmost 20 cm lie between 51.8 and 77.8 t/ha.

Compared with the unfertilized control, fertilization helps to maintain the total amount of organic carbon. In the case of winter wheat, the results do not differ significantly for wheat in rotation with or without fertilizer but for continuous winter wheat, application of fertilizer contributed to significantly higher total organic carbon stocks. The same trend is observed under sugar beet, where fertilizer application to continuous sugar beet contributes to significantly higher soil organic carbon stocks. Comparing the effects of crop rotation and fertilization, it appears that organic carbon levels are more influenced by fertilization than by crop rotation.

The meadow has maintained its original stocks of soil organic carbon and, with fertilization, even increased them. In contrast, there has been a significant depletion

[1] Haplic chernozem in *World reference base for soil resources 2006.*

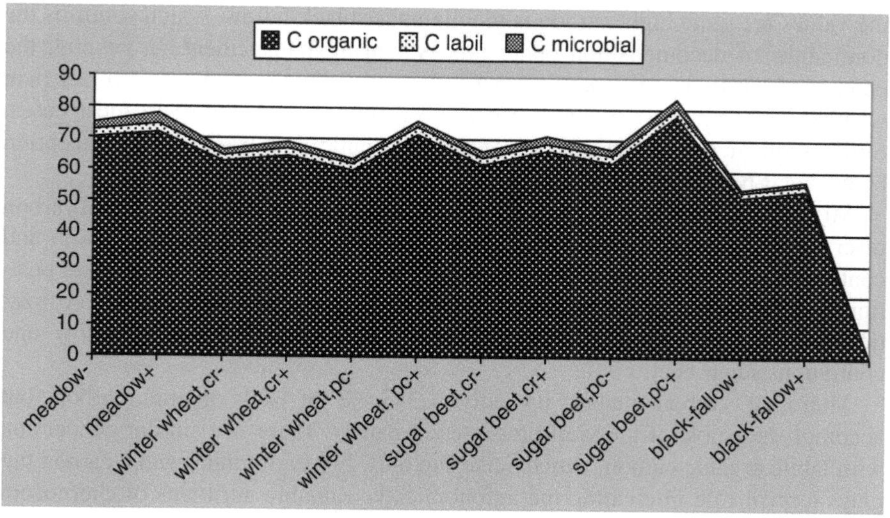

Fig. 25.1 Organic, labile, and microbial carbon (t/ha 0–20 m) in Typical chernozem in the Selectia long-term experiments, means 2005–2007 (*cr* crop rotation, *pc* continuous culture, – unfertilized background, + fertilized)

Table 25.1 Total organic carbon, labile, and microbial carbon (t/ha and % in the 0–20 cm layer) in Typical chernozem under different cropping systems, means 2005–2007

Variant	Total organic carbon	Labile carbon	Microbial carbon
Unfertilized			
Meadow	70.1	2.4 (3.42 %)	2.34 (3.33 %)
Winter wheat in crop rotation	62.6	2.0 (3.19 %)	1.71(2.73 %)
Continuous winter wheat	60.0	2.1 (3.5 %)	1.52 (2.53 %)
Sugar beet in crop rotation	61.9	2.2 (3.55 %)	2.03 (3.27 %)
Continuous sugar beet	62.6	2.4 (3.81 %)	2.0 (3.19 %)
Black fallow	51.8	1.7 (3.28 %)	0.92 (1.77 %)
Fertilized			
Meadow	71.8	2.7 (3.76 %)	3.42 (4.76 %)
Winter wheat in crop rotation	64.6	2.4 (3.71 %)	1.96 (3.03 %)
Continuous winter wheat	71.5	2.5 (3.49 %)	1.81 (2.53 %)
Sugar beet in crop rotation	66.5	2.2 (3.30 %)	2.50 (3.75 %)
Continuous sugar beet	77.8	2.7 (3.47)	2.84 (3.65 %)
Black fallow	54.0	1.9 (3.51)	1.13 (2.09 %)

of carbon stocks under black fallow which may be attributed to increased oxidation, exacerbated by tillage, without any counterveiling additions of fresh organic matter (Boincean 1999; Likov et al. 2004; Corceaghin and Terentiev 2007).

The quantity and quality of crop residues remaining in the soil after harvest influences the quantity and quality of labile organic matter (Boincean 1999; Likov et al. 2004). Next to the meadow, the greatest absolute values of labile organic carbon occurred under sugar beet, both unfertilized and fertilized. The lowest values are under black fallow but, relative to the total amount of organic carbon, the values are quite high; carbon is mobile under black fallow, which confirms the dominance of decomposition under this system of management. In general, the content of labile organic matter depended more on applications of fertilizer than whether the crop was grown in rotation or continuous culture. In most cases, fertilization increased the content of labile organic matter – the sole exception being sugar beet in crop rotation.

Microbial biomass carbon follows the same trends. The highest microbial carbon reserves were in the meadow, the lowest in black fallow. Crops in rotation and continuous cultures, both fertilized and unfertilized, occupy an intermediate position – with significantly more under sugar beet than under winter wheat. Fertilization increases microbial biomass in all cases, especially under meadow and continuous sugar beet.

Microbial carbon makes up only 1.3–3 % of soil organic carbon but accomplishes most of the vital processes in nature. There is a similar connection with labile organic carbon, which makes up only 2–4 % of total organic carbon but plays a great role in creating the agronomically valuable attributes of chernozem and, usually, determines crop productivity.

25.4 Conclusions

Total, labile, and microbial soil carbon reserves change according to the systems of cultivation and fertilization of arable crops. In the Balti chernozem, total organic carbon reserves in the topmost 20 cm ranged between 51.8 t/ha under black fallow and 77.8 t/ha under meadow. Other variants occupied an intermediate position.

Trends of labile organic matter and microbial biomass carbon were similar: reserves of labile organic matter ranged from 2.4 to 2.7 t/ha under unfertilized and fertilized meadow, respectively, down to 1.7–1.9 t/ha under unfertilized and fertilized black fallow, respectively. Microbial biomass ranged from 142.8mgC/100 g soil under meadow to 38.6mgC/100 g soil under black fallow. Again, soil under arable crops occupied an intermediate position, with fertilization having more effect than whether the crops were in rotation or continuous cultures.

Labile soil organic matter makes up only 2–4 % of soil organic carbon but determines the stability of soil organic matter transformation process. The share of microbial biomass is less: 1.3–3 % but this fraction accomplishes the most vital processes in soil.

References

Andrieş S (1993) Regulating the soil's nutrient regimes under the planned harvest of winter wheat and corn. Stiinta, Chisinau (Romanian)

Andrieş S, Arhip O, Filipciuk V et al (2008) Environment-friendly agricultural practices as guarantee of soil fertility restoration. Moldovan Agric 7–8:14–21 (Romanian)

Boincean BP (1999) Ecological agriculture in the Republic of Moldova (crop rotation and soil organic matter). Stiinta, Chisinau (Russian)

Boincean BP, Nica L, Bugaciuc L et al (2007) Long-term productivity and fertility of arable chernozem soils of Moldova. In: Boincean BP (ed) Sustainable, including ecological, agriculture: problems, perspectives. Materials of the international scientific/practical conference. Presa Universitara Balteana, Balti (Romanian), pp 22–30

Cambardella CA (1992) Particulate soil organic matter change across a grassland cultivation sequence. Soil Sci Soc Am J 56:777–783

Corceaghin VA, Terentiev OV (2007) On the reproduction of soil fertility. Agric Sci 3:27–29 (Russian)

Likov AM, Esikov AN, Novikov MN (2004) The organic matter of arable non-chernozem soils. VNIIPTIOU Academy of Agricultural Sciences of Russia, Moscow (Russian)

Ştefanic G, Săndoiu D, Gheorghiţă N (2006) The biology of agricultural soils. Elisavaros, Bucharest (Romanian)

Tiurin IV (1946) The quantitative part of the life of soil organic matter. Pochvovedenie 1:3–15 (Russian)

Ursu A (2000) Soil degradation and desertification. Bons Offices SRC, Chisinau (Romanian)

Zveaghintsev DG (1991) Methods of soil microbiology and biochemistry. MGU Lomonosov State University, Moscow (Russian)

Chapter 26
Primary Soil Tillage in Rotations of the Main Field Crops in Moldova

I.V. Boaghii and L.I. Bulat

Abstract Research on primary soil tillage for the main field crops of Moldova has been conducted on Typical chernozem at the Research Institute of Field Crops in Balti, since the foundation of the institute. On the basis of earlier data, a long-term field experiment on methods, depth and frequency of primary soil tillage for crop rotations was initiated in 1977. Different methods, depth and frequency of primary soil tillage produced much the same proportion of water-resistant structural aggregates, soil porosity and spring stocks of productive moisture. Different effects of tillage by mouldboard plough, as opposed to ploughless tillage, are evident in the greater contents of mobile nitrogen and phosphorus and, especially, in biological indices of soil fertility. Annual losses of soil organic matter were 1.5 times higher under ploughing than for ploughless tillage.

26.1 Introduction

Research on methods and the depth of primary soil tillage for the main field crops cultivated in Moldova has been conducted at the Research Institute of Field Crops in Balti for more than 50 years, since the foundation of the institute. On the basis of earlier data, a new long-term field experiment was initiated in 1977 on methods, depth and frequency of primary soil tillage for crop rotations. Besides primary soil tillage, other crop management technologies have been studied, particularly for row crops. Here, we describe research initiated and designed by Dr PT Chibasov in 1977 and continued by NF Malahovskaia, AG Socaliuc, CI Cebotari and IV Boaghii.

I.V. Boaghii (✉) • L.I. Bulat
Selectia Research Institute of Field Crops, Balti, Republic of Moldova
e-mail: selectia3@gmail.com

D. Dent (ed.), *Soil as World Heritage*, DOI 10.1007/978-94-007-6187-2_26,

26.2 Experimental Site and Methods

The soil is Typical chernozem: soil texture is silty clay over clay loam, the humus layer is 80–100 cm thick with 4.3–4.9 % humus in the plough layer, pH $_{water}$ 6.8–7.1, sum of exchangeable bases 32–33 mg/100 g and total nitrogen 0.25 %, phosphorus 0.14 % and potassium 1.3 %. Mean annual precipitation is 481 mm with up to 60–70 % of the annual total falling during a short period of the spring and summer; over the 19 years of the experiment, precipitation exceeded the mean in 9 years, 4 years were close to the mean and five of them were distinctly lower.

Investigations were carried out under a seven-field crop rotation of oats and vetch for green mass – winter wheat – maize for grain – maize for grain – spring barley – sunflower. The recommended rates of mineral fertilizer, in kg of active ingredient per hectare, were $N_{60}P_{60}$ for the oats and vetch, $N_{60}P_{90}K_{60}$ for winter wheat, $N_{120}P_{120}K_{120}$ for sugar beet, $N_{60}P_{90}K_{30}$ for maize, $N_{60}P_{60}K_{45}$ for spring barley and $N_{60}P_{60}K_{60}$ for sunflower. Fertilizer application at 150 % of the recommended rates was also used. Manure was applied for the sugar beet at 40 t/ha (equivalent to an annual rate of 5.7 t/ha).

Various systems of primary soil tillage were studied over two crop rotations: annual tillage with the mouldboard plough at a different depths for each crop in the rotation (variant no.1); in the third variant, the mouldboard plough was used three times during the rotation; in the fifth variant, twice; and the sixth, only once, for sugar beet. Also for sugar beet, beginning in 1985, the soil was loosened to a depth of 32–35 cm without inverting the topsoil. In the eighth variant, deep mouldboard ploughing was employed for all crops except winter wheat. The 50 % higher dose of fertilizer was applied in variants 2, 5 and 7. The experiment was carried out in three replicates. Each individual plot was 400 m^2.

After two cycles of the crop rotation, beginning with the year 1991, the variants with 50 % increase in the rates of mineral fertilizers (variants 2, 5 and 7) were discontinued, and for the second variant, all the principal tillage was done by para-plough, for the fifth variant by chisel plough, and for the seventh by rotary cultivator. Beginning in 1991, a sodded strip was included alongside the experimental fields to monitor the changes of bulk density without cultivation.

26.3 Results and Discussion

26.3.1 *Influence of Tillage on Agro-physical, Agrochemical and Biological Indices of Soil Fertility*

Soil bulk density takes pride of place amongst agro-physical indicators of soil fertility. Tillage is a means to change bulk density to suit the growth of different crops. Analysis of the dynamics of soil bulk density in the crop rotation (Table 26.1)

Table 26.1 Soil bulk density (g/cm^3) in the seven-field rotation with different methods of tillage

	1st rotation		2nd rotation		3rd rotation		
Depth (cm)	Mouldboard plough	Ploughless tillage	Mouldboard plough	Ploughless tillage	Mouldboard plough	Ploughless tillage	Sod since 1991
0–10	1.09	1.10	1.10	1.11	1.14	1.20	1.16
10–20	1.15	1.21	1.20	1.23	1.18	1.24	1.31
20–30	1.19	1.24	1.22	1.24	1.21	1.24	1.31
30–40	1.20	1.20	1.24	1.21	1.22	1.25	1.26
0–40	1.16	1.19	1.19	1.19	1.18	1.23	1.26

Table 26.2 Water-stable structural aggregates and soil porosity under different methods of tillage, means 1977–1995

	Water-resistant structural aggregates (%)		Soil porosity (%)	
Depth (cm)	Mouldboard plough	Ploughless tillage	Mouldboard plough	Ploughless tillage
0–10	66.2	65.9	57.1	56.1
10–20	69.6	71.7	54.6	52.8
20–30	70.9	72.6	53.7	52.2
30–40	72.6	73.7	53.2	53.0
0–40	69.8	70.9	54.7	53.5

shows similar values and trends for different methods tillage. Comparison with the bulk density under sod indicates how effective tillage has been.

A favourable soil structure in the upper layer of the root zone depends on the stability and porosity of the soil aggregates (Table 26.2).

Different methods of primary soil tillage produced much the same content of water-resistant structural aggregates and soil porosity. Moreover, for most crops in each of the three rotations, the stocks of productive moisture in spring were the same, despite differences in the methods, depth and frequency of tillage (Table 26.3).

Systematic application of the recommended rates of mineral and organic fertilizers caused substantial variation in the agrochemical indices of soil fertility (Table 26.4). The balances for the main nutrient elements show big deficits of nitrogen and potassium, and a surplus of phosphorus, regardless of the methods and frequency of primary soil tillage. Increasing the rates of mineral fertilizer by 50 % diminished, but did not eliminate, the loss of nitrogen and potassium – and increased the surplus of phosphorus.

Over three cycles of the crop rotation, the content of mobile phosphorus in the 0–30 cm layer of soil increased by 1.8–1.9 times (Table 26.5); this amounts to 160–168.2 mg/kg of soil. Despite the negative nitrogen balance, the content of soluble nitrogen also increased 2.1 times from the first to the third rotations under the mouldboard plough and by 1.9 times for ploughless tillage.

The different methods of tillage also have a different influence on the biological activity of the soil as expressed by the rates of decomposition of flax tissue (Table 26.6), soil respiration and the count and mass of earthworms (Table 26.7).

26.3.2 Influence of Soil Tillage on Soil Organic Matter

Soil organic matter is the integral index of soil fertility. Soil tillage, in itself, not only redistributes the existing organic matter in the soil but can change the ratio between synthesis and decomposition of soil organic matter. As a rule, without manuring, soil tillage accelerates the decomposition of the existing stock of organic matter. The decomposition rates are higher for soil tillage by the mouldboard plough than for other methods of tillage that do not invert the soil (Table 26.8).

Table 26.3 Spring stocks of available water (mm) under different methods of primary soil tillage

	Vetch + oats for green mass		Winter wheat		Sugar beet		Maize for grain		Maize for grain		Spring barley		Sunflower	
Cycle of rotation	1	2	1	2	1	2	1	2	1	2	1	2	1	2
1st (1977–1983)	329	315	376	357	349	350	331	327	338	339	365	369	376	360
2nd (1984–1990)	317	312	326	325	328	338	280	268	307	316	310	300	327	320
3rd (1991–1995)	324	326	344	350	317	321	273	263	338	318	332	340	337	320

1 mouldboard plough; *2* non-inverting soil tillage

Table 26.4 Balance of nutrients (kg/ha) in crop rotations with different systems of primary soil tillage, 1976–1985 means

	Mouldboard ploughing						Non-inverting soil tillage					
	Recommended rates of mineral fertilizer			Rates of fertilizer increased by 50 %			Recommended rates of mineral fertilizer			Rates of fertilizer increased by 50 %		
Indices	N	P	K	N	P	K	N	P	K	N	P	K
Applied as fertilizer	14.1	14.1	12.6	17.3	18.2	15.1	14.1	14.1	12.6	17.3	18.2	15.1
Nutrients removed by crops	18.4	7.3	23.1	18.2	7.2	22.8	18.1	7.2	22.9	18.4	7.3	23.0
Balance	−4.3	+6.8	−10.5	−0.9	+11.0	−7.7	−4.0	+6.9	−10.3	−1.1	+10.9	−7.9

Table 26.5 Dynamics of soluble nutrients (mg/kg) with recommended rates of fertilizers by methods of primary tillage, averages for seven-field crop rotation

	1st rotation				2nd rotation				3rd rotation				Average			
	1		2		1		2		1		2		1		2	
Depth (cm)	N	P	N	P	N	P	N	P	N	P	N	P	N	P	N	P
0–10	50	107	59	140	64	152	73	180	70	204	82	229	61	155	71	183
10–20	51	109	54	103	65	147	66	156	90	185	85	69	147	68	148	148
20–30	40	78	48	82	59	95	54	100	106	150	101	155	68	108	68	112
30–40	44	50	41	54	49	66	53	70	117	102	109	102	70	73	67	75
0–40	46	86	50	95	60	115	61	126	96	160	94	168	67	120	69	130

1 mouldboard plough; *2* ploughless tillage

Table 26.6 Decomposition of flax strips % with different methods of soil tillage, average 1987–1988

		Soil layers (cm)		
	Exposure (days)	0–20	20–40	0–40
Mouldboard ploughing to 25–27 cm	31	42.8	53.7	48.3
	66	87.1	71.3	79.2
Ploughless tillage to 8–12 cm	31	27.3	53.8	40.6
	66	65.9	66.8	66.3

Table 26.7 Earthworm activity and soil respiration in the 0–20 cm layer according to the methods of tillage, average 1991–1995

	Earthworms		
	pcs/m^2	gm	Soil respiration (mgCO_2/24 h)
Mouldboard	17	6.6	90.9
Non-inversive soil tillage	27	10.8	106.5

Table 26.8 Stocks of soil organic matter in crop rotation with recommended rates of mineral fertilizers for different methods of tillage, average for seven-field rotation

Depths (cm)	Stock of soil organic matter (t/ha) Mouldboard ploughing 1977	1990	Ploughless tillage 1977	1990	Loss of soil organic matter (t/ha)Mouldboard ploughing	Ploughless tillage
0–10	51.6	51.0	52.9	52.2	−0.6	−0.7
10–20	55.2	53.7	58.0	57.2	−1.5	−0.8
20–30	54.7	54.7	56.5	56.6	−0.2	+0.1
0–30	161.7	159.4	167.5	166.0	−2.3	−1.4

The loss of soil organic matter for the 0–30 cm soil layer over 14 years was 2.3 t/ha under tillage with the mouldboard plough and 1.4 t/ha for the same intensities of non-inversive tillage. Soil organic matter is being lost under both systems but the mineralization of soil organic matter is more than 50 % greater under the mouldboard plough.

26.3.3 Effects of the Method of Tillage on Potential and Actual Weediness

The different methods of primary soil tillage had different effects on both on the potential weediness of the soil, created by weed seed, and on the actual weediness of crops. Estimates of the stocks of weed seeds in the 0–40 cm soil layer were made every year in the ploughed fields and 6 years out of seven in the fields tilled by non-inversive techniques. There was more weed seed in the ploughed fields with 133.7 and 102.2 million seeds/ha, respectively, but estimates of emerged weeds during the growing period showed more in the unploughed fields (Table 26.9).

26.3.4 Effects of Tillage on Crop Yields

Regardless of the depth, methods and frequency of the primary soil tillage, much the same crop yields were achieved in each rotation (Table 26.10). Decreasing yields were observed for sugar beet and the second maize crop without ploughing. During the third rotation, the same tendency was observed for the second maize crop in comparison with maize planted after sugar beet. This was an effect of sugar beet on the soil moisture during these 3 dry years. The absolute grain yields of maize were also much lower in the third rotation than in the first two.

Table 26.9 Weediness per square metre of crops in rotation during the harvest period for different systems of soil tillage without herbicides. Weeds recorded by number (first column) and mass (second column)

Methods of primary soil tillage	Crops in crop rotation													
	Vetch and oats		Winter wheat		Sugar beet		Maize for grain		Maize for grain		Spring barley		Sunflower	
Mouldboard plough	7	89	16	348	31	904	11	147	13	–	–	181	–	–
Ploughless tillage	5	50	11	–	13	–	18	452	20	1,095	–	343	6	195

Table 26.10 Crop yields (t/ha) in rotation at recommended rates of fertilizers with different systems of tillage

Crops	Method of primary tillage	Crop rotation I (1977–1983)	Crop rotation II (1984–1990)	Crop rotation III (1991–1995)	Coefficient of variation 1977–1995 (%)
Vetch and oats for green mass	1	23.0	32.6	27.9	26.6
	2	23.6	33.0	28.7	28.0
Winter wheat	1	4.51	4.61	5.62	20.1
	2	4.66	4.75	5.51	17.7
Sugar beet	1	44.4	47.5	43.1	15.1
	2	–	45.3	41.8	13.7
Maize	1	7.87	6.67	4.16	38.5
	2	7.76	6.76	4.22	41.2
Maize	1	6.68	7.22	5.80	26.2
	2	6.46	6.74	5.60	29.2
Spring barley	1	3.86	4.19	4.84	24.0
	2	3.99	4.17	4.85	21.0
Sunflower	1	2.30	3.17	2.08	30.2
	2	2.30	3.12	2.02	29.4

For the third rotation, yields are reported as the mean of 5 years
1 mouldboard plough; *2* ploughless tillage

Application of an extra 50 % of mineral fertilizer did not increase crop yields on the ploughed plots in comparison with ploughless tillage. Indeed, the increased rates of mineral fertilizers diminished the yields of maize and winter wheat; only the mixture of vetch and oats for green mass produced more above-ground biomass in the first and second rotations (by 1.0–1.5 and 2.3–3.7 t/ha, respectively). The variability of crop yields is determined much more by the weather than by methods of primary soil tillage – provided that the tillage is timely and of good quality. Late or poor tillage leads to crop yield losses.

The picture is the same for the crop rotation as a whole (Table 26.11); output was the same, regardless of the method of primary soil tillage and increased rates of mineral fertilizer did raise yields in comparison with the recommended level. Variability was lower for the whole rotation than for each separate crop.

Table 26.11 Productivity of crop rotation with recommended rates of mineral fertilizers but different methods of tillage (t/ha grain units)

	Crop rotation				
	1st	2nd	3rd	Average	Coefficient of variation for 1977–1995 (%)
Mouldboard plough	5.55	6.00	5.38	5.64	12.1
Ploughless tillage	5.52	5.96	5.40	5.63	11.6

Table 26.12 Energy efficiency for the whole crop rotation with different systems of soil tillage, rates of mineral fertilizer and methods for weed control, average 1985–1990

Rates of mineral		Mouldboard plough		Ploughless tillage	
fertilizers		Without herbicides	With herbicides	Without herbicides	With herbicides
Recommended	1	17,739	18,594	17,038	17,897
	2	74,589	75,906	73,466	76,831
	3	4.2	4.1	4.3	4.3
50 % extra	1	21,114	21,972	20,421	21,280
	2	71,154	73,334	71,822	74,364
	3	3.4	3.3	3.5	3.5

1 Energy used for crop cultivation, Mdj/ha; *2* Content of energy in the harvested crops, Mdj/ha; *3* Coefficient of energy efficiency

26.3.5 Energy Efficiency

With limited and continually increasing cost of non-renewable energy, the rational use of energy has become a crucial issue. The energy efficiency of different systems of soil tillage with different rates of mineral fertilizers and methods for weed control is summarized in Table 26.12.

The data show a sharp decrease in the coefficient of energy efficiency at higher rates of mineral fertilizers in comparison with recommended rates, irrespective of the method of tillage. The energy expenditure required for extra mineral fertilizer is not repaid by the gain of energy in economically useful production. Herbicides increased the total productivity of the crop rotation, irrespective of the methods of soil tillage but, in terms of energy efficiency, there is no advantage with mouldboard ploughing and surprisingly little with ploughless tillage. However, ploughless tillage is more energy efficient than the mouldboard plough.

In view of the high energy costs of soil tillage, the manufacture of mineral fertilizers (especially nitrogen fertilizer) and chemicals for controlling pests, weeds and disease, we have to reconsider our stance on soil tillage. The search for alternative sources of nitrogen and new methods of weed control is now urgent and this will change the role of soil tillage, especially in interaction with other agricultural practices. There is pressing need for multidisciplinary research linked with long-term experiments with low-input (sustainable) farming systems to determine the action and interaction of the various components of each farming system: crop rotation, system of soil tillage, and fertilization. On-farm research will also increase as these experiments proceed.

26.4 Conclusions

1. The bulk density of tilled chernozem soils is close to that of untilled soils under sod. Different methods of soil tillage did not change the content of the water-stable structural aggregates, soil porosity and spring stocks of soil moisture.
2. The recommended rates of mineral fertilizers and manure did not compensate for the nitrogen and potassium removed by crops but contributed excess phosphorus to the soil. The content of mobile phosphorus increased from the first to the third rotation of crops.
3. Differences were found between the content of mobile nutrients in the soil under ploughless tillage compared with mouldboard ploughing. The differences were more pronounced for biological indices of fertility.
4. The seven-field crop rotation with cereals and sugar beet did not prevent annual losses of soil organic matter. The annual loss of soil organic matter was 1.5 times higher under the mouldboard than under ploughless tillage.
5. The methods, depth and frequency of tillage in crop rotation hardly influenced the yields of individual crops or the output of the rotation as a whole, although there were issues with sugar beet and second maize.
6. Higher than recommended rates of mineral fertilizer significantly decrease energy efficiency.

Chapter 27
Humus Dynamics and Efficiency of Crop Rotations on Calcareous and Common Chernozem

D.M. Indoitu and D. Indoitu

Abstract The dynamics of soil organic matter is evaluated in long-term experiments under different fertilization regimes on Calcareous and Common chernozem. Cropping for 50 years without fertilizer lowered the humus content of Calcareous chernozem by 15–19 % and of Common chernozem by 10–14 %. Combined organic and mineral fertilization had the most favourable effect on humus content. Regular application of 12–14 t/ha of farmyard manure or 12 t/ha of manure + $N_{60–90}P_{60}$ to the crop rotation maintains soil fertility and promotes high fertilizer-use efficiency by the crops. Mineral fertilizer does not prevent the decomposition of humus but optimum use significantly increases crop yields.

27.1 Introduction

Long-term field experiments enable measurement and monitoring of soil fertility and crop yields in relation to known inputs of nutrients and organic matter under specific soil-climatic conditions. In the Central Zone of Moldova, the first long-term experiments were begun in 1950 on the Ketros field experimental station at Anenii Noi under the direction of AG Timoshenko and PA Curcheatov (1946–1954) and continued by KL Zagorchea (1962–2005), MI Dumitrashko (1963–1970), MF Stratulat (1964–1969) and DM Indoitu (from 1968).

D.M. Indoitu (✉) • D. Indoitu
State Agrarian University of Moldova, Chisinau, Republic of Moldova
e-mail: indoitudumitra@rambler.ru

D. Dent (ed.), *Soil as World Heritage*, DOI 10.1007/978-94-007-6187-2_27,

27.2 Experimental Site and Method

The Central Zone of Moldova experiences a continental climate with a mean annual rainfall of 449 mm, varying between 246 and 550 mm. As a rule, drought occurs 1 year in three or four. The experiment on Calcareous chernozem[1] was founded in 1950. The soil is a light loam with 2.5–3.0 % humus (Tiurin), 0.8–1.5 mg/100 g mobile phosphate (Machigin), 18–22 mg/100 g exchangeable potassium and 1.8–2.2 % carbonates in the 0–20 cm layer. The original ten-field crop rotation was maize for silage – winter wheat – maize for grain – peas – winter wheat – maize for grain – maize for grain – peas – winter wheat – sunflower, all with low doses of fertilizer. After two complete cycles (1950–1970), an additional eight-field rotation was introduced: maize for grain – peas – winter wheat – winter wheat – maize for grain – peas – winter wheat – sunflower, with a system of fertilization for the planned yield: annual applications of 150, 200, 300 and 370 kg NPK/ha, including variants with paired combinations of nutrient elements and an equivalent annual application of farmyard manure at 24 t/ha and combined manure at 12 t/ha + mineral $N_{120}P_{90}K_{90}$.

The experiment on Common chernozem[2] began in 1963–1965. The soil has 2.8–3.2 % humus, 1.3–3.2 mg/100 g mobile phosphate, 26–30 mg/100 g exchangeable potassium and 0–0.1 % carbonates in the 0–20 cm layer. There is a six-field rotation of maize for grain – peas – winter wheat – maize for grain – barley – sunflower.

27.3 Results and Discussion

27.3.1 Humus Dynamics

Humus is an integrated index of soil fertility; soil physical attributes and the provision of nutrients depend mainly on humus content. Changes in the humus content of Calcareous chernozem under the eight-field rotation are depicted by Figs. 27.1, 27.2 and 27.3 for the 0–20, 20–40 and 40–60 cm layers, respectively. Compared with the initial soil condition, cultivation of crops without fertilization (absolute control) lowered the humus content of the plough layer by 0.40–0.42 %; a reduction of the humus content may also be seen in the underlying layers. Systematic fertilization in cycles I and II of the ten-field rotation (control after-effect) stabilized the humus content. Subsequently, variants of the experiment with organic and organic-mineral fertilizers showed an increase of humus content in the plough layer from 2.65 to 3.08 %; manuring at the 24 t/ha produced a significant increase in the underlying layers.

[1] Calcic chernozem in the *World reference base for soil resources 2006*.

[2] Haplic chernozem in WRB.

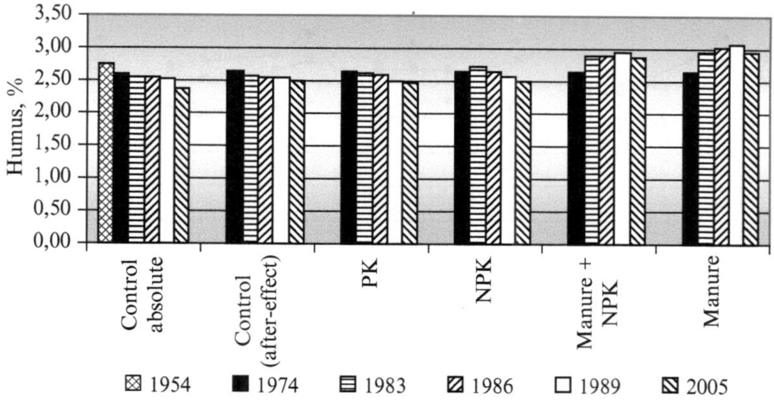

Fig. 27.1 Dynamics of the humus content in 0–20 cm layer in the eight-field crop rotation

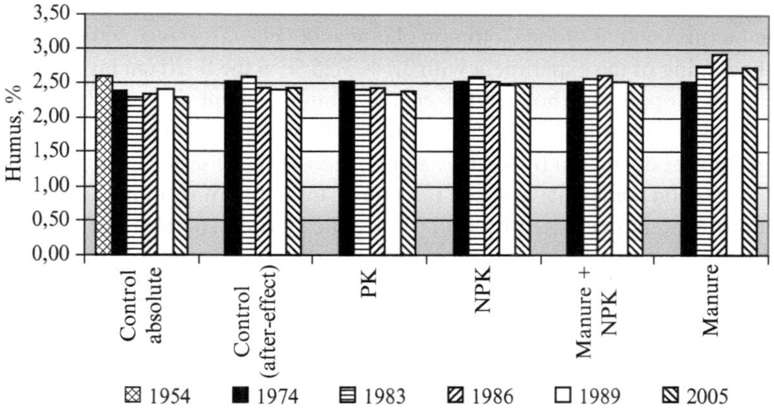

Fig. 27.2 Dynamics of the humus content in 20–40 cm layer in the eight-field crop rotation

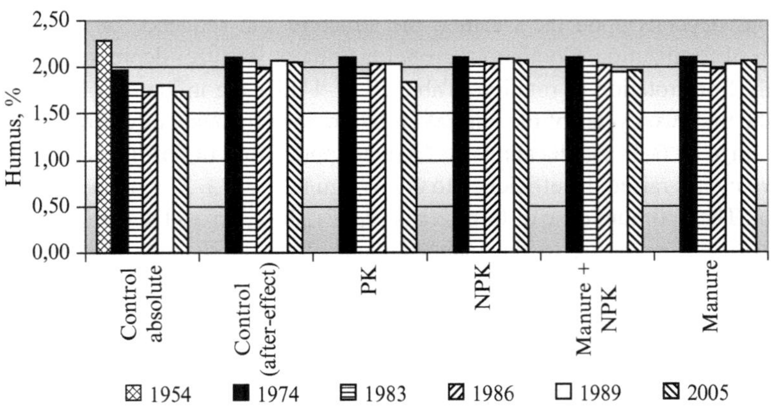

Fig. 27.3 Dynamics of the humus content in 40–60 cm layer in the eight-field crop rotation

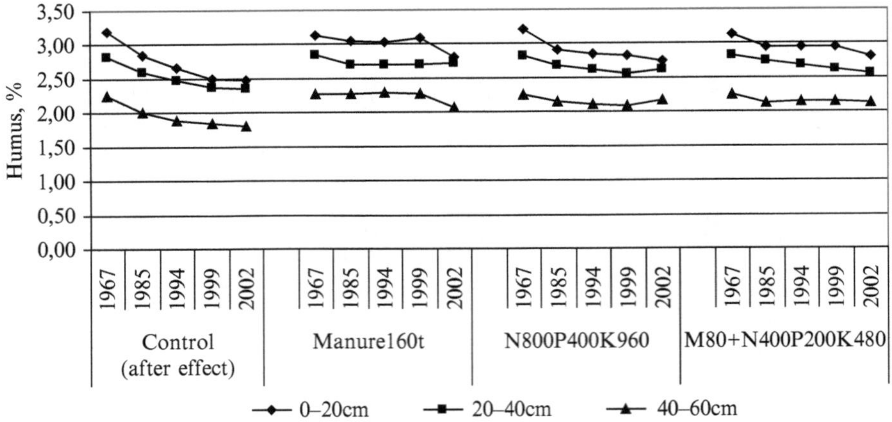

Fig. 27.4 Dynamics of the humus content in layers (six-field crop rotation)

The humus content of the Common chernozem (Fig. 27.4) was somewhat greater at the beginning of the experiment with 3.14–3.25 % in the 0–20 cm layer, but, within 36 years of cropping without fertilizer, the humus content decreased to 2.47 % in 0–20 cm layer and from 2.80 to 2.35 % in 20–40 cm layer; for the 0–60 cm layer, on average, humus decreased by 0.35 %. Application of $N_{400}P_{200}K_{480}$ mineral fertilizers over the rotation somewhat reduced the rate of loss of humus; in the 0–60 cm layer, the humus content decreased by 0.17 %. Application of 160 t/ha farmyard manure and 80 t/ha manure + $N_{400}P_{200}K_{480}$ to the crop rotation maintained the humus content at 3.05–3.10 %.

27.3.2 Efficiency of Crop Rotations

The relative efficiency of the crop rotations is measured in centner[3] grain units (cgu) per kg active substance of the fertilizers applied. The annual efficiency of the crop rotations depends upon the weather, the structure and sequence of crops and the system of fertilization. Continuous cropping without fertilizer depressed crop yields steadily from rotation to rotation (Table 27.1). Under the influence of the fertilizer over the first two cycles of the ten-field rotation, the efficiency of crops increased on the average by 6–14 cgu/ha annually. Greater applications of fertilizers in the following two cycles raised the efficiency to 64–69 cgu/ha, i.e. 23–27 cgu/ha more than the control. The optimum dose of fertilizer is $N_{480}P_{240}K_{240}$ across the complete rotation.

Fertilizing according to the planned yield of the field crops in the eight-field rotation increased the crop productivity significantly: annual application of $N_{120}P_{90}K_{90}$ increased the harvest by 65 cgu/ha but higher doses lowered the

[3] 1 centner = 100 kg.

Table 27.1 Annual productivity of ten-field crop rotation (cgu/ha) by treatment

Rotation	Years	Norm of fertilizers in rotation						
		Control	M*20 + $N_{45}P_{120}K_{105}$		M20 + $N_{45}P_{90}K_{90}$	M20 + $P_{120}K_{60}$		
I	1951–1960	28.8	35.7		34.9	34.7		
		Control	M30 + $N_{90}P_{220}K_{60}$		M30 + $N_{90}P_{220}K_{60}$	M40 + $N_{120}P_{250}K_{60}$		
II	1961–1970	35.2	46.1		48.8	49.1		
		Control	Control 2 after-effect	$N_{120}P_{240}K_{120}$	$N_{480}P_{240}K_{240}$	$N_{480}P_{480}K_{240}$	$N_{720}P_{720}K_{240}$	M40 + $N_{480}P_{480}K_{240}$
III	1971–1980	39.7	46.1	54.5	66.6	67.7	67.0	69.4
IV	1981–1990	36.5	42.1	51.7	64.2	64.8	63.2	69.3
		After-effect						
V	1991–2000	32.6	37.0	37.7	52.0	47.9	43.7	52.2
VI	2001–2010	29.7	36.1	53.6	57.5	54.1	48.5	49.5

M* manure, t/ha

efficiency of the crop rotation (Table 27.2). The efficiency of the six-field crop rotation was somewhat less than that of the eight and ten-field rotations (Table 27.3).

Increases in yield over the control in years without fertilizing characterize the total level of the after-effect of the treatments and the level of the soil fertility in different variants fertilized in past.

Overall, organic, mineral and organic-mineral fertilizer treatments had much the same effect on the efficiency of the crop rotations, with the organic-mineral combination having some advantage. The efficiency of the crop rotations varied from rotation to rotation and according to the dosage of fertilizers (Tables 27.4 and 27.5). The return on applied fertilizers in I and II rotations was 13–17 kg gu per 1 kg active substance applied. Rotations III and IV recouped 20–23 kg gu per kg active substance from the applied $N_{480}P_{240}K_{240}$. The return generally decreased with the increase of the doses of fertilizers. The six-field crop rotation yielded a rather low return on the applied fertilizers (Table 27.6).

Cropping without fertilizer in cycle VII reduced the efficiency of the six-field rotation on all variants but the after-effects of the treatments in the past cycles were high. The efficiency of crop rotations was the greater from application of 160 t/ha of farmyard manure, $N_{80}P_{400}K_{960}$, and manure + $N_{400}P_{200}K_{480}$ in past.

27.4 Conclusions

1. Cropping for more than 50 years without fertilizing lowered the humus content of Calcareous chernozem by 15–19 % and of Common chernozem by 10–14 % compared with the humus content at the outset of the experiment.
2. Regular annual application of 12–14 t/ha of farmyard manure or 12 t/ha of manure + $N_{60-90}P_{60}$ to the crop rotation maintains the fertility of the soil and leads to high fertilizer-use efficiency by the cultivated crops.
3. Application of mineral fertilizer does not prevent the decomposition of humus in soil but significantly increases the efficiency of the crop rotation.

Table 27.2 Annual productivity of the eight-field crop rotation (cgu/ha) by treatment

Rotation	Years	Norm of fertilizers on rotation						
		Control absolute	M20 + $N_{45}P_{120}K_{105}$			M20 + $P_{120}K_{60}$		
I	1955–1964	34.9	43,4			43.2		
		Control absolute	M30 + $N_{90}P_{220}K_{60}$		M30 + $N_{90}P_{220}K_{60}$		M40 + $N_{120}P_{250}K_{60}$	
II	1965–1974	35.2	46.1		48.8		49.1	
		Control absolute	Control 2 after-effect	$N_{480}P_{360}K_{360}$	$N_{960}P_{720}K_{720}$	$N_{1280}P_{720}K_{720}$	$N_{960}P_{960}K_{720}$	$N_{1280}P_{960}K_{720}$
III	1974–1981	38.6	48.4	62.5	65.0	65.8	63.2	63.6
IV	1982–1989	35.2	43.6	57.1	61.0	60.0	58.3	57.4
V	1990–1997	35.4	37.0	44.7	50.3	51.8	50.3	50.7

Table 27.3 Annual productivity of the six-field crop rotation (cgu/ha) by treatment

		Norm of fertilizers on rotation						
Rotation	Years	Control	M*40	$N_{200}P_{100}K_{240}$	M20 + NPK	M80	$N_{400}P_{200}K_{480}$	M40 + $N_{200}P_{100}K_{240}$
I	1967–1973	41.4	48.0	48.4	47.8	46.8	46.5	47.2
II	1973–1979	37.0	47.6	44.6	47.2	43.6	48.3	47.2
III	1979–1985	34.7	45.5	48.8	45.2	50.3	56.9	57.2
		Control	M80	$N_{400}P_{200}K_{480}$	M40 + $N_{200}P_{100}K_{240}$	M160	$N_{800}P_{400}K_{960}$	M80 + $N_{400}P_{200}K_{480}$
IV	1985–1991	36.0	53.4	52.6	52.1	59.7	60.3	60.8
V	1991–1997	29.9	39.4	43.4	41.0	43.9	44.2	46.5
VI	1997–2003	33.4	43.0	45.3	41.8	47.7	51.6	51.7
		After-effect						
VII	2003–2009	30.2	35.7	41.4	40.6	42.8	43.5	42.0

M*manure, t/ha

Table 27.4 Fertilizer efficiency of ten-field rotation (increase kg gu/kgNPK over control)

Rotation	Years	Norm of fertilizers per rotation						
		Control	M20 + $N_{45}P_{120}K_{105}$		M20 + $N_{45}P_{90}K_{90}$		M20 + $P_{120}K_{60}$	
I	1951–1960	28.8[a]	12.8		12.3		13.1	
		Control	M30 + $N_{90}P_{220}K_{60}$		M30 + $N_{90}P_{220}K_{60}$		M40 + $N_{120}P_{250}K_{60}$	
II	1961–1970	35.2[a]	14.1		17.5		14.3	
		Control	Control 2 after-effect	$N_{120}P_{240}K_{120}$	$N_{480}P_{240}K_{240}$	$N_{480}P_{480}K_{240}$	$N_{720}P_{720}K_{240}$	M40 + $N_{480}P_{480}K_{240}$
III	1971–1980	39.7[a]	46.1[a]	17.5	21.4	18.0	12.4	13.4
IV	1981–1990	36.5[a]	42.1[a]	20.0	23.0	18.9	12.6	15.6
		After-effect						
V	1991–2000	32.6[a]	37.0[a]	1.5	15.6	9.1	4.0	8.7
VI	2001–2010	29.7[a]	36.1[a]	17.5	21.4	18.0	12.4	13.4

[a]cgu/ha

Table 27.5 Fertilizer efficiency of eight-field rotation (increase kg gu/kgNPK over control)

Rotation	Years	Norm of fertilizers per rotation						
		Control absolute	M20 + $N_{45}P_{120}K_{105}$			M20 + $P_{120}K_{60}$		
I	1955–1964	34.9*	12.6			14.8		
		Control absolute	M30 + $N_{90}P_{220}K_{60}$		M30 + $N_{90}P_{220}K_{60}$		M40 + $N_{120}P_{250}K_{60}$	
II	1965–1974	35.2*	11.3		14.0		11.5	
		Control absolute	Control 2 after-effect	$N_{480}P_{360}K_{360}$	$N_{960}P_{720}K_{720}$	$N_{1280}P_{720}K_{720}$	$N_{960}P_{960}K_{720}$	$N_{1280}P_{960}K_{720}$
III	1974–1981	38.6*	48.4*	9.4	5.5	5.1	4.5	4.1
IV	1982–1989	35.2*	43.6*	9.0	5.8	4.8	4.5	3.7
V	1990–1997	35.4*	37.0*	5.1	4.4	4.4	4.0	3.7

*cgu/ha

Table 27.6 Fertilizer efficiency of six-field rotation (increase kg gu/kgNPK over control)

Rotation	Years	Norm of fertilizers per rotation						
		Control	M40	$N_{200}P_{100}K_{240}$	M20 + NPK	M80	$N_{400}P_{200}K_{480}$	M40 + $N_{200}P_{100}K_{240}$
I	1967–1973	41.4*	7.3	7.8	7.1	3.0	2.8	3.2
II	1973–1979	37.0*	11.8	8.4	11.3	3.7	6.3	5.7
III	1979–1985	34.7*	12.0	15.7	11.7	8.7	12.3	12.5
		Control	M80	$N_{400}P_{200}K_{480}$	M240 + $N_{200}P_{100}K_{240}$	M160	$N_{800}P_{400}K_{960}$	M80 + $N_{400}P_{200}K_{480}$
IV	1985–1991	36.0*	9.7	9.2	8.9	6.6	6.8	6.9
V	1991–1997	29.9*	5.3	7.5	6.2	3.9	4.0	4.6
VI	1997–2003	33.4*	5.3	6.6	4.7	4.0	5.1	5.1
		After-effect						
VII	2003–2009	30.2*	3.1	6.2	5.8	7.0	7.4	6.6

*cgu/ha

Chapter 28
Soil Conservation Capability of Crops and Crop Rotations: Data from Long-Term Studies in Belarus

A. Chernysh, A. Ustinova, I. Kas'yanenko, and S. Kas'yanchik

Abstract In Balarus, soil erosion can be acute under intensive arable crops. 45 years of run-off-plot data on washout of soil and associated humus and nutrients during snowmelt and downpours are used to derive coefficients of *erosion-preventing capability* for individual crops and soil rotations. Ranking the main crops in descending order of the coefficient: perennial grasses used for 2 or more years, 0.98; perennial grasses used for 1 year, 0.92; winter cereals, 0.89; spring cereals, 0.67 (0.36); annual grasses, 0.62 (0.36); and row crops, 0.15 (0.08). Values in brackets take account of the annual soil washout during the thaw.

Studies of physical properties under a soil-protective crop rotation alongside continuous (19 years) Caucasian goat's rue (*Galega orientalis* L.) demonstrated substantial benefits in restoring soil fertility and protecting against erosion. Continuous *Galega* increased soil resilience as expressed in lesser soil bulk density, increased permeability and improved soil structure. It was also associated with a substantial increase of biological activity.

28.1 Introduction

In Belarus, characteristics of the landscape, soil parent materials and intensive land use contribute to a serious problem of soil erosion. According to the data from the third round of soil and geobotanical studies (Kuznitsov et al. 2011), erodible land makes up nearly one third of the arable – and one tenth has already suffered significant damage. Soil erosion causes substantial economic losses: sloping arable land is losing up to 100 t/ha soil annually (Sidorchuk et al. 2006). As much as 150 kg/ha of humus, 10 kg/ha of nitrogen, 4–5 kg/ha of phosphorous and potassium and 5–6 kg of calcium and magnesium are being lost every year with run-off and

A. Chernysh (✉) • A. Ustinova • I. Kas'yanenko • S. Kas'yanchik
Institute of Soil Science and Agrochemistry, Minsk, Belarus
e-mail: chernysh-erosion@yandex.ru

D. Dent (ed.), *Soil as World Heritage*, DOI 10.1007/978-94-007-6187-2_28,

leaching (Institute of Soil Science and Agrochemistry 2000). Depending on the degree of erosion, this means losses of 12–40 % of cereal yields, 20–60 % for row crops, 15–40 % for flax and 3–30 % for perennial grasses (Chernysh 2005). Erosion also causes ecological damage; in undulating country with arable close to streams and reservoirs, eroded sediment and biogenic elements cause siltation and contamination of rivers and lakes with nitrates, phosphates and pesticides. In Belarus, radionuclides released by the Chernobyl accident have been relocated and concentrated where soil eroded by run-off has accumulated (lower slopes and basins), as well as in forest and scrub where windblown silt has accumulated.

Data from run-off plots of the Institute for Soil Science and Agrochemistry, opened in 1967 near Minsk, have been used to estimate the soil conservation capability of crops and crop rotations.

28.2 Experimental Site and Methods

The experimental site encompasses a transect across a watershed to footslopes of both southerly and northerly aspect. The slopes are convex with an average gradient of 5–7° and length up to 120 m. The soils are Haplic luvisols (sod podzolic soils) on light-textured loess. Non-eroded soils are located on the flat watershed, slightly eroded soils on in the upper slopes, moderately eroded and severely eroded on the middle and lower slopes, and gleys in the bottomland.

Run-off plots characterize water flows and erosion process under different kinds of ground cover in conditions similar to the natural ones. The rectangular plots are isolated from the surrounding area by partitions (25–30 cm high) running along the contour and up-and-down slope. Each plot is equipped to measure the volume of run-off and suspended material during the spring thaw and periods of intense rainfall. Agrochemical properties of soils and run-off, as well as structural and aggregate composition of soils of different erodibility were determined according to currently accepted Belarus methods. The soil-protective capability of cultivated crops was evaluated using 25 years of observational data; comparisons of agro-physical, agrochemical and biological properties and structural and aggregate composition are drawn from recent 5-year observations.

28.3 Results and Discussion

28.3.1 Assessment of Soil-Protective Capability

Vegetation of all kinds protects the soil from erosion. The longer and more completely the soil is protected by above-ground vegetation and bound together by roots, the less the likelihood of erosion. It is of prime importance that the soil be well protected during the spring thaw and during downpours. Long-term data show

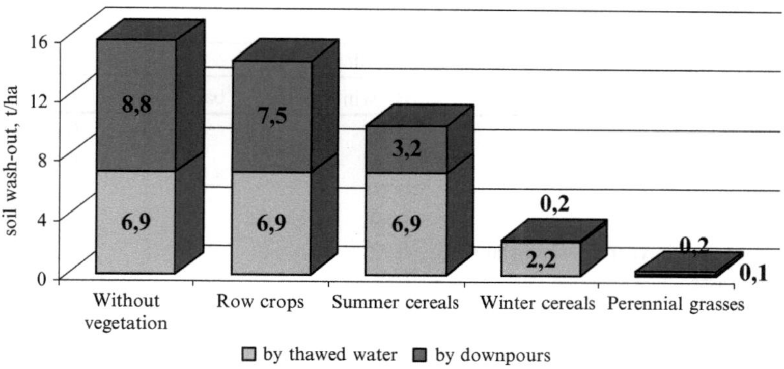

Fig. 28.1 Total soil washout under different cultivated crops, t/ha per year

Table 28.1 Losses of humus and major nutrients by water erosion (kg/ha/year) according to land use

Element		Row crops (including winter tillage)	Summer crops (including winter tillage)	Winter crops	Perennial grasses
Humus	1[a]	4.7	3.3	4.1	3.0
	2	235.3	103.5	30.1	3.1
	Total	240.0	106.8	34.2	6.1
N_{total}	1	0.4	0.3	0.3	0.2
	2	12.8	7.4	1.6	0.1
	Total	13.2	7.7	1.9	0.3
P_2O_5	1	0.2	0.2	0.2	0.1
	2	5.7	3.4	0.7	0.2
	Total	5.9	3.6	0.9	0.3
K_2O	1	2.3	1.3	0.9	0.6
	2	3.1	1.6	0.4	0.5
	Total	5.4	2.9	1.3	1.1

[a]*1* losses in solution, *2* losses as suspended material

that average annual washout of fine-grained soil from bare ground was 16 t/ha, of which 44 % was accounted for by thaw water and 56 % run-off from downpours (Fig. 28.1).

The intensity of erosion is somewhat reduced under row crops (average annual losses of 14.5 t/ha of soil) and substantially reduced under continuously sown crops, especially winter cereals. Under summer cereals, the average run-off was 10.1 t/ha per year (36 % less than from bare soil and 30 % less than from row crops); this represents only a reduction of washout during summer rains – erosion in the spring thaw is undiminished. Annual soil washout under winter cereals averaged 2.4 t/ha, hardly more than the acceptable level of 2.0 t/ha and nearly all during the spring thaw. Under perennial grass, there is hardly any soil loss – no more than 0.3 t/ha per year.

Losses of humus, phosphorous and potassium in solution and suspension are in line with the washout of soil under different land uses (Table 28.1). Under row crops and summer cereals (including losses from bare soil over winter), annual washout

Table 28.2 Coefficients of soil-protective capability for arable crops

Crops	Coefficients	
	By winter tillage + bare fallow	By bare fallow
Row crops (potato)	0.08	0.15
Annual grasses	0.36	0.62
Spring cereals and grain legumes	0.36	0.67
Winter crops	0.89	–
Perennial grasses in year of establishment	0.92	–
Perennial grasses in the second year	0.98	–

comprises 240 and 106.8 kg/ha of humus, respectively; 13.2 and 7.7 kg/ha total nitrogen; 5.9 and 3.6 kg/ha labile phosphorus; 5.4 and 2.9 kg/ha labile potassium. Because winter cereal crops protect soil for most of the year and perennials all year long, the annual washout of nutrients was substantially less: under winter cereals, 2.4 times less than under summer cereals and comprising 34.2 kg/ha humus, 1.9 kg/ha nitrogen, 0.9 kg/ha phosphorous, and 1.3 kg/ha potassium. Under perennial crops, the macroelement losses did not exceed 1 kg/ha and humus losses 6.1 kg/ha.

Long-term studies of soil loss enable a ranking of crops according to their effectiveness in preventing erosion (in descending order): perennial grasses – winter crops – summer crops – row crops. For planning of measures to prevent erosion, it is also useful to have indices that characterize the soil-protective capability of different agricultural practices. Amongst those for field crops, the most efficient is that suggested by Konstantinov (1977):

$$C_p = (W_t - W_f)/W_t$$

where C_p the coefficient of soil-protective capability; W_t the soil washout from winter tillage and/or bare fallow, t/ha; and W_f the soil washout from a field in crop, t/ha.

Table 28.2 presents the coefficients determined on the basis of long-term data on soil washout under different crops.

The highest coefficients are for perennial grasses (0.92–0.98) and winter cereals (0.89) the lowest for row crops (0.08–0.15). However, the coefficients of summer cereals and row crops are halved if we take account of total annual soil losses, rather than just losses over the growing period and the soil-protective capabilities of summer cereals, and row crops are better represented by considering soil losses relative to bare fallow during the growing season and, also, winter tillage so that the total annual soil losses are accounted for.

Knowing the soil-protective coefficients of individual crops, standards of erosion-preventive capability of crop rotations may be calculated as weighted means:

$$S_{ea} = (Cp_1 \times S_1 + Cp_2 \times S_2 + \cdots + Cp_n \times S_n)/100$$

Table 28.3 Soil physical properties under a protective crop rotation and *Galega*

		Density, kg/m^3			Porosity, %		
Crop	Erodibility[a]	2006	2010	2006 ± 2010	2006	2010	2006 ± 2010
Galega	1	1.09	1.03	−0.06	59	61	+2
	2	1.14	1.11	−0.03	57	58	+1
	3	1.18	1.19	+0.01	55	55	0
Soil-protective rotation	1	1.30	1.23	−0.07	50	53	+3
	2	1.40	1.32	−0.08	47	50	+3
	3	1.44	1.37	−0.07	47	48	+1
$HCP_{0.05}$ factor A (soil)				0.08			1
factor B (crop)				0.11			2

[a]*1* non-eroded soil, *2* slightly eroded, *3* moderately eroded

Table 28.4 Soil biological indices under the soil-protective rotation and continuous *Galega*

Crop	Degree of erosion[a]	Dehydrogenase, mg triphenylformazan/kg of soil	Polyphenol oxidase, mg quinone/kg soil	Peroxidase, mg quinone/ kg soil	Invertase, mg glucose/ kg soil
Galega	1	857	46.8	42.5	3,358
	2	797	44.9	40.4	2,895
	3	505	43.5	36.7	2,557
Soil-protective crop rotation	1	642	30.3	31.2	2,099
	2	392	28.2	27.4	1,948
	3	255	25.0	21.6	1,420
$HCP_{0,05}$ factor A (soil)		138	1.32	2.16	318
factor B (crop)		113	1.08	1.76	260

[a]*1* non-eroded soil, *2* slightly eroded soil, *3* moderately eroded soil

where S_{ea} is a standard of erosion-preventive capability of a crop rotation; Cp_1, Cp_2, Cp_n the coefficients of soil-protective capability of individual crops; and S_1, S_2, S_n is the saturation of a crop rotation with certain crops (%)

The greater the value of the standard, the greater is the erosion-preventive capability. For common crop rotations in Belarus, the standard varies from 0.52 to 0.96, depending on the composition and frequency of the component crops.

28.3.2 Erosion-Protective Rotations

Some of the more interesting studies over the 45 years since the establishment of the run-off plots have focused on the effects of a soil-protective crop rotation (peas-and-oats for green mass – winter barley – lucerne and clover – perennial grasses for 1–3 years) and a continuous (19 years) culture of Caucasian goat's rue *(Galega orientalis* L.) – a valuable fodder crop. Over the last 5 years, measurements have been made of agro-physical properties (Table 28.3) and biological activity (Table 28.4).

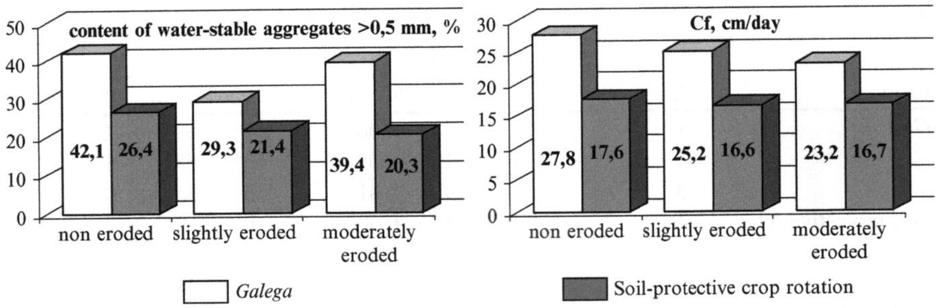

Fig. 28.2 Erosion resistance under a soil-protective rotation and *Galega*

At the beginning of the experiment in 2006, the bulk density of the plough layer under the soil-protective crop rotation was 1.30–1.44 kg/m^3 with the higher densities on more-eroded soils. In the course of 5 years, it was reduced to 1.23–1.37 kg/m^3. Compared with the crop rotation, the density of the plough layer under *Galega* was 11–26 % lower throughout the study – close to optimal – thanks to the well-developed root system and elimination of the compacting effects of farm traffic. Porosity is inversely related to the bulk density; under *Galega*, it is characterized as optimal, even on the eroded variants. There was also a measurable improvement of soil structural resilience measured by the content of water-stable aggregates >0.5 mm and filtration coefficient (Fig. 28.2).

Galega increased erosion resistance and lessened the differences between non-eroded and eroded variants; the non-eroded soil contained 42.1 % of water-stable aggregates and the moderately eroded soil was almost equal with 39.4 %. Under the crop rotation, the content of water-stable aggregates was 26.1 % in non-eroded soil and 20.3 % in moderately eroded soil – 1.4–1.9 times less than under *Galega*. The soil's infiltration capacity (Cf) also improved, largely through better physical characteristics as well as the strong root system; the calculated infiltration coefficient under *Galega* was 23.2 cm/day on moderately eroded soil up to 27.8 cm/day on non-eroded soil – 6.5–10.2 cm/day higher than under the protective crop rotation. Furthermore, *Galega* is an efficient accumulator of organic matter and nitrogen in soil.

Dehydrogenase output indicates the number and activity of the soil microbial community. *Galega* increased biological activity throughout the soil profile and catena, most significantly in slightly and moderately eroded soils, in which biological activity doubled compared with the soil-protective crop rotation.

Invertase output indicates available carbohydrate substrate. *Galega* increased available carbohydrates throughout the catena by 1.6 times in non-eroded soil and by 1.5–1.8 times in eroded variants.

Polyphenol oxidase output under *Galega* was 43.5–46.8 mg quinone/kg of soil, 1.5–1.7 times higher than under the crop rotation, and the influence of the soil's erosion status was less. Similarly, peroxidase output under *Galega* was 36.7–42.5 mg quinone/kg of soil, 1.4–1.7 times higher than under the crop rotation, and decreased by 5–16 %, depending on the degree of erosion, whereas under the protective crop rotation, it decreased by 14–44 % (from a lower standard) according to soil erosion status.

28.4 Conclusions

1. Washout of soil, humus and nutrient elements during the spring snowmelt and during spring and summer rainstorms depends upon soil preparation and the protection of the soil surface by vegetation. Total losses of humus and macroelements under row crops and spring cereals were in 2–4 times higher than under winter crops and in 20–40 times higher than under perennial grasses.
2. Data on total annual soil washoff under different crops made it possible to calculate coefficients of their erosion-preventive capability: perennial grasses over 2 or more years, 0.98; perennial grasses in their first year, 0.92; winter cereals, 0.89; summer cereals, 0.67 during the growing season but 0.36 over the whole year; and row crops, 0.15 during the growing season but 0.08 over the whole year.
3. Continuous Caucasian goat's rue *(Galega orientalis* L.) provided increased soil resilience against erosion measured in terms of bulk density and increase in permeability to optimal values, improved soil structure and water stability and a substantial increase of biological activity. Furthermore, it is a productive forage crop!

References

Chernysh AF (ed) (2005) Recommendations for the construction of anti-erosion complexes and use of the erodible lands in different Belarus landscape zones. Institute for Soil Science and Agrochemistry, Minsk (Russian)

Institute of Soil Science and Agrochemistry (2000) Environmentally safe use of eroded land in Belarusian Poozer'ye landscapes: recommendations. ISSA, Minsk (Russian)

Konstantinov I (1977) Soil erosion in crop rotations on slopes and its control. In: Water erosion and its control. Kolos, Moscow, pp 156–161 (Russian)

Kuznitsov G, Smeyan N, Tsytron G et al (eds) (2011) Soils of agricultural lands of Republic of Belarus: practical guide. Orgstroj, Minsk (Russian)

Sidorchuk A, Litvin L, Golosov V, Chernysh A (2006) Soil erosion in European Russia and Belarus. In: Soil erosion in Europe. Wiley, Chichester, pp 72–95

Chapter 29
Crop Yield and Quality Depending on Fertilization in Crop Rotation on Sod-Podzolic Soil

V. Lapa and M. Lomonos

Abstract On the basis of field experiments, optimal fertilization was established for various crops in rotation on Albeluvisol (sod-podzolic) soil near Minsk, Belarus. Peas with oats receiving $N_{60}P_{60}K_{120}$ at pre-sowing yielded 5.9 t/ha of green fodder; Voltario winter triticale receiving 40 t/ha farmyard manure and $P_{60}K_{120}$ at autumn pre-sowing, N_{90} during spring growth and N_{30} at the first-node stage yielded 9.8 t/ha grain and more than 10 centner/ha of crude protein; under a cover crop of winter triticale, application of $P_{60}K_{120}$ and early spring application of $P_{60}K_{140}$ on clover yielded 11.65 t/ha green mass; for Toma spring wheat, pre-sowing application of $N_{60}P_{60}K_{120}$ with additional N_{30} at first node and N_{30} at last leaf yielded 7.1 t/ha grain; Yantar spring rape with $N_{90}P_{60}K_{120}$ applied at pre-sowing yielded 2.8.t/ha of seed with a crude protein content of 662 kg/ha. For the crop rotation as a whole, $N_{24-72}P_{60}K_{124}$ mineral fertilizers with 8 t/ha of manure provided a yielding capacity of 102–104 c/ha fodder units. However, under this rotation, the soil's humus content diminished from 1.98 to 1.78 % and soil mobile potassium lost 61 mg/kg (12.2 mg/kg/year), while mobile phosphate increased by 18 mg/kg (3.6 mg/kg/year).

29.1 Introduction

Scientifically proven crop rotations and agronomic practices are the basis of sustainable, profitable, arable farming. The issue of increasing farm productivity is dealt with by application of fertilizers in conjunction with other technical practices – so efficient fertilizer use is a priority. Attention is also paid to the quality of food crops, animal feed and industrial raw materials which, in many ways, measures their economic efficiency and competitiveness.

V. Lapa (✉) • M. Lomonos
RUE Institute for Soil Science and Agrochemistry, Minsk, Belarus
e-mail: brissagro@gmail.com

D. Dent (ed.), *Soil as World Heritage*, DOI 10.1007/978-94-007-6187-2_29,

The goal of agrochemistry is to obtain high crop yields of optimum quality, maintain or raise soil fertility, and meet the needs for ecological security and protection of the environment.

29.2 Experimental Site and Methods

The effects of mineral fertilizer on crop yield and quality were investigated at the Shchemyslitsa co-operative, in the Minsk region, over the period 2006–2011 in a crop rotation of peas with oats > winter triticale undersown with clover > clover meadow > spring wheat > spring rape. The soil is sod-podzolic[1] light loamy on thick loess with a plough layer of pH_{KCl} 5.8–6.0, P_2O_5 229–422 mg/kg and K_2O 63–222 mg/kg and humus content of 1.62–1.78 %.

Farmyard manure was applied at 40 t/ha before sowing peas with oats. Plant protection included control of weeds, disease and lodging of cereal crops. Analysis of soil samples followed standard practice: available phosphorus and potassium were determined by Kirsanov's method (0.2 M HCl) and humus by Tiurin's method (0.4 M $K_2Cr_2O_7$). In plant samples, after wet oxidation in a mixture of sulphuric acid and hydrogen peroxide, nitrogen and phosphorous were determined by colorimetry and potassium by flame photometry. Crude protein was estimated by multiplying the total nitrogen content by 6.25 (Vildflush et al. 1998).

29.3 Results and Discussion

The productivity of the rotation depended on fertilization (Table 29.1). Yield without fertilizer was 64.6 centner fodder units per hectare and 103.9 c/ha in the variant receiving $N_{96}P_{60}K_{124}$ with split application of nitrogen fertilizer.

Field Peas with Oats: In comparison with the control, manuring prior to sowing yielded an increase of 13.5 c fodder units/ha. Against a background of $P_{30}K_{62}$ in variants receiving complete mineral fertilizers, application of $N_{24–72}$ yielded an additional 3.4–6.4c fodder units/ha. Application of $N_{24–96}$ against a background of $P_{60}K_{124}$ yielded only an extra 2.6–4.9 c fodder units/ha. Optimal nitrogen application for peas with oats was N_{60} before sowing against a background of $P_{60}K_{120}$, yielding 538 c/ha green mass. There was little or no benefit from increased doses of nitrogen. The small benefit from mineral nitrogen was underscored by the negative effect of carried-over nitrogen on the subsequent clover crop.

Farmyard manure applied at 40 t/ha and ploughed in before sowing proved to be an efficient fertilizer, yielding an additional 77 c/ha green mass, equivalent to 8.5 c/ha

[1] Albeluvisol in *World Reference Base for Soil Resources 2006.*

Table 29.1 Productivity of grain-and-grass rotation under different fertilizer treatments

Variant: Σ per 1 ha	Variant: Ø per 1 ha	Peas with oats, green mass, c/ha	Winter triticale grain, c/ha	Clover, green mass, c/ha	Spring wheat, grain, c/ha	Spring rape, seed, c/ha	Mean yield fodder units, c/ha/year
–	–	286	57.5	687	40.4	21.4	64.6
	40 t manure	363	66.7	862	46.3	24.4	78.1
N_{120}	N_{24}	419	74.4	840	51.3	28.1	82.5
N_{240}	N_{48}	484	78.1	822	56.2	31.2	86.1
N_{360}	N_{72}	498	80.2	767	59.1	34.8	86.3
$N_{240}P_{150}$	$N_{48}P_{30}$	489	79.4	857	57.1	31.4	88.3
$N_{240}K_{310}$	$N_{48}K_{62}$	490	79.6	875	57.8	31.8	89.4
$P_{150}K_{310}$	$P_{30}K_{62}$	422	73.8	1,035	50.9	28.7	90.7
$N_{120}P_{150}K_{310}$	$N_{24}P_{30}K_{62}$	453	78.6	1,008	57.4	32.7	94.1
$N_{240}P_{150}K_{310}$	$N_{48}P_{30}K_{62}$	489	81.7	983	61.2	35.8	96.3
$N_{360}P_{150}K_{310}$	$N_{72}P_{30}K_{62}$	517	84.6	931	64.6	38.7	97.1
$P_{300}K_{620}$	$P_{60}K_{124}$	442	81.0	1,165	51.5	31.8	99.3
$N_{120}P_{300}K_{620}$	$N_{24}P_{60}K_{124}$	504	85.8	1,109	58.7	34.4	101.9
$N_{240}P_{300}K_{620}$	$N_{48}P_{60}K_{124}$	538	89.3	1,072	62.8	37.3	103.8
$N_{360}P_{300}K_{620}$	$N_{72}P_{60}K_{124}$	547	92.8	1,009	65.3	40.4	103.6
$N_{360}P_{300}K_{620}$	$N_{72}P_{60}K_{124}$	546	94.1	1,015	67.8	38.3	104.2
$N_{480}P_{300}K_{620}$	$N_{96}P_{60}K_{124}$	556	98.4	959	70.6	38.9	103.9

of fodder units. The most efficient system was application of complete mineral fertilizer $N_{60}P_{60}K_{120}$ at preplanting, yielding 538 c/ha (59.2 c/ha fodder units).

Winter Triticale: Grain yield of Voltario winter triticale varied from 57.5 to 98.4 c/ha, according to fertilization. Compared with the unfertilized control, application of 40 t/ha of farmyard manure to the bare ground before sowing yielded an increase of 9.2 c grain/ha. The maximum yield was achieved with split application of N_{120} (N_{90} during spring growth + N_{30} at the first-node stage) against the background of pre-sowing application of $P_{60}K_{120}$ and 40 t/ha of manure, yielding 98.4 c/ha (5.6 c/ha attributable to the additional nitrogen N_{30} dose and 31.7 c/ha attributable to the complete mineral fertilizer).

Clover Meadow: The productivity of clover depends a lot on the weather, especially moisture supply in the soil, as well as on the availability of nutrients. The green mass harvested in two cuts, 687–1,165 c/ha (Table 29.1), benefitted from the after-effect of manure applied earlier in the rotation, yielding an additional 175 c/ha compared with the control.

The condition and fertilization of the cover crop also affected the growth of the clover meadow in its first year – a big cover crop means worse conditions for the interplanted perennial grasses, which were severely thinned. Excess nitrogen caused lodging of the cover crop, resulting in heavy shading and death of the clover (Kazantsev 2002, Kidin 2009). In our trials, application of 30 kgN/ha to the triticale increased the triticale yield by 7–10 % but decreased the yield of clover by 2.5–5 %, depending on background conditions, and application of 90kgN/ha raised the yield of triticale by 15–21 % but decreased the yield of clover by 11–13 %. The optimal fertilization for triticale was $N_{90+30}P_{60}K_{120}$ but, in this case, the yield of clover was 959 c/ha (206 c/ha lower than the highest clover productivity). The highest yield of clover (1,165 c/ha) was obtained against the background of the after-effects of manure and application of $P_{60}K_{120}$ to the triticale and $P_{60}K_{140}$ the following year to the clover.

Spring Wheat: Mineral fertilizers substantially enhance the yield and grain quality of Toma spring wheat (Table 29.1). Grain yield varied from 40.4 c/ha without fertilizer up to 70.6 c/ha with $N_{60+30+30}P_{60}K_{120}$. The after-effect of manure (8 t/ha of crop rotation area) yielded an increase of 5.9 c/ha compared with the control. Application of phosphate and potassium fertilizers provided additional grain yields of 4.6–5.2 c/ha but increasing the dose from $P_{30}K_{60}$ up to $P_{60}K_{120}$ gave little benefit. Nitrogenous fertilizers played the main role producing big yields of spring wheat; increasing doses of nitrogen yielded increases of 5.0–19.1 c/ha and split application of N_{60} at pre-sowing and N_{30} at the first-node stage achieved much the same as a single application of N_{90}. The highest yield (70.6 c/ha) was achieved with a pre-sowing application of $N_{60}P_{60}K_{120}$ in combination with additional N_{30} at first node and N_{30} at last leaf; in this variant, 19.1 c/ha yield increase was attributable to the nitrogen fertilizer and 24.3 c/ha to the complete fertilizer.

Spring Rape: Yields of rape seed increased from16.6 c/ha without fertilizer to 28.2 c/ha with $N_{90}P_{60}K_{120}$. Nitrogen fertilizer proved to be of prime importance in yield formation; increasing doses yielded an additional 2.6–4.6 c/ha. Phosphate and potassium also had a positive effect: $P_{30}K_{60}$ at pre-sowing yielded an additional 2.5 c/ha in the background variant and increasing doses up to $P_{60}K_{120}$ yielded a further 1.3 c/ha. The after-effect of 40 t/ha of farmyard manure yielded an increase of 3.2 c/ha. The highest yield of rape seed (28.2 c/ha) was achieved with pre-sowing application of $N_{90}P_{60}K_{120}$. In this variant, yield increase due to nitrogen fertilizer was 4.6 c/ha and to complete mineral fertilizer 8.4 c/ha – with cost recovery of 1 kg NPK with 3.1 kg of seed. Further increase of nitrogen dosage actually decreased the yield and split application of 90 kgN/ha showed no benefit over a single application.

29.3.1 Dynamics of Agrochemical Indices

Over the period of the rotation, the humus content of the soil diminished by 0.18–0.37 %, depending on fertilization (Table 29.2). Even with the addition of farmyard manure and complete mineral fertilization, the humus content of the plough layer declined from 1.94–1.96 to 1.72–1.78 %.

Against the background application of 8 t/ha farmyard manure, an average annual application of 30 kg/ha phosphate (active ingredient) in combination with nitrogenous fertilizer decreased labile phosphate by 44 mg/kg. In combination with potassic fertilizers, there was no significant change. With nitrogenous and potassic

Table 29.2 Change of agrochemical indices of light loamy sod-podzolic soil under grain-and-grass rotation, 2006–2010

	Agrochemical indices								
	Humus (%)			P_2O_5 (mg/kg)			K_2O (mg/kg)		
Annual average fertilizer doses	2006/ 2007	2010/ 2011	±	2006/ 2007	2010/ 2011	±	2006/ 2007	2010/ 2011	±
8 t/ha farmyard manure	1.87	1.63	−0.24	251	229	−22	121	63	−58
Manure + $N_{24–72}$	1.93	1.70	−0.23	281	247	−34	147	80	−67
Manure + $N_{48}P_{30}$	1.99	1.62	−0.37	322	278	−44	153	76	−77
Manure + $N_{48}K_{62}$	1.98	1.62	−0.36	282	258	−24	213	139	−74
Manure + $P_{30}K_{62}$	1.90	1.69	−0.21	357	358	1	223	154	−69
Manure + $N_{24–72}P_{30}K_{62}$	1.94	1.72	−0.22	356	346	−10	208	151	−57
Manure + $P_{60}K_{124}$	2.02	1.76	−0.26	400	417	17	297	248	−49
Manure + $N_{24–72}P_{60}K_{124}$	1.96	1.78	−0.18	404	422	18	283	222	−61
Student's HCP_{05}	0.10	0.14		17	24		10	20	

fertilizers, labile phosphate decreased by 10 mg/kg. Increasing the annual dose of phosphate from 30 to 60 kg/ha increased the content of labile phosphate by 17–18 mg/kg of soil. Application of potassic fertilizers at 62 kg/ha of active ingredient per year resulted in a decrease of labile potassium by 69 mg/kg soil per rotation cycle; doubling the annual dose of potassium resulted in an annual diminution of 12.2 mgK/kg of soil.

Alongside crop yields, attention is also given to indices of quality such as protein content. In the grain-and-grass rotation, clover meadow is the dominant protein producer yielding with 1,536–2,433 kg/ha of protein (Table 29.3). Spring rape produced the lowest protein harvest as a result of its low seed yield.

Over the whole grain-and-grass crop rotation, the total protein yield was 3.4–6.1 t/ha with the annual average yield of 682–1,220 kg/ha. The greatest average annual protein yield was obtained with application of complete mineral fertilizers ($N_{96}P_{60}K_{124}$) against the background of 8 t/ha average annual application of farmyard manure over the crop rotation area.

29.4 Conclusions

On light loamy sod-podzolic soil with a background application of 40 t/ha of farmyard manure, optimal fertilization for the constituent crops of a grass-and-grains rotation was:

- For field peas with oats, $N_{60}P_{60}K_{120}$ at pre-sowing, yielding 5.9 t/ha of green fodder.
- For Voltario winter triticale, $P_{60}K_{120}$ at autumn pre-sowing, N_{90} during spring growth and N_{30} at the first-node stage, yielding 9.8 t/ha grain and more than 10 centner/ha of crude protein.

Table 29.3 Crude protein yield of grain-and-grass rotation, kg/ha

Variant: Σ per 1 ha	Variant: Ø per 1 ha	Peas with oats, green mass	Winter triticale, grain	Clover, green mass	Spring wheat, grain	Spring rape, seed	Annual average protein yield
–	–	651	492	1,536	410	319	682
40 t	8 t	862	588	1,959	479	379	853
N_{120}	N_{24}	1,002	698	1,811	542	439	898
N_{240}	N_{48}	1,166	762	1,856	617	495	979
N_{360}	N_{72}	1,260	805	1,638	674	562	988
$N_{240}P_{150}$	$N_{48}P_{30}$	1,208	768	1,963	636	498	1,015
$N_{240}K_{310}$	$N_{48}K_{62}$	1,156	771	1,938	649	506	1,004
$P_{150}K_{310}$	$P_{30}K_{62}$	1,012	670	2,313	540	435	994
$N_{120}P_{150}K_{310}$	$N_{24}P_{30}K_{62}$	1,118	736	2,266	623	513	1,051
$N_{240}P_{150}K_{310}$	$N_{48}P_{30}K_{62}$	1,285	798	2,134	702	565	1,097
$N_{360}P_{150}K_{310}$	$N_{72}P_{30}K_{62}$	1,342	848	1,942	759	610	1,100
$P_{300}K_{620}$	$P_{60}K_{124}$	1,081	734	2,565	563	476	1,084
$N_{120}P_{300}K_{620}$	$N_{24}P_{60}K_{124}$	1,246	828	2,433	653	534	1,139
$N_{240}P_{300}K_{620}$	$N_{48}P_{60}K_{124}$	1,348	879	2,319	719	604	1,174
$N_{360}P_{300}K_{620}$	$N_{72}P_{60}K_{124}$	1,426	942	2,133	803	662	1,193
$N_{360}P_{300}K_{620}$	$N_{72}P_{60}K_{124}$	1,412	964	2,132	843	639	1,198

- Under winter triticale as a cover crop, application of $P_{60}K_{120}$ and early spring application of $P_{60}K_{140}$ on clover meadow (Vitebchanin variety) yielded 11.65 t/ha green mass.
- For Toma spring wheat, pre-sowing application of $N_{60}P_{60}K_{120}$ with additional N_{30} at first node and N_{30} at last leaf, yielding 7.1 t/ha grain.
- For Yantar spring rape, application of $N_{90}P_{60}K_{120}$ at pre-sowing, yielding 2.8 t/ha of seed with crude protein yield of 662 kg/ha.
- The regime of $N_{24–72}P_{60}K_{124}$ mineral fertilizers with 8 t/ha of manure provided a crop rotation yielding capacity of 101.9–104.2 c/ha fodder units. The highest annual average protein yield of 1,220 kg/ha was obtained after application of a complete mineral fertilizer ($N_{96}P_{60}K_{124}$) against the background of application of manure at 8 t/ha of the crop rotation area.

Under this rotation, humus content of the topsoil diminished from 1.98 to 1.78 %; mobile potassium diminished by 61 mg/kg soil (12.2 mg/kg/year), while mobile phosphate increased by 18 mg/kg (3.6 mg/kg/year).

References

Kazantsev VP (2002) Meadow fodder production. Nauka – Russian Academy of Science, Novosibirsk (Russian)

Kidin VV (2009) Special features of cultivated crop nutrition and fertilizer application: educational guidance. RSAU – MTAA KA Timiryazev Agricultural Academy, Moscow (Russian)

Vildflush IR et al (1998) Practical course in agrochemistry. Uradzhai, Minsk

Chapter 30
Potassium Effects on Wheat Yield and Quality in Long-Term Experiments on Luvisol in Romania

G. Ciobanu, C. Domuţa, R. Brejea, Cornelia Ciobanu, L. Bara, Camelia Bara, Ramona Albu, Maria Şandor, G. Bunta, and A. Vuscan

Abstract Wheat needs a substantial amount of potassium for optimal growth; adequate fertilization improves the efficiency of photosynthesis, water-use efficiency and resistance to disease. In Romania, potassium fertilizers have been less used and less well researched than nitrogen and phosphate fertilizers. We discuss the effect of potassium on wheat yield and quality in a long-term field experiment on Haplic luvisol at the Agricultural and Development Research Station at Oradea, Romania. Four levels of potassium were applied on four nitrogen/phosphate backgrounds. Under these conditions, wheat yield and quality are maximized by balanced fertilization with optimum potassium management; different levels of potassium produced significant variation in wheat yield, protein and gluten content; the best performance was achieved with 80 kgK/ha.

30.1 Introduction

Potassium fertilizer is required on soils that are not naturally well supplied with potassium, especially when large amounts of nitrogen (N) and phosphorus (P) fertilizer are applied. However, across Romania, the response to potassium fertilizers is weaker than to N and P. Several researchers have shown that the efficacy of potassium fertilizer depends on optimal N and P; efficacy is less when potassium fertilizer is applied alone, or with P only, but more when applied with N (Burlacu 2007). In recent years, there has been speculation in Romania about the yield and quality of wheat and the technical capacity of the farmers (Marinciu and Săulescu 2008), fuelled by various issues, especially the variability of grain quality.

G. Ciobanu (✉) • C. Domuţa • R. Brejea • C. Ciobanu • L. Bara •
C. Bara • M. Şandor • A. Vuscan
University of Oradea, Oradea, Romania
e-mail: scdaoradea@yahoo.com

R. Albu • G. Bunta
Agricultural Research and Development Station, Oradea, Romania

D. Dent (ed.), *Soil as World Heritage*, DOI 10.1007/978-94-007-6187-2_30,

This has been attributed to climate change and the lack of suitably adapted new Romanian varieties but research points to inadequate technology and unbalanced fertilization (Hera et al. 1986; Tabără 2008; Semun 2010).

Győri and Sipos (2006) define wheat-grain quality according to physical properties (hectolitre mass, thousand-kernels weight, grain hardness), protein-linked properties (total protein and gluten, expansiveness, sedimentation volume, protein and amino acid composition), rheological properties (farinograph, valorigraph, alveograph and extensograph tests), enzymatic properties (Hagberg falling number, amilograph test) and other examinations (baking test, mycotoxin content, residues of pesticides). Wheat quality, especially protein content and bread-making quality, depends on soil, weather, nitrogen fertilization, plant protection and genotype (Pepó 2002; Tanács et al. 2004; Szentpétery et al. 2004). Appropriate fertilization can increase the protein content by 26–42 % compared with control experiments without fertilizer. Recent research highlights the dependence of grain quality on rates of NPK fertilizer; in the case of unbalanced fertilization, the influence on yield and quality is more pronounced (Wagner and Tabără 2007; Sandor et al. 2010).

Here, we analyse the effects of long-term potassium fertilization against different NP backgrounds on the baking quality of winter wheat of the A2(B1)variety Crişana, bred at Oradea, which is recommended for the hilly region of Crişana, Maramureş, Transylvania and Bucovina and the western Romanian plain.

30.2 Experimental Site and Method

The data are from long-term field experiments with fertilizers at the Agricultural Research and Development Station, Oradea, beginning in 1974, using the unique design of the Fundulea Research Institute (Hera, this symposium). The location is lat. 47°03′N, long. 21°56′E, on a flat terrace of the Crisul Repede River at an elevation of 136 m.

The soil is Haplic luvisol (IUSS 2008, local classification pre-luvisol) on clay loam alluvium; the water table is at 6–8 m. The soil has a dense, slowly permeable subsoil clay pan; a weakly acid plough layer – so mobile aluminium in the topsoil restricts the growth of some crops, such as clover – and no free carbonates; but is well provided with potassium and phosphorus (Table 30.1).

Table 30.2 presents the main agrochemical properties under the conditions of the experiment. Long-term application of fertilizers at different rates produced different trends in nutrients and reaction, depending on the fertilizer rate, and we might expect this to influence wheat yield and its quality.

The experiment with potassium fertilizers was set up in 1974 using a crop rotation of pea – winter wheat – maize – sunflower. Two factors were compared:

A – K fertilization with a_1 K_0, a_2 K_{40}, a_3 K_{80} and a_4 K_{120}
B – NP fertilization with b_1 N_0P_0, b_2 $N_{80}P_{40}$, b_3 $N_{80}P_{80}$, b_4 $N_{160}P_{80}$

Table 30.1 Properties of Haplic luvisol, Oradea

Soil depth cm	Sand (%)	Silt (%)	Clay (%)	SOC (%)	Humus (%)	$CaCO_3$ (%)	Al mobile (mg/100 g soil)	pH 1:2 H_2O	N Total (%)	P mobile (ppm)	K mobile (ppm)
0–5	43.5	28.3	28.2	1.25	2.32	0.00	3.68	6.3	0.12	21.8	83.0
5–15	41.8	28.4	29.8	1.12	2.28	0.00	2.32	6.4	0.11	22.7	102.1
15–30	40.0	28.5	31.5	1.02	1.91	0.00	0.52	6.3	0.09	5.7	112.1
30–60	32.0	28.0	40.0	0.99	1.93	0.00	0.77	6.6	0.09	6.1	117.9
60–90	24.1	36.7	39.2	0.29		0.00	0.32	6.6			
90–50	35.1	27.3	37.6	0.17		0.00	0.59	6.5			

Table 30.2 Soil agrochemical properties under the long-term field experiment, August 2010

		Nutrient elements			
K Rates	NP Rates	pH_{water}	Humus (%)	Mobile P (ppm)	Mobile K (ppm)
K_0	N_0P_0	6.1	1.59	48.2	95.8
	$N_{80}P_{40}$	5.92	1.66	35.6	127.2
	$N_{80}P_{80}$	5.65	1.83	36.3	173.1
	$N_{160}P_{80}$	5.05	1.71	45.1	208.3
Average		5.68	1.69	41.3	151.1
K_{40}	N_0P_0	5.75	1.75	81.3	87.3
	$N_{80}P_{40}$	5.78	2.04	72.4	124.9
	$N_{80}P_{80}$	5.52	2.28	71.8	150.3
	$N_{160}P_{80}$	4.91	1.87	68.1	161.3
Average		5.49	1.98	73.4	130.8
K_{80}	N_0P_0	5.62	1.98	128.3	88.9
	$N_{80}P_{40}$	5.55	2.5	120.1	117.1
	$N_{80}P_{80}$	5.45	2.37	101.5	140.6
	$N_{160}P_{80}$	4.68	1.87	95.2	127.3
Average		5.32	2.09	111.2	118.5
K_{120}	N_0P_0	5.48	1.75	82.5	69.7
	$N_{80}P_{40}$	5.45	2.04	73.4	107.1
	$N_{80}P_{80}$	5.21	2.28	70.2	126.1
	$N_{160}P_{80}$	4.55	1.87	65.2	147.6
Average		5.17	1.98	72.8	112.6

The application of N, as ammonium nitrate, was split equally between sowing time and the beginning of stem elongation in spring; P as superphosphate and K as KCl were applied at sowing. In the case of $N_{160}P_{80}K_{80}$ rates, the experimental results were obtained in the period 2008–2010.

The experiment was designed as a randomized block with four replicates. At the end of October, all plots were sown with approximately 550 wheat seeds/m^2. Weeds were controlled with RIVAL and STAR herbicides, but no treatment was given to control crop diseases. Crops were harvested mechanically at full maturity at the beginning of July. Crop samples were taken from each experimental variant

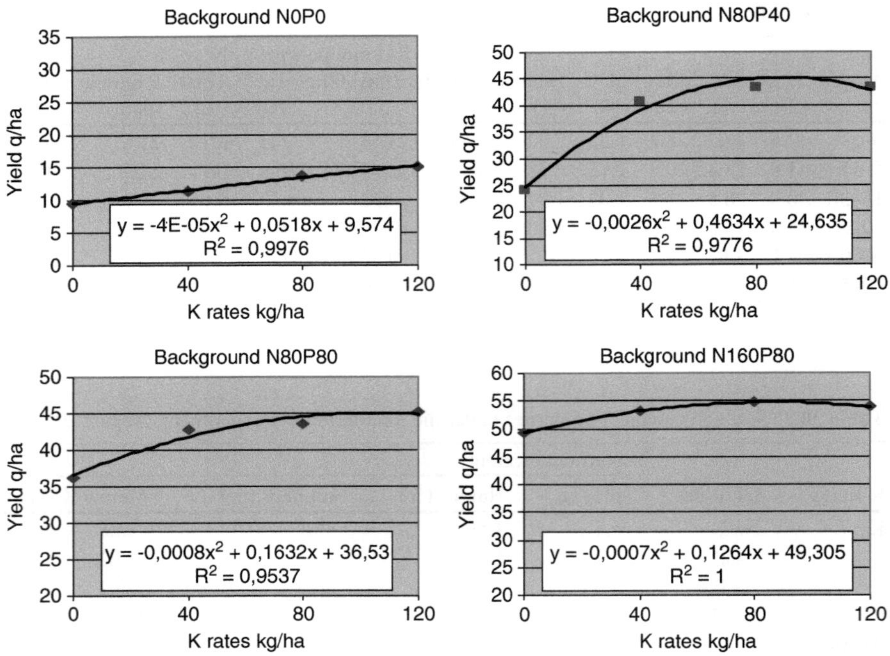

Fig. 30.1 Influence of K × NP fertilizer on yields of winter wheat at Oradea, 2008–2010

for biometric measurements and grain samples from three replicates of each variant were tested to determine the main bread and milling characteristics: TGW (1,000 grain weight) and HM (hectolitre mass) at ARDS Oradea; grain protein was determined by the Kjeldahl method. Soil samples from 0 to 20 cm were collected from each plot in August 2010, after the harvest, and subjected to routine chemical analysis.

30.3 Results and Discussion

30.3.1 Effect of Potassium on Wheat Yield

Long-term application of KxNP fertilizers influenced the main agrochemical properties of the soil and was clearly reflected in grain yields. On this soil, the effect of potassium fertilizer on grain yield depended on the background level of N and P (Fig. 30.1); the optimal application of potassium was 80 kg K_2O/ha for every background level except zero.

30.3.2 Values of Quality Indices

Technical specifications for bread wheat specify a *hectolitre mass* (HM) greater than 75 kg/hl; above 78kg/hl is considered good. HM is significantly influenced by the background nutrient levels created in the long-term experiments. Potassium fertilizer increased HM by 3–5 kg/hl; the highest values were registered in variants $N_{80}P_{80}K_{80}$ and $N_{160}P_{80}K_{80}$. HM values greater than 80 kg/hl and as much as 83 kg/hl show that potassium fertilization facilitates grain filling (Fig. 30.2).

High *protein content* is associated with good bread wheat. Taking values between 8.4 and 13.4 %, the protein content depends on the NP background and the rate of K applied. Lower values, between 8.41 and 8.57, were registered in the case of N_0P_0 backgrounds; the highest values, 12.82–13.41 %, were achieved from the $N_{160}P_{80}$ background. Potassium fertilizer increased protein values by 0.1–0.6 % (Fig. 30.3).

Moist gluten content depends mainly on the dosage of N (Fig. 30.4). The effect of K fertilization is slight in the case of N_0P_0 and $N_{80}P_{40}$ backgrounds, (0.3–1 % increase of moist gluten value). With $N_{80}P_{80}$ and $N_{160}P_{80}$ backgrounds, application of 80 kgK_2O/ha yielded a 3–4 % increase in moist gluten.

Falling number (FN) characterizes the activity of amidine in wheat grain on alpha-amylase activity. The optimal value for bread making is between 180 and 260 s. The lowest FN values (146–149 s) were registered in the case of N_0P_0 background where potassium application brings a slight decrease. The best FN

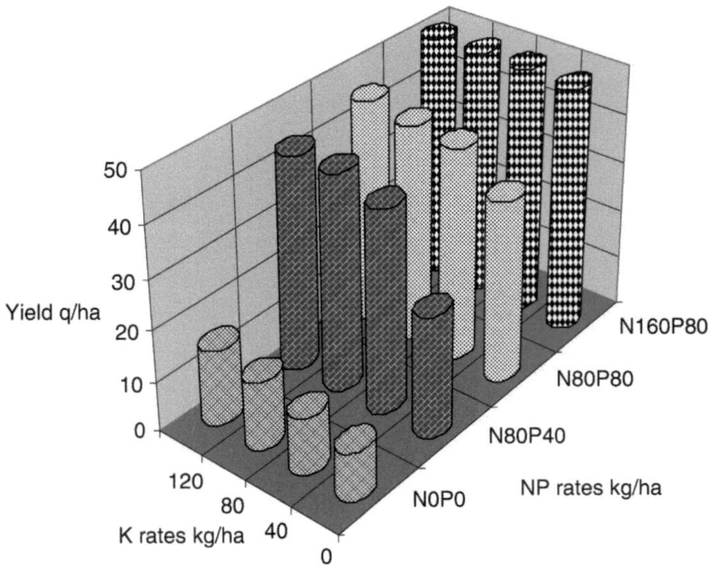

Fig. 30.2 Influence of K × NP fertilizer on HM of wheat grain at Oradea, 2008–2010

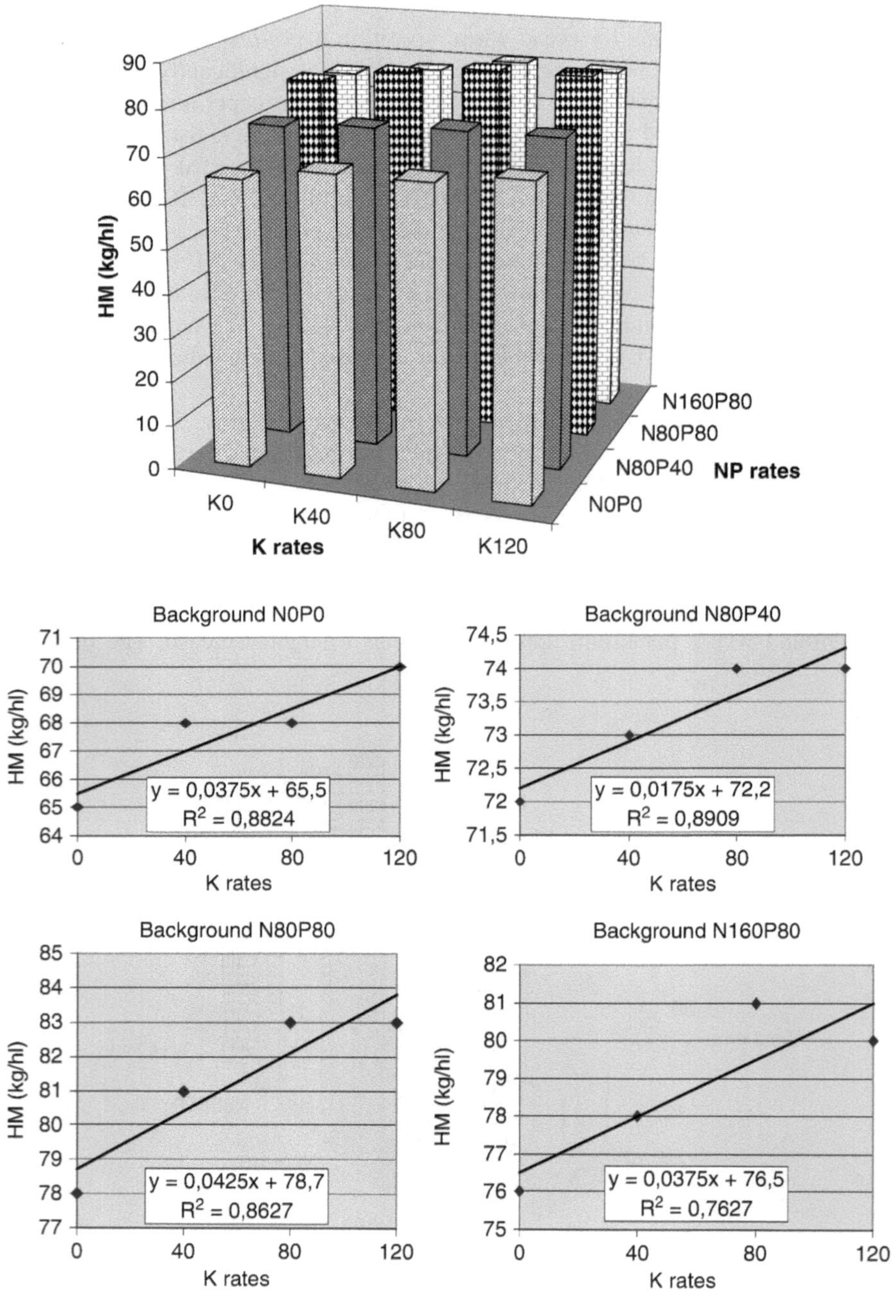

Fig. 30.2 (continued)

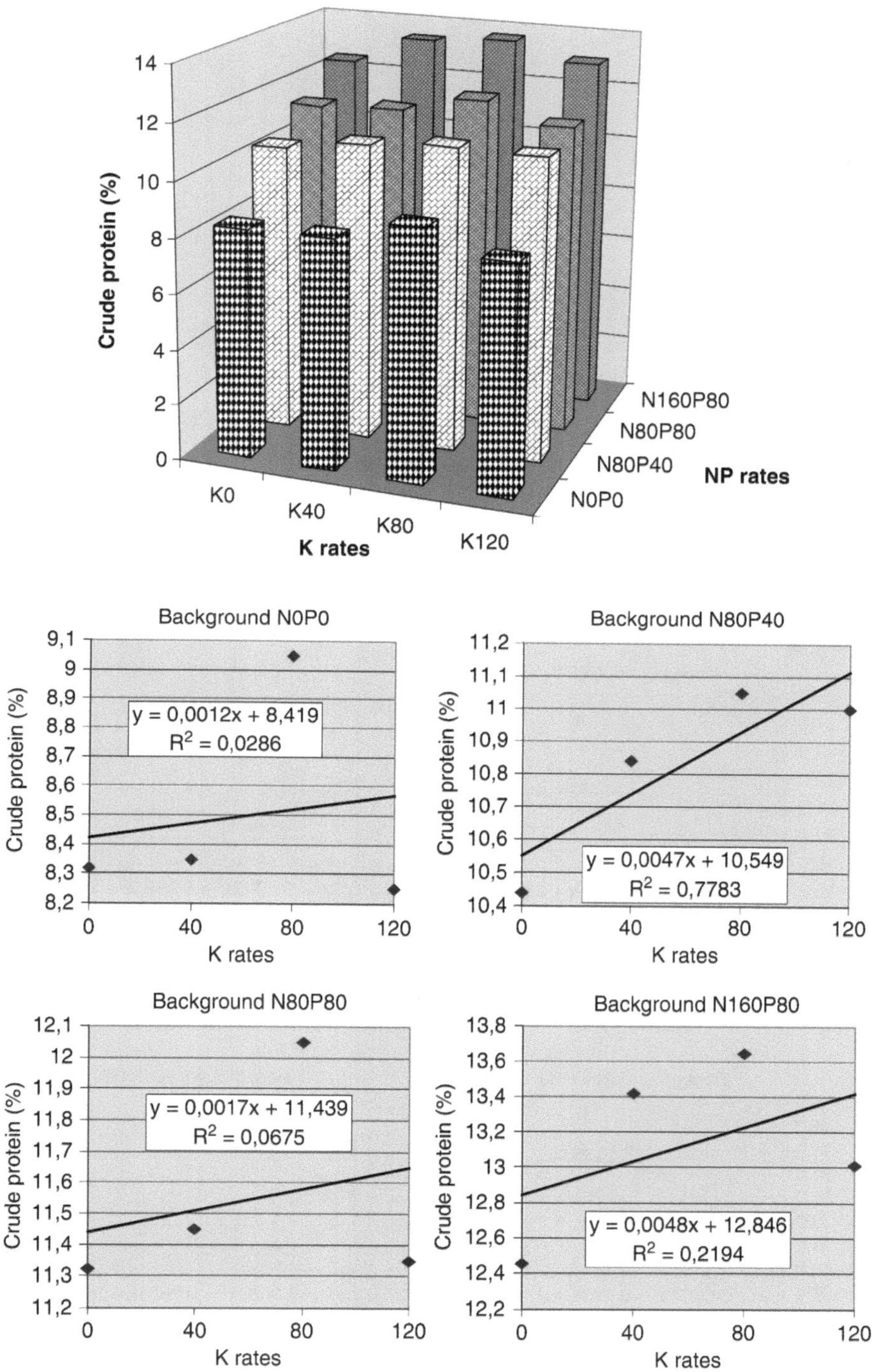

Fig. 30.3 Effect of K × NP fertilizer on crude protein of wheat grain, Oradea 2008–2010

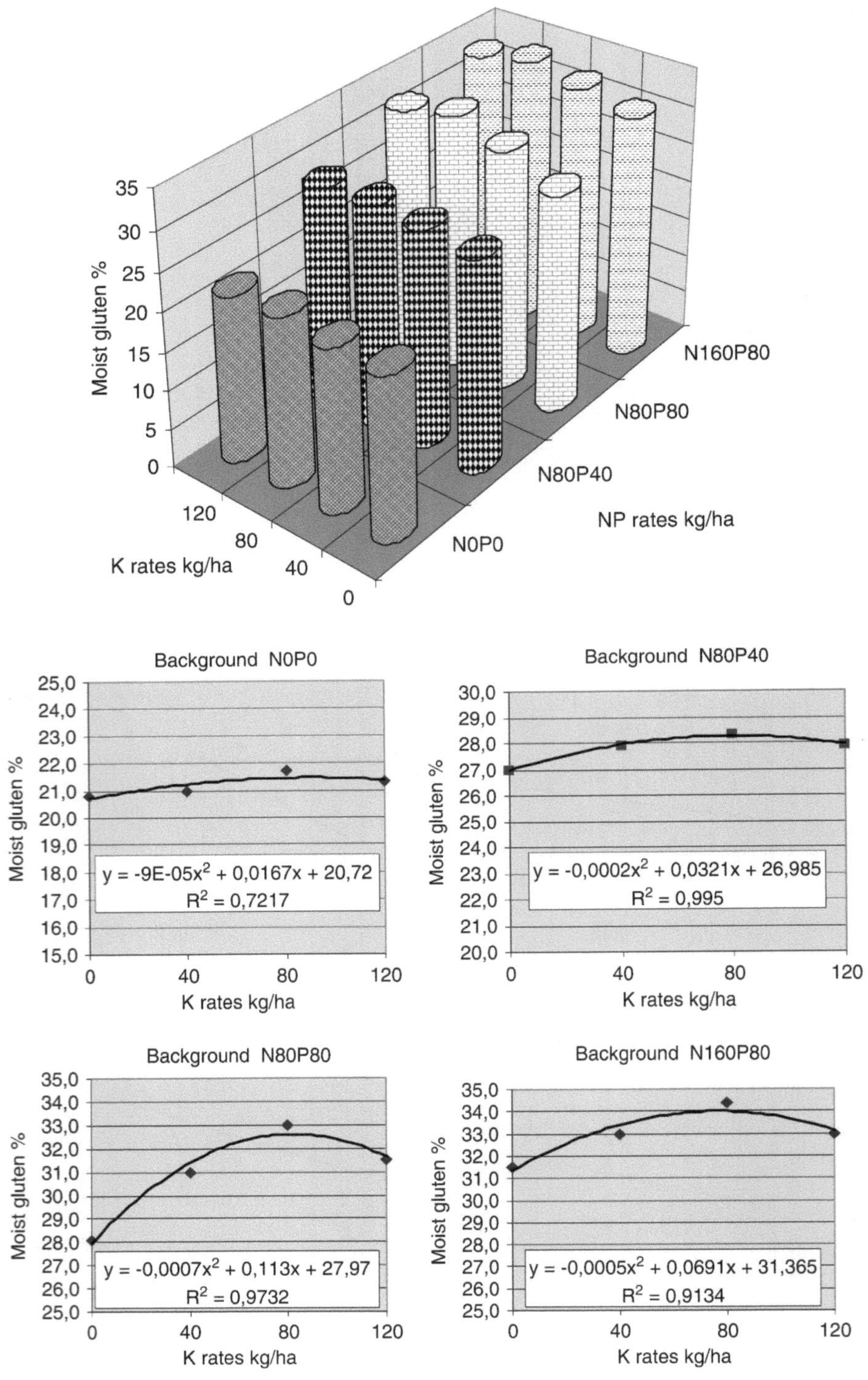

Fig. 30.4 Effect of K × NP fertilizer on moist gluten % of wheat at Oradea, 2008–2010

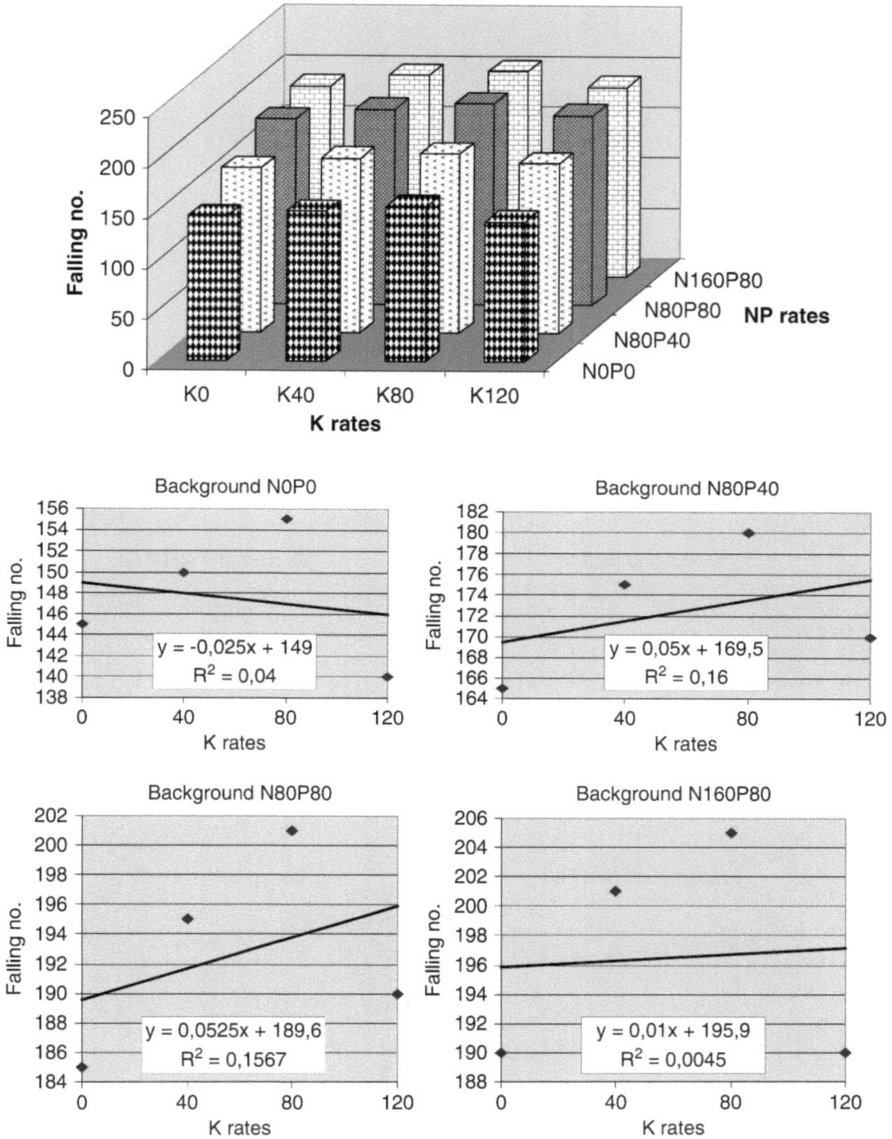

Fig. 30.5 Effect of K × NP fertilizer on the FN of wheat grain, Oradea 2008–2010

was registered in the case of $N_{80}P_{80}$ and $N_{160}P_{80}$ backgrounds, when potassium application brings an increase of these values of 1–6 s (Fig. 30.5).

High values of *deformation index* (DI) in the range 15.8–16.7 mm, indicating a poor baking quality, were registered in the case of N_0P_0 background; the lowest values (6–8 mm) were achieved from the $N_{160}P_{80}$ background. Application of potassium fertilizer had a slight positive influence on deformation index (Fig. 30.6).

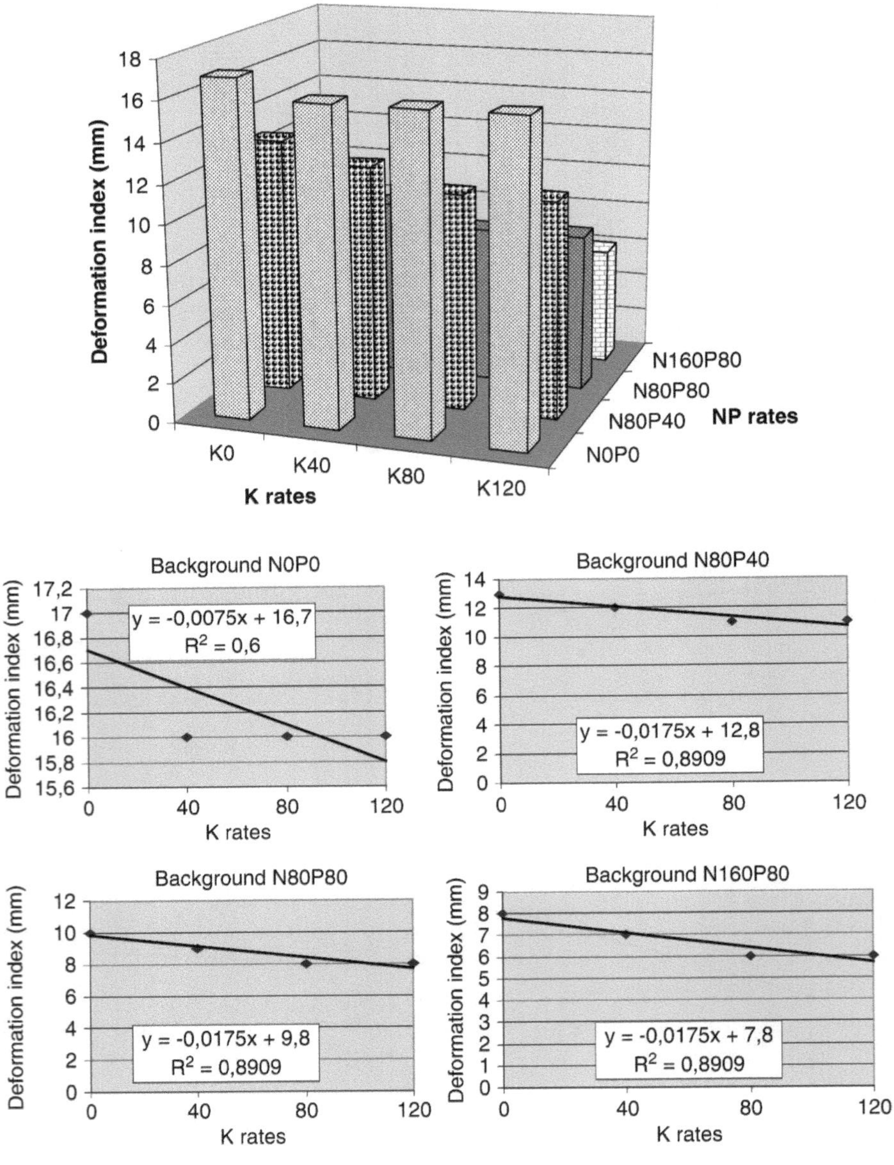

Fig. 30.6 Effect of K × NP fertilizer on the DI of wheat grain at Oradea, 2008–2010

30.4 Conclusions

Potassium promotes the synthesis of the carbohydrates and contributes to the crop's resistance to drought and disease. It is positively correlated with crop yield and hectolitre mass, protein content, moist gluten and falling number of wheat grain and inversely correlated with deformation index. Long-term balanced NPK fertilization maintains the nutrient status of Haplic luvisol at a level that enables high yields of wheat of good bread-making quality.

Acknowledgment The research was carried out under project PN-II-ID-PCE-2008 690/2009: *The study of influences of some technological elements on wheat yield quality under the conditions of the North-Western part of Romania.*

References

Burlacu G (2007) Research results on the rational use of fertilizers. Analele INCDA Fundulea LXXV, Volum jubiliar:287–307 (Romanian)
Győri Z, Sipos P (2006) Investigation on wheat quality in different samples. Buletin USAMV Cluj-Napoca 62:258–263 (Romanian)
Hera C, Alina I, Mihăilă V, Popescu S, Vassiliki P (1986) Influence of fertilizers on wheat quality indices. Probleme Agrofitotehnice 8(2):23–29 (Romanian)
IUSS Working Group WRS (2008) World reference base for soil resources 2006. World Soil Resources Report 103. FAO, Rome
Marinciu C, Săulescu NN (2008) Cultivar effects on the relationship between grain protein concentration and yield in winter wheat. Rom Agric Res 25:19–27 (Romanian)
Pepó P (2002) The behaviour of winter wheat variety depending on seeding period. Nővénytermelés 51:189–198 (Romanian)
Sandor M, Domuţa C, Bunta G et al (2010) Cultivar influence on wheat quality on the Crişurilor Plain. Res J Agric Sci 42(1):264–268 (Romanian)
Semun T (2010) Variation in grain yield and quality of Romanian bread wheat varieties compared to local varieties in north-western Turkey. Rom Biotechnol Lett 15(2):5189–5196 (Romanian)
Szentpétery Z, Hegedűs Z, Jolánkai M (2004) Impact of agrochemicals on yield quality and pesticide residues of winter wheat varieties. Cereals Res Commun 32(2):635–642
Tabără V (2008) Baking quality of winter wheat varieties under ARDS Lovrin conditions. Buletinul AGIR 1–2:3–6 (Romanian)
Tanács L, Matuz J, Gerő L, Petróczi IM (2004) Effect of NPK fertilizers and fungicides on the quality of bread wheat in different years. Cereals Res Commun 32(2):627–634
Wagner L, Tabără V (2007) Studies on milling quality traits in winter wheat (*T. aestivum vulgare*) under the influence of fertilization. Res J Agric Sci 39(1):237–242 (Romanian)

Chapter 31
Effect of Systematic Mineral Fertilization on Available Potassium in Pellic Vertisol

V. Koteva

Abstract The long-term field experiment at the Institute of Agriculture in Karnobat, Bulgaria, encompasses 47 years of systematic fertilization of field crops in rotation on Pellic vertisol with low, moderate and high levels of NPK fertilizer and, in a control experiment, without fertilizer. Cropping without addition of fertilizer decreased available potassium by 5.7 kgK_2O/decare(daa)/year. Annual application of 4 and 8 kg/daa of potassium did not compensate for potassium removed by the crops – there remains a deficit of 5.5 and 3.5 kg/daa, respectively; only annual application of 12 kg/daa achieves a positive balance in the rotation. Despite the deficit, the levels of available potassium in the soil remain high under all treatments. Low and moderate fertilization levels do not significantly change the soil's reserve of available potassium compared to the unfertilized control; under the high fertilization level, available potassium increases in the 0–40 cm horizon compared to the unfertilized control, but changes in the deeper layers are insignificant and reserves remain high.

For our 47-year trial, available potassium determined by Milcheva's method gives better idea of the potassium nutrient for plants than exchangeable potassium determined by Ivanov's method.

31.1 Introduction

Potassium is essential to plants for its participation in biochemical processes throughout the life cycle (Beringer 1980; Mengel 1982; Munson 1985; Mengel and Kirkby 1987). It is taken up mainly by the roots. Most soils have large potassium reserves but only the water-soluble, exchangeable and some of the non-exchangeable potassium – known as *available potassium* – is accessible to plants. Usually, the natural supply of available potassium is quickly depleted

V. Koteva (✉)
Institute of Agriculture, Karnobat, Bulgaria
e-mail: vilikoteva@yahoo.com

D. Dent (ed.), *Soil as World Heritage*, DOI 10.1007/978-94-007-6187-2_31,

by cropping and has to be replenished by application of mineral fertilizers, so it is important to know the changes in soil potassium reserves under long-term cultivation (Milcheva 1980; Berestov et al. 1986)

Agrochemistry depends on long-term field experiments to investigate the dynamics of available potassium (Dospehov et al. 1976; Koteva et al. 2001; Seculic et al. 1998). The long-term trial with mineral fertilization of field crops has been maintained at the Institute of Agriculture at Karnobat since 1963, directed in turn by Philipov (1963–1970), Dimov (1971–1979), Marchev (1979–1982) and, since 1983, by Koteva. The soil is *Pellic vertisol* (IUSS Working Group WRB 2006) derived from andesite. Measurements of water-soluble, mobile and reserve potassium were conducted after the eighth and the 27th year (Koteva 1993) of the trial. The negative potassium balance found after 8 years in some variants of the experiment suggested that there may be a continuing decrease in available potassium; the present study reports on the effects on the potassium balance of 47 years of mineral fertilization of field crops.

31.2 Experimental Method

The 30 cm plough layer of the soil at the trials site has a total humus content of 2.5–2.8 %, acid reaction (pH_{KCl} 5.5–6.2), low reserves of mineral nitrogen (20–40 mg/1,000 g) and mobile phosphorus (3–5 mg/100 g), but a good reserve of mobile potassium (>30 mg/100 g) (Koteva 1993). Below 30 cm, the contents of humus, mineral nitrogen, mobile potassium and phosphorus decrease sharply and the reaction becomes alkaline.

The long-term field experiment involves cultivation of field crops in rotation: maize > wheat > sunflower > barley, according to the traditional system of the region. Nitrogen, phosphorus and potassium fertilizers are applied at 3 levels, *low* T_1, *moderate* T_2 and *high* T_3, and compared with an unfertilized *control* T_0, each with four replicates on trial plots of 200 m^2. According to the biological needs of the plant, the inherent nutrient status of the soil and agronomic recommendations, fertilizer applications over the period 1963 to 2010 were, for T_1, T_2 and T_3, respectively, 156, 320 and 484 kg/decare (daa) K_2O (as potassium chloride); 188, 376 and 564 kg/daa P_2O_5 (as triple superphosphate), 274, 544 and 816 kg/daa N (as ammonium nitrate).

In the early years, very low rates of potassium were applied: from 1963 to 1966, the annual application for T_1, T_2 and T_3, respectively, is 1, 2 and 3 kg/daa K_2O; in 1967–1974, 1, 3 and 5 kg/daa K_2O; increasing to 4, 8 and 12 kg/daa K_2O after 1975. The fertilizer regime has remained unchanged for the 27-year period of our investigation (1983–2010) (Table 31.1).

Soil samples from each variant and replicate were analyzed in 2010 (47th year of the trial). Exchangeable potassium was determined by the acetate-lactate method of Ivanov (1984) and available potassium by extraction with 2 N HCl (Milcheva 1986). Statistical measures of variation and dispersion were applied to the data.

Table 31.1 Fertilizing scheme in long-term trial in the period 1983–2010

Crops	Fertilizing levels			
	Control T_0	Low T_1	Moderate T_2	High T_3
Maize	$N_0P_0K_0$	$N_5P_2K_2$	$N_{10}P_6K_6$	$N_{15}P_8K_8$
Wheat	$N_0P_0K_0$	$N_7P_5K_5$	$N_{14}P_{10}K_{10}$	$N_{21}P_{15}K_{15}$
Sunflower	$N_0P_0K_0$	$N_5P_2K_2$	$N_{10}P_6K_6$	$N_{15}P_8K_8$
Barley	$N_0P_0K_0$	$N_7P_5K_5$	$N_{14}P_{10}K_{10}$	$N_{21}P_{15}K_{15}$
Total 1983–2010	$N_0P_0K_0$	$N_{162}P_{108}K_{108}$	$N_{324}P_{216}K_{216}$	$N_{486}P_{324}K_{324}$
Annual mean	$N_0P_0K_0$	$N_6P_4K_4$	$N_{12}P_8K_8$	$N_{18}P_{12}K_{12}$

Table 31.2 Potassium balance for the period 1983–2010, kgK_2O/daa

Fertilizer levels	K_2O input with fertilizer	K_2O output with biomass	Balance	
			For period	For 1 year
T_0	0	153.9	−153.9	−5.7
T_1	108	257.7	−149.7	−5.5
T_2	216	310.2	−94.2	−3.5
T_3	324	298.9	+ 25.1	+ 0.9

To explain the changes in potassium status in the soil, a provisional potassium balance was calculated as the difference between K_2O input from fertilizer and output as harvested biomass.

31.3 Results and Discussion

The influence of fertilization on soil-available potassium depends mainly on the balance between the input with the fertilizer and the output as harvested biomass. The calculated potassium balance (K_2O/daa) for the last 27 years of the trial (1983–2010) shows an output of 153.9 kg (Table 31.2) with cropping of maize, wheat, sunflower and barley in rotation without fertilization; an annual loss of 5.7 kg from the soil layer exploited by crop roots. The amounts of potassium imported with mineral fertilization at low and moderate levels did not compensate for removals by the crops, leaving annual deficits of −5.5 and −3.5 kg. Only the high rate of fertilizer (with an annual input of 12 kgK_2O/daa) gave a positive balance of 25.1 kg for the period, averaging + 0.9 kg/year.

The potassium output by the crops, established in our previous investigation (Koteva 1993), is significant: at the low and moderate potassium fertilizer levels of 2 and 6 kg/year, the annual deficit under maize is −5.5 and −3.0 kg (Table 31.3); under sunflower, the deficit is bigger, −10.4 and −13.4 kg. In the same variants, winter cereal crops before the sowing received higher potassium inputs (5 and 10 kg), but the balance remained negative (−4.3 and −5.7 kg in wheat and −3.3 and 3.5 kg in barley). At the high fertilizer level, the annual balance for the whole crop

Table 31.3 Provisional potassium balance in crop rotation in period 1983–2010, kgK_2O/daa

Crops	Fertilizer levels T_0	T_1	T_2	T_3
Maize	−6.9	−5.5	−3.0	−1.8
Wheat	−4.5	−4.3	−5.7	+ 3.4
Sunflower	−6.1	−10.4	−13.4	−11.2
Barley	−4.1	−3.3	−3.3	+ 4.2

Table 31.4 Available and exchangeable potassium in the long-term experiment, 2010 (mg/100 g soil)

Forms of potassium	Horizon of the soil, cm 0–20	20–40	40–60
T_0			
Available K_2O[a]	43.6	42.6	41.6
Exchangeable K_2O[b]	34.1	32.5	28.9
T_1			
Available K_2O	44.5	43.6	42.8
Exchangeable K_2O	35.4	32.9	32.4
T_2			
Available K_2O	46.2	45.0	44.2
Exchangeable K_2O	36.0	35.9	31.7
T_3			
Available K_2O	52.4	48.4	44.2
Exchangeable K_2O	36.3	35.8	29.8

[a]Available K_2O determined by Milcheva's method
[b]Exchangeable K_2O determined by Ivanov's method

rotation is positive (+0.9 kg) with clear differentiation of the crops: under wheat +3.4 kg, barley +4.2 kg, but under maize −1.8 kg and sunflower −11.2 kg. In summary, the potassium requirements of the crops in the unfertilized, low fertilizer and moderate fertilizer treatments were met mainly from supply by the soil. We should expect this to be reflected in the measured soil reserves of exchangeable and available potassium.

For scientific investigations and practical fertilization recommendations, plant-available potassium is determined by acetate-lactate extraction (Ivanov 1984), which measures mainly exchangeable K^+, and by extraction in 2 N HCl (Milcheva 1986) which also extracts additional non-exchangeable potassium that is available to the plants. Table 31.4 compares the results of the two extractions for the different variants of the long-term experiment.

The present study uses extraction by 2 N HCl (Milcheva 1986) to represent available potassium, which includes exchangeable and part of the non-exchangeable potassium. The results (Table 31.5) demonstrate that, despite the negative balance, the quantity of available potassium remains high in all treatments and soil layers to a depth of 100 cm. Compared with the control, the low and moderate variants of fertilization do not significantly affect soil reserves of available potassium. The positive potassium balance in the T_3 variant is reflected in an increase in available potassium in the 0–20 cm soil layer from 43.6 mg/100 g in T_0 to 52.4 mg/100 g and in the 20–40 cm

Table 31.5 Available K_2O content in the soil in 2010, after 47 years of mineral fertilization

	Soil layer, cm				
Fertilizer levels	0–20	20–40	40–60	60–80	80–100
T_0					
Content, mg/100 g	43.6	42.0	41.6	40.4	37.6
Variation, %	6.92	6.74	6.07	4.88	3.99
T_1					
Content, mg/100 g	44.5	43.6	42.8	42.1	38.5
Variation, %	8.67	7.21	6.28	5.02	4.17
T_2					
Content, mg/100 g	46.2	45.0	44.2	42.0	38.3
Variation, %	9.16	8.00	6.97	5.66	4.99
T_3					
Content, mg/100 g	52.4*	48.4*	44.2	42.6	38.8
Variation, %	9.90	8.12	7.09	4.91	4.57

$*p = 0.01$ %

layer from 42.0 mg/100 g in T_0 to 48.4 mg/100 g. In all treatments, the variation in potassium content is greater in the humose topsoil (0–60 cm) than in the subsoil (60–100 cm): CoV 6.74–9.90 % compared with 3.99–5.66 %. A clear tendency for variability increase is marked from the unfertilized variant to the variant receiving the high level of mineral fertilization.

31.4 Conclusions

Cropping maize, wheat, sunflower and barley in rotation on Pellic vertisol without potassium fertilizer leads to annual loss of 5.7 kgK_2O/daa. Applications of 4 and 8 kg/daa of potassium are not enough to compensate for the removal of potassium by the crops; there are annual deficits of 5.5 and 3.5 kg/daa. Potassium fertilizer application of 12 kg/daa leads to a positive balance for the rotation as a whole, made up by positive balances for wheat (+3.4 kg/daa) and barley (+4.2 kg/daa) and negative balances for maize (−1.8 kg/daa) and sunflower (−11.2 kg/daa).

Despite the deficit balances, available potassium remains high in all fertilization variants and throughout the soil profile. The low and moderate levels of fertilization hardly change the reserve of available potassium compared with the control. The high level of fertilizer increases available potassium in the 0–40 cm layer; changes in the deeper layers are insignificant.

However determined, there are big reserves of plant-available potassium in the Pellic vertisol derived from andesite. Comparing the results of determination by Milcheva's method and Ivanov's method, changes induced by the different treatments over 47 years are reflected by both, but the quantities of available potassium determined by Milcheva's method are 25–44 % higher and may give a better estimation of the plant-available nutrient.

References

Berestov I, Rinks G, Rampe H (1986) Balance of nutrients in crop rotation at optimal nutrition of plants. Agrochem 6:37–42 (Russian)

Beringer H (1980) The role of potassium in yield formation. From international conference on role of potassium in crop production. Bull Indian Soc Soil Sci, New Delhi, 10:241–258

Dospehov B, Kiriuchina B, Bratskaia A (1976) Effect of 60-year fertilization, regular liming and crop rotation on the agro-chemical properties of soil. Agrochem 4:3–14 (Russian)

IUSS Working Group WRB (2006) World reference base for soil resources, vol 103, World soil resources report. FAO, Rome

Ivanov P (1984) Determination of plant-available potassium. Soil Sci Agrochem 4:88–98 (Russian)

Koteva V (1993) Some leached smonitza soil fertility parameters change under long fertilization in crop-rotation. Dissertation, University of Sofia (Bulgarian)

Koteva V, Stoichev D, Stoicheva D (2001) Soil Sci Agrochem Ecol 4–6:140–144

Mengel K (1982) Factors and processes affecting potassium requirements of crops. Potash Rev 9:1–12

Mengel K, Kirkby E (1987) Principles of plant nutrition, 4th edn. International Potash Institute, Bern

Milcheva M (1980) Soil Sci Agrochem 1:3–10 (Russian)

Milcheva M (1986) Soil Sci Agrochem 6:4–10 (Russian)

Munson RD (ed) (1985) Potassium in agriculture. American Society of Agronomy, Madison

Seculic P, Ubavic M, Dozet M (1998) Effect of fertilizer application on chemical soil properties. In: 2nd Balkan symposium on field crops, Novi Sad, part II, pp 303–308

Part III
Different Ways of Doing Things

Chapter 32
Towards Sustainable, Self-Supporting Agriculture: Biological Nitrogen Factories as a Key for Future Cropping Systems

E. Triboi and A.-M. Triboi-Blondel

Abstract The LOME concept of self-sufficient agricultural systems encompasses legumes supplying nitrogen and proteins, oil-seed crops producing fuel, and methanization of biomass producing renewable energy. The benefits of introducing a legume into a crop rotation have been widely discussed but there has been no comparison of long-term yield trends and few long-term experiments on the interaction between biological and mineral sources of nitrogen (N).

During five 6-year cycles of a long-term experiment comparing cropping systems with and without lucerne over 2 years followed by four identical crops, the 2 years of lucerne produced 689 kgN as above-ground biomass and, also, had a big residual effect. The four subsequent crops after the lucerne exported a further 202 kgN and, in four rotations out of five, wheat grown in the first and third years after lucerne without mineral N fertilizer achieved more than 80 % of the maximum wheat yields under non-N limited conditions. There is a negative interaction on N uptake from different N sources. These results suggest that a self-sufficient cropping system based on biological N fixation can be a real alternative to conventional intensive systems.

32.1 Introduction

Current cropping systems must change – for several pressing reasons:

1. Mineral nitrogen (N) fertilizers contributed in large measure to the spectacular increase in crop yields over the last 50 years. Cheap and easy to use, they replaced other sources of N, especially biological sources. Consequently, the area under legumes diminished dramatically, especially in developed countries (Crews and Peoples 2004); in France, for example, the area under lucerne has

E. Triboi (✉) • A.-M. Triboi-Blondel
Department of Environment and Agronomy, INRA UR-Agrononie, Clermont-Ferrand, France
e-mail: eugene.triboi@neuf.fr

D. Dent (ed.), *Soil as World Heritage*, DOI 10.1007/978-94-007-6187-2_32,

been reduced by about one million hectares, which means a loss of about 300 000 tonnes of nitrogen. Moreover, excess mineral fertilizer has often been applied in the hope of greater profit; for France, Pointerau (2001) estimated an excess of 835 000tN, which has brought about pollution of rivers and groundwater. What the new agriculture gained in efficiency, it lost in self-sufficiency because intensification depends on products that are external to the system of production (fertilizer, pesticides, mechanization powered by fossil fuels).

2. Having sacrificed self-sufficiency in energy and other inputs, N in particular, agriculture is now vulnerable to price volatility – especially the price of energy. Synthetic N fertilizer is one of the principal energy and financial outlays, and it should be remembered that symbiotic production of N by legumes requires two to three times less energy than chemical synthesis of mineral N fertilizer (Thiébeau et al. 2003). Thus, the replacement of synthetic N fertilizer by symbiotic N can be a big contribution to a more self-sufficient agriculture and, since it is renewable energy, the system also becomes more sustainable Tonitto et al. (2006).
3. To arrest the pollution of soils, streams and groundwater, European agricultural policy encourages ecological agriculture within a framework of socially sustainable development. Renunciation of monoculture in favour of crop rotation is a premise of movement towards a more ecological agriculture in which crop management is no longer annual but, rather, multi-annual and in which evaluation is no longer based on immediate profit but, rather, on several criteria.

In this context, we may envisage the evolution of agriculture towards self-sufficiency in nitrogen and energy based on the LOME concept (Triboi 2010):

L for *legumes* to ensure self-sufficiency in nitrogen and in proteins
O for *oil-seed crops* to produce the energy/fuel
ME for *methanization* (anaerobic digestion of biomass) to produce methane as renewable energy.

The three pillars of the system are interdependent: legumes produce not only nitrogen but also carbon for energy, oil crops produce not only fuel but also proteins, and both may be used as substrate for methane production. The main exported element of this agricultural system will be C energy (30–50 % of C in the biomass) because the other elements (N, P, K, Ca, Mg, etc.) will be returned to the soil with the digest of methanization – a high-quality organic fertilizer. So, this new agriculture will be of high productivity, *environmentally friendly and self-sufficient in energy*.

The capacity of legumes to replace mineral N fertilizers has been widely reviewed. Comparing legume-based systems (LBS) with fertilizer-based systems (FBS), Crews and Peoples (2004) and Tonitto et al. (2006) concluded that N from legumes is potentially more sustainable than from industrial sources and LBS are environmentally superior because of synchrony between crop N demand and N release from organic matter. They comment that 'the N-difference method, where available N in soil or N uptake by crops following legume incorporation is compared with control plots that received no legume residues, is the most useful for evaluating the agronomic utility of a particular crop rotation, and can be used to interpret N synchrony in a LBS when the rotation has been followed long enough to reach an equilibrium in soil organic matter dynamics' (Crews and Peoples 2005).

We need long-term experiments (LTEs) to develop sustainable cropping systems because 'agricultural systems that undermine long-term productive capacity through ... over application of chemical fertilizers and biocides and other forms of mismanagement, although possibly displaying higher gross productivity in the short term, are less likely to remain productive in the long-term than agricultural systems ... that include greater reliance on locally available energy sources and greater nutrient cycling to maintain their productivity' (Rydberg and Haden 2006). Unfortunately, there are neither LTEs where legume and fertilizer sources of N have been directly compared (Crews and Peoples 2005) nor long-term studies on the dynamics of legume N in cropping systems. Here we present the results of an LTE where legume and fertilizer sources of N have been directly compared over a 30-year period. To our knowledge, this is the only LTE which enables quantification of the contribution and the use over time of legume and mineral N sources (Triboi and Triboi-Blondel 2004, 2008a, b). We consider (1) the productivity of a 2-year lucerne crop, (2) its residual effects on annual crops during 4 subsequent years, (3) the apparent nitrogen use of organic N (lucerne, vetch) versus mineral N fertilizer and (4) the interaction between organic and mineral N.

32.2 Materials and Methods

32.2.1 Experimental Design

The experiment compares the direct effects and interactions of:

- *The nature of rotations.* Two 6-year rotations were compared: A, comprising only annual crops, and L, including 2 years of lucerne in place of sugar beet and wheat. The four other crops were identical: winter wheat, maize or oil-seed rape, winter wheat and winter wheat or barley.
- *Green manure (G).* Each of the two rotations was subdivided: with and without vetch as a cover crop following a cereal and used as green manure. In addition, the green residue of sugar beet in rotation A and the last cut of lucerne from rotation L were applied as green manure.
- *Crop residues (R).* With or without return of the wheat and maize straw.
- *Mineral nitrogen (N).* With or without application of mineral N to annual crops.

Finally, on the elementary 24 × 4.2 m plots, the experiment comprised 3 replications of 16 treatments which allowed measurement of the simple and combined effects of four factors, each present at two levels (with or without):

- Eight treatments without mineral N: C (control) (1), R (2), G (3) and GR (4) in rotation A; and L (5), LR (6), LG (7) and LGR (8) in rotation L

– Eight treatments with mineral N: N (9), NR (10), NG (11) and NGR (12) in rotation A and LN (13), LNR (14), LNG (15) and LNGR (16) in rotation L

Five 6-year rotations were completed over a 30-year period (1969–1999).

32.2.2 Nitrogen Inputs

Only the annual crops received nitrogen fertilizer. In rotation A, the R, G and N_{min} inputs were applied over a 6-year period instead of four in rotation L (lucerne did not receive mineral or organic inputs). Thus, for an average annual application of mineral N of 120 kg over 30 years, the total quantity applied was 720 kg for rotation A and 480 kg for rotation L.

N inputs with green manure (G) varied in timing and quantity. Their frequency diminished over time: 4, 3, 2, 1 and 0 inputs, respectively, for the rotations 1–5. In addition, in the second year of rotation L, the last cut of lucerne was used as green manure. Over the 6 years, the total amounts applied in the two rotations averaged 187 and 164 kgN, respectively, for rotations A and L, corresponding to an annual average of 31 and 27 kgN/ha, respectively.

The quantity of R harvest residues, depended on the productivity of the treatment – nearly double in the treatments with mineral N (NR, NRG) compared with treatments without mineral N (R and GR). Thus, during the first four rotations, the average annual quantity was roughly 20 kgM/ha without N and 35 kgN/ha with N. The crops were protected by specific fungicide, herbicide and insecticide.

32.2.3 Soil and Weather Conditions

The trial was carried out on the Limagne plain on a thick, clay soil well provided with P and K. However, to maintain the natural nutrient status of the soil, phospho-potassic fertilizer was applied over the whole experiment to compensate for the export through harvests of the most productive treatments. The climate is continental with oceanic influences. Mean annual rainfall over the 30 years was 600 mm with a 20 % coefficient of variation; the minimum of 349 mm was recorded in 1991 and the maximum of 826 mm in 1992. The seasonal minimum is in winter and the maximum in late spring and summer, so in association with soils of high water retention capacity, leaching is usually less than 20 kgN/ha/year. The soil has a substantial capacity for mineral nitrogen storage, especially in the 60–90 cm layer (Gachon 1973; Perrier et al. 2000). Over the 30 years, the mean annual temperature was 11.3 °C, with a minimum of 9.9 °C in 1971 and a maximum of 12.6 °C in 1997.

Monthly maximum temperatures are about 25 °C in July and August when daily temperatures are often above 30 °C, so heat stress, associated with droughts, is possible. Late spring frosts and relatively mild autumns are often recorded.

32.2.4 Analysis and Statistics

Crop biomass was measured at harvest from 20 to 40 m^2 per plot. A sample of grain, green biomass, stems or roots was tested for moisture and N content (N determined by the Kjeldahl or Dumas method) and N uptake calculated as the product of N concentration and biomass.

A multifactor ANOVA was carried out for grain and N yields over all the results to determine the effects of the different experimental factors. To highlight any interaction between the different sources of nitrogen (mineral, organic) and/or the residual and cumulative effects, the analysis was complemented by two synthetic variables. First, the additional quantity of N in the above-ground biomass following a direct input of N (mineral or organic) compared with treatments without inputs; since the use of nitrogen is a function of the total bioavailable quantity in the soil, this additional effect was calculated for several levels of N availability of the treatments that did not receive a direct mineral or organic input. The second synthetic variable is the apparent utilization coefficient of the N applied as fertilizer (N-5C), calculated as the ratio between the difference of N exported by the above-ground biomass in the treatment with fertilizer and the N exported by the treatment without N input (expressed as a percentage of the quantity of N applied as fertilizer). In order to take account of cumulative effects, Nc-AUC was calculated from the accumulation over time of the inputs and outputs. The cumulative N balance, calculated by the difference between the N applied by fertilization and N exported by the above-ground biomass, was used as an explicit variable of the evolution in time of Nc-AUC. Finally, a relative yield was calculated as the percentage of the yield in an ***i*** treatment in relation to the maximum yield of the year in non-limiting nitrogen conditions (average of the LNG and LNGR treatments). Statistical analyses were carried out with Statgraphics Plus.

32.3 Results

To establish a cropping system that is self-sufficient in N, we need to quantify the symbiotic sources of N that can be produced by the cropping system and their direct and residual effects compared with mineral N sources. The Clermont-Ferrand long-term experiment allows us to analyse the productivity of 2 years of lucerne and, also, vetch grown as cover crop; their direct effect, in the case of vetch; and residual effects compared with mineral N fertilizer.

32.3.1 *Productivity of a 2-Year Lucerne Crop: Dry Matter and Nitrogen Yield*

The two 6-year rotations differed only by the presence or absence of 2 years under lucerne, so we can quantify the productivity of lucerne and its additional effect in relation to annual crops (sugar beet and wheat) over four succeeding years and five cycles (Table 32.1).

The cumulative effect of lucerne is underscored by statistical analysis of above-ground biomass, its nitrogen content and the quantity of nitrogen present in the biomass over five rotations. The mean N exported by the above-ground biomass over the 2 years was 689 kgN/ha, 268 kg (39 %) in the first year and 421 kg (61 %) in the second. Inter-annual variations of N output ranged from 199 to 347 kgN/ha for the first year and from 355 to 515 kgN/ha in the second year. This variability may be explained by variation in rainfall.

32.3.2 *Residual Effect of Lucerne*

Residual Effect on the Grain and Nitrogen Yield of Winter Wheat

The yield of wheat grown in first and third years after lucerne, and without mineral N, attained the maximum yield of the year achieved where nitrogen was not limiting (average of LNG and LNGR treatments) in at least two of five rotations (third and fourth), while the relative N yield was less than the maximum N yield, even in the treatments which received green manures (Fig. 32.1). So, our results suggest that the residual N effect of lucerne constrains the protein content rather than the grain yield.

- The relative yield represents the percentage of yield in the ***i*** treatment compared with the maximum yield of the year in non-N-limiting conditions (average of the LNG and the LNGR treatments)
- 1, 2, 3, 4, 5 = rotation number;
- Treatments: 5 = L, 6 = LR, 7 = LG, 8 = LGR, 13 = LN, 14 = LNR, 15 = LNG, 16 = LNGR as in Table 32.1

Residual Effect on Nitrogen Uptake Over 4 Years

Table 32.2 summarizes the beneficial effect of lucerne in succeeding crops as the difference between the quantity of N contained in the above-ground biomass of C&R, G&GR, N&NR and NG&NGR treatments with and without lucerne as the preceding crop.

Table 32.1 Lucerne productivity over 2 years

	Year 1			Year 2			Total 2 years	
Years	Biomass Mg/ha	%N	N yield kgN/ha	Biomass Mg/ha	%N	N yield kgN/ha	Biomass Mg/ha	N yield kgN/ha
1969–1970	11.6	2.98	347	16.7	3.08	515	28.3	863
1975–1976	8.5	3.31	281	14.2	3.20	453	22.7	734
1981–1982	10.3	2.82	289	14.9	2.58	385	25.2	674
1987–1988	6.5	3.07	199	12.0	2.96	355	18.5	554
1993–1994	7.2	3.12	225	12.4	3.19	396	19.6	621
Mean	8.8	3.06	268	141	3.00	421	22.9	689
ANOVA, P value								
Treatment		*0.0000*						
Year	*0.0000*	*0.0000*	*0.0000*	*0.0000*	*0.0000*	*0.0000*		
Interaction		*0.0000*			*0.0005*			

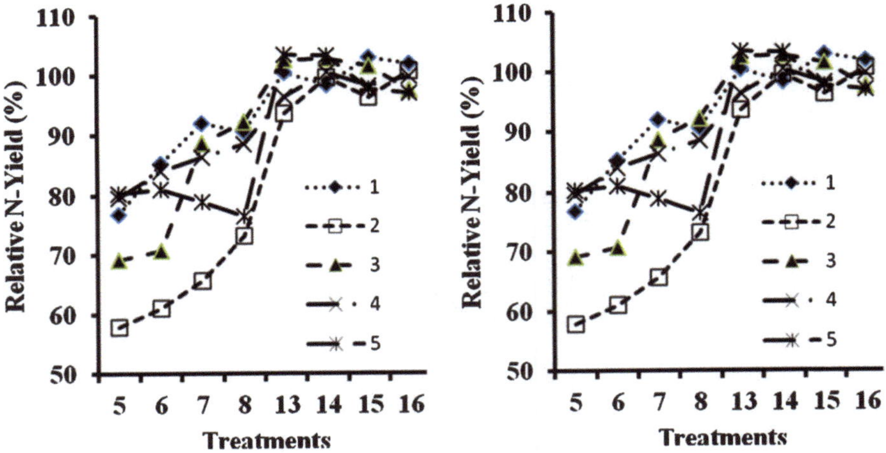

Fig. 32.1 Relative yield and relative N yield (%) of wheat following lucerne

Table 32.2 Residual effect of lucerne over 4 years at 4 levels of N inputs (treatments): additional N in above-ground biomass of annual crops

	Beneficial effects of lucerne (KgN/ha)					
	Year 1	Year 2	Year 3	Year 4		
Nitrogen treatments*	Wheat	Maize or rapeseed	Wheat	Barley or wheat	Total	Mean
C&R	79	58	47	24	208	52a
G&GR	87	35	48	25	195	49a
N&NR	48	31	29	22	130	32b
NG&NGR	34	9	8	13	64	16c
Mean	62a	34b	33b	21b		Lsd = 8.0

Each value is the mean of five rotations

Interaction year *nitrogen treatments is not significant

This effect was evaluated with four crops: two winter cereals (wheat and barley) in the third, fifth and sixth years of the rotation and maize and rapeseed in the fourth year. The residual effect of lucerne was calculated at four different levels of N bioavailability created mainly by green manure inputs (G) and mineral N: C&R, G&GR, N&NR and NG&NGR treatments.

In the absence of mineral fertilization (C&R, G&GR), the cumulative residual effect over 4 years was 200 kgN whereas, with added N, it was 130 kgN (N&NR) and 64 kgN (NG&NGR). The residual effect of lucerne in first year (62 kgN) was markedly higher than the other 3 years (34–21 kgN). The weak effect of years 2 and 4 may be attributed to the test crops – maize and rapeseed in second year and wheat in third year. The general trend was a higher level of benefits associated with weak nitrogen bioavailability and a decline over the 4 years after ploughing-in lucerne. The interaction between these two factors was not significant. The averages over the five rotations for the treatments without mineral N (C&R and G&GR) are presented in Table 32.3.

Table 32.3 N exported by lucerne and beneficial effect on following cereal crops

	N export (kgN/ha)								
	Lucerne cuts			Beneficial effects of lucerne[a]					
Years of lucerne stands	Year 1	Year 2	Total	Year 3	Year 4	Year 5	Year 6	Total	Total
1969–1970	347	515	863	67	13	43	28	150	1,013
1975–1976	281	453	734	53	60	19	22	152	886
1981–1982	289	385	674	86	98	46	30	258	932
1987–1988	199	355	554	112	18	107	24	260	814
1993–1994	225	396	621	99	44	25	24	190	811
Average	268	421	689	83	47	48	25	202	891
%	39	61	100	41	23	24	12	100	

[a]The beneficial effects of the 2-year lucerne on the following cereal crops were assessed as the difference of above-ground N export by the crops following lucerne and the rotation without lucerne in treatments without N mineral input (C&R and G&GR)

Out of the 3 years with maize, only in 1978 (rotation two) was there a significant response to N fertilization. In the third and fifth rotations, rapeseed used as test crop proved to be a better indicator of the nitrogen bioavailability after lucerne (98 and 44 kgN). With rapeseed, the second-year residual effect was higher than the third and even higher than the first year during the very favourable year 1984. Finally, during the fourth year with a barley crop, the average residual effect of the lucerne decreased with nitrogen bioavailability of 24–13 kgN/ha, and it was lower than the third year, especially with weak bioavailability of N (Table 32.2).

32.3.3 Nitrogen Balance and Utilization from Different Sources

The negative interactions between the different nitrogen sources may be due to the limited capacity of the crops to store N in their biomass, in which case the evolution of the soil N balance could explain much of the variation of N-AUC. To test this hypothesis, we compared the evolution of cumulative N_c-AUC over 30 years with the evolution of cumulative nitrogen balance. The balance, calculated from 1969, varied greatly according to the system used, resulting in balances between +371 and −2176 kgN/ha after 30 years (data not shown). The regression analysis of Nc-AUC as a function of the N balance reveals two types of situations (Fig. 32.2):

- Use of only crop residues R and green manure G compared with mineral N alone or with G and R
- R, G or N inputs with or without lucerne.

Without mineral N fertilizer, the nitrogen balance in treatment R (receiving only harvest residues) became more and more negative, ending up at -1,813 kg (Fig. 32.2a). For the balances from 0 to 1,000 kg, Nc-AUC decreased considerably,

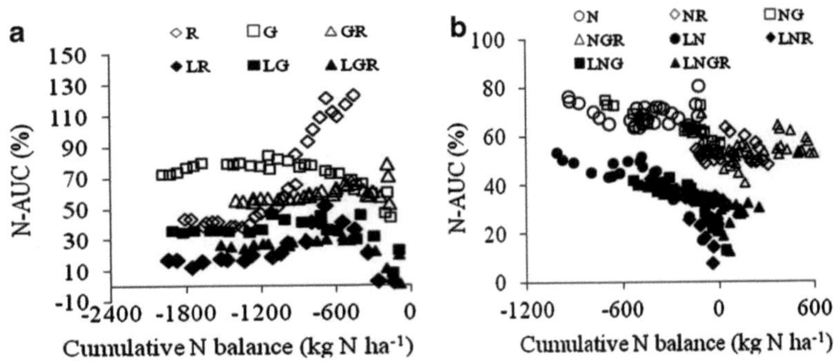

Fig. 32.2 Apparent N-use coefficient (N-AUC %) relative to cumulative N balance (kgN/ha) (Treatments as in Table 32.1: (**a**) without mineral N, (**b**) with mineral N)

from more than 100 to roughly 40 %; then it stabilized despite the continued decrease of the N balance. This *S*-shaped curve expresses, in reality, the arrival at a steady state with a very weak use of crop residues – because the N outputs by the time of the harvest were practically identical to C control, without any N input.

In the presence of relatively small organic nitrogen inputs (G, GR, LR, LG, LGR, with total input of 825–855 kg/ha[1]) and despite the continuous increase in the N deficit, Nc-AUC increased, at first, before diminishing and then stabilizing around variable values according to the treatment (Fig. 32.2a). The G and GR treatments had higher Nc-AUCs than the R treatment, 70 and 55 %, respectively. In the presence of lucerne, the N balance–AUC relationship was parallel with that of the G and GR treatments, and even with LR for the balances lower than −500 kgN, but was out of synchrony towards the bottom by about 30–40 %. The equilibrium values of AUC were weaker in the presence of lucerne: 35, 25 and <20 %, respectively, for LG, LGR and LR. This decline could be due to the increase in the available nitrogen pool connected with a greater export of nitrogen with the above-ground biomass.

Mineral fertilization (Fig. 32.2b), alone or with other nitrogen sources, considerably increased the pool of available nitrogen and, thus, the yield and the nitrogen content in the biomass. The nitrogen balance improved but remained mostly negative. The continuous decrease in nitrogen balance was accompanied by a proportional increase in Nc-AUC. The N, NR, NG and NGR treatments came together in two linear relations, according to the presence or absence of lucerne. The slope of these straight lines, −0.0165 without lucerne and −0.0272 with lucerne, shows that a variation in nitrogen balance of 100 kgN ha/ha leads to a variation in the AUC from 1.7 to 2.7 %. In the absence of lucerne, the N-use coefficient of mineral nitrogen, calculated according to Fig. 32.1b for a zero balance, was 58 % – which value is often found in studies with ^{15}N-marked fertilizer. However, as previously seen in the case of a small input of nitrogen, the presence of lucerne diminished the Nc-AUC by roughly 30 %. This might reflect the presence of an overall pool of bioavailable nitrogen which was higher by 30/1.65 = 182 kgN following lucerne.

32.4 Discussion

The experimental results, quantifying the effect of different nitrogen sources, complement our current knowledge of the fate of nitrogen at the level of a cropping system. Now we may think about new cropping systems that can be both sustainable and self-sufficient in nitrogen. Of course, the results presented are specific to the soil and climate of the Limagne in the middle of France with a moderate winter rainfall and low N leaching; we need to consider how far they may be applicable to conditions of significant leaching during intercropping period.

32.4.1 *Generic Character of Observations*

First, we may note that the overall conclusions agree with and complete previous results obtained in short-term experiments elsewhere. Our results indicate a negative interaction between the different nitrogen sources: a decrease in the use efficiency of nitrogen applied as mineral or organic fertilizer. According to the principles of the *A-value* method of isotopic dilution (Fried and Dean 1952; Triboi 1987), the quantity of an element taken up by the crop from several sources ***i*** ($NDFS_i$) is proportional to the quantity of each source (S_i). Then, the utilization coefficient by the crop of source ***i*** (UC_i) which represents the proportion of N from the source ***i*** recovered in the above-ground biomass is

$$UC_i = \frac{N_b * NDFS_i i}{S_i} = \frac{N_b * \frac{S_i}{S_t}}{S_i} = \frac{N_b}{S_t}$$

where N_b = total N in above-ground biomass and S_t = total mineral N pool in soil.

The result is that at time *t*, the different N sources have the same UC and the change of ratio N_b/S_t by the variation of the N biomass or of soil N pool changes the UC. So any increase in S_t, by the introduction of new N inputs as organic residues of lucerne or vetch biomass or by increase of rate, raises the total pool of N mineral in soil (S_t) and decreases the UC of all N sources. The UC at harvest integrates the dynamic of mineral N pool in soil during the growing period of the crop; hence, the negative interaction between N sources demonstrated in our experiment.

32.4.2 *Nitrogen Self-Sufficient Cropping System*

Our results over 30 years, summarized in Table 32.3, reveal lucerne as a veritable nitrogen factory. On average over five rotations, a 2-year lucerne crop produced 689 kgN, 40 % in the first year. This could still be improved by paying particular attention to crop establishment, which determines productivity in the first year, and by irrigation. During the 4 years after lucerne, another 202 kgN was recovered in above-ground biomass from treatments not receiving mineral N, bringing the gross exported quantity due to the lucerne to 891 kgN. Considering that the 202 kgN exported in annual crops during the 4 years after the lucerne originated from a pool that has an uptake efficiency of 30–60 %, we calculate a total soil available pool of 337–673 kgN. This is an approximate calculation, but a gross production of more than 1,000 kgN/ha for the 2 years' lucerne seems realistic, ensuring a supply of more than 200 kg of organic N/ha/year that can satisfy the needs of annual crops for 5 years!

Comparable results have been recorded in other soil and climate conditions. With irrigated lucerne in Australia, Gault et al. (1995) obtained 3–4 tonnes/ha of biomass in the first year and 10–13 tonnes in years 2 and 3, containing 340–410 kgN; the input to the soil was estimated at 800 kgN, with a flux of 250–300 kgN/ha/year to a depth of 1.4 m. For a 3-year crop, Kelner et al. (1997) estimated the contribution of above-ground biomass at 167, 407 and 404 kgN and that of the root system at 107, 259 and 223 kgN/ha; the root contribution could be higher because the authors took into account only the top 50 cm. At the landscape scale, Peterson and Russelle (1991) estimated the average input of lucerne over four million hectares of the American Corn Belt at 252 kgN/ha/year, of which 56 kg comes through the soil; similar values have been recorded with other leguminous fodder crops, in particular clover (Strong et al. 1996; Kumar and Goh 2000).

A great deal of published data shows significant residual effects from legumes, going so far as to claim complete satisfaction of nitrogen needs (Committee on Alternative Farming Systems 1989; Boawn et al. 1963; Gault et al. 1995; Strong et al. 1996; Drinkwater et al. 1998; Bachinger and Zander 2007). However, our results suggest that the residual N effect of lucerne is not enough to produce high yields with a high level of protein; supplementary N input is needed to fully satisfy crop demand.

The nitrogen generation of the rotation might be improved by an additional, intercropped legume, which could produce 50 to over 100 kgN, depending on to the available water during the summer and autumn. Lucerne can be used as green manure, but in that case, it must be grown for a full year, halving the number of cash crops – and growing legumes as nitrogen fertilizer seems be uneconomic or, rather, dependent of price of the energy. However, this negative perception will change if the lucerne biomass and other crop residues were to be used in a biogas plant (producing of methane) instead using the crop directly as green manure. This new energy represents supplementary income, making good the loss from the decreased number of cash crops in rotation, and the digested effluent (which contains most of the nutrients) can be returned to the soil at the time of the crops' highest N uptake. Thereafter, both stockless and livestock systems can be self-sufficient in N and energy.

We conclude that legumes in crop rotation can satisfy the nitrogen demand for the entire system and that self-sufficient cropping systems can be a real alternative to the systems based on maximum productivity and large synthetic input.

Acknowledgements This chapter is dedicated to Dr Louis Gachon (1926–1999), Director of INRA-Agricultural Research Station at Clermont-Ferrand and Chief of Agronomy of INRA-France who, in 1968, transformed the field experiment initiated by Dr D Collier in 1948 to this long-term experiment. Special thanks to INRA technical staff who contributed to the project by their enthusiastic assistance throughout the trial over 30 years.

References

Bachinger J, Zander P (2007) ROTOR, a tool for generating and evaluating crop rotations for organic farming systems. Eur J Agron 26:130–143

Boawn LC, Nelson JL, Crawford CL (1963) Residual nitrogen from NH_4NO_3 fertilizer and from alfalfa plowed under. Agron J 55:231–235

Committee on the Role of Alternative Farming Methods in Modern Production Agriculture, National Research Council (1989) Crop rotation, research and science. In: Alternative agriculture. National Academy Press, Washington, DC, p 147

Crews TE, Peoples B (2004) Legumes versus fertilizer sources of nitrogen: ecological tradeoffs and human needs. Agric Ecosyst Environ 102:279–297

Crews TE, Peoples B (2005) Can the synchrony of nitrogen supply and crop demand be improved in legume and fertilizer-based agroecosystems? A review. Nutr Cycl Agroecosyst 72:101–120

Drinkwater LE, Wagoner P, Sarrantonio M (1998) Legume-based cropping systems have reduced carbon and nitrogen losses. Nature 396:262–265

Fried M, Dean IA (1952) "A" concept concerning the measurement of available soil nutrient. Soil Sci 73:263–271

Gachon L (1973) Effect of farming system on the nitrogen economy of calcareous clay soil in Limagne. Sci du Sol Bull AFES 3:177–197 (French)

Gault RR, Peoples MD, Turner GL et al (1995) Nitrogen fixation by irrigated lucerne during the first three years after establishment. Aust J Agr Res 46:1401–1425

Kelner DJ, Vessey JK, Entz HM (1997) The nitrogen dynamics of 1-, 2- and 3-year stands of alfalfa in a cropping system. Agric Ecosyst Environ 64:1–10

Kumar K, Goh MK (2000) Biological nitrogen fixation, accumulation of soil nitrogen and nitrogen balance for white clover (*Trifolium repens* L.) and field peas (*Pisum sativum* L.) grown for seed. Field Crop Res 68:49–59

Perrier MJ, Triboi E, Pépin D (2000) Measurement of nitrate leaching on the Black Soils of Limagne. In: Banton O, Porel G (eds) ESRA 2000 international symposium on groundwater in agriculture, 13–15 Sept, Poitiers (French), pp 67–70

Peterson TA, Russelle PM (1991) Alfalfa and the nitrogen cycle in the corn belt. J Soil Water Conserv 46:229–235

Pointereau P (2001) Legumes – what is their ecological role? Courrier de l'environnement de l'INRA 44:69–72 (French)

Rydberg T, Haden AC (2006) Energy evaluations of Denmark and Danish agriculture: assessing the influence of changing resource availability on the organization of agriculture and society. Agric Ecosyst Environ 117:145–158

Strong WM, Dalal RC, Cahill MJ et al (1996) Sustaining productivity of a vertisol at Warra, Queensland, with fertilizers, no-tillage or legumes: 3. Effects of nitrate accumulated in fertilized soil on crop response and profitability. Aust J Exp Agric 36:675–682

Thiébeau P, Parnaudeau V, Guy P (2003) What future for lucerne in Europe? Le Courrier de l'environnement de l'INRA 49:1–14 (French)

Tonitto C, David BM, Drinkwater EL (2006) Replacing bare fallow with cover crops in fertilizer-intensive cropping systems: a meta-analysis of crop yield and N dynamics. Agric Ecosyst Environ 112:58–72

Triboi E (1987) Field measurement of the amount of nitrogen fixed symbiotically by vetch intercropped with oats. In: Nitrogen fixation by legumes, INRA symposia 37. Paris (French), pp 265–270

Triboi E (2010) Cropping systems self-sufficient in nitrogen and energy: LOME. In: Conférence Festival de Non Labour et Semis Direct NLSD, Ferme des Manets, Semalle, Orne, 15 septembre 2010 (French)

Triboi E, Triboi-Blondel A-M (2004) Cropping system using lucerne as nitrogen source: effect on yield and nitrogen content. In: VIII ESA Congress: European agriculture in a global context. LVL Copenhagen, Denmark, 11–15 July 2004, pp 683–684

Triboi E, Triboi-Blondel A-M (2008a) Cropping systems self-sufficient in nitrogen and energy, reality or Utopia? Alter Agric 89:17–18 (French)

Triboi E, Triboi-Blondel A-M (2008b) Cropping systems self-sufficient in nitrogen and energy, reality or Utopia? In: International conference on organic agriculture and climate change, ENITA, France, 17–18 Apr, (French)

Chapter 33
Legumes as an Alternative Source of Nitrogen for Modern Agriculture

B.P. Boincean, G.T. Rusnac, I.V. Boaghii, D.I. Pasat, and S. Gavrilas

Abstract The capacity of legumes to fix biological nitrogen is, increasingly, recognized as an alternative to synthetic nitrogen which requires a lot of nonrenewable energy and presents hazards to the environment and public health. The amounts of nitrogen accumulated by annual and perennial legumes were determined in a long-term field experiment on Typical chernozem: 3 years of lucerne in a seven-field rotation accumulated 39.7–43.5 kgN per hectare of crop rotation per year as above-ground biomass; over 3 years, the following crops took up from the soil another 14.9 kgN. Calculation of the nitrogen balance for crop rotations with and without perennial legumes shows that even perennial legumes cannot reinstate all the nitrogen taken up by other crops in the rotation. However, the legume crop can also provide a substantial contribution of energy from the synthesis of methane from biomass.

33.1 Introduction

Current industrial farming is not sustainable, either economically, environmentally or socially. Nitrogen fertilizer, not so long ago a panacea, is now a problem – not only because of its soaring cost to the farmer but also its hazard to the environment and public health. Self-sufficiency in nitrogen and energy is now a strategic issue for farming, and scientists and farmers all over the world are revisiting biological nitrogen from leguminous crops, especially perennial legumes (Berestechi 1984; Lupascu 2004; Misustin and Cerepcov 1991; Triboi and Triboi-Blondel Chap. 32 in this symposium).

Replacement of perennial legumes by cash crops led to a drastic loss of soil fertility and, in the long term, declining crop yields. Reintroducing perennial

B.P. Boincean (✉) • G.T. Rusnac • I.V. Boaghii • D.I. Pasat • S. Gavrilas
Selectia Research Institute for Field Crops, Balti, Republic of Moldova
e-mail: bboincean@gmail.com

D. Dent (ed.), *Soil as World Heritage*, DOI 10.1007/978-94-007-6187-2_33,

legumes in rotation can improve fertility and reduce dependence on synthetic nitrogen, but better nitrogen-use efficiency can be achieved only by an optimal balance of industrial and biological nitrogen (Trepacev 1980). Perennial legumes can also provide an alternative source of energy – as methane from anaerobic digestion of the harvested biomass – clearing the way to growing perennial legumes even on farms without livestock (Triboi and Triboi-Blondel 2008).

The long-term poly-factorial field experiment on Typical chernozem[1] on the Balti steppe was established by Dr B Boincean in search of a practical alternative to conventional farming systems. It studies the actions and interactions between three main components of farming systems: crop rotations with and without perennial legumes, two systems of soil tillage and three systems of fertilization. Within this framework, the nitrogen balance was calculated for each rotation and, also, for continuous winter wheat.

33.2 Experimental Method

The sequences of crops in the two rotations are:

I	II
1. Lucerne + rye grass, third year after first cutting	1. Maize for silage
2. Winter wheat	2. Winter wheat
3. Sugar beet	3. Sugar beet
4. Maize for grain	4. Maize for grain
5. Winter barley	5. Peas for grain
6. Maize silage undersown with lucerne and rye grass	6. Winter wheat
7. Lucerne + rye grass for green mass	7. Sunflower

The design of the experiment specifies the methods and depth of soil tillage for each crop in each rotation and, also, the rates of application of organic and mineral fertilizers. No chemicals are applied for control of pests, diseases and weeds. The two systems of tillage are (1) alternation of mouldboard plough with ploughless tillage and (2) ploughless tillage. The three systems of fertilization are (1) the control, without fertilization, (2) composted farmyard manure and (3) farmyard manure + mineral fertilizers where the average application per hectare of crop rotation is, for crop rotation I, 10 t/ha of farmyard manure + $N_{12.8}$ $P_{21.4}$ $K_{24.2}$ and, for crop rotation II, 10 t/ha of farmyard manure + $N_{38.6}$ $P_{24.2}$ $K_{24.2}$.[2]

There are three replicates. Each plot is 264 m^2, making a total experimental area of 8.7 ha. Simultaneously, research is conducted under continuous winter wheat, winter barley, sugar beet and maize under the same conditions as the poly-factorial experiment but without replication.

[1] Haplic chernozem in *World reference base for soil resources 2006.*

[2] Fertilizer application in kg active ingredient per hectare.

33.3 Results and Discussion

33.3.1 *Perennial Legumes*

The nitrogen accumulated by lucerne was estimated as both above-ground biomass (Table 33.1) and below-ground reserves available to the succeeding crops (Table 33.2). The amount of nitrogen in the above-ground biomass of lucerne and perennial grasses was 278.1 kgN/ha/year for unfertilized plots, 301.9 kgN/ha/year for plots receiving farmyard manure and 304.7 kgN/ha/year for plots receiving farmyard manure + NPK fertilizer. Over the full rotation, this is equivalent to 39.7, 43.3 and 43.5 kgN/ha/year, respectively.

The amount of nitrogen accumulated by the roots was estimated by the difference in yields of the same crops grown in the crop rotation with lucerne and in the rotation without lucerne on unfertilized plots – the difference in yields being attributed to the residual nitrogen. The supplementary nitrogen taken up from the soil by three crops following lucerne in the rotation was 103.7 kg/ha/year or 14.9 kg per hectare of crop rotation per year (Table 33.2).

The total nitrogen accumulated by lucerne in the crop rotation was estimated as the sum of nitrogen in above-ground biomass and the amount of nitrogen recovered by the following crops (Table 33.3). On unfertilized plots, it amounted to 381.8 kgN/ha/year or 54.6 kgN per hectare of crop rotation per year. Fertilization somewhat increased the amount of nitrogen accumulated as above-ground biomass but cut the amount of nitrogen recovered by the following crops from 103.7 to 26.2 kg/ha – so the total amount of nitrogen accumulated by lucerne on fertilized plots was lower than on unfertilized plots.

Table 33.4 presents the nitrogen balance calculated for the crop rotations with and without lucerne and also for continuous winter wheat. In all cases, there remained an uncompensated loss of nitrogen – which was not avoided even by

Table 33.1 Nitrogen in above-ground biomass of lucerne in crop rotation in the Selectia polyfactorial experiment, average over two full rotations, 1996–2009, kgN/ha

	Unfertilized	Manured	±kg N/ ha and %	Manure + NPK	±kg N/ ha and %
Lucerne, second year, t/ha	26.5	28.4	+1.9	28.8	+2.3
Lucerne, third year, t/ha	14.1	15.7	+1.6	15.7	0
Total green mass, t/ha	40.6	44.1	+3.5/ 8.6 %	44.5	+3.9/ 9.6 %
Dry mass, t/ha	10.15	11.02		11.12	
Nitrogen in the above-ground biomass, kg N/ha/year[a]	278.1	301.9		304.7	
Nitrogen per ha of crop rotation, kg/ha/year	39.7	43,3		43.5	

[a]2.74 % of total dry matter

Table 33.2 Nitrogen used by crops following lucerne on unfertilized plots in the Selectia poly-factorial experiment, average for 1996–2009, kgN/ha

Crops	Yields in rotation with lucerne (t/ha)	Yields in rotation without lucerne (t/ha)	Difference (t/ha)	Supplementary N taken up by crops following lucerne (kg/ha)
Winter wheat	4.21	2.41	+1.80	59.4
Sugar beet	35.2	28.9	+6.30	31.5
Maize for grain	5.54	4.90	+0.64	12.8
Total, kg/ha/year				103.7
Per ha of crop rotation, kg N/ha/year				14.9

Table 33.3 Total nitrogen accumulated by lucerne in rotation, average for 1996–2009

	kgN/ha/year	kgN per 1 ha of crop rotation per year
Unfertilized		
Nitrogen accumulated by lucerne as above-ground biomass	278.1	39.7
Nitrogen recovered by following crops on unfertilized plots	103.7	14.9
Total	381.8	54.6
Fertilized (farmyard manure + NPK)		
Nitrogen accumulated by lucerne as above-ground biomass	304.7	43.5
Nitrogen recovered by following crops on unfertilized plots	26.2	3.7
Total	330.9	47.2

Table 33.4 Nitrogen balance for the crop rotation and for continuous winter wheat with above-ground lucerne biomass returned to the soil, average for 1996–2009

	N input from above-ground lucerne biomass (kg/ha)	Nitrogen taken up by crops	Difference (kgN/ha)
Crop rotation with lucerne	54.6	88.5	−33.9
Crop rotation without lucerne	0	99.8	−99.8
Continuous winter wheat	0	68	−68.0

Table 33.5 Nitrogen balance with farmyard manure + NPK including input of N in manure

	Input of nitrogen (kg/ha)				
	As mineral fertilizer	As farmyard manure	As lucerne biomass and residue	Nitrogen taken up by crops	Difference (kgN/ha)
Crop rotation with lucerne	12.8	50.0	47.2	98.5	+11.5
Crop rotation without lucerne	47.1	50.0	–	123.7	−26.6
Continuous winter wheat	60	50.0	–	81.8	+28.2

application of mineral fertilizer (Table 33.5). Only the further application of 10 tonnes of farmyard manure per hectare of crop rotation achieved a positive nitrogen balance under the rotation with lucerne and under continuous wheat.

However, nitrogen-use efficiency is poor: 23.1–24.6 % by the crop rotations and only 11 % by continuous wheat (for Table 33.6, we determined the extra yield relative to the unfertilized control and the plots to which only manure was applied). It is evident that further measures are needed to reduce the losses of nitrogen by volatilization and leaching.

33.3.2 Annual Legumes

The amount of nitrogen accumulated by an annual legume – peas was estimated in the same way. The above-ground biomass of peas in the rotation under different systems of fertilization amounted to 64.0–68.5 kgN/ha/year or 9.1–9.8 kgN per hectare of the crop rotation per year (Table 33.7), and the amount of supplementary nitrogen taken up by the winter wheat crop that followed the peas (relative to the crop following maize silage) was 37.6 kg/ha or 5.4 kg per hectare of the crop rotation per year (Table 33.8). The total amount of nitrogen accumulated by peas in rotation, averaged over 14 years, was 101.6 kg/ha on unfertilized plots and 80.4 kg/ha on fertilized plots or 14.5 and 11.5 kgN per hectare of crop rotation per year, respectively (Table 33.9).

The nitrogen balance was recalculated in respect of the complete crop rotations and for continuous winter wheat on unfertilized and fertilized plots but including the contribution of the pea crop (Tables 33.10 and 33.11). The amount of nitrogen accumulated by the peas was too small to significantly improve the nitrogen balance of the whole crop rotation (compare Tables 33.4 and 33.5).

Table 33.6 Nitrogen-use efficiency in crop rotations and continuous wheat on fertilized plots, 1996–2009

	Input (kgN/ha)		Nitrogen taken up by extra yield of crops (kg/ha)		Nitrogen-use efficiency (%)	
	Farmyard manure	Manure + NPK	Farmyard manure	Manure + NPK	Farmyard manure	Manure + NPK
Crop rotation with lucerne	50	62.8	9.8	14.5	19	23.1
Crop rotation without lucerne	50	97.1	11.2	23.9	22.4	24.6
Continuous winter wheat	66.5	126.5	–	13.9	–	11.0

Table 33.7 Nitrogen in above-ground biomass of peas in rotation under different systems of fertilization, average for 1996–2009

Indices	Unfertilized	Farmyard manure	±, t/ha	Farmyard manure + NPK	±, t/ha
Yield, t/ha	1.28	1.34	+0.06	1.37	+0.09
Nitrogen in above-ground biomass, kgN/ha/year[a]	64.0	67.0		68.5	
Amount of nitrogen per hectare of crop rotation per year, kg	9.1	9.6		9.8	

[a]Amount of nitrogen in 1 tonne of peas (main and secondary production) is 50 kg

Table 33.8 Nitrogen taken up by winter wheat following peas for grain and maize silage on unfertilized plots, average for 1996–2009

	Yield after peas (t/ha)	Yield after maize silage (t/ha)	Difference (t/ha)	Additional N taken up (kg/ha)
Winter wheat	3.55	2.41	1.14	37.6
Nitrogen per hectare of crop rotation per year				5.4

Table 33.9 Total nitrogen accumulated by peas in crop rotation, average for 1996–2009

	kgN/ha/year	kgN per hectare of crop rotation/year
Unfertilized		
Nitrogen in above-ground biomass	64.0	9.1
Nitrogen used by the following winter wheat crop	37.6	5.4
Total	101.6	14.5
Fertilized		
Nitrogen in above-ground biomass	68.5	9.8
Nitrogen used by the following winter wheat crop	11.9	1.7
Total	80.4	11.5

Table 33.10 Nitrogen balance for the unfertilized crop rotation when the above-ground biomass of lucerne is returned to the soil and for continuous winter wheat, kg/ha, average for 1996–2009

	Nitrogen input with above-ground lucerne biomass and crop residues	Nitrogen uptake by crops	Difference (kg N/ha)
Crop rotation with lucerne	54.6	88.5	−33.3
Crop rotation without lucerne + peas	5.4	99.8	−94.4
Continuous winter wheat	0	68	−68.0

Table 33.11 Nitrogen balance for the crop rotation with manure and NPK with the return of above-ground biomass of lucerne to the soil and for continuous winter wheat, kg/ha, 1996–2009

	Nitrogen input			Nitrogen taken up by crops	Difference (kg N/ha)
	As mineral fertilizer	Farmyard manure	With lucerne biomass and crop residues		
Crop rotation with lucerne	12.8	50	47.2	98.5	+11.5
Crop rotation without lucerne + peas	47.1	50	4.3	123.7	−22.3
Continuous winter wheat	60	50	–	81.8	+28.2

33.4 Conclusions

1. The total amount of nitrogen accumulated by lucerne in crop rotation, averaged over two complete rotations (14 years), was 381.8 and 330.9 kgN/ha/year on unfertilized and fertilized plots, respectively. This amounts to 54.6 and 47.2 kgN per hectare of crop rotation per year on unfertilized and fertilized plots, respectively.
2. The total amount of nitrogen accumulated by peas, averaged over two complete rotations, was 101.6 and 80.4 kg/ha/year on unfertilized and fertilized plots. That is14.5 and 11.5 kgN per hectare of crop rotation per year, respectively. Most of the nitrogen accumulated by the pea crop is harvested as green peas.
3. Application of fertilizer decreases the amount of nitrogen accumulated by both annual and perennial legumes.
4. Returning nitrogen to the soil by ploughing in the above-ground biomass lucerne does not compensate for the amount of nitrogen taken up by the other crops in the rotation. Supplementary sources of nitrogen are required to make good the deficit.
5. Perennial legumes in crop rotation and integrated with animal husbandry are the best way of restoring soil fertility, including the soil's capacity to provide the nitrogen requirements of arable crops. However, the reintroduction of perennial legumes and grasses in rotation is beneficial for all farms, including those without animals. The loss of a proportion of the cash crops may be compensated by anaerobic digestion of the harvested grass-legume mixture to produce methane as an energy source as well as organic fertilizer. This would restore soil fertility and achieve a significant reduction in the outlay for industrially produced fertilizers and, also, an alternative source of energy.

References

Berestechi OA (ed) (1984) Biological basis of soil fertility. Kolos, Moscow (Russian)

Lupascu MTT (2004) Lucerne. Ecological and forage importance. Stiinta, Chisinau (Romanian)

Misustin EN, Cerepcov N (1991) DN Prianishnicov and the problem of biological nitrogen in agriculture. In: Misustin EN (ed) The modern development of scientific ideas of DN Prianishnicov. Nauka, Moscow, pp 50–64 (Russian)

Trepacev EP (1980) Biological and mineral nitrogen in agriculture – the scales and problems. Agric Biol XV(2):178–189 (Russian)

Triboi E, Triboi-Blondel A-M (2008) Cropping systems self-sufficient in nitrogen and energy. Reality or Utopia? Alter Agric 89:17–18 (French)

Chapter 34
Resource-Conserving Agriculture: Undersowing and Mixed Crops as Stepping Stones Towards a Solution

Hans Ramseier and Valentin Crismaru

Abstract In today's agricultural environment, the conservation of water, soil and energy is of utmost importance. In the past, various approaches have been tested in a practical environment. Two practical approaches that were, once, commonplace have been neglected, namely, undersowing and mixed cultivation where various species are simultaneously grown with or between plantings of a main crop. These have been researched in multiyear field studies.

White clover and lucerne have proven to be suitable for undersowing in Moldova where undersowing of these legumes significantly reduced weediness. In Moldova, spring barley grown with white clover or lucerne produced significantly higher yields than a pure crop. In Switzerland, the seeding of an undersown crop was found to produce only slightly less yield of winter barley without application of herbicide compared with barley alone and treated with herbicide.

Field studies with mixed crops were primarily conducted with false flax (*Camelina sativa*) within a field pea crop. Camelina effectively inhibits weeds, so herbicide treatment is no longer necessary. But it is hard to calculate the amount of seed needed to prevent weed infestation; well-developed false flax will compete with the peas, reducing their yield and resulting in diminished profit. The problem is of no concern if the Camelina can be marketed as oil – which can be used for human consumption, cosmetics, paints and fuel.

H. Ramseier (✉)
Bern University of Applied Sciences HAFL, Zollikofen, Switzerland
e-mail: hans.ramseir@bfh.ch

V. Crismaru
Institute of Ecology and Geography of the Academy of Sciences of Moldova, Chisinau, Republic of Moldova

D. Dent (ed.), *Soil as World Heritage*, DOI 10.1007/978-94-007-6187-2_34,

34.1 Introduction and Objectives

The ever-increasing stress imposed on our planet by mankind is a growing concern. The emergence of a free-market economy in Europe at the end of the seventeenth century instigated an economic expansion culminating in today's global industrial economy. Science and technology has aided and abetted a doubling of the world's population in the last 50 years, and scientists predict that the upheavals experienced in the last three centuries will become even more common in future. Today's population is beyond seven billion and growing at a rate of 80 million people per year; if we are to feed the 8–12 billion people predicted for 2050, we may have to quadruple agricultural productivity and use 7–8 times the amount of energy that we are using today (Wallimann and Dobkowski 2003). Other estimates are not so high but all point in the same direction; for instance, FAO expects a 70 % increase in global demand for agricultural production (FAO 2011). Can the industrial economy support this growth without destroying itself or future generations' quality of life? It is hard to say whether the situation will evolve quite as predicted by Wallimann and Dobkowski, but this is the way the world is heading. Conservation of water, soil and energy is of utmost importance; various approaches have been tested in a practical environment, and from our evidence, undersowing and mixed cropping might play an important role.

Undersowing: Undersowing is the practice of seeding one or more crops at the same time as the main crop, where only the main crop is harvested. Experiments were conducted in rape, winter and spring barley, winter and spring wheat, maize, linseed and soya (Table 34.1). For example, undersowing of rape with clover:

- Suppresses weeds through competition and, possibly, through root exudations by the undersown crop, reducing or, even, eliminating the need for herbicides.
- Fixes nitrogen through the symbiotic *Rhizobium* of legumes. Some nitrogen may be available for the main crop, but much more will be available for the following crops.
- Increases ground cover and root penetration, arresting erosion and increasing trafficability.
- Produces humus to enhance soil fertility and soil structure.

Mixed Crops: In this case, two species of crops are sown together and both are harvested. Well-matched crops achieve a more efficient utilization of light, metabolizable energy, nutrients and soil water, expressed in the land equivalent ratio (LER) where LER greater than 1 indicates that the resources are more efficiently used in the mixed crop compared with the principal crop grown as a monoculture (Fig. 34.1).

Table 34.1 Undersown crops tested in field studies

Common name	Linnaean name
Legumes	
White clover (three varieties)	*Trifolium repens*
Sub clover	*Trifolium subterraneum*
Lucerne (alfalfa)	*Medicago sativa*
Sainfoin	*Onobrychis viciifolia*
Black medic	*Medicago lupulina*
Persian clover	*Trifolium resupinatum*
Berseem clover	*Trifolium alexandrinum*
Common vetch	*Vicia sativa*
Fenugreek	*Trigonella foenum-graecum*
Blue fenugreek	*Trigonella caerulea*
Non-legumes	
False flax	*Camelina sativa*
Buckwheat	*Fagopyrum esculentum*
Niger	*Guizotia abyssinica*
Mixtures	
White clover + Berseem clover + Phacelia	*Trifolium repens, Trifolium alexandrinum, Phacelia tanacetifolia*
White clover + buckwheat	*Trifolium repens, Fagopyrum esculentum*
White clover + Niger	*Trifolium repens, Guizotia abyssinica*

Fig. 34.1 Complementary use of water, nutrients (roots) and metabolizable energy (above ground) in mixed crops

34.2 Field Research

Undersowing and mixed-crop studies have been conducted in Switzerland since 2006. Most field experiments with mixed crops used a mixture of field peas and false flax in typical block designs. On-farm studies were conducted using strip plots. In order to answer specific questions, some pot trials were conducted in a climate chamber. Field studies have been conducted in Moldova since 2008.

34.3 Selected Results and Discussion

34.3.1 Republic of Moldova

Field trials were conducted with spring barley by Dr Valentin Crismaru at the Institute for Plant Protection and Ecological Agriculture in Chisinau.

Undersowing significantly reduces the number of weeds. On average, weed dry matter is 34–49 % less in the undersown crop (Fig. 34.2). Yields were significantly higher with undersown crops throughout all three trial years (Table 34.2). This may be attributed to the drastic decrease in weed competition; the subsequent crop also profited from the nitrogen fixed by the legume (Fig. 34.3).

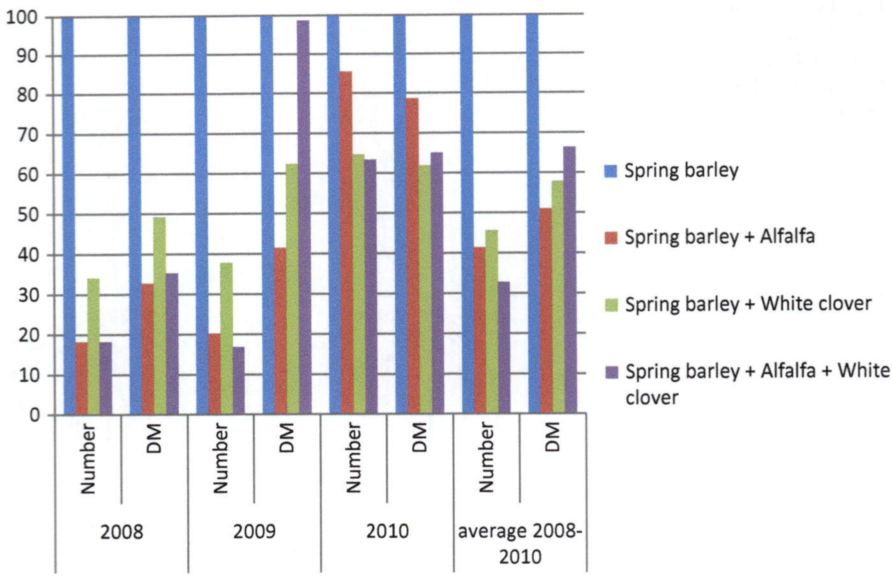

Fig. 34.2 Number and dry matter (*DM*) of weeds in spring barley with and without undersowing, 2008–2010 (% compared with scores without undersowing)

Table 34.2 Spring barley yields 2008–2010, Chisinau

	Yield (kg/ha)				Increase of yield 2008–2010	
Procedure	2008	2009	2010	2008–2010	kg/ha	%
Spring barley alone	1,970[a]	1,368[a]	2,265[a]	1,868		
Spring barley + alfalfa	2,268[b]	1,713[b]	2,560[b]	2,180	+312	+17
Spring barley + white clover	2,145[b]	1,640[b]	2,445[b]	2,077	+209	+11
Spring barley + white clover + alfalfa	2,295[b]	1,628[b]	2,600[b]	2,174	+306	+16

Superscripts mark statistically significant differences (ANOVA, $p < 0.05$)

Fig. 34.3 Spring barley undersown with white clover, Chisinau 2008. Competition from the quickly emerging clover is quite evident but weeds are rare

34.3.2 Switzerland

Undersown Rape

As a specific example, we may take the trials on the Schaedeli farm at Uettligen, 2008/2009. Trial 1 was a randomized block design with six treatments: control (hoeing), undersown white clover (10 kg/ha), undersown alfalfa (20 kg/ha), undersown sainfoin (100 kg/ha), undersown buckwheat (40 kg/ha) and white clover (5 kg/ha), and without undersowing or weed control. Trial 2 was a strip plot with two treatments: control (hoeing) and undersown white clover (10 kg/ha), also hoed.

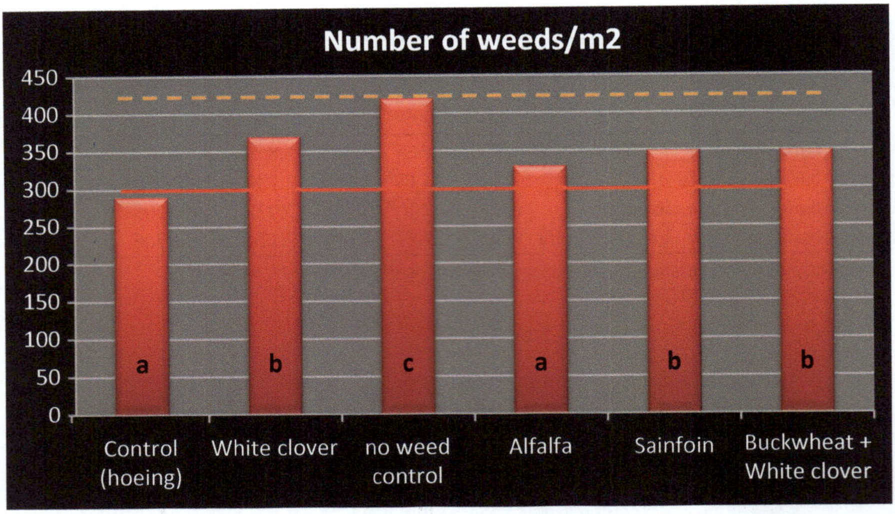

Fig. 34.4 Uettligen block trial, total weeds/m^2 in autumn 2008 (Different letters indicate significant differences (ANOVA, $p \leq 0.05$))

Weed Suppression

There were fewer weeds on the undersown plots compared with the control; alfalfa (lucerne) proved to be most effective in suppressing weeds. The effect of undersowing on specific weed varieties varied greatly: white clover suppressed shallow-rooted weeds such as annual meadow grass (*Poa annua*) and chickweed (*Stellaria media*) but not forget-me-not (*Myosotis arvensis*) which thrived in this trial (Fig. 34.4).

In the spring of 2009, the biomass of the rape and the undersown plants in the strip plot was measured (Fig. 34.5). The biomass of rape showed a small benefit from the white clover. The biomass of accompanying plants (white clover and weeds) showed no difference.

Soil Cover

White clover provided a thick ground cover. Even in the strip plots, where both the control and undersown strips were hoed, clover quickly refilled gaps in the rows. After harvest, there was a visible difference between the treatments with and without undersowing; the stubble was mown after harvest and transported together with the clover to a dryer and made into pellets. Afterwards, the clover was allowed to regrow and harvested and pelleted once more before sowing winter wheat.

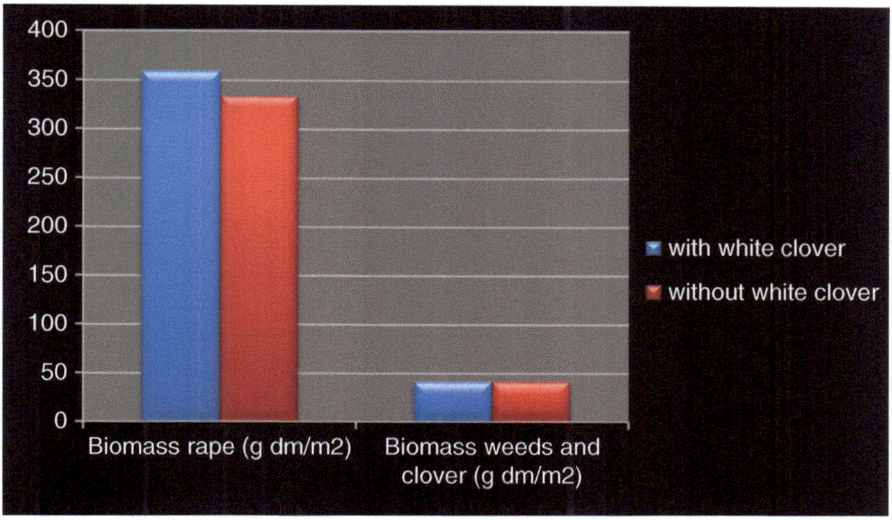

Fig. 34.5 Uettligen strip trial, biomass of rape and accompanying plants (weeds and white clover) in spring 2009

Table 34.3 Harvested fodder after the rape harvest, Uettligen, 2009

	Total pellets (tonnes)	Pellet yield (t/ha)	Dry matter (t/ha)
First cut	2.40	1.71	1.57
Second cut	2.50	1.79	1.60

Forage Value of Undersown Crops (Tables 34.3 and 34.4)

The undersown white clover is valuable in its own right, and the profit greatly exceeds the additional costs for seed and extra work involved. The presence of rape stems in the first cut of forage gave it a low energy level and high amounts of cellulose, hemicellulose and lignin. However, it contained a lot of protein, even more than hay from an extensive meadow, and also much calcium from the white clover. The second cut contained more net energy as well as a high concentration of protein (typical for white clover) – ideal as a concentrate for dairy cattle in early lactation. High-producing cows need a more quickly metabolized source of energy, but this is not important for organic farms where high production is not the priority. Once again, the high calcium content makes this a useful feed for mineral absorption.

Nitrogen Production from Undersown White Clover

Symbiotic *Rhizobium* in the clover roots fixes atmospheric nitrogen. This nitrogen is used for the above-ground parts of the plants, but much remains in the soil and

Table 34.4 Values of dry white clover pellets, Uettligen, 2009

Investigated value	First cut after rape harvesting	Second cut after rape harvesting
Dry matter (DM, %)	92	90
Neutral detergent fibre (g/kg DM)	651	335
Hemicellulose (g/kg DM)	207	91
Lignin	90	38
Crude protein (g/kg DM)	131	212
Soluble protein as % of crude protein	26	21
Net energy (MJ/kg DM)	3.8	6.8
Calcium (g/kg TS)	129	114

Table 34.5 Effect of white clover on following wheat yield, Uettligen, 2010

Practice	Yield (t/ha)	Ears/m^2	Grains/spike	WTG (g)	Weight (kg/hl)
Without white clover	5.19	419	31.2	39.8	81.1
With white clover	5.78	461	32.7	38.3	81.1

becomes available for future crops; yield studies of subsequent crops have shown an increased level of nitrogen. After rape, the field was planted with winter wheat, and measurements were made at harvest (Table 34.5).

Because of the nitrogen inherited from the white clover, more nitrogen was available to the wheat, which produced more tillers and thus more spikes per m^2. The crop following white clover yielded 11 % more than the crop without this benefit.

34.3.3 *Field Peas and False Flax* (Camelina sativa) *as a Mixed Crop*

Effective suppression of weeds was observed in all trials of peas and false flax as a mixed crop. As an example, in trials at Zollikofen in 2009, 4 kg/ha *Camelina* was seeded. The field without weed control produced an exceptionally high yield of 4.7 t/ha (Table 34.6). The planting of *Camelina* resulted in some decrease of yield of peas, though not statistically significant, but the protein yield remained the same. In addition, a bonus was achieved with the yield of 88–162 l/ha of *Camelina* oil (Figs. 34.6 and 34.7).

In 2010, a field trial was conducted to ascertain the optimal rate of seeding *Camelina* with peas; previous results suggested an ideal amount between 2 and 2.5 kg/ha. In 2011, another block trial was undertaken with 1.8 kg/ha *Camelina* (Table 34.7).

The trial confirmed the weed-suppressive capabilities of *Camelina* but proved that it is hard to determine the ideal amount of seed. The peas were seeded at the end

Table 34.6 Grain yield, crude protein and calculated oil yield of Camelina in the Zollikofen block trial, 2009

Yield component	Peas without weed control	Peas, hoed	Peas with herbicide treatment	Peas + Camelina no weed control	Peas + Camelina hoed	Peas + Camelina + clover, hoed
Peas (dt/ha)	47.5[a]	41.6[a, b]	52.5[a]	42.0[a]	30.0[b]	34.2[a, b]
Camelina (dt/ha)	0	0	0	4.4	8.1	7.1
Crude protein (dt/ha)	11.8[a]	10.4[a]	13.1[a]	12.1[a]	10.5[a]	11.1[a]
Camelina oil (l/ha)	0	0	0	88	162	142

Different letters indicate significant differences (ANOVA, $p \leq 0.05$)

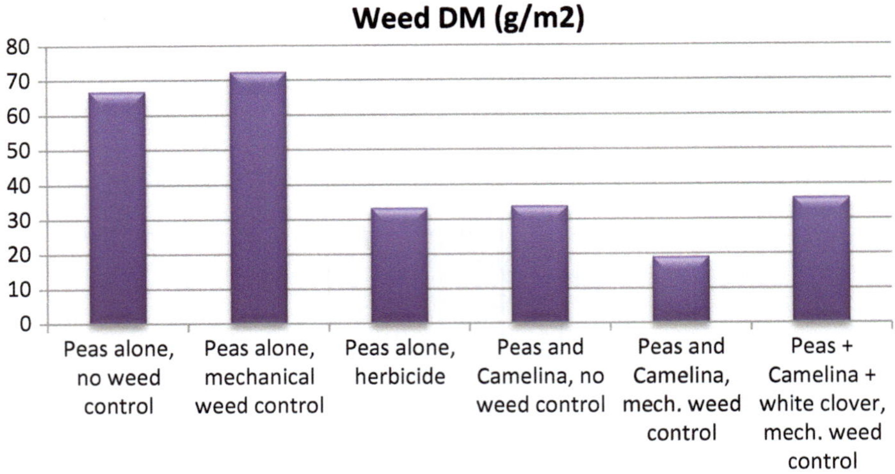

Fig. 34.6 Weed dry matter/m^2, block trial Zollikofen, 2009

Fig. 34.7 Camelina–pea mixture, Zollikofen, June 15, 2009

Table 34.7 Yield components in block trial of field pea–false flax, Zollikofen 2011

Factor	Peas alone, herbicide	Peas alone, no weed control	Peas + *Camelina*
Dry matter, weeds g/m^2	59.5	93.8	31.9
Yield, peas (dt/ha)	47.2	44.6	18.8
WTG (g)	116.0	119.7	100.3
Yield, *Camelina* (dt/ha)	–	–	16.34
Approximate oil yield of *Camelina* (l/ha)	–	–	327

of February. Because of the early seeding, the amount of *Camelina* was reduced to 1.8 kg/ha, but, even so, it competed with the peas. However, the oil yield of 327 l/ha was very high, and if there is a market for the oil, the competition will not be an issue.

34.4 Conclusions

Soil is the very foundation of agriculture; maintaining good soil structure, high humus levels and biological activity are as important as erosion minimization. Resource-conserving farming systems are vital for the future, and practices such as good crop rotation, selection of resistant crop varieties and seed quality are indispensable. The above results show that undersowing and mixed crops can be a valuable component of conservation farming.

Sustainable agriculture is a productive, competitive and efficient way to produce safe agricultural products while at the same time protecting and improving the natural environment and social/economic conditions for local communities. However, our objectives can only be achieved through international cooperation; research partnerships and international knowledge exchanges – as in the Sustainable Agriculture (SAI 2010) – are of vital importance.

Acknowledgement The Resource-Conserving Agriculture project is supported by KFH – Rectors' Conference of the Swiss Universities of Applied Sciences.

References

FAO (2011) The state of the world's land and water resources for food and agriculture – managing systems at risk. Food and Agriculture Organization of the United Nations/Earthscan, Rome/London

SAI (2010) Sustainable agriculture initiative platform. www.saiplatform.org

Wallimann I, Dobkowski MN (2003) Das zeitalter der knappheit. Verlag Paul Haupt, Bern (German)

Chapter 35
Rationale for Maintaining Humus in Arable Soils of Moldova

A.L. Rusu and V. Plămădeală

Abstract A humus balance sheet was calculated for the 1,786 million ha of arable land in Moldova. Overall and under most crops, the humus balance is strongly negative. Only perennial herbs, especially when sown as mixtures with grasses, produce a large surplus of crop residues; they also contribute to the nitrogen balance. The current structure of cropping, with 68 % of arable under clean-weeded crops, viticulture and orchards, mineralizes 2,992 thousand tonnes of humus but returns only 718 thousand tonnes of crop residues – only one quarter of the necessary return of soil organic matter.

Various model cropping structures may be proposed to close the gap by reducing the area of clean-weeded crops and substituting perennial grasses and legumes in suitable areas, which could increase the input of humus by one third and, at the same time, reduce the mineralization of soil reserves. This would compensate for 60 % of the current humus deficit. Beyond this, perennial grasses and legumes used as fodder could increase the amount of manure to provide 4.6 t/ha/year, compensating for the remaining 40 % of the humus deficit. As well as achieving a balance of humus, the proposed modification of cropping patterns would provide an annual turnover of about 2,900 lei ($US242)/ha.

35.1 Introduction

A fertile soil is, primarily, a soil rich in humus. Humus improves nearly every physical, chemical and biological attribute of soil and is directly correlated with plant productivity. It is estimated that, over the course of a year, 60–80 % of soil organic matter is mineralized, 10–30 % is converted into humus, 3–8 % is taken up in the biomass

A.L. Rusu (✉) • V. Plămădeală
Nicolae Dimo Institute of Pedology, Agrochemistry and Soil Protection,
Chisinau, Republic of Moldova
e-mail: Alexandru-ipaps_dimo@mtc.md

D. Dent (ed.), *Soil as World Heritage*, DOI 10.1007/978-94-007-6187-2_35,

of soil organisms and about the same amount remains unconverted (Bucur and Lixandru 1997). Although relatively stable, humus is itself subject to continuous decay, serving as food and energy source for microorganisms. Humus accumulates under perennial grasses which annually produce a huge amount of plant debris: under hoed row crops and black fallow, mineralization predominates, including the previously accumulated humus, and soil fertility is diminished. Sustainable agriculture requires replacement of the humus consumed, either by crop residues or manure. The objective of this chapter is to describe quantitatively the balance of humus in tilled soils in Moldova and rationale for measures to improve it.

35.2 Method of Calculation

Humus balance was calculated as the difference between income and loss. Income equals the residue from crops grown and the mass of manure produced annually, considering that the entire mass is incorporated into the soil. Normative indices are borrowed from Donos and Andries (2001) and Banaru (2002) with further additions and clarifications (Table 35.1). Within the total loss of humus, the loss of nitrogen was calculated as losses in harvested crops, ammonification, denitrification and leaching; allowance made for fixation of symbiotic and non-symbiotic nitrogen; and net consumption adjusted by a crop factor and indices for soil texture and tillage practices. The following factors for soil texture follow Lakov (1985): clays 0.8, clay loam 1.0, medium loam 1.2–1.4 and sand 1.8. For groups of crops and farm practice, we used the following factors: perennial grasses 1.0, thickly sown crops 1.2, hoed row crops 1.8 and black fallow 2.2. Since 60 % of Moldovan arable land is clayey, nitrogen consumption was calculated and multiplied by coefficient 0.8; then nitrogen consumption was recalculated in terms of humus. The total loss of humus mineralized was decreased to take account of the intake of humus synthesized from crop residues. Further, it is taken that a tonne of grass, semi-fermented, forms 0.78 t of compost or manure.

The *Statistical Yearbook* (National Bureau of Statistics 2011) presents means for the last 8 years for the cultivated area, areas sown to different crops and areas of perennial crops and also average production by crop, costs of production and farm-gate prices.

35.3 Results and Discussion

Calculation confirms that the humus balance is negative for most crops: the crops with the greatest consumption of fertility and humus deficit (over 1 t/ha/year) are, in decreasing order, sugar beet, maize, oil seed rape, potatoes, vineyards and orchards and black fallow. With the exception of rape, they are hoed row crops.

Table 35.1 Normative indices used in calculating the humus balance

	Crop	Main production with natural humidity, t/ha		Coefficient of dry mass of crop residues	Coefficient of humification of crop residues	Coefficient of mineralization of humus
		Mean yield 2003–2010	Adjusted to norm under fertilization in rotation with perennial legumes			
1	Wheat	2.14	2.48	1.40	0.12	0.66
2	Barley	1.70	1.97	1.40	0.12	0.64
3	Maize grain	2.83	3.28	1.00	0.12	0.82
4	Leguminous grains	1.11	1.29	1.14	0.21	0.14
5	Millet, sorghum and other cereals	2.00	2.32	1.10	0.17	0.66
6	Sunflower	1.26	1.46	4.24	0.14	1.38
7	Soya	1.50	1.74	1.14	0.21	0.38
8	Sugar beet	26.24	30.44	0.09	0.16	0.13
9	Tobacco	1.47	1.71	3.88	0.13	0.99
10	Rapeseed and other industrial crops	1.70	1.97	4.24	0.13	1.59
11	Potatoes	9.06	10.51	0.11	0.15	0.21
12	Field vegetables	9.01	10.45	0.11	0.15	0.12
13	Fodder roots	23.44	27.19	0.07	0.17	0.08
14	Maize silage	10.21	11.84	0.14	0.13	0.11
15	Green perennial herbs	14.00	16.24	0.34	0.21	0.02
16	Clean-weeded orchards	4.53	5.25	0.80	0.06	0.29
17	Grassed orchards	5.25	6.09	1.23	0.13	0.16
18	Clean-weeded vineyards	3.51	4.07	1.11	0.07	0.48
19	Grassed vineyards	4.07	4.72	1.90	0.11	0.21
20	Herbs in orchards and vineyards	10.00	11.60	0.29	0.21	0.03

Tobacco incurs a relatively low humus deficit, producing about 6 t/ha plant debris, which is converted to more than 700 kg of humus. Actually, rapeseed produces more crop residues (over 7 t/ha), but it also consumes 2.7 times as much nitrogen and, therefore, causes more wasting of humus. Legumes have modest requirements of nitrogen and humus, thanks to their capacity to fix nitrogen symbiotically, consuming only 8–30 kg N/ha and 158–564 kg/ha humus. Under annual legumes, the humus balance ranges from minus 200 kg/ha to plus 100 kg/ha. Perennial legumes and, especially, mixtures of grasses and legumes not only add a lot of organic residues, but also more than 100 kg/ha nitrogen. Humus accumulation by perennial grasses and legumes is about three times their consumption, accruing an annual benefit of 700 kg/ha.

Two diametrically opposed systems operate under vineyards and orchards. Most are clean weeded, so the annual intake of organic residues is low (217–273 kg/ha). The alternative, with grassy ground cover between the rows, protects the soil against erosion, increases the mass of organic residues by 1.8–2.0 times and humus accrual by 3.1–3.9 times, and increases the yield of the main crop by more than 15 quintals per hectare. Thus, grassed orchards and vineyards maintain a neutral or positive humus balance, whereas clean weeding produces a profoundly negative balance, characterized by an annual deficit of 1,100–1400 kg/ha.

Countrywide, the current structure of the arable is dominated by the 68 % hoed crops, which conduct an immense destruction of humus. In an annual arable cycle of 1.786 million ha, 2,992 thousand tonnes of humus is mineralized. In compensation, only 718,000 t of humus is synthesized from crop residues, just a quarter of what is needed to balance the budget (Table 35.2).

Robust measures are needed to improve the fertility of arable soils. With this objective, we have drawn up some model cropping structures that would improve soil fertility but not diminish food security and farm incomes (Table 35.2). Compared with the current crop structure, it is proposed to reduce the area sown to maize for grain by 70,000 ha and, in the longer term, cut its area to 200,000 ha and gradually reduce the land under sunflowers, from the present 270,000 ha to about 80,000 ha. In their place, leguminous crops, including soya, would be expanded by about 30,000 ha per year.

The sustainable model pays special attention to the improvement of fertility by growing perennial legumes in rotation with field crops and in vineyards and orchards. As prescribed by our agronomic patriarch, Academician Lupaşcu (1996), the model includes perennial legumes in rotation over 200–230 thousand ha instead of clean-weeded crops, which will increase the mass of crop residues and the humus synthesized by some 34 % (Table 35.3). At the same time, the new model will reduce the intensity of mineralization of humus from 2,992 thousand tonnes at present to 2,021 thousand tonnes per year. So, modification of crops could cut the annual deficit of humus by 1,336 thousand tonnes (2,274 – 938) or 0.74 t/ha, making good 60 % of the current deficit.

Since grasses and legumes are fodder for animals, the humus status of the soil will be improved both by crop residues and by application of manure generated from the grass. It is estimated that 1 t of sown grasses is equivalent to 0.63 t of farmyard manure. So, a crop rotation with 26 % of sown grass (468,000 ha), even at the level of average harvests, will produce 7.6 million tonnes of forage which, fed to livestock, will produce 4,728 thousand tonnes of farmyard manure – more than 1.3 as much again as the present production of manure. Together with humus synthesized from crop residues, the manure would balance the stock of humus (a deficit of 120,000 t, negligible for the whole area of 1,876 thousand ha of arable, averaging 0.07 t/ha).

Table 35.2 Humus balance on arable soils of Moldova under different crop-structure models, thousand tonnes

	Crop	Actual structure: Clean weeded 68 %, perennial herbs 3 %					Sustainable structure: Clean weeded 35 %, perennial herbs 41 %				
		1	2	3	4	5	1	2	3	4	5
1	Wheat	324	18.1	117	454	−337	324	18.1	117	454	−337
2	Barley	122	6.8	35	133	−98	122	6.8	35	133	−98
3	Maize grain	471	26.4	160	1,097	−937	200	11.2	68	466	−399
4	Leguminous grains	38	2.1	10	6	+4	50	2.8	13	8	+5
5	Millet, sorghum and other cereals	16	0.9	6	21	−15	10	0.6	4	13	−9
6	Sunflower	266	14.9	199	462	−263	80	4.5	60	139	−79
7	Soya	41	2.3	15	23	−8	60	3.4	22	34	−12
8	Sugar beet	32	1.8	12	111	−99	32	1.8	12	111	−99
9	Tobacco	4	0.2	3	6	−3	4	0.2	3	6	−3
10	Rapeseed and other industrial crops	30	1.7	28	81	−53	40	2.2	37	108	−71
11	Potato	33	1.9	5	63	−58	40	2.2	6	76	−70
12	Vegetables	49	2.7	7	52	−45	49	2.7	7	52	−45
13	Fodder roots	6	0.3	2	11	−9	6	0.3	2	11	−9
14	Maize silage	45	2.5	8	49	−41	15	0.9	3	16	−13
15	Green perennial herbs	23	1.3	23	7	+16	468	26.2	468	137	+331
16	Clean-weeded orchards	120	6.7	26	160	−134	13	0.8	3	17	−14
17	Grassed orchards	13	0.8	11	11	0	120	6.7	101	99	+2
18	Clean-weeded vineyards	138	7.7	38	232	−194	15	0.9	4	25	−21
19	Grassed vineyards	15	0.9	13	13	0	138	7.7	118	116	+2
Total		*1,786*	*100.0*	*718*	*2,992*	*−2,274*	*1,786*	*100.0*	*1,083*	*2,021*	*−938*

Note: *1* cultivated area, thousand ha; *2* cultivated area, %; *3* humus synthesized from crop residues; *4* humus mineralized; *5* humus balance

Soil fertility will be improved both by altering the crop structure and by addition of manure. The value of manure is well known (Rusu et al. 2012); less is known about the effectiveness of different crop structures, so Table 35.4 provides an economic analysis of the model crop structures.

The models propose a gradual reduction of 301,000 ha of the area sown to maize for grain and a reduction of 186,000 of the area sown to sunflowers, which will incur an annual loss of 852,000 t and 234,000 t of grain and seed, respectively. The production foregone has a value of 2.584 million lei or 155,000 euros. In the place of these field crops will be 445,000 ha of perennial grasses and legumes. Furthermore, perennial grasses and legumes will be established between the rows of vineyards and orchards which, as a rule, occupy sloping land. This entails an increase to 26 % of the arable area for perennial forage crops in rotation and up

Table 35.3 Sources of humus recovery in arable soils of the Republic of Moldova

		Annual structure		Long-term structures			
		Clean weeded 68 %, perennial grasses 3 %		Clean weeded 56 %, perennial grasses 14 %		Clean weeded 35 %, perennial grasses 41 %	
	Specification	thousand tonnes	t/ha	thousand tonnes	t/ha	thousand tonnes	t/ha
1	Crop residues	6,134	3.43	6,461	3.62	7,130	3.99
2	Humus synthesized from crop residues	718	0.40	822	0.46	1,083	0.61
3	Mineralization of humus and crop residues	2,992	1.68	2,701	1.51	2,021	1.13
4	Balance without manure (row 2 − row 3)	−2,274	−1.27	−1 879	−1.05	−938	−0.53
5	Area under perennial forage crops in rotation (thousand ha)	23	0.01	154	0.09	468	0.26
6	Forage yield (row 5 × with 14.00 t/ha for present structure and with 16.24 t/ha for model structures)	322	–	2 501	–	7,600	–
7	Farmyard manure from forage production (row 6 × 0.63)	203	0.11	1,576	0.88	4,728	2.65
8	Total national manure production (row 7 + 3,389 thousand tonnes)	3,592	2.01	4,965	2.78	8,177	4.58
9	Quantity of synthesized humus incorporated as fertilizer (row 8 × 0.10)	359	0.20	497	0.28	818	0.46
10	Input of humus to the soil from crop residues and farmyard manure (row 2 + row 9)	1,077	0.60	1,319	0.74	1,901	1.06
11	Humus balance with better use of farmyard manure (row 10 − row 3)	−1,915	−1.07	−1,382	−0.77	−120	−0.07

to 90 % of the area of vineyards and orchards. The production of forage will amount to 7.2 million tonnes, providing an increase in meat production of over 200,000 t, valued at 4.337 million lei. The total worth of agricultural production obtained by modifying the crop structure is estimated at 7.77 million lei (446 million euros). Compared with the current situation, the new model crop structure is expected to increase total revenue by 5.185 million lei/year or 2,900 lei/ha – a return of 200 %.

35.4 Conclusions

1. In Moldova, the present arable area of 1,786 million ha mineralizes 2,992 thousand tonnes of humus annually. This is compensated by only 718 thousand tonnes of humus synthesized from crop residues – only one quarter of the

Table 35.4 Economic analysis of model crop structures at 2010 prices (1 lei MD = 0.08 US \$ = 0.06 euro)

	Specification	Clean weeded 56 %, perennial grasses 14 %	Clean weeded 35 %, perennial grasses 41 %
Losses from arable under maize and sunflower			
1	Reduction of area under maize, thousand ha	101	301
2	Quantity of grain foregone (row 1 × 2.83 t/ha), thousand tonnes	286	852
3	Worth of grain production foregone (row 2 × 1,783 lei/t), thousand lei	509,938	1,519,116
4	Reduction of area under sunflower, thousand ha	78	186
5	Quantity of seed foregone (row 4 × 1.26 t/ha), thousand tonnes	98	234
6	Worth of seed production foregone (row 5 × 4,552 lei/t), thousand lei	446,096	1,065,168
7	Total annual loss of production (row 3 + row 6), thousand lei	956,034	2,584,284
Benefits from modification of crop structure			
8	Extension of area under perennial grasses and legumes in rotation, thousand ha	131	445
9	Additional production (row 8 × 16.24 t/ha), thousand tonnes	2,127	7,227
10	Meat production, live weight, obtained from forage (row 9 × 0.028 t/ha), thousand tonnes	60	202
11	Worth of livestock produced (row 10 × 21,469 lei/t), thousand lei	1,288,140	4,336,738
12	Extension of grassed orchards and vineyards, thousand ha	68	230
13	Additional fruit and grape production (row 12 × 4.66 t/ha), thousand tonnes	317	1,072
14	Worth of additional production (row 13 × 2,790 lei/t), thousand lei	884,430	2,990,880
15	Worth of additional production from extension of area under rape, potato and legumes, thousand lei	442,005	442,005
16	Total worth of extension of grassland and other crops (row 11 + row 15), thousand lei	2,614,575	7,769,623
17	Annual benefit from change of crop structure (row 16 − row 7), thousand lei	1,658,541	5,185,339
18	Total annual benefit (row 17: 1,786), lei/ha	929	2 903
19	Efficiency of structure modification (row 17: row 7 × 100), %	173	201
20	Efficiency of structure modification (row 17: row 7), lei	1.73	2.01

amount required to balance the humus budget. As a consequence, the annual balance is negative to the tune of 1.27 t/ha.

2. Increasing the area sown to perennial grasses and legumes to 26 % in rotation with annual crops and up to 90 % of the area of orchards and vineyards will increase the amount of crop residues, and the humus synthesized from these crop residues, by 34 %. At the same time, humus mineralization will be abated. In total, this will redress 60 % of the current annual humus deficit.
3. Perennial grasses and legumes in rotation with annual crops will also provide fodder for livestock which will generate enough farmyard manure to make good the remaining 40 % of the humus deficit.
4. Besides achieving a balance of humus in the soil, reducing the area under clean-weeded row crops and expanding the area under perennial grasses and legumes would provide an annual financial benefit of about 2,900 lei/ha.

References

Banaru A (2002) Improved methodological guidance for determining humus balance in arable soils. N Dimo Institute, Chisinau (Romanian)

Bucur N, Lixandru G (1997) Fundamental principles of soil science. Dosoftei, Iasi (Romanian)

Donos A, Andries S (2001) Improved procedures for determining and adjusting the balance of biological elements in soils of Moldova. N Dimo Institute, Chisinau (Romanian)

Lakov AM (1985) Humus and soil fertility. Moskovskii Rabochy, Moscow (Russian)

Lupaşcu M (1996) Agriculture of Moldova and its ecological amelioration. Stiinta, Chisinau (Romanian)

National Bureau of Statistics (2011) National accounts 2011, Chisinau (Romanian)

Rusu A, Plămădeală V, Siuris A et al (2012) Guide for use of organic fertilizers. Pontos, Chisinau (Romanian)

Chapter 36
Phyto-amelioration of Degraded Chernozem

V. Cerbari and T. Ciolacu

Abstract The prevailing farming system in Moldova has brought about a loss of humus, degradation of soil structure and compaction of the plough layer of chernozem soils. The most effective remedy is steppe vegetation. Fifteen years under grass fallow restored 80 % of the humus and 95 % of structural quality. Five years under lucerne had little effect; but growing a legume-and-grass mixture for 5 years restored tilth, the degraded plough layer was enriched with organic matter and plant residues (by 0.45 % or 0.1 % annually), soil structure improved and a sod began to form. Large-scale implementation of this method requires restoration of the livestock sector and allocation of 15–20 % of the land to perennial grasses.

36.1 Introduction

Chernozem soils are characterized by great natural fertility. *Typical chernozem*[1] is the gold standard for soil quality but, under continuous arable, it has lost many of its favourable attributes. The fragmentation of land holdings through land reform makes it hard to apply crop rotation and the protective measures essential to sustainable use, and destruction of the livestock sector has arrested the cycling of organic matter – leaving the soil without its main source of natural fertilizer. The result has been accelerated loss of soil organic matter, degradation of soil structure and compaction.

Investigations by the N Dimo Institute of Pedology, Agrochemistry and Soil Protection show that the balance of organic matter is deeply negative; Moldovan soils are losing 0.01 % of humus per year (Andries 2007). In the absence of manure,

[1] Haplic chernozem in *World reference base for soil resources 2006*.

V. Cerbari (✉) • T. Ciolacu
Nicolae Dimo Institute of Pedology, Agrochemistry and Soil Protection,
Chisinau, Republic of Moldova
e-mail: Valerian-valerian-cerbari@rambler.ru

D. Dent (ed.), *Soil as World Heritage*, DOI 10.1007/978-94-007-6187-2_36,

Fig. 36.1 Experimental plot with grassland established 15 years ago

Fig. 36.2 (**a**) Profile of *Typical chernozem* after 15 years under grass. (**b**) Soil structure restored by grass vegetation

alternative sources of organic matter are needed and further work, in collaboration with the Institute of Practical Phyto-technical Science (Balan 2007a, b; Balan et al. 2007; Cerbari and Balan 2010), demonstrated a significant improvement of soil quality in Typical chernozem under grass fallow (Figs. 36.1 and 36.2). Over the 15 years, no organic matter was removed from the site and, in comparison with

Fig. 36.3 Experimental plot used for 5 years for lucerne forage

chernozem under natural steppe, humus content was restored up to 80 % and soil structure to 95 % in the 0–25 cm layer and up to 78 % in the 25–35 cm layer.

The cardinal factor in soil remediation is the steppe vegetation that created chernozem in the first place; however, it is obvious that no one can afford to leave the land fallow for 15 years to restore soil quality. Lucerne is valuable forage that accumulates nitrogen but even 5 years under lucerne hardly improved soil structure (Fig. 36.3); its taproot system does not have the same beneficial effect as a dense mass of grass roots (Fig. 36.4), and the humus content of the soil remained almost the same because the crop was harvested for forage. Long-term field experiments have shown that even 30 % of lucerne in crop rotation does not achieve a positive humus balance (Sidorov et al. 2006).

Seeking phyto-amelioration techniques that will regenerate tilth and enable sustainable arable farming, we established an experimental plot on *Leached chernozem*,[2] sown every 5 years with a forage mixture of perennial grasses and legumes.

[2] Haplic chernozem in *World reference Base for soil resources 2006*. Leached chernozem are close to Typical chernozem, distinguished mainly by deep leaching of carbonates.

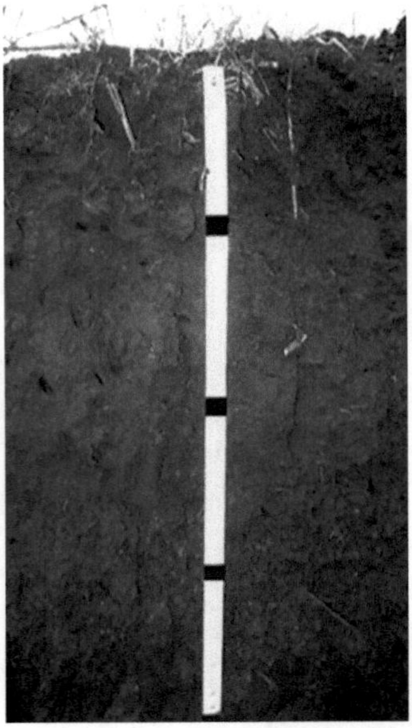

Fig. 36.4 Profile of arable *Typical chernozem* after 5 years under lucerne

36.2 Experimental Site and Methods

The experiment was established at the N Dimo Experimental Station at Ivancea in Orhei District on a 10 × 200 m plot, characterized at the outset by a 2 m soil profile and four 50 cm profiles. Bulk density was measured in the field and soil samples collected from each horizon for laboratory analysis. A 50:50 mixture of lucerne (*Medicago sativa*) and tall oat grass (*Arrhenatherum elatius*) was sown in autumn 2007 (Fig. 36.5). The fibrous root system of the grass was supposed to restore soil structure and the legume to provide nitrogen and improve forage quality. Unfortunately, drought from mid-July to mid-September 2009 killed the oat grass (its roots were confined to the 0–15 cm layer which dried completely). Perennial rye-grass (*Lolium perenne*) was sown instead; it is drought resistant and has a strong fibrous root system but, unfortunately, it is less productive. The biomass was harvested annually for forage.

36.3 Results and Discussion

Leached chernozem has an Ahp1-Ahp2-Ah-Bhw1-Bhw2-BCk1-BCk2-Ck profile. The solum (A + B horizons) is leached, effervescence with dilute acid appears at a depth of about 95 cm, visible carbonates occur as *pseudomycelia* below 100 cm and occasional *loess dolls* below 150 cm.

Fig. 36.5 Experimental plot with perennial legume and grass

Humus content varies between 3.2 and 3.5 % in the plough layer, 2.2–2.5 % in the Bhw1 horizon, 1.0–2.0 % in the Bhw2 and 0.7–1.0 % in the BCk1. The C–N ratio lies between 8.4 and 8.9, indicating a low annual input of fresh organic matter. Available phosphorus and potassium are moderate, cation exchange capacity ranges from 31 me/100 g soil in the Ahp1 to 22 me/100 g in Ck horizon and hydrolytic acidity is quite low at 1.7–2.5 me/100 g. Arable chernozem generally has good chemical attributes but degradation appears as a loss of humus, wasting of natural structure, compaction, and decreasing nutrient reserves.

In 2011, we assessed changes over 4 years produced by the perennial root system and organic residues left on the surface after harvest (about 25 % of the above-ground biomass). Field observation indicated enrichment of organic matter and significantly improved the soil structure in the topmost 12 cm and formation of a sod 4–5 cm thick (Fig. 36.6); this improvement was confirmed in the laboratory (Table 36.1).

Over 4 years, the soil was enriched by 9.1 t/ha dry mass of organic residues (2.2 t/ha annually) with average nitrogen content of 2.3 %, as well as 15.1 t/ha of roots (4.1 t/ha annually) with nitrogen content about 1.4 %, a total of 24.2 t/ha of organic residues (about 6 t/ha annually) (Table 36.2). This increased soil organic matter in the 0–12 cm layer by 0.45 %, which was reflected in restoration of soil structure, increased quantity and stability of agronomically favourable aggregates and lower bulk density. A deeper and more stable recovery of the plough layer can be achieved by this means over a period of 15–20 years.

Fig. 36.6 Plough layer (**a**) before and (**b**) after 4 years under legume and grass

Table 36.1 Changes in Leached chernozem after 4 years under perennial grass-legume vegetation

Horizon, depth (cm)	Bulk density (g/cm^3)	Total porosity (% v/v)	Sum of aggregates 10–0.25 mm (%)	Water stability of favourable aggregates (%)	Soil organic matter (%)	Total nitrogen (%)	Total phosphorus (%)
Ahp1 0–12	1.29	50.8	66.5	65.3	3.43	0.212	0.148
	1.22	53.4	68.1	79.8	3.88	0.241	0.157
Ahp1 12–20	1.41	46.4	51.5	68.7	3.22	0.203	0.136
	1.40	46.8	61.7	80.6	3.28	0.224	0.130
Ahp2 20–35	1.48	44.5	50.8	73.3	3.06	0.195	0.124
	1.45	45.5	57.3	83.6	3.12	0.208	0.123
Ah 35–50	1.43	46.5	79.3	75.7	2.86	0.187	0.102
	1.40	47.8	–	–	2.84	0.179	0.095

Numerator initial characteristics, *denominator* changed characteristics

36.4 Conclusions

1. The existing farming system causes continuous loss of humus, deterioration of soil structure and diminishing soil resistance to compaction.
2. Over 4 years, a mixture of perennial grasses and legumes, grown for forage, improved the quality of Leached chernozem. The plough layer was enriched with organic matter by 0.45 %, soil structure improved significantly and a sod 4–5 cm thick was formed.
3. The proposed technology is inexpensive and provides forage equivalent to 4 t of grain/ha/year. Its implementation on a large scale in Moldova requires restoration of the livestock sector and allocation of 15–20 % of the land to perennial grasses.

Table 36.2 Yields of forage obtained during the 4 years of the experiment

Mowing dates	Green mass (t/ha)	% water	Dry mass (t/ha)	Equivalent grain units (t/ha)	Ash	Total N	P_2O_5	K_2O	C
					% dry mass				
Year 2008 (absolute domination of *Arrhenatherum elatius* $K = 0.12$)									
13.05.2008	55	79.7	11.2	6.6	–	–	–	–	–
26.06.2008	5	60.2	2.0	0.6	–	–	–	–	–
Total	60	78.1	13.2	7.2	–	–	–	–	–
Roots in 0–30 cm layer (calculated)			10.6/4.2	–	–	–	–	–	–
Year 2009 (*Medicago sativa* and *Arrhenatherum elatius* 50:50, $K = 0.16$)									
26.05.2009	26	79.5	5.3	3.1	–	–	–	–	–
Drought, *Arrhenatherum elatius* killed									
Total	26	79.5	5.4	4.2	–	–	–	–	–
Roots in 0–30 cm layer (calculated)			4.2/4.2	–	–	–	–	–	–
Year 2010 (*Medicago sativa* and *Lolium perenne* ratio 50:50, $K = 0.16$)									
09.06.2010	25	73.2	6.7	4.0	9.4	2.18	0.68	2.20	37.1
05.08.2010	14	72.1	3.9	2.2	10.8	2.15	0.70	2.39	35.3
Total	39	72.8	10.6	6.2	9.9	2.16	0.69	2.27	36.2
Roots in 0–30 cm layer (determined)			8.9/3.6	–	18.9	1.60	0.21	0.45	29.5
Year 2011 (*Medicago sativa* and *Lolium perenne* ratio 50:50, $K = 0.16$)									
31.05.2011	19	71.9	5.3	3.0	9.8	2.11	0.47	1.54	37.2
12.07.2011	5	65.4	1.7	0.8	10.1	3.40	0.64	1.61	36.0
Total	24	70.5	7.0	38	9.9	2.38	0.51	1.56	36.6
Roots in 0–30 cm layer (determined)			7.8/3.1	–	–	1.22	0.25	0.25	30.5
Total yield in 4 years/ha	149	Mean	34.5	21.4	9.9	2.27	0.60	1.92	36.4
Mean annual yield/ha	37	water content 75.7	8.6	5.3					
Annual organic residues t/ha (25 % of yield)		Total in 4 years	9.1	–					
		Annual average	2.2						
Roots in 4 years: total mass (numerator) and recycled (denominator) in the 0–30 cm layer			$\frac{31.5}{15.1}$	–	–	1.41	0.23	0.35	30.0
Roots, total average annual mass (numerator) and recycled (denominator) in the 0–30 cm layer			$\frac{7.9}{3.8}$						
Net yield 4 years	112	75.7	25.6	16.0	–	–	–	–	–
1 year	28		6.4	4.0	–	–	–	–	–

References

Andries S (2007) Optimization of nutritional regimes and crop productivity. Pontos, Chisinau (Romanian)

Balan T (2007a) Compaction of Typical chernozem on the Balti steppe by agricultural use. In: Teodor R, Petru G (eds) Soil compaction – processes and consequences. Risoprint, Cluj-Napoca, pp 117–121 (Romanian)

Balan T (2007b) Degradation and remediation of properties of Typical chernozem from the Balti steppe. Mold Agric 7–8:37–39

Balan T, Cerbari V, Boincean BP, Boagii I (2007) Anthropogenic aspects of transformation of Typical chernozem on the Balti steppe and the problem of their sustainable use. In: Boincean B (ed) Proceedings of the international scientific-practical conference sustainable, including ecological, agriculture: results, problems, perspectives. Presa Universitaria Baltiana, Balţi, pp 14–17 (Romanian)

Cerbari V, Balan T (2010) Typical chernozem from the moist warm area of Nothern Moldova. In: Andries SV (ed) Soil quality monitoring (database, conclusions, prognosis, recommendations). Nicolae Dimo Research Institute for Pedology and Agrochemistry, Chisinau, pp 98–171 (Romanian)

Sidorov M, Vanicovici G, Coltun V et al (2006) Agrotechnics. Presa Universitaria Balteana, Balti (Romanian)

Chapter 37
Phyto-technology for Remediation of Chernozem in the South of Moldova

V. Cerbari, V. Scorpan, M. Ţaranu, and I. Bacean

Abstract Experiments were established on Common chernozem in Cahul District to repair the degraded plough layer by enhancing its productivity. After establishing baseline conditions, strips were established with sulla + rye-grass, lucerne + rye-grass and winter vetch as a catch crop for green manure. In parallel, a plot was established with application of 50 t/ha fermented sheep manure. Over 4 years, a mixture of perennial legumes and grasses improved the physical state of the plough layer. A significant improvement of humus content and soil structure was achieved by using vetch as green manure within a 4/5 field crop rotation; with just one harvest of vetch as green manure, the content of soil organic matter increased by 0.19 %, with a parallel improvement of the physical state of the degraded arable layer and a 20–30 % benefit to the following crop.

37.1 Introduction

Soil quality is unsatisfactory across most of the arable land of Moldova (Cerbari et al. 2010a) and maintenance of long-term soil fertility and productive capacity has become a priority. As a consequence of continual, intensive tillage, chernozem soils that make up 80 % of the arable have lost 40 % of their initial humus content; soil structure has deteriorated accompanied by compaction at the plough sole.

V. Cerbari (✉)
Nicolae Dimo Institute for Pedology, Agrochemistry and Soil Protection, Chisinau, Republic of Moldova
e-mail: Valerian-valerian-cerbari@rambler.ru

V. Scorpan • M. Ţaranu
Climate Change Office, Ministry of Environment of the Republic of Moldova, Chisinau, Republic of Moldova

I. Bacean
Agrarian State University of Moldova, Chisinau, Republic of Moldova

D. Dent (ed.), *Soil as World Heritage*, DOI 10.1007/978-94-007-6187-2_37,

This condition may be improved only by systematic soil conservation (Cerbari 2010; Boincean 2011) and by enhancing inputs of organic matter (Shein and Milanovsky 2002; Cerbari et al. 2010b). Regretfully, farmyard manure is hardly ever applied nowadays and the task is to replace it with an alternative source of organic matter in the shape of green manure. Here we discuss the possibility of improving arable soil quality by growing vetch as a high-nitrogen green manure, in rotation with grain crops.

37.2 Experimental Site and Methods

The aim was to develop and test methods for enhancing the quality of the degraded plough layer and improve its productive capacity (Cerbari 2011). The experimental plots are located on Common chernozem[1] on the property of Tarsal-Agro Ltd. at Tartaul de Salcie in Cahul District, lying on a broad, almost flat upland (Fig. 37.1).

Experimental plots of sulla (*Hedysarum coronarium*), rye-grass + lucerne, and rye-grass were established in the spring of 2010. The vetch plot was sown in the autumn of 2009 and the yield of 28 t/ha of green matter ploughed in a green manure in August 2010; the following month, a seedbed was prepared and sown with winter wheat. Baseline soil conditions were determined in autumn 2009, prior to establishing the experiment (Tables 37.1, 37.2, 37.3 and 37.4).

Compared with the Typical chernozem of Central Moldova, Common chernozem of Southern Moldova have more favourable physical attributes, thanks to a coarser texture; on the other hand, they have less humus and are poorer in nutrients.

Fermented sheep manure was incorporated in a 50 m $\times$ 10 m plot adjacent to the vetch strip; 2.5 t (50 t/ha) was spread, disked to break up the lumps, and incorporated into the soil by ploughing to 25 cm early in September 2010 (Fig. 37.2). The composition of the manure is presented in Table 37.5: 46.5 t/ha of dry matter was incorporated, comprising 5.1 t/ha carbon, 0.81 t/ha nitrogen, 0.57 t/ha phosphorus and 0.34 t/ha potassium. Early in the following April, the plot was seeded with sunflower.

37.3 Results and Discussion

Here we present preliminary data on the impact of the various treatments on soil properties and productive capacity; of course, the impact on soil quality of perennial legumes and grasses seeded for hay after the first year is comparatively less than after several successive years. From visual observation, the sulla + rye-grass mixture

[1] Common chernozem and Typical chernozem are both *Haplic chernozem* in the World Reference Base. In Common chernozem, carbonates occur immediately beneath the topsoil, in the 30–60 cm layer, whereas Typical chernozem is leached of carbonates to below 50–90 cm.

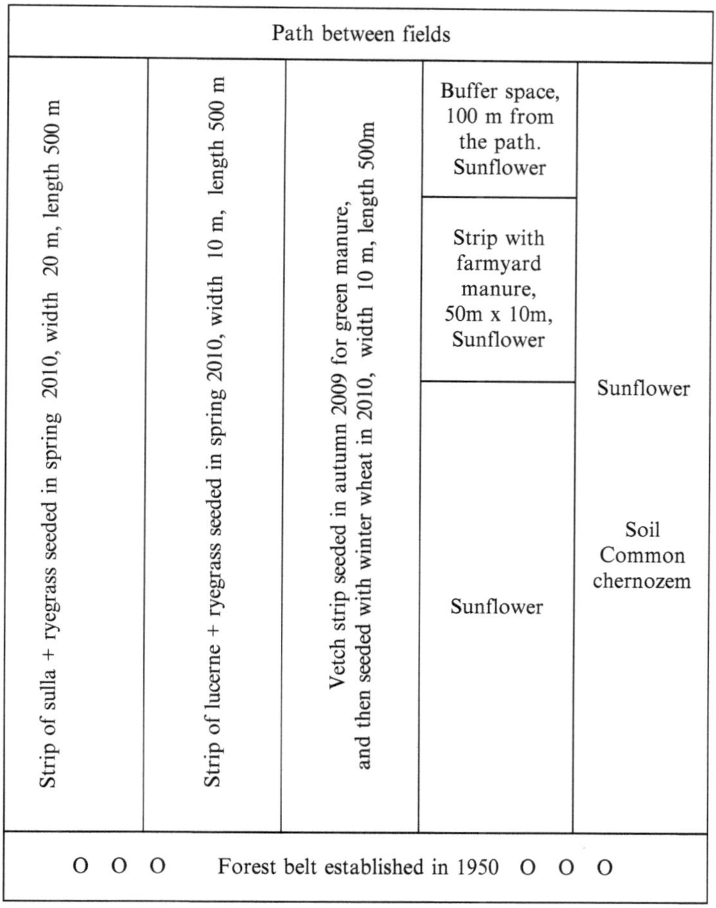

Fig. 37.1 Layout of the field experiment at Tartaul de Salcie, 2011

Table 37.1 Particle size distribution

Horizon, depth (cm)	Size of fractions (mm), % by mass						
	1.00–0.25	0.25–0.05	0.05–0.01	0.01–0.005	0.005–0.001	<0.001	<0.01
Ahp1 0–25	0.8	16.6	33.5	8.1	10.0	31.0	49.1
Ahp1 25–34	0.8	16.6	33.5	7.3	11.1	30.7	49.1
Ahk 34–49	0.6	15.5	33.9	8.2	11.0	30.8	50.0
Bhk1 49–71	0.5	9.9	38.2	8.9	11.5	31.0	51.4
Bhk2 71–96	0.5	8.1	40.0	8.2	11.8	31.4	51.4
BCk 96–120	0.4	7.0	41.3	8.1	12.2	31.0	51.3
Ck 120–150	1.1	7.1	40.2	9.2	12.6	29.8	51.6

Table 37.2 Soil structure, data for dry sieving/data for wet sieving

Horizon, depth (cm)	Diameter of structural units (mm), % by mass: >10	< 0.25	Sum (10 – 0.25)	Sum (>10 + <0.25)	Structure quality (dry sieving)	Structure water stability
Ahp1 0–25	35.5/–	14.0/70.2	50.5/29.8	49.5/70.2	Medium	Low
Ahp2 25–34	57.0/–	2.9/75.2	40.1/24.8	59.9/75.2	Medium	Low

Table 37.3 Physical properties

Horizon, depth (cm)		Ahp1 0–25	Ahp1 25–34	Ahk 34–49	Bhk1 49–71	Bhk2 71–96	BCk 96–120	Ck 120–150
Horizon thickness (cm)		25	9	18	23	25	19	30
Clay <0.001 mm	% mass	31.0	30.7	30.8	31.0	31.4	31.0	29.8
Clay <0.01 mm		49.1	49.1	50.0	51.4	51.4	51.3	51.6
Hygroscopicity		5.1	5.1	8.8	4.7	4.6	3.9	3.5
Hygroscopicity index		6.8	6.0	5.8	5.8	5.4	4.7	4.3
Density	g/cm^3	2.59	2.60	2.63	2.65	2.67	2.68	2.70
Apparent density		1.25/1.35[a]	1.45	1.34	1.38	1.4	–	–
Total porosity	% volume	49.4	44.2	49.0	47.9	47.6	–	–
Compaction[b]		3	13	4	7	7	–	–

[a]Apparent density 0–10 cm/apparent density 10–25 cm
[b]Compaction = 100(PMN-PT)/PMN, where PMN is minimum required porosity (% volume) and PT is actual porosity (% volume)

Table 37.4 Chemical properties

Horizon/depth (cm)		Ahp1 0–25	Ahp1 25–34	Ahk 34–49	Bhk1 49–71	Bhk2 71–96	BCk 96–120	Ck 120–150
$pH_{(water)}$		7.1	7.2	7.3	7.6	7.8	7.9	8.0
$CaCO_3$	% mass	0	0	0	1.4	4.0	6.6	8.0
Total P_2O_5		0.139	0.111	0.080	–	–	–	–
Humus		3.16	3.11	2.85	2.60	1.84	1.00	0.61
Total N		0.208	0.202	0.190	–	–	–	–
C–N		8.8	8.9	8.7	–	–	–	–
P_2O_5	(mg/100 g sol)	1.6	1.0	0.8	–	–	–	–
K_2O		21	18	14	–	–	–	–

was much more productive than lucerne + rye-grass. This is confirmed by the data for green matter: the yield of sulla + rye-grass (20 % dry matter) was 35 t/ha, while the yield of lucerne + rye-grass (24 % dry matter) was 17 t/ha. The impact

Fig. 37.2 Spreading and ploughing in manure

Table 37.5 Chemical composition of manure applied at the establishment of the experimental plot at Tartaul de Salcie

	Water	Carbon	Ash	Total N	Total P	Total K		
	% by mass						C:N	pH
Sheep manure	7.3	10.2	38.8	1.61	1.14	0.68	6.3	8.2

Table 37.6 Changes in the plough layer of Common chernozem following incorporation of 50 t/ha of manure (initial state/changed state)

Horizon depth (cm)	Apparent density (g/cm^3)	Total porosity (% v/v)	Sum of favourable aggregates (10–0.25 mm %)	Water stability of favourable aggregates (%)	Humus (% mass)	Mobile forms, mg/100 g P_2O_5	Mobile forms, mg/100 g K_2O
Ahp1 0–10	1.25	51.7	50.5	29.8	3.16	1.6	21
	1.23	52.5	57.5	31.8	3.37	1.9	22
Ahp1 10–25	1.35	47.9			3.16	1.6	21
	1.33	48.6			3.35	1.8	21
Ahp2 25–34	1.45	44.2	40.1	24.8	3.11	1.5	18
	1.43	45.0	42.1	25.8	3.16	1.5	19
Ahk 34–49	1.34	49.0	–	–	2.85	0.8	14
	1.35	48.7			2.87	1.0	16

of organic manure on soil properties after the first year of its incorporation, as measured in May, is presented in Table 37.6.

Incorporation of 50 t/ha of manure in the 0–25 cm layer increased the humus content by 0.2 %, improved soil structure and enhanced mobile phosphorous, translating to an increase of 0.4 t/ha in the sunflower yield compared with the control.

The vetch strip, 10 m wide and 1,100 m long, was established in autumn 2009. Late in July 2010, the plot was disked to a depth of 12 cm. During the first 10 days of September 2010, the plot was ploughed, 259 kg/ha Amofos phosphate fertilizer applied and winter wheat was sown. Baseline soil condition was established prior to seeding the vetch strip, and measurements were repeated on the field seeded with

Table 37.7 Changes in the plough layer of Common chernozem following incorporation of 28 t/ha vetch green manure (initial state/changed state)

Horizon depth (cm)	Apparent density (g/cm^3)	Total porosity (% v/v)	Sum of favourable aggregates (10–0.25 mm %)	Hydrostability of favourable aggregates (%)	Humus (% mass)	Mobile forms mg/100 g	
						P_2O_5	K_2O
Ahp1 0–10	1.25/1.21	51.7/53.3	50.5/68.8	29.8/43.4	3.16/3.36	1.6/1.9	21/21
Ahp1 10–25	1.35/1.30	47.9/49.8			3.16/3.34	1.6/1.8	21/21
Ahp2 25–34	1.45/1.43	44.2/45.0	40.1/41.5	24.8/38.6	3.11/3.06	1.5/1.4	18/18
Ahk 34–49	1.34/1.35	49.0/48.7	–	–	2.85/2.90	0.8/0.9	14/14

Fig. 37.3 Plough layer of Common chernozem, (**a**) prior to and (**b**) after incorporation of vetch green manure and sowing winter wheat

wheat after vetch in the second half of May 2011. Table 37.7 and Fig. 37.3 present data on the impact of the vetch on soil properties.

Incorporation of 28 t/ha of vetch green manure (6.1 t/ha dry mass, 4.2 % nitrogen) increased the organic matter content of the 0–25 cm plough layer by 0.19 %, improved soil structure and enhanced the level of mobile phosphorous. Table 37.8 presents data for the crop of winter vetch.

The succeeding winter wheat crop yielded 4.2 t/ha (8 % moisture), compared with the unfertilized control yield of 2.9 t/ha.

Table 37.8 Yield of winter vetch in the Tartaul de Salcie experiment

Yield in 2010	Green manure (t/ha)	Dry mass (%)	Dry mass (t/ha)	Grain units (t/ha)	Ash	N	P_2O_5	K_2O	C
					% of dry mass				
The vetch strip was established in September 2009									
Main yield	28	21.8	6.1	5.6	11.4	4.12	1.00	2.38	35.1
Roots, 0–30 cm layer (determined)			5.6	–	19.7	1.51	0.27	0.85	28.7

37.4 Conclusion

The current farming system is mining soil organic matter, reducing the resilience of soil structure and compacting the topsoil. Soil quality is certainly improved by incorporation of farmyard manure – but this will only be possible in Moldova through the restoration of stockbreeding and allocation of 15–20 % of the land to perennial grasses and legumes. Lucerne alone, with its taproot system, hardly improves soil structure or, even, total soil organic matter (Cerbari et al. 2010a). Soil improvement on a broad scale is possible by seeding vetch as a catch crop, to be used as green manure in crop rotations. Preliminary results on Common chernozem demonstrate that even a single green manure crop of vetch increased the humus content of the plough layer by 0.19 % and improved soil structure – while raising production by 20–30 %.

References

Boincean BP (2011) Soil tillage – trends and prospects. Akademos (J Sci Innov Cult Art) 22 (3):61–67 (Romanian)

Cerbari V (2010) No-till – a system of agriculture which protects the soil. Mold Agric 8–9:9–14 (Romanian)

Cerbari V (2011) Program for development and implementation of soil conservation technologies in agriculture. Mold Agric 4–5:7–9 (Romanian)

Cerbari V (coordinator) and others (2010a) Monitoring of soil quality. Database, conclusions, prognoses, recommendations. Pontos, Chisinau

Cerbari V, Scorpan V, Taranu M (2010b) Decreasing CO_2 emissions from arable land of the steppe zone of the Republic of Moldova. Environment (Sci J Inf Ecol Cult) 49(1):6–13 (Russian)

Shein EV, Milanovsky EY (2002) Soil organic matter and aggregate stability. In: Proceedings of the international conference on soils under global change, Constanta, Romania, vol 1, 3–6 September 2002

Chapter 38
Organic Manuring to Restore the Fertility of Eroded Soils

A. Siuris

Abstract In an agrarian economy like Moldova's, many economic and social problems may be mitigated by protection and amelioration of the soil. The arable is not in good health; some 865,000 ha, more than 25 % of the country, is eroded or humus-depleted and the degraded area is expanding. Starting from a firm, anti-erosion foundation, the productive capacity of degraded lands may be rebuilt by the rational use of manure which contributes biologically available energy and increases heterogeneity, humus stocks, and crop yields.

38.1 Introduction

Moldova's farmland is not in good health (Krupenikov 2004, 2008) and erosion is insidiously removing the soil itself. The affected area amounts to 864,600 ha (25.5 % of the country); 537,200 ha is weakly eroded, 268,700 ha moderately eroded, and 58,750 ha strongly eroded. The most fertile soils are the worst affected: chernozem comprise 91 % of the eroded soils and some 788,200 ha of chernozem is eroded to some degree (Rozloga 2010).

The damage to the national economy is tremendous. Loss of crop harvests costs €37 million annually (Andrieş et al. 2008) but the greatest loss is of fertile soil itself. Every year, on average, 30 t/ha is washed off the eroded land of Moldova (including Gagauz ATU and the territory on the left bank of the Dniester). That is 26 million tonnes of soil, containing 700,000 t of humus, 50,000 t of nitrogen, 34,000 t phosphate, and 587,000 t of potassium – equivalent to the complete destruction of 2,000 ha of chernozem. This chapter highlights the possibilities of restoring the fertility of eroded soils by application of farmyard manure as an organic fertilizer.

A. Siuris (✉)
Nicolae Dimo Institute of Pedology, Agrochemistry and Soil Protection, Chisinau, Republic of Moldova
e-mail: ipaps_dimo@mtc.md

D. Dent (ed.), *Soil as World Heritage*, DOI 10.1007/978-94-007-6187-2_38,

38.2 Experimental Site and Methods

The research was conducted between 1996 and 2009 at the N Dimo Experimental Station in the village of Lebedenco, in Cahul district. The experimental field slopes at 5–6° with a northeastern aspect; the soil is a Common chernozem,[1] dusty clay loam texture on loess, and moderately eroded. The plough layer was weakly alkaline (pH_{water} 7.5–7.8) with 2.07–2.54 % humus, 1.54–1.93 mg/100 g mobile phosphorus, and 15.3–16.8 mg/100 g exchangeable potassium.

The experiment was carried out in three replicates; individual 6 × 40 m plots were laid out in a single row across the slope with their long sides oriented up-and-down slope. Winter barley was cultivated in 1977, followed by maize, Hungarian vetch (oats + peas), winter wheat, maize, winter barley, maize, sunflower, and winter wheat. The farmyard manure applied was cattle manure with bedding. Soil samples were collected from fixed points for determination of humus (Tiurin's method), mobile phosphorus (Macighin's method), exchangeable potassium by flame photometry, particle size distribution (pipette method with dispersal in $Na_2P_2O_7$ solution), structural composition by dry and wet sieving (Savinov's method), particle density (pycnometer), and apparent density (cylinder) to calculate porosity.

38.3 Results and Discussion

Manure is a valuable product that constitutes some 80 % of all local fertilizer. Its composition depends on the way the beasts and poultry are housed, whereby the manure is obtained with or without bedding materials which modify the nutrient content and mechanical properties (Ţurcan et al. 1993). The Organic Fertilizers and Soil Fertility Laboratory at the N Dimo Institute has measured wide variations in organic matter and nutrient content (Table 38.1).

The worth of manure as a fertilizer is unquestionable. Innumerable experiments and practical experience have demonstrated its benefits on all crops (Batula 1988; Mineev 1988; Ţurcan 1985). Our research quantifies its potential for improving the status of humus, mobile phosphate, and exchangeable potassium (Table 38.2).

By the tenth year of application, the humus content of manured variants increased by 0.41–0.72 % compared with the control. At the same time, mobile phosphate increased from 1.44 to 2.03 mg/100 g soil and exchangeable potassium increased from 2.9 to 8.3 mg/100 g compared with the initial state.

The clay loam texture may be regarded as favourable since it provides good conditions for the growth of crops; clay loams are easy to work over a wide range of moisture content. However, dustiness means that structural aggregates formed by

[1] Haplic chernozem in the *World reference base for soil resources 2006.*

Table 38.1 Organic matter and nutrient content of manures at natural moisture content

	Moisture	Organic matter	N	P_2O_5	K_2O	NPK sum
Type and form of manure	%		kg/t			
Cattle manure with bedding	53	17.3	5.6	4.3	10.4	20.3
Cattle manure without bedding	82	11.2	3.9	2.7	4.6	11.2
Pig manure with bedding	57	18.3	8.2	7.4	7.4	23.0
Pig manure without bedding	84	11.7	5.7	2.9	2.4	11.0
Poultry droppings with bedding	61	29.0	16.3	14.5	13	43.8
Poultry droppings without bedding	49	29.2	22.2	7.4	9.9	39.5
Sheep manure with bedding	40	23.1	9.5	4.5	17.7	31.7
Sheep manure without bedding	53	21.3	9.2	3.6	10.7	23.5
Horse manure	55	16.7	7.3	4.8	8.4	20.5

Table 38.2 Change of humus, mobile phosphorus, and exchangeable potassium in the plough layer of eroded Common chernozem

	Total humus		Mobile phosphate		Exchangeable potassium	
	%		mg/100 g soil			
Manure application	Content	Increase	Content	Increase	Content	Increase
1996 baseline, before the application of manure						
Unmanured (control)	2.07	–	1.89	–	16.7	–
50 t manure/ha – once in 2 years	2.09	–	1.54	–	16.1	–
100 t manure/ha – once in 4 years	2.54	–	1.80	–	16.5	–
150 t manure/ha – once in 6 years	2.44	–	1.85	–	17.8	–
200 t manure/ha – once in 8 years	2.17	–	1.78	–	16.8	–
2006, tenth year of experiment						
Unmanured (control)	2.11	0.04	2.04	0.15	16.8	0.13
50 t manure/ha – once in 2 years	2.56	0.47	3.27	1.73	19.0	2.9
100 t manure/ha – once in 4 years	2.95	0.41	3.34	1.54	21.3	4.8
150 t manure/ha – once in 6 years	3.07	0.63	3.88	2.03	24.3	6.5
200 t manure/ha –once in 8 years	2.89	0.72	3.22	1.44	25.1	8.3

tillage are not water stable, compact readily, and are prone to erosion by run-off. Both the content of clay (<0.01 mm) and fine clay (<0.001 mm) remained constant, but the increased content of organic matter in the manured variants results in a decrease in particle density and, importantly, decreased apparent (bulk) density. Manuring with 50–100 t/ha reduced the proportion of clods >10 mm by 26.6 % while increasing crumbs with a diameter less than 0.25 mm from 6.0 to 11.2 % (Table 38.3).

These changes are associated with an increase of the void space, which now attains up to 55 %, and a decrease of penetration resistance by about 10 kg F/cm^2 or 43 % compared with the control (Table 38.4).

The improvement of agro-physical and agrochemical indicators by means of manuring has brought about increased of crop yields (Tables 38.5 and 38.6).

Table 38.3 Structural improvement in the plough layer of eroded Common chernozem after 10 years of manuring, average statistical data (X ± s)

Manuring variant	Percentage of structural elements by diameter (mm) >10	<0.25	∑ 10 – 0.25	>10 + <0.25	Quality of structure (dry sieving)	Water stability (wet sieving)
Unmanured	49.5[a]	3.6	47.0	53.1	Average	Low
	–[b]	72.5	27.5	72.5		
Manure, 50 t/h – once in 4 years	22.9	14.8	62.3	37.7	Good	Low
	–	71.6	28.4	71.6		
Manure, 100 t/ha – once in 4 years	25.5	9.6	62.9	37.1	Good	Low
	–	71.0	29.0	71.0		

[a]*Numerator* total aggregates by dry sieving
[b]*Denominator* water-stable aggregates by wet sieving

Table 38.4 Impact of manuring on the physical condition of the plough layer of moderately eroded Common chernozem, mean statistical parameters (X ± s)

Manuring variant	Fractions % <0.001 mm	<0.01 mm	Density (g/cm^3)	Apparent density (g/cm^3)	Porosity (%)	Penetration resistance (kgF/cm^2)
Control, no manure	25.9	45.9	2.66	1.26	52.6	23.4
Manure, 50 t/ha – once in 4 years	26.3	45.4	2.64	1.22	53.8	20.1
Manure, 100 t/ha – once in 4 years	25.8	45.7	2.63	1.18	55.1	13.3

Table 38.5 Effect of manuring on crop production from moderately eroded Common Chernozem (1997–2005), q/ha cereal units

Manure application	Total production	Total increase	Annual average increase
Control, unmanured	198.7	–	–
50 t manure/ha – once in 2 years	277.2	78.5	8.7
100 t manure/ha – once in 4 years	295.1	96.4	10.7
150 t manure/ha – once in 6 years	306.4	107.7	12.0
200 t manure/ha – once in 8 years	302.0	103.3	11.5

Over 9 years, manured variants with different doses yielded 78.5–107.7 q/ha cereal units more than the control. The greatest increase in yield was registered on the variant treated with 150 t manure/ha applied 1 year in six. Lucerne was grown on the experimental field for 4 years (2006–2009) during which no manure

Table 38.6 Effect of manuring on lucerne yield on moderately eroded Common chernozem (2006–2009), kg/ha dry mass (hay)

	Total yield per 4 years		Total increase per 4 years		Annual average increase from organic manure	
Manure application	Hay	Nutritive units	Hay	Nutritive units	Hay	Nutritive units
Control, no manure	13,432	7,388	–	–	–	–
50 t manure/ha – once in 2 years	27,731	15,259	14,299	7,871	3,535	1,968
100 t manure/ha – once in 4 years	23,625	12,994	10,193	5,606	2,548	1,402
150 t manure/ha – once in 6 years	26,718	14,695	13,286	7,307	3,322	1,827
200 t manure/ha – once in 8 years	25,462	14,004	12,030	6,616	3,008	1,654

was applied. The after-effect of manure was manifest in the yield of lucerne: an increase of dry mass (hay) of 10,193–13,286 kg/ha, equivalent to 5,606–7,811 nutritive units.

38.4 Conclusions

1. Manuring moderately eroded Common chernozem increased the contents of humus and mobile phosphate and potassium. In the course of 10 years, humus increased by 0.47–0.72 % or 0.05–0.07% per year; mobile phosphate and exchangeable potassium increased by 3.3–3.9 and 2.9–8.3 mg/100 g soil, respectively.
2. Manuring also promoted a 24.0–26.6 % reduction of clods >10 mm and, simultaneously, a 6.0–11.2 % increase of crumb structure and improved mechanical properties. The penetration resistance decreased by 43 %.
3. Manuring over a period of 9 years led to a yield increase of 78–108 q/ha cereal units or by 39–54 %, compared with the control. On this moderately eroded soil, the best results were achieved by manuring at 50 t/ha once in 4 years.

References

Andrieş S, Cerbari V, Rusu A et al (2008) Natural and anthropic conditions of soil deterioration and technological processes for the minimization of the consequences of unfavourable eco-pedological factors. In: Toma S (ed) Diminution of the impact of extreme pedoclimatic factors on cultivated plants. Academy of Sciences, Chisinau (Romanian), pp 44–46

Batula AA (1988) Organic fertilizers. Harvest, Kiev (Russian)

Krupenikov IA (2004) Biosphere and ecological consequences of erosion processes. Assessment of fertility of eroded soils. In: Krupenikov IA (ed) Soil erosion. Pontos, Chisinau, pp 72–97 (Romanian)

Krupenikov IA (2008) The origin of Chernozem, improvement, the tragedy of deterioration, protection and rehabilitation. Pontos, Chisinau (Russian)
Mineev VG (1988) The ecological problems of agrochemistry. Moscow University Publishing House, Moscow (Russian)
Rozloga I (2010) Structure of soil cover on sloping fields in the Republic of Moldova. Agricultural Science 2:7–11 (Romanian)
Ţurcan MA (1985) Agrochemical basis of application of organic fertilizers. Stiinta, Chisinau (Russian)
Ţurcan MA, Serjentu EP, Banaru AZ et al (1993) Recommendations for the use of organic fertilizers in Moldova. Agroinformreclama, Chisinau (Romanian)

Chapter 39
Structural Indices of Typical Chernozem Under Various Methods of Tillage

G. Jigau, M. Motelica, M. Lesanu, I. Chircu, E. Tofan, L. Georgescu, and C. Iticescu

Abstract Systems analysis identifies the need for a more ecological as well as more efficient approach to issues in modern agriculture. Soil structure of Typical moderately humic chernozem was monitored under mouldboard ploughing and three variants of conservation tillage under production conditions. Different kinds of tillage bring into play different mechanisms of soil structural development. As compared with ploughing, deep loosening and disking produce a more favourable, more resilient soil structure and improved soil water status.

39.1 Introduction

Modern agriculture faces problems that have been brought about largely by farming practices, such as the degradation of soil structure that adversely affects crop yields, and, also, overarching socio-economic problems – not least continuously increasing energy consumption and the soaring cost of fuel and fertilizers. On either hand, the problems stem from the application of industrial technologies. To mitigate these problems, we must look to alternatives that embrace the rational use and maximum efficiency of all components of the farming system.

In the case of soil management, we must ensure the optimal functioning of soil ecosystems in relation to all the other components of agro-ecosystems. We look to conservation technologies that (a) provide optimal conditions for soil functioning and crop growth, (b) conserve and improve soil resources and fertility, (c) conserve

G. Jigau (✉) • M. Motelica • M. Lesanu • I. Chircu
State University of Moldova, Chisinau, Republic of Moldova
e-mail: pedologieaplicativa@gmail.com

E. Tofan
Al. I. Cuza University, Iasi, Romania

L. Georgescu • C. Iticescu
Dunărea de Jos University, Galati, Romania

D. Dent (ed.), *Soil as World Heritage*, DOI 10.1007/978-94-007-6187-2_39,

water resources and promote maximum water-use efficiency to reduce drought risk and (d) restore degraded land and soil cover. The guiding principles of a conceptual framework might be:

1. Bringing biological processes back into soil management, as opposed to industrial processes
2. Adapting farm management to the specific conditions of the landscape
3. Integrated management of the various processes within agro-ecosystems, as summarized in Table 39.1

39.2 Experimental Site and Methods

The research was conducted over the period 2008–2010 at the SRL Vatmol-Agro experimental farm at Scaieni, in the Donduseni District of Moldova. The soil is Typical chernozem,[1] moderately humic and fine loamy, with combined A_m + A_mB horizons thicker than 65 cm; the underlying B_1 is calcareous below 70 cm. Soil texture is clay loam throughout. The plough and sub-plough layers are clearly visible; the average humus content of the plough layer is 4.57 % and of the sub-plough layer 3.64 %, declining to 3.31 and 1.93 % in the upper and lower A_mB horizons, respectively, and to 1.38 % in the B_1. Reaction is pH 7.2 in the plough layer and 7.25 in the B_1.

Within a crop rotation of wheat, barley and sunflower, four kinds of soil tillage were practised: traditional mouldboard ploughing to 22–25 cm, para-ploughing to 40 cm, disking to 18 cm and disking to 14–16 cm. The measured soil physical attributes were bulk density, using Petinov's method; water-stable aggregates, by wet sieving; total porosity (Et) calculated using the relation $Et = (1 - \rho b/\rho s)$, where ρb is bulk density and ρs is the density of solid phase; and differential porosity, using Kachinski's method based on the soil water potential curve.

39.3 Results and Discussion

Comparison of the soil structural aggregates created under different kinds of tillage reveals some common features but, also, a general reduction in the proportion of agronomically favourable aggregates (10–0.25 mm) during the growing season (Table 39.2). This tendency is most pronounced under the mouldboard plough where, compared with soil structure in May, in June, favourable aggregates were less by almost 10 % in the 0–10 cm layer; by July, the loss was almost 17 %. The same trend, though less pronounced, was observed in the deeper soil layers and, by July, the layering observed at the beginning of the growing season was hardly noticeable.

[1] Haplic chernozem in *World reference base for soil resources 2006.*

Table 39.1 Hierarchical levels of adaptive management in agro-ecosystems

Hierarchical level	Technical factors	Soil-landscape functions	Intrinsic factors	Evaluated parameters
0	Biological characteristics of the plant	Physical support; root system development; supply of air, water, nutrients and warmth	Soil texture, bulk density, cohesiveness, penetrability	Soil texture, bulk density, infiltration and permeability
00	Physical properties determining the biological potential of the plant			
1	*Idem* + solar radiation + temperature	Soil surface which receives solar radiation – as heat transformer and supplier	Thermal absorption capacity, specific heat, thermal conductivity	Content, distribution and composition of humus, soil texture and permeability
10	Physical proprieties favouring effectiveness of solar radiation			
2	*Idem* + water resources (rainfall, irrigation)	Soil as a partitioning surface and reservoir in the water cycle	Infiltration capacity, permeability-hydraulic conductivity, total and plant-available water capacity	Soil texture, bulk density, permeability, differential porosity (available water capacity)
20	Physical properties determining water-use efficiency			
3	*Idem* + biogeochemical cycles; transformation and formation of materials	Storage and transformation of organic matter, interaction of geological and biological cycles	Hydrothermal, air-water and redox regimes	Porosity and permeability
30	Physical properties determining direction and intensity of transformations			
4	*Idem* + practised technologies	Space for accumulation and translocation of materials; receptor of fertilizers and amendments; material manipulated by tillage and traffic	Pore space (volume, size, stability, continuity)	Soil texture and bulk density, soil structure and stability
40	Physical properties affecting practised technologies			
401	Irrigation			
402	Fertilization			
4021	Mineral fertilization			
4022	Manuring			
403	Tillage system			
404	Soil protection and conservation measures			

Table 39.2 Percentage of 10–0.25 mm soil aggregates under different kinds of tillage, means 2008–2010

Tillage	May				June				July			
	0–10 cm	10–20 cm	20–30 cm	30–40 cm	0–10 cm	10–20 cm	20–30 cm	30–40 cm	0–10 cm	10–20 cm	20–30 cm	30–40 cm
Ploughing	78.4	73.1	71.5	71.3	67.6	68.4	67.3	67.0	61.5	63.7	64.5	61.0
Loosening to 40 cm	79.5	77.6	78.0	79.3	75.3	74.7	74.3	75.6	73.8	73.8	74.6	72.5
Disking to 18 cm	79.3	78.5	78.0	77.1	76.8	75.9	74.9	75.0	73.9	75.0	74.8	74.3
Disking to 14–16 cm	80.7	81.3	79.9	78.7	79.1	79.7	78.0	78.3	76.2	76.8	76.8	78.4

Deep soil loosening creates a more resilient structure that hardly degrades between May and July: the 0–10 cm layer lost only 2.5 % of favourable aggregates by June and 5.7 % by July. Disking, which maintains about a third of crop residues on the soil surface, creates yet more favourable conditions for the preservation of soil aggregates and an almost homogeneous arable layer with little degradation during the growing season. We conclude that, compared with ploughing, conservation tillage favours natural processes of structure development.

Changes in soil structure are synchronized with changes in soil moisture (Table 39.3). Under the plough, soil water reserves in the 0–20 cm layer in May are more than 85 % of field capacity, with plant-available water held within the 10–0.25 mm aggregates. By June, water reserves are reduced to 72–73 % of field capacity and, by July, to 55–57 %. As a result, the topsoil shrinks and fissures. In the deeper layers, soil moisture remains optimal (more than 75 % of field capacity) so, by June, the 0–40 cm arable layer separates into two structural regimes: 0–20 cm deeply fissured (the agro-modified regime) and 20–40 cm where the soil structure is developed under a more natural regime.

By July, three layers are differentiated. Drying of the 0–20 cm layer creates strong fissuring and discrete, compact aggregates >10 mm in size. Water is lost by evaporation from the topmost 10 cm; the dry soil heats strongly during the day and cools quickly overnight, fragmenting the soil mass to a dry mulch in which organic matter is intensively oxidized. Beneath the mulch, temperatures are less extreme and soil aggregates are created mainly by shrinkage fissures. In the 20–30 cm layer, water is depleted to about 65 % of field capacity by crop roots and upward flux to the dry surface. Fissuring is less pronounced in the 20–30 cm where shrinkage reinforces the natural granular soil aggregates. The sub-plough layer at 30–40 cm remains at optimum moisture (75–80 % of field capacity) even in July. Structural development takes place through alternate shrinking and swelling and mechanical pressures from farm operations are attenuated so that, even in July, bulk density hardly exceeds the critical threshold for root development.

With deep loosening without inversion of the furrow, water reserves in the 0–10 cm layer in May are 87 % of field capacity, increasing to 99 % in the 30–40 cm layer. In July, soil moisture is still 81–88 % of field capacity – the optimum for crop growth. The soil water regime is relatively stable in the face of variable weather so crops are not exposed to water stress. Under the alternative disc cultivations, soil water regimes are identical. At the same time, neither deep loosening nor disking stratifies the soil profile, the entire arable layer is in an optimal condition for root ramification, and the slight increase in water availability at depth encourages deep rooting so conservation tillage reduces the drought hazard; all in all, favourable conditions for the functioning of the soil ecosystem.

Table 39.4 presents data on the water stability of soil aggregates under the different systems of tillage. The water stability of the aggregates varies between treatments – even early in the growing season. Under the plough, water-stable aggregates comprise 61 % of the 0–10 cm layer and somewhat less in the deeper layers. In June, the values are 53 % at 0–10 cm, 51 % at 10–30 cm and 49 % at 30–40 cm and in July, 46–49 % of the 0–30 cm layer. In the deep-loosening variant,

Table 39.3 Moisture dynamics and bulk density of fine loamy Typical chernozem according to tillage, means 2008–2010

	May				June				July			
Depth (cm)	Bulk density (g/cm^3)	Water (%)	Field capacity (%)	% of FC	Bulk density (g/cm^3)	Water (%)	Field capacity (%)	% of FC	Bulk density (g/cm^3)	Water (%)	Field capacity (%)	% of FC
Ploughed												
0–4	1.09	23.7	27.8	85	1.21	20.1	27.8	72	1.17	15.3	27.3	55
10–20	1.13	24.5	28.3	87	1.28	20.7	28.3	73	1.37	16.2	28.0	57
20–30	1.16	26.8	28.0	96	1.29	21.0	28.0	75	1.41	17.8	26.8	64
30–40	1.20	26.8	27.4	98	1.30	22.7	26.8	85	1.36	19.9	26.8	75
Loosened to 40 cm												
0–4	1.07	24.3	27.8	87	1.12	23.6	27.8	85	1.14	22.4	27.8	81
10–20	1.10	25.5	28.3	90	1.16	24.7	28.3	87	1.24	23.5	28.3	83
20–30	1.12	27.1	28.0	97	1.18	25.0	28.0	89	1.26	23.6	28.0	84
30–40	1.16	27.0	27.4	99	1.18	25.4	27.4	93	1.29	23.6	26.8	88
Disked to 18 cm												
0–4	1.03	24.1	27.8	87	1.08	23.7	27.8	85	1.16	22.8	27.8	84
10–20	1.09	24.9	28.3	88	1.13	24.7	28.3	87	1.21	23.8	28.3	84
20–30	1.12	26.7	28.0	95	1.17	25.3	28.0	90	1.24	23.8	28.0	85
30–40	1.13	27.2	27.4	99	1.21	25.9	27.4	95	1.25	24.1	26.8	90
Disked to 14–16 cm												
0–4	1.03	25.3	27.8	91	1.09	24.0	27.8	86	1.12	22.7	27.8	82
10–20	1.07	25.7	28.3	91	1.14	24.9	28.3	88	1.20	23.9	28.3	84
20–30	1.11	27.2	28.0	97	1.17	26.0	28.0	93	1.23	24.0	28.0	86
30–40	1.14	27.4	27.4	100	1.20	26.3	27.4	96	1.27	24.0	26.8	90

Table 39.4 Percentage of water-stable 10.0–0.25 mm aggregates under different systems of tillage, means 2008–2010

Dry sieving / Wet sieving

	May				June				July			
Tillage	0–10 cm	10–20 cm	20–30 cm	30–40 cm	0–10 cm	10–20 cm	20–30 cm	30–40 cm	0–10 cm	10–20 cm	20–30 cm	30–40 cm
Ploughing	78.4/61.0	73.1/58.0	71.5/58.0	71.3/57.0	64.6/53.0	68.4/51.0	67.3/51.0	67.0/49.0	61.5/48.0	63.7/46.0	64.5/49.0	61.0/52.0
Deep loosening (40 cm)	79.5/63.0	77.6/65.0	78.0/65.0	79.3/61.0	75.3/64.0	74.7/62.0	74.3/60.0	75.6/62.0	73.8/62.0	73.8/59.0	74.6/59.0	72.5/54.0
Disking (18 cm)	79.3/67.0	78.5/67.0	78.0/65.0	77.1/64.0	76.8/68.0	75.9/70.0	74.9/65.0	75.0/66.0	73.9/67.0	75.0/61.0	74.8/63.0	74.3/66.0
Disking (14–16 cm)	80.7/70.0	81.3/72.0	79.9/72.0	78.7/68.0	79.1/72.0	79.7/74.0	78.0/73.0	73.8/73.0	76.2/69.0	76.8/71.0	76.8/70.0	78.4/73.0

Table 39.5 Porosity in Typical chernozem at the end of the 2009 growing season under different systems of tillage

Parameter	Depth, cm	Tillage system Ploughing	Deep loosening to 40 cm	Disking to 18 cm	Disking to 14–16 cm
Total porosity, %	0–10	52.4	57.5	55.4	55.2
	10–20	49.6	53.8	52.6	52.7
	20–30	48.3	50.1	50.3	50.1
	30–40	46.7	48.6	49.4	48.9
Sum aggregate porosity, %	0–10	32.3	36.8	35.7	35.9
	10–20	30.8	38.4	38.9	39.4
	20–30	29.1	36.7	35.5	36.3
	30–40	28.3	32.8	33.6	32.9
Inter-aggregate porosity, %	0–10	20.1	20.7	19.7	19.3
	10–20	18.8	15.4	13.7	13.9
	20–30	19.2	13.4	14.4	13.8
	30–40	18.4	15.8	15.8	16.0
Water-conducting pores, %	0–10	20.1	21.4	20.7	20.4
	10–20	19.3	23.4	22.4	22.9
	20–30	18.7	21.7	20.4	20.8
	30–40	18.4	18.3	14.3	18.6
Water-retaining pores, %	0–10	12.2	15.4	15.0	15.5
	10–20	11.1	15.0	16.5	16.5
	20–30	10.9	15.0	15.3	16.0
	30–40	9.9	14.5	15.3	14.3

aggregates are more water stable and remain more resilient throughout the growing season. Water-stable aggregates in the 10–0.25 mm range make up three quarters of all aggregates greater than 0.25 mm and the proportion of coarse aggregates (>10 mm) is much less than under the plough. The improvement is even more pronounced with disc tillage.

The stability of soil aggregates is inversely proportional to the seasonal changes in soil moisture and temperature. In ploughed soil, extremes of soil moisture and temperature produce deep fissuring and shattering of aggregates. In contrast, the more temperate regime under disc cultivation promotes maximum water stability. Based on these observations, we consider that soil water management can be used to optimize the structural-aggregate condition during the growing season – this is an important benefit of conservation tillage. Retention of crop residues on the soil surface, as opposed to burying them with the mouldboard plough, is beneficial in avoiding stratification of the cultivated profile and in moderating the hydrothermal regime. The situation is nearer to the natural ecosystem of the steppe where humification throughout the soil profile reinforces aggregate stability.

The pore space is the seat of (and an index of) biophysical activity in the soil. Different systems of tillage have very different impacts on the pore space (Table 39.5). Under the plough, the pore space of the 0–10 cm layer is 52 % and less than 50 % in the deeper layers. Most of this pore space is within aggregates: 32.3 % at 0–10 cm, falling to 28.3 % in the 30–40 cm layer. The volume of water-retaining pores falls from 12.2 % at 0–10 cm to only 9.9 % at 30–40 cm. This distribution of pore space gives rise to an oxidant-pulsating air-water regime caused by the ratio between pores with different sizes and functions.

Conservation tillage leaves the soil with more favourable pore space: in all variants, the pore space of the 0–10 cm layer is excellent; in the10–30 cm layer, it is satisfactory (taking 50 % total pore space as the cut-off point). However, although the bulk density of the 30–40 cm layer is rated satisfactory, total pore volume is not – which we attribute to some silting up of the pore space in this layer. Under all three variants of conservation tillage, most of the pore space is within the soil aggregates and falls within the optimal range. Significantly, all variants have an increased proportion of water-retentive pores, establishing a balanced air-water regime and thereby an optimal ratio between reduction processes (humus accumulation) and oxidation (mineralization).

39.4 Conclusions

Conservation of chernozem soils and their soil-forming environment is inversely proportional to the intensity of soil tillage. Within a conceptual agricultural system, cultivation with the mouldboard plough diminishes positive integration of the processes of soil aggregate formation, humification and air-water relationships. As a result, the formation of agronomically and environmentally useful soil aggregates is diminished. Systems of conservation tillage favour the integration of processes that create more favourable, resilient soil aggregates and soil water – soil air conditions that better maintain the ecosystem.

Bibliography

Jigau G (2009) Soil genesis and soil physics. CEP Moldova State University, Chisinau (Romanian)
Jigau G (2011) Conservative technologies, applied aspects. Agric Mold 4–5:10–15 (Romanian)
Jigau G (2012) Farming system: past, present and future. Lider AGRO 1–2:12–17 (Russian)

Chapter 40
Direct and Residual Effects of Phosphate-Enhanced Organic Fertilizers on Soil Fertility and Crop Production

V. Plămădeală and L. Bulat

Abstract Full recycling of phosphate with presently available farmyard manure can satisfy only one fifth of Moldova's annual phosphate requirement. Crop nutrients from other sources are needed. Data from long-term experiments with different doses of sewage sludge and sugar-factory waste in crop rotation have been studied to establish their direct and residual effects on soil phosphate status. In the first 8 years after applying 40 t/ha of sludge, the P_2O_5 content was increased from 6.0 to 70 mg/100 g soil – a very high level of assurance; after 23 years, P_2O_5 decreased to 2.4–2.8 mg/100 g soil – corresponding to a moderate level of assurance. Doubling the dose of sludge gave higher P_2O_5 levels but not in proportion to the amount of sludge applied; treatment with 80–160 t/ha of sludge produced background levels of 4.9–6.2 mg P_2O_5/100 g soil over 23 years. Fertilization increased crop yields by 16–42 %, depending on the dose.

40.1 Introduction

Mobile phosphate is a prime limiting factor on agricultural production. In 1969, 78 % of topsoil in Moldova was low or very low in phosphate; soils with a high level of assurance made up not more than 8 % of the total area (Burlacu 2000). During the era of intensive fertilizer use (1966–1990), incorporation of about 900 kg/ha of phosphate as chemical fertilizer and annual application of 5.6–6.6 t/ha of manure (supplying a further 15–16 kg of phosphate) raised much of the farmland to the moderate category so that, over the period 1985–1990, only about 30 % of soils remained low or very low in phosphate (National complex program 2004; Cerbari 2004). Virtual cessation of nitrogen fertilizer use after 1992 prevented

V. Plămădeală (✉) • L. Bulat
Nicolae Dimo Institute for Agrochemistry and Soil Protection,
Chisinau, Republic of Moldova
e-mail: vaileon.1954@yahoo.com

D. Dent (ed.), *Soil as World Heritage*, DOI 10.1007/978-94-007-6187-2_40,

uptake of the accumulated phosphate – which has undergone gradual immobilization so that, at the present time, the phosphate supply is much the same as in unfertilized soil (Cerbari 2004).

Present circumstances emphasize the value of other sources of nutrients, especially those with high phosphate content. The traditional local resources are farmyard manure and crop residues that can be returned to the soil. The annual quantity of manure is 4.0–4.5 million tonnes; wheat straw and sunflower stalks make up the equivalent of 1.5 million tonnes of manure with bedding, making a grand total of 5.5–6.0 million tonnes of manure containing about 19,000 tonnes of phosphate. However, to ensure optimal soil phosphate status, annual application of 91,000 tonnes is needed (National complex program 2004). Even complete recycling of phosphate with presently available traditional manures can satisfy only a fifth of the country's needs, so it is necessary to bring into circulation nutrients from other sources (Cerbari 2004). Sewage sludge and sugar-factory waste are of particular interest and have been investigated by the Organic Fertilizers and Soil Fertility Laboratory of the Nicolae Dimo Institute to establish their effects on crop yields and long-term effects on soil phosphate status (Ţurcan 1985; Rusu et al. 2012).

40.2 Experimental Site and Methods

Research was carried out at the State Experimental Station of the N Dimo Institute at Ivancea, in Orhei District, during the period 1986–2010. The soil is *Leached chernozem*[1] of clay loam texture with topsoil humus content of 3.8–4.0 %, phosphate (Macighin method) 1.8–2.0 mg/100 g soil, exchangeable potassium 27 mg/100 g, pH 6.7 and hydrolytic acidity 2.65 me/100 g.

Sewage sludge and sugar-factory waste were applied in various doses to a crop rotation of 3 years lucerne-wheat-sunflower-maize-peas-wheat-maize-maize, i.e. with 40 % of row crops and 30 % of perennial legumes. Manured plots provided a comparison with those receiving wastes. The applied materials had the following composition:

Manure – moisture 52 %, organic matter 17.3 %, total nitrogen 0.5 %, total phosphorus 0.23 %, total potassium 0.7 %, pH 8.2

Sewage sludge – moisture 64 %, organic matter 21.3 %, total nitrogen 0.7 %, total phosphorus 0.9 %, total potassium 0.5 %, pH 7.5

Sugar-factory waste – moisture 25.4 %, organic matter 4.8 %, total nitrogen 0.3 %, total phosphorus 0.5 %, total potassium 0.5 %, pH 8.5

For soil analysis, humus was determined by Tiurin's method, mobile phosphorus colourimetrically after Macighin and exchangeable potassium by flame photometry. For waste analysis, moisture was determined according to GOST 26713-85, organic matter by GOST 27980-88, total nitrogen by GOST 26715-75, total phosphorus by GOST 26717-85 and total potassium by GOST 26718-85 (Methodological directions on the analysis of organic fertilizers 1984).

---

[1] Haplic chernozem in the *World reference base for soil resources 2006.*

Table 40.1 Humus content of plough layer depending on applied fertilizer, 1987–2009

	Humus (%)						Mean values (%)	
Variant	1987	1991	1996	2001	2004	2009	Content	Efficiency
Control	4.0	3.7	3.9	3.9	3.9	3.5	3.8	–
Manure, 40 t	–	3.7	3.8	4.0	4.0	3.6	3.8	–
Manure, 80 t	–	3.9	4.0	4.1	3.8	3.6	3.9	0.1
Manure, 160 t	–	4.0	4.1	4.2	4.1	3.9	4.1	0.3
Sewage sludge, 40 t	3.9	3.7	4.1	4.0	4.1	4.0	4.0	0.2
Sewage sludge, 80 t	4.0	3.8	3.8	3.9	3.9	3.8	3.9	0.1
Sugar waste, 40 t	3.9	3.9	4.0	4.0	3.9	3.8	3.9	0.1
Sugar waste, 80 t	3.8	3.9	4.0	4.1	4.2	3.8	4.0	0.2
Sugar waste, 160 t	4.1	–	4.1	4.2	4.3	3.8	4.1	0.3
Sugar waste, 40 t + sewage sludge, 40 t	4.2	3.9	4.0	4.1	4.2	3.9	4.1	0.3
Manure, 40 t + sewage sludge, 40 t	4.3	4.4	4.2	4.2	4.1	4.0	4.2	0.4
Manure, 40 t + sugar waste 40 t + sludge, 40 t	4.3	4.0	4.3	4.3	4.3	4.2	4.2	0.4
Manure, 40 t + sugar waste, 40 t	4.3	3.9	4.1	4.2	4.3	4.2	4.2	0.4

40.3 Results and Discussion

Monitoring humus content in the long-term experiment provides objective information on the evolution of soil fertility depending on management practice and fertilization (Table 40.1). At the outset, the plough layer contained 4.0 % humus. After 23 years, the humus content of the control was 3.5 %, which we interpret as a slow but significant decline of humus content in the absence of any manure application.

This is in line with other long-term experiments in the same environment: according to the N Dimo Laboratory of Agrochemistry and Plant Nutrition, Leached chernozem with an initial humus content of 4 % lost 0.8 % humus in 25 years; Calcareous chernozem and Grey forest soils lost 0.1 and 0.3 %, respectively (Cerbari 2000; Zagorcea 1990). Discrepancies between our data and some others, especially regarding the Leached chernozem, may be related to the crop rotation, but general experience is that even 30 % of perennial legumes and grasses supplemented by modest doses of farmyard manure cannot maintain the humus status of chernozem. This is confirmed by the long-term field experiments on Typical chernozem at Balti (Boincean 1999, Boincean in this symposium). Humus content did increase in variants receiving high doses of organic fertilizers, especially from multiple sources, but clear differences in humus accumulation depending on the type of soil and dosage of fertilizer were not established; this may be attributed to the levelling effect of 10 years of lucerne.

Data for 23 years allow us to establish the long-term effects of sewage sludge and sugar-factory waste on soil phosphate status (Table 40.2).

Leached chernozem is characterized by a low level of mobile phosphate – 1.0–2.1 mg/100 g soil with a mean of 1.4 mg/100 g. Applying 40 t/ha of farmyard

Table 40.2 Mobile phosphate in plough layer depending on fertilization 1987–2009

Variant	Mobile phosphate (mg/100 g soil)						Means (mg/100 g soil)	
	1987	1991	1996	2001	2004	2009	Content	Efficiency
Control	2.1	1.5	1.4	1.7	1.0	0.94	1.4	–
Manure, 40 t	–	2.9	2.5	1.7	1.0	1.25	1.9	0.5
Manure, 80 t	–	3.3	2.4	1.5	1.0	1.25	1.9	0.5
Manure, 160 t	–	5.7	3.5	3.3	2.7	1.65	3.4	2.0
Sewage sludge, 40 t	7.8	7.3	3.5	3.8	3.5	2.40	4.7	3.3
Sewage sludge, 80 t	8.0	7.5	4.5	4.3	3.9	2.81	5.2	3.8
Sugar waste, 40 t	4.9	4.3	3.1	4.8	3.2	1.46	3.6	2.2
Sugar waste, 80 t	6.2	8.4	5.1	4.0	3.2	2.56	4.9	3.5
Sugar waste, 160 t	7.5	–	6.4	7.4	5.4	4.48	6.2	4.8
Sugar waste, 40 t + sludge, 40 t	8.0	6.8	5.0	3.9	3.9	1.95	4.9	3.5
Manure, 40 t + sludge, 40 t	8.2	8.4	6.5	4.1	4.8	2.64	5.8	4.4
Manure, 40 t + sugar waste 40 t + sludge, 40 t	12.0	8.6	5.4	4.8	3.9	3.14	6.3	4.9
Manure, 40 t + sugar waste, 40 t	14.0	9.7	6.2	3.5	3.5	3.04	6.7	5.3

manure increased mobile phosphate by 1.9 mg/100 g soil; the highest value (3.4 mg/100 g) was obtained with the application of 160 t/ha of manure. The positive effect was strong during the first 5 years but then fell back, whereas the effects of phosphate-rich wastes were both stronger and longer-lasting. Application of 40 t/ha of sludge raised mobile phosphate to 6.0–7.0 mg/100 g soil during the first 8 years, falling back after 23 years to 2.4–2.8 mg/100 g soil, corresponding to a moderate level of assurance. Doubling the dose of sludge somewhat increased the soil's phosphate level but by no means proportionately; the mean values were 4.7 mg/100 g soil with application of 40 t/ha of sludge and 5.2 mg/100 g soil with 80 t/ha. Application of sugar-factory waste substantially increased soil phosphate levels. A dressing of 40 t/ha increased mobile phosphate by as much as 160 t/ha of manure; higher doses of 80 and 160 t/ha raised mobile phosphate levels over 23 years to means of 4.9 and 6.2 mg/100 g soil, which correspond to *relatively optimal* and *high* levels of assurance. Strong effects were also demonstrated for mixed dressings of wastes and manure.

Post-action soil exchangeable potassium was also measured. The highest values were recorded in variants with manure and sugar-factory waste, sugar waste at 160 t/ha and the mixture of manure + sugar waste + sludge, which raised exchangeable potassium levels to 33, 30 and 31 mg/100 g soil, respectively – higher than the control by 6, 3 and 4 mg/100 g soil, respectively.

Crop yields were raised significantly (Table 40.3).

Over the 23-year period, crop yields increased, compared with the control, by 5 q/ha grain units for a 40 t/ha application of farmyard manure to 13 q/ha for the maximum dose of 160 t/ha, an increase of 16–32 %. Production increases in variants with high doses of farmyard manures and waste were in the range of 10–13 q/ha grain units, exceeding the control by 32–42 % (Krupenikov and Boincean 2004).

Table 40.3 Effect of organic manures on crop yield, q/ha grain equivalents

Variant	1987–1990	1991–1995	1996–2000	2001–2005	2006–2010	Mean	Gain q/ha	Gain %
Control	41	46	24	27	16	31	–	–
Manure, 40 t	43	56	28	32	19	36	5	16
Manure, 80 t	45	63	32	34	20	39	8	26
Manure, 160 t	46	64	32	37	24	41	10	32
Sewage sludge, 40 t	44	67	33	34	26	41	10	32
Sewage sludge, 80 t	50	68	34	39	18	42	11	35
Sugar-factory waste, 40 t	50	70	31	35	17	41	10	32
Sugar waste, 80 t	46	69	32	38	22	41	10	32
Sugar waste, 160 t	52	73	33	41	21	44	13	42
Sugar waste, 40 t + sludge, 40 t	47	70	29	37	20	41	10	32
Manure, 40 t + sludge, 40 t	44	69	31	39	27	42	11	35
Manure, 40 t + sugar waste 40 t + sludge, 40 t	53	73	29	37	22	43	12	39
Manure, 40 t + sugar waste, 40 t	51	72	30	38	27	44	13	42
DL, 5 %	3.3	4.3	2.3	2.7	1.5			

The long-term effects of fertilizers on crop production stem from two main factors. The first is the stabilization of soil organic matter by a combination of crop rotation and manuring. The second, which is closely related, is the optimization of crop-available phosphate thanks to the substantial addition of phosphate from the various sources. In the first 4–5 years, yields benefitted from the nitrogen content of the fertilizers; production increases in later years were influenced most by the residual action of phosphate (Andreis 2006).

The long-term experiment clearly demonstrates that the use of organic fertilizers and nutrient-rich waste can bring considerable benefits. Local sources of organic matter and nutrients can be effectively and economically applied in field crop rotations. Their implementation costs are recovered by production increases in 2–3 years and substantial profits accrue over more than 10 years (Rusu et al. 2012; Methodological directions on the analysis of organic fertilizers 1984). Sewage sludge is rich in nitrogen and phosphorus, providing an average annual benefit equal to 50 euro/ha; with application of 80 t/ha, this index increases to 56 euro/ha.

Similar benefits may be obtained from dressings of sugar-factory waste. The best returns were associated with a combination of manure and sewage sludge and the heaviest dressing of sugar-factory waste (160 t/ha) which gave an average annual income over 23 years of 59–64 euro/ha, which compares well with the conventional artificial fertilizers.

40.4 Conclusions

1. Dressings of organic fertilizers on Leached chernozem produced a significant residual effect on the contents of soil organic matter and mobile phosphate.
2. Applications of sewage sludge and sugar-factory waste, having higher phosphate content than farmyard manure, produced a greater and more long-lasting effect. Sewage sludge and sugar-factory waste increased the content of mobile phosphate two-and-a-half to threefold compared with the control; application of sludge or sugar waste along with manure increased mobile phosphate sixfold.
3. Manuring raised crop yields significantly. Maximum yields (43–44 q/ha grain units) were obtained from application of manure + sewage sludge + sugar-factory waste.

References

Andrieş S (2006) Regime of phosphorus in the soils of Moldova and application of fertilizers with phosphorus. Pontos, Chisinau (Romanian)

Boincean BP (1999) Organic farming in the Republic of Moldova. Ştiinţa, Chisinau (Russian)

Burlacu I (2000) Agricultural service for agriculture in the Republic of Moldova. Pontos, Chisinau (Romanian)

Cerbari V (2000) Bulletin of ecopedologic monitoring (agrochemistry), 7th edn. Agroinformreclama, Chisinau (Romanian)

Cerbari V (2004) National program for soil fertility; database, conclusions, prognosis, recommendations. Pontos, Chisinau (Romanian)

Krupenikov IA, Boincean BP (2004) Black soil and arable farming. Typography, Balti (Russian)

VV Dokuchaev Soil Science Institute (1984) Methodological directions on the analysis of organic fertilizers. Kolos, Moscow (Russian)

National complex program (2004) National complex program for recovery of degraded lands and soil fertility increase. Part II: soil fertility increase. Pontos, Chisinau (Romanian)

Rusu A, Plămădeală V, Siuris A et al (2012) Guide to the use of organic fertilizers. Pontos, Chisinau (Romanian)

Ţurcan MA (1985) Agrochemical basis for application of organic fertilizers. Ştiinţa, Chisinau (Russian)

Zagorcea CL (1990) Optimization of fertilizers and crop field rotation. Ştiinţa, Chisinau (Russian)

Chapter 41
Worm Compost to Improve Soil Fertility

L. Cremeniac and T. Boclaci

Abstract As the main natural wealth of Moldova, soil warrants special care. Agriculture, in particular, should pay attention to the soil's humus and nutrient status – and restore losses of humus and the nutrients used by crops. This requires measures to improve soil fertility. Research at Maximovca ETS highlights the value of compost produced from organic wastes by earthworms; soils to which worm compost was incorporated in doses of 3–4 t/ha over 3 months showed increased humus content of 15.2–24.2 % compared with the control.

41.1 Introduction

Soil degradation means loss of biological and economic productivity, whether through natural causes or injudicious management. In Moldova, exploitative land use over the last 30 years has increased soil erosion and landslides, salinity and sodicity, deterioration of soil structure and compaction, and has exacerbated a negative nutrient budget through inadequate application of manure and fertilizer. The amount of humus in soil is one of the main indices of fertility; it has a big influence on soil chemistry, hydro-physical and biological activity, holding 98 % of nitrogen reserves, 60 % of phosphorus, 80 % of sulphur and essential quantities of other micro- and macronutrients (Toma 2008). Already, about half of the initial humus content of the black earth has been lost under cultivation. With business as usual, it is likely that the humus content of arable soils will decrease by a further 10–25 % over the next few decades; this would have devastating effects on soil fertility, biodiversity, structural stability and resilience (National Inventory 2009)

L. Cremeniac (✉) • T. Boclaci
Scientific and Practical Institute for Biotechnology in Animal Husbandry and Veterinary Medicine, Anenii Noi, Republic of Moldova
e-mail: kremeneak@yandex.com

D. Dent (ed.), *Soil as World Heritage*, DOI 10.1007/978-94-007-6187-2_41,

so there is pressing need to apply procedures for maintaining soil fertility and to stabilize and increase the humus content.

The annual loss of humus is 0.5–0.7 t/ha. To compensate for this loss requires the annual addition of about 6.3 t/ha of farmyard manure – which is not, at present, available. Bioconversion of organic waste by earthworm cultivation is a relatively recent development (Kosolapov 1996). Research worldwide on reducing harmful substances entering the environment and production of green organic fertilizer from organic wastes using the California red hybrid (*Eisenia andrei*) indicates that the resultant compost is an effective natural fertilizer – dark brown, granular, odourless, hygroscopic and long acting (Condîreva et al. 1994; Cremeneac 2001, 2003; Cremeneac and Boclaci 2008). It is a well-balanced mixture of macro- and micronutrients with a C:N ratio less than 4, so its nitrogen-effectiveness is very high (Gorodnii 1996) and it is effective at doses 8–12 times less than traditional compost or manure (Condîreva et al. 1994).

Research at Maximovca ETS demonstrated that incorporation of 3–4 t/ha worm compost increased nutrient status, soil moisture and soil biological activity, reduced soil bulk density and diminished the content of nitro compounds. A worm culture with an initial stage of 30,000 individuals (rhymes) can produce 1,200 kg of compost annually. The production unit can be upscaled depending on the purpose.

Application of ordinary compost or farmyard manure to the soil is costly and not particularly effective. One tonne of compost yields only 20 kg of humus, whereas a tonne of worm's compost contains 270–300 kg of humus – which offers a significant reduction in the time required to restore the soil's humus deficit. With this in mind, an experiment was set up under field conditions at STE Maximovca, in the centre of the Republic of Moldova.

41.2 Experimental Site and Methods

The site at STE Maximovca was located on *Typical chernozem* with a mean humus content of 3.3 %. Experimental plots and the control were each 100 m^2.

Worm compost was produced by bioconversion of vegetable waste. The experiment consisted of two treatments and a control (Table 41.1).

For determination of soil organic matter and humus in the first year of action of fertilizer, soil samples were taken from the surface and from a depth of 15 cm from each group prior to incorporation of compost and, again, 3 months after incorporation of fertilizer. Analytical methods followed the manual of E. Petuhova (1989) and Standard GOST 26213-84 Soils (1998).

Table 41.1 Scheme of the experiment

No.	Groups	Conditions of experiment
1	I – experimental	Worm compost – 3 t/ha
2	II – experimental	Worm compost – 4 t/ha
3	III – control	Natural background

Table 41.2 Content of humus and organic matter in soil fertilized with worm's compost

	Elements	Sample	Conditions of experiment: Experimental group I worm compost 3 t/ha: Initial	After adding compost	% of initial	Experimental group II worm compost 4 t/ha: Initial	After adding compost	% of initial
1	Organic matter %	Surface	4.59 ± 0.26	4.84 ± 0.10	105.4	5.08 ± 0.10	5.37 ± 0.02	105.7
		15 cm	5.17 ± 0.10	5.49 ± 0.02	106.2	4.91 ± 0.10	5.28 ± 0.11	107.5
2	Humus %	Surface	3.5 ± 0.08	4.00 ± 0.15	114.3	3.30 ± 0.12	3.80 ± 0.01	115.2
		15 cm	3.5 ± 0.01	4.10 ± 0.07	117.1	3.40 ± 0.07	4.00 ± 0.15	117.6

Table 41.3 Content of humus and organic matter in soil fertilized with worm's compost compared with the control

No.	Nutritive elements	Sample collection	Conditions of experiment: Control group: Initial state	After 3 months	% compared to initial state	Compared to control group, %: Group I: After adding compost	Group II: After adding compost
1	Organic matter %	Surface	3.58 ± 0.07	4.40 ± 0.03	107.5	110.0	122.0
		15 cm	4.43 ± 0.09	4.45 ± 0.02	100.5	123.4	118.7
2	Humus %	Surface	3.20 ± 0.12	3.30 ± 0.07	103.1	121.2	115.2
		15 cm	3.30 ± 0.15	3.30 ± 0.07	100.0	124.2	121.2

41.3 Results and Discussion

At the outset, the quality of the compost was determined. Its reaction was neutral (pH 7.23 ± 0.002), water content 61.08 ± 0.05 % and organic matter/humus content 30.85 ± 1.65 and 36.40 ± 4.20 %, respectively (Table 41.2).

After 3 months, the organic matter content of surface samples from experimental groups I and II exceeded the initial values by 5.4 and 5.7 %, respectively. For samples from a depth of 15 cm the gain was 6.2 and 7.5 %. The humus content of surface soil from experimental groups I and II 3 months after incorporation of worm's compost surpassed the initial content by 14.3 and 15.2 %, respectively; the gains at 15 cm were 17.1 and 17.6 %.

The values of soil organic matter and humus from samples collected from the experimental groups were also compared with those collected from the control group at the same time (Table 41.3). Compared with the control, surface samples collected from experimental group I, 3 months after incorporation of compost, surpassed that of the control in terms of organic matter and humus by 10.0 and

21.2 %, respectively, and those from depth 15 cm by 23.4 and 24.2 %. The same trend was found with experimental group II.

41.4 Conclusions

Bio-compost produced by conversion of organic waste by worms, used as an organic fertilizer, significantly improved soil fertility, increasing the amount of organic matter and humus during the first year of application.

References

Condîreva M, Cremeneac L, Bahcivanji M (1994) Biological processing technology for manure using worm cultivation. Recommendations. Agroinformreclama, Chisinau (Romanian)

Cremeneac L (2001) Technology of worm cultivation and bioconversion of organic waste. In: Conference on fruitful collaboration between researchers and farmers in the third millennium. Central Typography, Chisinau (Romanian)

Cremeneac L (2003) Bioconversion of organic waste in Moldova. Summary information. INEI, Chisinau (Romanian)

Cremeneac L, Boclaci T (2008) Agri-environmental assessment of worm's compost and its role in delivering agricultural production. In: Proceedings of the international scientific conference on durable agriculture in the context environmental changes, Iasi, Romania, 16–18 October 2008

Gorodnii N (1996) Bioconversion of agro-systems management. Monography, Kiev (Russian)

GOST 26213-84 (1988) Soils. Determination of humus by the method of Tiurin, modified by ŢINAO. State Committee for Standards, Moscow (Russian)

Kosolapov A (1996) Worm culture and its capabilities. Recommendations – Ryazan Regional Dept Agric and Ryazan Regional Inst for Improving Qualifications and Agribusiness, Ryazan (Russian)

National Inventory Report (2009) Emission sources and greenhouse gases sequestration in Moldova 1990–2005. Bons Offices, Chisinau (Romanian)

Petuhova E (1989) Zootechnical feed analysis. Textbooks and teaching aids for students in higher education. Agropromizdat, Moscow (Russian)

Toma S (2008) Application of fertilizers in sustainable agriculture. SAM, Chisinau (Romanian)

Chapter 42
Efficiency of Grass Strips and Sodded Waterways

L. Popov

Abstract Soil erosion is the most severe hazard of land and ecosystem degradation – it is irreversible and has far-reaching environmental and social consciences. Loss of crop production, alone, costs Moldova US$ 200 million a year; the cost of infrastructure damage is undoubtedly much greater. Erosion can be minimized by mechanical and biological methods, both of which always involve considerable investment. Simple, effective biological controls include alternation of crops and perennial grasses in strips and sodding of waterways – which may reduce soil loss by 20 t/ha/year and safely dispose of the remaining runoff.

42.1 Introduction

In Moldova, privatization and fragmentation of land holdings by the 1995–2000 land reforms has complicated land use and management, crop rotations and the design and implementation of measures to combat erosion, and has noticeably reduced soil fertility. As a result, the proportion of arable has declined, especially in land units of only 1–4 ha (Krupenikov 2004; Ursu 2006; Agency for Land Relations 2010).

The ecological principle of arresting soil erosion is that the land should be permanently covered by vegetation. Obviously, that is not the case in the present system of agriculture. Periodically, the soil is bare and subject to aggressive action by natural and man-made agents; great volumes of water from heavy rains are concentrated in overland flow that carries away the topsoil. Soil erosion is most prevalent under annual crops, especially row crops; furrows that remain on the surface after various agricultural operations serve as conduits for runoff, especially where they run downslope. Unwise placement of private plots has led to deluges of soil

L. Popov (✉)
Nicolae Dimo Institute of Pedology, Agrochemistry and Soil Protection,
Chisinau, Republic of Moldova
e-mail: ipaps_dimo@mtc.md

D. Dent (ed.), *Soil as World Heritage*, DOI 10.1007/978-94-007-6187-2_42,

and accumulation of sediment at the base of the slope, silting up houses and surrounding fields, and damaging property and morale. The annual cost of lost agricultural production is 2.5 billion lei or US$ 200 million (Andrieş 2007; Constantinov 2004); the cost of damage to infrastructure is undoubtedly much greater.

Soil protection is a precondition of economic and environmental improvement. In Moldova, it is a problem that requires action at state level to strengthen the efforts of all participants in the administrative, scientific, industrial and farming sectors.

42.2 Experimental Site and Methods

Research was undertaken at the N Dimo Institute experimental station at Ursoaia, in Cahul district. The site is part of a catchment of about 1,000 ha that includes sloping ground under arable crops. The experimental strips were each 100 m wide; the upper slope includes five plots, the lower four, separated by a 12 m-wide strip of oak woodland.

The erosion-protection measures were assessed on conventionally ploughed plots under various arable crops and, also, meadow using simulated rainfall (Rowell 1994; Kuznetov and Glazunov 2004). The spray was applied at 2 mm/min over 30 min; at this intensity and with a fall of 1.8 m, the kinetic energy is comparable with a rainfall intensity in the upper 10 % of expected downpours. Simulated rainfall was applied in three replications on runoff-controlled plots and over unprotected plots as a control. Samples (500 cm^3) of runoff water and sediment were collected every 5 min and the amount of soil erosion determined by turbidity values, weighing and volume measurements.

For safe discharge of runoff, a sodded waterway 150 m long and 30 m wide, serving a sub-catchment of 270 ha, was established downslope of the woodland. A mixture of grasses and clover (*Trifolium repens* (10 %), *Poa pratensis* (10 %), *Lolium perenne* (47.5 %), *Phleum pratensis* (15 %) and *Festuca pratensis* (17.5 %) was sown at 20 kg/ha on the 5th of March. The effectiveness of the outlet was monitored under a natural rainfall of average intensity of 0.6 mm/min and a duration of 50 min, amounting to 25.4 mm.

42.3 Results and Discussion

Field research has established that, on slopes of less than 2°, clean-weeded crops can handle the simulated rainfall when they occupy not more than 50–60 % of the ground (Fig. 42.1). As the slope gradient increases, the share (or width of the clean-weeded crop strip) should decrease and the alternating strips of dense cover, like perennial grasses, should be wider. On gradients more than 7°, clean-weeded crops should not be grown at all. Following these guidelines will ensure better conditions for the arable crops and a high degree of soil protection: 69–100 % with soil loss decreased by about 20 t/ha/year.

Fig. 42.1 Alternating strips of arable crops

Fig. 42.2 Sodded waterway

Allowing for the nil cash-crop harvest from the alternating strips, the yield of cash crops was winter wheat, 3.5 t/ha; maize, 3.8 t/ha; and sunflower, 2.4 t/ha.

An effective sodded waterway (Fig. 42.2) needs a dense sward of perennial grasses that will protect the soil surface against erosion and clogging sediment

Table 42.1 Quantity and cost of soil erosion under strip cropping and monoculture

Slope aspect	Black fallow		Maize		Winter wheat		Peas		Perennial grass		Alternating crop strips	
	t/ha	lei/ha	t/ha	lei/ha	t/ha	lei/ha	t/ha	lei/ha	t/ha	lei/ha	t/ha	lei/ha
West I	33.6	2,113	27.5	1,732	6.7	422	13.1	826	1.0	64	12.8	791
East II	25.8	2,225	21.1	1,822	5.1	443	10.0	858	0.8	67	12.1	1,122
West III	51.7	4,022	42.4	3,298	10.3	804	17.8	1,525	1.7	121	24.1	1,793
East IV	30.9	1,887	24.6	1,567	6.0	265	9.8	780	0.9	58	10.7	709

carried by the runoff. Its efficiency depends on the density and longevity of the grass cover and its capacity to regenerate on the silted surface; its value is determined by its high protecting capacity, of the order of 86 %, limiting soil loss to 0.3–0.9 m^3/ha (Constantinov 2004; Neamţu 1996). In our case, sodding the water outlet decreased the usual flow rate to 0.5–1.4 m/s and, also, allowed for the passage of a torrent of 2.5 m/s without suffering erosion.

Table 42.1 compares the quantity and cost of soil erosion by runoff under strip cropping and conventional monoculture. In general terms, the biological measures cut soil erosion by half, yielding a substantial saving in terms of maintaining crop yields over the long term.

42.4 Conclusions

Alternating strips or blocks of arable crops and perennial grasses provide a high degree of soil protection – in the range of 69–100 % and contain soil loss within the range of 10–24 t/ha, depending on the erosion hazard. In money terms, this represents a saving of 1,192–2,053 lei/ha in improved crop yield.

Providing that a perennial sod can be maintained, well-engineered sodded waterways provide safe disposal of surplus rainwater.

References

Agency for Land Relations and Cadastre (2010) Land cadastre of Republic of Moldova 2010. Agency for Land Relations and Cadastre, Chisinau (Romanian)

Andrieş S (2007) Optimization of soil nutrient regimes in crop productivity. Pontos, Chisinau (Romanian)

Constantinov I (2004) Agro-technical and phyto-technical measures to mitigate soil erosion processes. In: Nicolae Dimo Institute of Pedology, Agrochemistry and Soil Protection (ed) Soil erosion. Pontos, Chisinau, pp 148–195 (Romanian)

Krupenikov IA (2004) Biosphere and ecological consequences of erosional processes. Assessing of the fertility of eroded soil. In: Nicolae Dimo Institute of Pedology, Agrochemistry and Soil Protection (ed) Soil erosion. Pontos, Chisinau, pp 72–85 (Romanian)

Kuznetov MS, Glazunov GP (2004) Erosion and soil protection. Kolos, Moscow (Russian)

Neamţu T (1996) Ecology, erosion and anti-erosional agro-technics. Ceres, Bucharest (Romanian)
Rowell DL (1994) Soil science: methods and applications. Longman, London
Ursu A (2006) The consequences of privatization and pedological principles for consolidation of agricultural land. In: The current state, use problems and soil protection. Proceedings of the scientific and practical conference. National Society of Soil Science of Moldova, Chisinau (Romanian), pp 133–139

Chapter 43
Sown and Natural Grassland for Soil Protection and Productivity in the Forest Steppe of Ukraine

V. Olifirovich, S. Makoviychuk, M. Kolenchuk, and G.V. Cossack

Abstract A 3- and 4-year study of sown and natural grassland in the foothills of the Carpathians demonstrates the enhancement of productivity by legumes. Oversowing with bird's-foot trefoil more than doubled dry matter yield and improved the botanical composition.

43.1 Introduction

In the forest-steppe zone of Ukraine, sustainability demands that ploughland should be limited to 40–45 % of the landscape. However, the last century witnessed a huge expansion of extensive arable, at the expense of natural grassland, as the easiest means of increasing crop production. Overall, 60 % of the land is now arable but, in some regions, ploughland occupies as much as 82–96 %. This has unleashed unprecedented soil erosion; in the Chernivtsi region, sloping land makes up 57 % of the territory, arable 72 %, and every other ploughed acre suffers some degree of erosion. Compared with the situation in 1960, the area of eroded farmland has increased 1.7 times and the area of eroded arable by 2.4 times, greatly degrading soil fertility (Cherniavsky 2003). Moreover, according to UNEP (2007), fertilizers occupy fourth place in the big league of polluters.

Planting perennial vegetation is the most effective way to conserve the soil cover. Grasses and legumes can produce 500–700 kg/ha of humus annually, equivalent to 20–30 t/ha of farmyard manure, so the return of a significant proportion of the landscape to grassland will improve the environment and arrest water pollution by nitrates and agrochemicals. Natural grasslands, along with forests and marshes, serve as a natural water filter and purifier; essentially they guarantee

V. Olifirovich (✉) • S. Makoviychuk • M. Kolenchuk • G.V. Cossack
Bukovyna Institute for Agro-industrial Production,
NAAS, Chernivtsi, Ukraine
e-mail: bukselek_kuk@rambler.ru

D. Dent (ed.), *Soil as World Heritage*, DOI 10.1007/978-94-007-6187-2_43,

the quantity and quality of the freshwater resources of Ukraine – and steppe everywhere (Bogovin 1994). They are an integral part of the cultural landscape and meadows are custodians of biodiversity in the landscape, the habitat of dozens of valuable food and medicinal herbs and myriad beneficial insects – not least bees and, hence, both honey and pollination of farm crops and orchards.

One of the most promising means of intensifying the management of grassland is to create a seeded legume-grass sward. Even the partial replacement of mineral nitrogen by symbiotically fixed nitrogen is an important factor in reducing energy costs and the environmental contamination associated with mineral fertilizers (Kurgak 2006–2007; Triboi and Triboi-Blondel Chap. 32 in this symposium). On natural pastures, the introduction of legumes is an effective way to increase the production of high-quality forage while reducing human impact on the environment – taking advantage of the potential of perennial legumes as a cheap source of biological nitrogen. According to Mashchak and others (2006–2007), natural pastures can be effectively improved by direct seeding with legumes, employing minimum tillage (Veklenko 2008; Efremova 2007).

43.2 Experimental Site and Methods

Studies were conducted in grassland on a south-western slope of 5–7° in the forest-steppe zone of the Carpathian foothills in the country around Chernivtsi during 2004–2007.

The legume components of legume-grass mixtures are short-lived. Our studies were conducted on timothy-clover and timothy-lucerne mixtures sown in 2001, and at our experimental sites, the red clover had been lost completely, while the lucerne was much diminished. One way to maintain the proportion of legumes at the desired level is to reseed them into the turf in the second or third year (Kugak 2006–2007). In 2005, the established grasslands were over-sown with bird's-foot trefoil (*Lotus corniculatus*); the scheme of experiments is shown in the tables. Observations, measurements, surveys and analyses were conducted in accordance with methods generally accepted in forage and grassland science.

43.3 Results and Discussion

The yields obtained from the established grassland and following over-seeding with bird's-foot trefoil are shown in Table 43.1. The more successful over-seeding was into the clover-grass sward, where the resultant yield of dry matter averaged 5.67 t/ha over 3 years (Table 43.1). Thanks to its substantially greater legume component, the over-seeded clover grass sward was much more productive, achieving an increase of 45.8 % compared with 39.8 % achieved by sowing into the timothy-lucerne mixture.

However, in the third year of harvest (the fourth year after seeding bird's-foot trefoil), there was a sharp decrease in the proportion of legumes and an increase in

Table 43.1 Yield of dry matter from herbage with legume components, 2006–2008, t/ha

Culture, sown	2006	2007	2008	Average (2006–2008)
Red clover (15 kg/ha) + timothy (6 kg/ha) meadow sown (2001)	1.70	1.24	2.46	1.80
As above with bird's-foot trefoil (10 kg/ha) sown (2005)	8.37	4.06	4.59	5.67
Lucerne (15 kg/ha) and timothy (6 kg/ha) sown (2001)	3.43	1.23	2.30	2.32
As above with bird's-foot trefoil (10 kg/ha) sown (2005)	7.14	3.71	4.13	4.99
NIR_{05}	0.42	0.27	0.26	

Table 43.2 Yield of legume-rich natural grasslands in the Carpathian foothills, tonnes dry matter/ha

Culture, rate of seeding	2004	2005	2006	2007	Average (2004–2007)
Meadow grass without reseeding	2.12	3.45	3.30	3.29	3.04
Over-sown with bird's-foot trefoil, 8 kg/ha	2.75	4.23	5.33	4.02	4.08
Over-sown with clover hybrid, 8 kg/ha	3.13	4.53	4.81	3.91	4.10
Over-sown with bird's-foot trefoil, 4 kg/ha + clover hybrid, 4 kg/ha	3.21	4.92	5.02	4.46	4.40
NIR_{05}	0.47	0.26	0.27	0.23	

the proportion of forbs. It should also be noted that reseeding with bird's-foot trefoil is really successful only in years of favourable rainfall.

Increasing the productivity of natural grasslands of the foothills of Carpathians can be achieved not only by fertilizer but also by over-sowing with perennial legumes like clover, bird's-foot trefoil and hybrid clover-trefoil (Table 43.2). On average over the first 4 years of natural meadow hay, the best yield (4.4 t/ha) was obtained from over-seeding with a mixture of bird's-foot trefoil and hybrid clover.

43.4 Conclusions

1. On sown timothy grassland, over-sowing with bird's-foot trefoil substantially raises productivity for 3 years. The second year after over-sowing is the most productive.
2. Natural hayfields in the foothills of the Carpathians may also be enriched by over-sowing. The highest productivity was achieved by over-sowing with a mixture of bird's-foot trefoil and clover-bird's-foot trefoil hybrid.

References

Bogovin A (1994) Grasslands – an important factor in optimising the structural and functional organisation of agricultural landscapes and production of cheap food. In: Bogovin A (ed) Feed and forage protein: Ukrainian first international conference, Vinnitsa, 16–17 November 1994 (Ukrainian)

Cherniavsky A (2003) Effective and efficient use of degraded land. B Sivak – Green Bukovyna, Chernivtsi (Ukrainian)
Efremova G (2007) Effect of seeding legumes on the productivity of grasslands in the northern steppe of Ukraine: Abstract. Ph.D. thesis, Agricultural Science spec. 1 June 2012 (Ukrainian)
Kurgak V (2006–2007) Methods of enrichment of ecosystems with prairie legume components. In: Balian A (ed) Problems of the Carpathians agricultural complex. Inter-agency thematic research collection 15–16, JSC "Pat", Uzhgorod (Ukrainian), pp 147–150
Mashchak J, Timchisin S, Smetana S, Irshak R (2006–2007) Reserves of Carpathian natural grassland. In: Balian A (ed) Problems of the Carpathians agricultural complex. Inter-agency thematic research collection 15–16, JSC "Pat", Uzhgorod (Ukrainian), pp 143–146
UNEP (2007) The land, Chapter 3. In: Global environmental outlook GEO4. United Nations Environment Programme, Nairobi
Veklenko A (2008) The effectiveness of direct seeding of perennial grasses in turf grass stands in the dry valleys old grass stand central steppe of Ukraine. In: Petrichenko V (ed) Forage and fodder production. Inter-agency thematic research collection, vol 60, Vinnitsa, pp 82–89 (Ukrainian)

Chapter 44
Perennial Grasses Creating Soil Structure and Raising Fertility

V.D. Osadchuk, T.I. Gunchak, L.I. Miku, and G.V. Cossack

Abstract Under the conditions of the southwest Ukrainian forest-steppe, planting wind-eroded croplands with mixtures of perennial grasses significantly improves soil structure, stability and fertility. In field trials over the period 2001–2010, the structural coefficient (the proportion of agronomically favourable soil aggregates) depended on the grass mixture, its period of development and the weather. All mixtures under trial proved more effective than natural regeneration. In years with enough rainfall during the growing season, the best soil structure was achieved by a mixture of lucerne and timothy. However, the lucerne succumbed to severe drought whereas bird's-foot trefoil, which also achieved a good, water-stable structure, proved to be more drought resistant.

44.1 Introduction

Soil physical properties play an important part in soil fertility. Optimum agro-physical conditions provide a foundation for the provision of the nutrients required for crop growth, but no amount of nutrients will produce a good harvest under unfavourable physical conditions (Pechenyuk 2007). A good soil structure that creates favourable moisture, air and nutrient availability also needs to be resilient in the face of wind and water. The best way to re-establish a stable structure on eroded land is to sow a mixture of perennial legumes and grasses; the roots of perennial grasses ramify the soil, splitting and compressing it into granular aggregates (Chernyavskiy and Sivak 2003). Sowing perennial grasses not only provides good-quality forage but also increases soil fertility and, thereby, agricultural production (Rubin 1959).

V.D. Osadchuk (✉) • T.I. Gunchak • L.I. Miku • G.V. Cossack
Bukovyna Institute for Agro-Industrial Production, NAAS, Chernivtsi, Ukraine
e-mail: bukselek_kuk@rambler.ru

D. Dent (ed.), *Soil as World Heritage*, DOI 10.1007/978-94-007-6187-2_44,

44.2 Experimental Site and Methods

A field experiment testing various grass mixtures on wind-eroded croplands was carried out over 10 years (2001–2010) at the Bukovyna Institute of Agro-industrial Production. The trials were located on a 7° southwest-facing slope on moderately leached, dusty-loamy Grey forest soil of pH 5.1–5.3. Four experimental fields, each 1.5 ha, were located consistently across the slope. Each sown area measured 15 × 100 m with a border strip, making a grand total of 6.5 ha. Harvest measurements were conducted in six replicates.

44.2.1 Program

1. Seed mixture: clover meadow – 15 kg (7.5 million seeds) per hectare; as a control: timothy meadow – 6 kg (9.5 million)/ha
2. As above $+N_{90}P_{90}K_{90}$
3. Natural regeneration
4. As 3 + $N_{90}P_{90}K_{90}$
5. As 4 + 5 t CaO/ha
6. Seed mixture: bird's-foot trefoil – 8.5 kg (8.5 million)/ha; lucerne – 5.5 kg (2.7 million)/ha; timothy – 6 kg (9.5 million)/ha
7. As 6 + $N_{90}P_{90}K_{90}$
8. Seed mixture: bird's-foot trefoil – 12 kg(12 million)/ha; timothy– 6 kg (9.5 million)/ha
9. As 8 + $N_{90}P_{90}K_{90}$
10. As 9 + chisel ploughing to 55–60 cm every 5 m along the contour

3a. Seed mixture: lucerne (15 kg/ha) + timothy (6 kg/ha)
5a. As 3a + $N_{90}P_{90}K_{90}$

Soil aggregates were measured annually in spring/summer, in triplicate, by Savinov's method; water-stable aggregates were measured using Andrianovim's procedure as modified by Kachinskim (Vadyunina and Karchagin 1986); and a *structural coefficient* was calculated as $_{sum}$agronomically favourable structural aggregates (0.25–7 mm)/$_{sum}$fractions < 0.25, 7–10 and > 10mm.

44.3 Results

The sward created a water-stable, granular soil structure. Over the period 2001–2010, the structural coefficient depended on the grass mixture, its period of development and the weather (Table 44.1).

Compared with natural regeneration, all sown mixtures of perennial grasses and legumes created an increased structural coefficient (Table 44.2). In year 2008, the

Table 44.1 Composition of structural aggregates in the 0–40 cm soil layer, 2004–2010

Years	Size fraction (mm), structural coefficient	Variants				
		1	3	6	8	3a
2004	Sum of fractions 0.25–7 mm	59.5	62.0	65.7	61.6	66.0
	Sum of fractions <0.25, 7–10 and >10 mm	40.5	38.0	34.3	38.4	34.0
	Structural coefficient	**1.47**	**1.63**	**1.91**	**1.60**	**1.94**
2005	Sum of fractions 0.25–7 mm	65.0	67.0	62.0	62.8	69.7
	Sum of fractions <0.25, 7–10 and >10 mm	35.0	33.0	38.0	37.2	30.3
	Structural coefficient	**1.86**	**2.03**	**1.63**	**1.69**	**2.30**
2006	Sum of fractions 0.25–7 mm	–	64.3	62.8	59.6	70.4
	Sum of fractions <0.25, 7–10 and >10 mm	–	35.7	37.2	40.4	29.6
	Structural coefficient	–	**1.80**	**1.69**	**1.48**	**2.38**
2007	Sum of fractions 0.25–7 mm	58.0	58.1	55.2	52.0	58.4
	Sum of fractions <0.25, 7–10 and >10 mm	42.0	41.9	44.8	48.0	41.6
	Structural coefficient	**1.38**	**1.39**	**1.23**	**1.08**	**1.40**
2008	Sum of fractions 0.25–7 mm	57.2	56.2	61.2	60.3	–
	Sum of fractions <0.25, 7–10 and >10 mm	42.8	43.8	38.8	39.7	–
	Structural coefficient	**1.33**	**1.28**	**1.58**	**1.51**	–
2009	Sum of fractions 0.25–7 mm	55.2	58.5	60.9	57.8	–
	Sum of fractions <0.25, 7–10 and >10 mm	44.8	41.5	39.1	42.2	–
	Structural coefficient	**1.23**	**1.41**	**1.56**	**1.37**	–
2010	Sum of fractions 0.25–7 mm	56.0	56.8	58.8	60.1	–
	Sum of fractions <0.25, 7–10 and >10 mm	44.0	43.2	41.2	39.9	–
	Structural coefficient	**1.27**	**1.31**	**1.43**	**1.50**	–

structural coefficient on areas of natural regeneration (variant 3) was 1.28, whereas under the sown mixture of bird's-foot trefoil, lucerne and timothy meadow (variant 6), the index was 1.58. In 2009 the index was 1.41 under natural regeneration and 1.56 for the sown perennial legumes-timothy mixture. The lucerne-timothy mixture (variant 3a) achieved the highest structural coefficient (2.38) prior to the demise of the lucerne in 2008.

In the drought years 2004 and 2007, the structural index of the lucerne-timothy sward was 1.94 and 1.40, respectively, while in the better rainfall years 2005 and 2006, the index increased to 2.30 and 2.38, reflecting an increase in the proportions of aggregates in the favourable 0.25–7 mm range. The opposite situation prevailed in drought years when the structural coefficient was sharply reduced, falling to 1.23 in the case of the timothy-red clover meadow in 2009.

Wet strength, the ability of soil aggregates to resist dispersal by wetting, is agronomically valuable. It is known from the literature that water-stable soil aggregates depend on humic substances produced by fungi and bacteria (Chernyavskiy and Sivak 2005). By Savinov's scale (Vadyunina and Karchagin 1986), determinations of water-stable aggregates of size 3–5 mm in the 0–20 and 20–40 cm layers demonstrated that all the tested grass mixtures improved the index (Table 44.2). Water stability was consistently higher in the upper layer; interestingly, the highest ratio of water-stable aggregates in this layer in drought years was under natural regeneration.

Table 44.2 Percentage of water-stable 3–5 mm soil aggregates under long-term grassland

	Variant														
	1			3			6			8			3a		
	Soil layer (cm)			Soil layer (cm)			Soil layer (cm)			Soil layer (cm)			Soil layer (cm)		
Year	0–20	20–40	**0–40**	0–20	20–40	**0–40**	0–20	20–40	**0–40**	0–20	20–40	**0–40**	0–20	20–40	**0–40**
2004	98.8	87.4	**93.1**	98.7	93.6	**96.1**	97.0	88.6	**92.8**	99.0	87.7	**93.3**	97.4	90.8	**94.1**
2005	97.4	86.7	**92.0**	97.2	92.5	**94.9**	94.1	84.5	**89.3**	95.1	89.4	**92.2**	97.6	92.0	**94.8**
2006	–	–	–	97.1	95.5	**96.3**	97.2	94.3	**95.7**	98.0	95.5	**96.7**	97.6	94.5	**96.0**
2007	95.7	90.5	**93.1**	96.9	93.4	**95.1**	96.2	92.0	**94.1**	97.9	90.7	**94.3**	–	–	–
2008	96.5	87.3	**91.9**	96.5	91.1	**93.8**	94.5	89.6	**91.8**	96.6	89.7	**93.1**	–	–	–
2009	93.0	92.7	**92.9**	94.5	91.9	**93.4**	95.7	83.6	**89.7**	93.6	84.6	**89.1**	–	–	–
2010	98.0	98.0	**98.0**	95.8	95.0	95.4	95.6	94.3	**94.5**	96.8	96.2	**96.5**	–	–	–

44.4 Conclusions

All introduced seed mixtures under trial proved to be more effective than natural regeneration in rebuilding soil structure. The structural coefficient and wet strength of soil aggregates depended on the sown mixture and on sufficiency of rainfall. In years with enough rainfall during the growing season, the best structural coefficient was achieved by the mixture of lucerne and timothy but bird's-foot trefoil, which also achieved a good, water-stable structure, proved to be more drought resistant than lucerne.

References

Chernyavskiy OA, Sivak VK (2003) The effective and rational use of degraded soils. Green Bukovyna, Chernivtsi (Ukrainian)

Chernyavskiy OA, Sivak VK (2005) Rehabilitation of degraded land. Ruta, Chernivtsi (Ukrainian)

Pechenyuk VI (2007) Change of physical properties of soil under action of the systems of maintenance in a garden. In: Pechenyuk VI, Galychina GV (eds) Kamenets-Podilsky, Khmelnytsky region. Vidavets Zvoleyko DG, Kamenets-Podisky, pp 68–70 (Ukrainian)

Rubin SS (1959) Agrotechnics of perennial herbs in field crop rotations. State Publishing House of Agricultural Literature, Kiev (Ukrainian)

Vadyunina AF, Karchagin ZA (1986) Methods of investigation of soil physical properties. Agropromizdat, Moscow (Russian)

Chapter 45
Ecological Agriculture to Mitigate Soil Fatigue

L. Volosciuc and V. Josu

Abstract Soil fatigue may be defined as exhaustion of the soil through depletion of essential plant nutrients. Another situation, which might be termed *false exhaustion*, arises when the cause of the poor crop is the presence of injurious excretions from former crops. Solving the problem requires determining the causes of soil fatigue and establishment of soil quality indicators – in which a special role belongs to the community of microorganisms. The chapter discusses the assessment of the number of different microorganisms under various soil moisture conditions. To enhance soil quality and reduce the impact of root-zone fatigue caused by pathogens, a register of biological preparations for plant protection is proposed.

45.1 Introduction

No civilization may survive except under well-developed, sustainable agriculture. This is the most significant issue of all time and for all people. Even in 500 BC, Heraclitus observed that 'soil health reflects the health of the people' and this is still true; it might be adopted as the slogan of ecological agriculture. Across Europe, soil resources are being overexploited, degraded and irreversibly lost through inappropriate management practices (Jones et al. 2012), and we have much to learn from the Orientals who have preserved soil fertility for millennia without great material investments but with care and respect for both agriculture and nature (Volosciuc and Josu 2006; Volosciuc 2009b).

Soil fertility is a general indicator of ecological health, and we observe that modern agriculture is violating the natural cycling of matter and the resilience or self-regulating ability of the soil. Soil fatigue, as defined by Schreiner and Sullivan (1908), is exhaustion of the soil through depletion of essential plant nutrients: a

L. Volosciuc (✉) • V. Josu
Institute for Plant Protection and Ecological Agriculture, Ministry of the Environment, Chisinau, Republic of Moldova
e-mail: l.volosciuc@gmail.com

D. Dent (ed.), *Soil as World Heritage*, DOI 10.1007/978-94-007-6187-2_45,

different situation, which might be termed *false exhaustion*, arises when the cause of the poor crop is the presence of injurious excretions from former crops (Volosciuc 2009a, c; Burlakova et al. 2005). It often happens that crops do not thrive on soil recently cleared of another crop: growth is curtailed, root development is slow, yields are low and the cycle of growth and fruiting stalls – the syndrome is generally referred to as *soil fatigue*. It is very evident in intensive culture of stone fruits, especially peach and cherry, also in apples replanted in a short cycle and, even, with replanting of different species such as peaches after cherries, apricots or plums or cherries after peaches (Mai et al. 1994; Mazzola 2001).

The primary factors responsible for trouble with initial tree growth are nutritional and biotic (Zydlik et al. 2006):

- Lack of micro- and macronutrients due to depletion by the previous crop.
- Accumulation of toxins secreted by the roots of the previous tree crop. Allelopathic compounds synthesized by some species may have stressful effect on plants of other species or may increase their sensitivity to stressors; these chemicals include juglone, acid, hydrolysed or condensed tannins, phenols or folic acid, alkaloids, purines, coumarins, lactones, terpenes and steroids.
- Direct or indirect action of nematodes. For instance in peach trees, nematode damage to the roots promotes the formation of enzymes that can hydrolyze amygdalin, which is toxic to the roots of newly planted trees (Otto et al. 1994; Utkhede et al. 2001).

The present objective is to assess the roles of different pathogens in soil fatigue and to explore ways to mitigate the problem.

45.2 Materials and Methods

Studies have been undertaken on various microbial agents present in soils after monoculture, crop rotations and in orchards, depending on the genus of the crop and place in the rotation (Volosciuc 2010). Samples for microbiological analyses were taken from fresh soil and assayed for bacteria (on peptone nutrient agar 3 days after inoculation at 26–28°C), fungi (on Cheapec's nutrient agar 5 days after inoculation at 24°C) and actinomycetes (on chitin nutrient agar 7 days after inoculation at 24°C).

Inoculations were made in four replicates and the number of microorganisms calculated per gram of fresh soil. Identification of pathogens and useful microorganisms was performed in accordance with standard procedures (Zydlik et al. 2006). Procedures were also developed to produce microbiological plant-protection agents (Volosciuc 2009a, b; Volosciuc et al. 2011). Assessment of biological activity and statistical analysis followed Dospehov (1979).

45.3 Results and Discussion

Pathogenic fungi play a leading role in soil fatigue; they include the genera *Phytophthora* (Utkhede et al. 2001), *Rhizoctonia solani*, *Fusarium* or *Pythium* (Mazzola 2001). Several researchers emphasize the importance of *Actinomycetes* in replanting disease (Otto et al. 1994); others suggest that damage to the capillary root epidermis disturbs the uptake of water and nutrients, thereby stunting tree growth (Mazzola 2001; Burlakova et al. 2005; Butowskii 2004).

Several measures may be employed to combat the problem. Soil sterilization provides secure and quick relief from harmful substances and organisms such as bacteria, viruses, fungi, nematodes and other pests; it is an alternative to application of specific pesticides. Further positive effects of sterilization include the killing of all weeds and weed seeds and activation of biochemical reactions whereby blocked nutrients in the soil are tapped and made available for plants. Nearly always, soil fatigue is relieved and crop yields are increased (Volosciuc 2009b).

In contrast, ecological agriculture can *prevent* soil fatigue. Evolution of the soil and its biota over many millennia has created communities of organisms with their own ways and means of maintaining soil health. All kinds of symbiotic relationships, as well serving the immediate needs of the participants, accumulate considerable quantities of nutrients. Rhizosphere microorganisms lend natural resistance to plants by releasing antibiotics and insect repellents so that, under ecological agriculture, the soil biota regulate the populations of pathogenic microorganisms and noxious insects.

Painstaking research on relations between different components of the biota has isolated and identified various biological agents that are now the basis of biological preparations to control pests and diseases which may not be controlled in any other way. We have made a start on a register of these microbiological agents:

- Trichodermin-BL: a preparation from the fungus *Trichoderma lignorum* (Tode) Harz, isolated from soil and antagonistic to various pathogens. It reduces attacks by pathogens two- or threefold, thereby stimulating plant growth and increasing yields by 25–30 %.
- Trichodermin-F7: in liquid form, based on the fungus *Trichoderma harzianum* Refai. It is applied against root mould of vegetables and ornamental crops, decreasing pathogen attack 1.5 to 2-fold.
- NEMATOFAGIN-B: based on the fungus *Arthrobotrys oligospora* Fres. It used against gall nematodes on vegetable crops, decreasing meloidogenosis two to threefold and increasing the crop by 0.5–1 kg/m^2.
- Rizoplan: based on the bacterium *Pseudomonas fluorescens* AP-33. Applied against root moulds of cereals and legumes, decreasing the symptoms of disease by 30 %.

Several ecologically safe viral preparations have been developed from soil organisms to control pests that may not be controlled by other biological means:

- Virin-MB: based on the nuclear polyhedrosis virus of *Mamestra brassicae*. The titre is 3 billions/g; application rate is 0.1–0.2 kg/ha against cabbage moth on cabbage, tomatoes and other vegetable crops.

- Virin-OS-B: based on granulosis virus and nuclear polyhedrosis virus with synergetic action. The titre is 3 billions/g; application rate is 0.1 kg/ha against turnip moth and moths of the genus *Agrotis*.
- Virin-HS-2: based on nuclear polyhedrosis virus of a nonspecific host. Titre is 3 billions/g; application rate is 0.2 kg/ha against cotton bollworm and bollworms of the genus *Heliothis*.

45.4 Conclusions

Soil fatigue is widespread. Its causes are not yet completely understood but include specific deprivation of nutrients, accumulation of pests in the soil, metabolic excretions of roots, decline of the soil biota and changes in soil pH. It can be avoided by judicious crop rotation and regular application of organic fertilizers.

Analysis of the gross number of microorganisms in the soil under many and various treatments has demonstrated different ways to combat soil fatigue. The development and application of biological preparations can increase beneficial microorganisms and decrease the number of pathogens, especially phytopathogenic fungi and insect pests. Increasing the range and widening the application of microbiological plant protection will be a key to increasing the uptake and output of ecological and organic farming.

References

Burlakova LM, Kudreavtsev AE, Konontseva EV (2005) Soil degradation and problems of sustainable development of agricultural production in the Altai region. Polzunovsky Bull 4:28–34 (Russian)

Butowskii RO (2004) The problem of chemical contamination of soil and groundwater in the European Union. Agrochemistry 3:74–81 (Russian)

Dospehov BA (1979) Methods for field experiments. Kolos, Moscow (Russian)

Jones A, Panagos P, Barcelo S et al (2012) The state of soil in Europe. In: JRC reference report. Joint Research Centre of the European Commission, Ispra

Mai WF, Merwin IA, Abawi GS (1994) Diagnosis etiology and management of re-plant disorders in New York cherry and apple orchards. Acta Hortic 363:33–41

Mazzola M (2001) Suppression of specific apple root pathogens by *Brassica napus* seed meal amendment regardless of glucosinolate content. Phytopathology 91:673–679

Otto G, Winkler H, Szabó K (1994) Proof of actinomycetes in rootlets of species of *Rosaceae* from a SARD soil – a contribution to the specificity of re-plant diseases. Acta Hortic 363:43–48

Schreiner O, Sullivan MX (1908) Soil fatigue caused by organic compounds. Proc Am Soc Biol Chem 3:120

Utkhede RS, Sholberg PL, Smirle MJ (2001) Effects of chemical and biological treatments on growth and yield of apple trees planted in *Phytophthora cactorum* infested soil. Can J Plant Pathol 23:163–167

Volosciuc L (2009a) Biotechnology of production and application of baculoviral preparations in organic farming. Mediul Ambiant, Chisinau (Romanian)

Volosciuc L (2009b) Ecological problems in agriculture. Bons Offices, Chisinau (Romanian)
Volosciuc L (2009c) Organic farming – a reliable way of sustainable agriculture in Moldova. Quar Plant Prot (Kiev) 6:15–17 (Russian)
Volosciuc L (2010) Identification of problems and selection of baculoviruses. Bul AŞM Ştiinţele vieţii 310(1):96–106 (Romanian)
Volosciuc L, Josu V (2006) Soil – ways and means of improving plant health in organic farming. In: Modern pedological development of organic farming, vol 2. Centre for Applied Pedology, Chisinau, pp 161–171 (Romanian)
Volosciuc L, Lemanova NB, Gorbunova VV (2011) Use of soil-bacteria of different taxonomic groups in growing vegetables in the greenhouse. In: Sosnowska D (ed) Plant protection in greenhouses: 21st century perspectives. Bulletin EPWS/IOBC (Russian), pp 207–213
Zydlik Z, Rutkowski K, Pacholak E (2006) Effect of soil fatigue prevention methods on microbiological soil status in replanted apple orchards. Part III. Number of fungi and actinomycetes. Electron J Pol Agric Univ 9:4

Chapter 46
Mitigation of Western Corn Rootworm (*Diabrotica virgifera virgifera* LeConte) by Maize Varietal Selection

Y. Zaplitnyy, I. Mykulyak, M. Linska, T. Karp, Taras Matskiv, and A. Metelska

Abstract Maize parent material was assessed by the resistance of its root system to plucking. Highly productive hybrids created from this base were used to identify advanced hybrids that are most tolerant to infestation by western corn rootworm.

46.1 Introduction

Crop variety selection aims to create highly productive hybrids that are adapted to different soil and climatic conditions and which combine high yield and quality with resistance to pests and diseases, efficient response to fertilizers and compliance with advanced technology requirements – including the ecologically clean one (Molotskyy et al. 2006). The challenge is to combine productivity with resistance to environmental stress as well as increasing the crop's adaptive potential (Zhuchenko 1988). For maize, the main strategy is hybridization; researchers seek deeper knowledge of the way selectively valuable traits are inherited so as to create parent material for the selection of highly productive hybrids (Zozulya and Mamaliga 1993).

In the Ukrainian western forest steppe, the most injurious fungal pathogens of maize are stem rots and cob Fusarium; amongst the most dangerous pests is the stem moth which infests 11–14 % of cobs and 22–28 % of stems (Chernomyz et al. 2010). A serious threat has recently appeared in western regions of the country in the shape of a quarantine pest – western corn rootworm (*Diabrotica virgifera virgifera* LeConte), which is endemic in North America (Triebel et al. 2010). The adult beetles gnaw the anthers and stigmas of the maize cobs, damage grain at the

Y. Zaplitnyy (✉) • I. Mykulyak • M. Linska • T. Karp
Bukovyna Institute AIP of Ukraine, Chernivtsi, Ukraine
e-mail: bukselek_kuk@mail.ru

T. Matskiv • A. Metelska
SRPQS PPI NAAS of Ukraine, Kiev, Ukraine

D. Dent (ed.), *Soil as World Heritage*, DOI 10.1007/978-94-007-6187-2_46,

milk-ripe stage and gnaw the parenchyma between leaf veins; but the most severe damage is caused by the larvae, which attack the roots. Damaged plants have suspended development, decline and even die at an early stage. During strong wind and rain, the stems drop and take on a characteristic gooseneck form; such plants do not form cobs and yield losses are substantial (Kuner and Tsellner 2011). Crop rotation is the first line of defence but, in intensively farmed areas, the main treatment has been the use of pesticides. In the USA, more than 12 million ha are treated annually with insecticides at a cost of more than a million dollars – and the pest has quickly built up resistance. As an alternative to insecticides, resistant maize hybrids are being investigated and significant efforts are focused on genetic amendment of maize for resistance to *Diabrotica virgifera virgifera* larvae. The trait for resistance is a big root system (Riedell 1994).

Maize selection has been carried out since 1945 in Bukovyna, a traditional area of maize production with favourable climate and soils. Being within the humid zone, a principal direction in selection has been the adaptation for dropping resistance under wet soil conditions; this is also important for resistance to corn rootworm. The Bukovyna Institute AIP, at Chernivtsi, and the Ukrainian plant quarantine station PPI NAAS select and investigate parent material and create new, early-maturing hybrids that combine high productivity, dropping resistance and resistance to western corn rootworm.

46.2 Materials and Method

During 2006–2011, in a selection rotation on fine loamy black earths, selection was carried out by visual estimation of the plants' root dropping level during wet years. Furthermore, an apparatus has been constructed to measure the root system's resistance to vertical plucking from soil; a high resistance shows vigour and strength of the root system, which is important for resistance to corn rootworm (Melnik et al. 2005).

From the gene fund of inbred constant maize lines created in the maize selection laboratory, 17 lines have been selected as donors to root-system resistance to root dropping and a further 19 lines with high potential crop capacity. The research results have shown that the strongest root-system resistance to vertical plucking is amongst the lines Uch85, Uch44, Uch73, Uch53, Uch72, Uch19163, Uch37, Uch88, Lk18923, Lk14794, Lk18913, Uch22, Uch79, Uch57 and Uch56 which exhibited plucking resistance of 48.0–65.5 N/m. The resistance of foreign selection lines proved to be weak, particularly F_2, ДВ$_c$17, KL18, Ma45, CM24, Co125, LC1879, 3346, A203, Ma69 and A46 with values of 28.0–39.5 N/m.

In order to combine useful traits, we conducted pair crossing amongst the selected lines and determined the root-system resistance at various growth stages on the simple hybrids (Table 46.1).

All genotypes exhibit the highest root resistance to plucking at the milk-ripe stage, when they are best-protected against pest infestation. Particularly good resistance to plucking was shown by hybrid combinations Uch88 × Uch57,

Table 46.1 Root-system resistance of the best experimental simple maize hybrids, 2008–2010

	Root resistance to vertical plucking (N/m)			
Genotype	4–5 leaf phase	30 days after germination	Milk ripe	20 days after full ripeness
Uch85 × Lk18913	10.5	28.5	113.5	43.5
Uch44 × Uch22	10.0	23.5	79.0	35.0
Uch73 × R346	10.5	24.0	70.0	32.5
Uch53 × Uch22	9.0	23.0	72.0	34.0
Uch72 × Uch79	11.0	25.0	73.0	36.0
Lk19163 × Uch72	12.0	26.0	89.0	54.5
Uch37 × Uch57	14.5	36.5	141.5	56.5
Uch88 × Uch57	14.0	34.0	123.0	72.5
Lk18923 × Uch56	13.5	38.5	146.0	51.5
Lk14794 × Uch44	13.0	35.0	106.0	67.5

Table 46.2 Root-system vigour determination of registered maize hybrids

	Root resistance to vertical plucking (N/m)			
Hybrid	2009	2010	2011	Mean
Stizhok 192 CV	105	135	163	134
Sadgir	90	130	176	132
BM 281 ACV	132	154	195	163
Kitsmansky 215 CV	155	152	195	170

Lk18923 × Uch56, Lk14794 × Uch44 and, especially, Uch37 × Uch57 which exhibited high resistance at every development stage (14.5, 36.5, 141.5 and 56.5 N/m, respectively).

The best isolated inbred lines and experimental hybrids were used for further selection. In our region, one of the limiting factors for crop growth is accumulated temperature (2,400–2,600°C), so we selected early-ripening (FAO 150–199) and middle-early (FAO 200–299) hybrids. In cooperation with the Institute for Steppe-Zone Agriculture NAAS, at Dnepropetrovsk, new early-maturing hybrids Sadgir (FAO 180), Stizhok 192 (FAO 190), Kitsmansky 215 (FAO 215) and CV 281 ACV (FAO 280) have been created and successfully tested at the state level and registered in the Ukraine Register of Types. These hybrids are characterized by a combination of high productivity and resistance to root dropping (Table 46.2). The best root-system resistance was observed in middle-early hybrids (FAO 200–299), particularly Kitsmansky 215.

46.3 Conclusions

Increasing the resistance of maize hybrids by varietal selection assists the development of agriculture in an ecological direction – avoiding or, at least, minimizing the use of pesticides and maintaining soil health and ecological balance. Several years of research at the Bukovyna Institute AIP has selected maize parental material with a well-branched root system and high regenerative capability, which provides resistance to Diabrotica rootworm. Middle-early hybrids (FAO 200–299) have better resistance to root dropping and to Diabrotica than the early-ripening ones (FAO 150–199).

References

Chernomyz A, Mykulyak I, Linska M, Zaplitnyy J (2010) Methods for selecting corn hybrids in breeding for resistance to major diseases and pests in Bukovyna. Foothill Mt Agric Anim Husb 52:97–104, Ukrainian

Kuner A, Tsellner M (2011) Where lives and harms Diabrotica. Agroexpert 1:36–37 (Ukrainian)

Melnik P, Matsko T, Redko Y et al. (2005) Evaluation of damage to corn by *Diabrotica virgifera*. Plant protection and quarantine 10:48 (Russian)

Molotskyy M, Vasilkovsky P, Knyazyuk V, Vlasenko V (2006) Breeding and seed crops. Vishchashkola, Kiev (Ukrainian)

Riedell WE (1994) Root responses of maize hybrids following corn rootworm larval feeding damage. Cercal Res Commun 22(4):327–333

Triebel S, Hetman M, Bakhmut A (2010) Protection of maize against pests. Agronomy 2:58–62 (Ukrainian)

Zhuchenko A (1988) Ecological and genetic bases of adaptive potential of cultivated plants. Stiinta, Kishinev (Russian)

Zozulya A, Mamaliga V (1993) Breeding and seed production of field crops. Urozhay, Kiev (Ukrainian)

Part IV
Soil Policy and Communications to Decision Makers

Chapter 47
Abating Climate Change and Feeding the World Through Soil Carbon Sequestration

R. Lal

Abstract Two degrees Celsius was accepted by the Copenhagen Accord and the G-8 Summit as an acceptable upper limit of increase in global temperature. This requires identification and implementation of viable options to reduce emissions of CO_2 and other greenhouse gases and sequester carbon from the atmosphere: business as usual will mean a drastic increase in atmospheric CO_2 with dire consequences for the environment, ecosystem services and human well-being. However, net emissions can be reduced by enhancing terrestrial C pools: the soil (4,000 Pg to 3 m depth) and the biotic (620 Pg). The soil C pool is ~5 times the atmospheric pool (780 Pg) and 6.5 times the biotic pool. Most agroecosystems have severely depleted their soil organic carbon (SOC). The magnitude of depletion (30–40 MgC/ha, i.e. 25–75 % of the antecedent) depends on climate, soil type, land use history, farming systems and management.

In the long term, extractive farming practices can severely deplete SOC, exacerbate degradation and adversely affect agronomic productivity. Nonetheless, depleted and degraded soils have a large carbon sink capacity, and the SOC pool can be restored by restorative land use and adoption of management practices that create a positive soil carbon budget, reduce emissions from farming operations like tillage, and minimize risks of soil erosion and nutrient and SOC depletion. These practices include conservation agriculture with mulch farming and cover cropping, complex rotations including agroforestry, integrated nutrient management in conjunction with biological N fixation and recycling of plant nutrients fortified by rhizobial and mycorrhizal inoculations, biochar, fertigation with drip subirrigation, and creating disease-suppressive soils through improvement of rhizospheric processes. The SOC pool should be enhanced to above a threshold level of 1.5–2.0 % in the surface layer of most cultivated soils. Increase in SOC pool in the root zone by 1 Mg/ha can enhance total food production in

R. Lal (✉)
Carbon Management and Sequestration Center, Ohio State University, Columbus, OH, USA
e-mail: lal.1@osu.edu

D. Dent (ed.), *Soil as World Heritage*, DOI 10.1007/978-94-007-6187-2_47,

developing countries by 30–50 million Mg/year. The rate of SOC sequestration in most cropland soils ranges from 100 to 1,000 kgC/ha/year with a total global sequestration potential of 0.4–1.2 PgC over 50–100 years. The potential of C sequestration in the terrestrial biosphere is estimated to be equivalent to a drawdown of 50 ppm of atmospheric CO_2 over a century.

47.1 Introduction

Increase in atmospheric concentrations of carbon dioxide (CO_2) and other greenhouse gases (GHGs) is considered to be an important driver of abrupt climate change (ACC) and the attendant increase in intensity and frequency of extreme events. An increase in global temperature of 2 °C has been accepted by G-8 and the world community as the upper acceptable limit because there are numerous implications of climate change to crop production (Gornall et al. 2010; Hatfield et al. 2011) and the environment. The so-called Century Drought experienced in the USA during the summer of 2012 is a stark reminder of the adverse impacts of ACC and the vulnerability of agroecosystems. Global reduction in the yields of principal crops may be as much as 3–15 % for maize, 2–14 % for wheat, 1–3 % for rice and 2–7 % for soybean (Lobell et al. 2011). There may also be slight gains in yields of maize, wheat and soybeans (1–2 %) due to the CO_2 fertilization effect (Ainsworth and McGrath 2010). However, enrichment of CO_2 can also inhibit NO_3 assimilation in wheat and other plants (Bloom et al. 2010). Therefore, the impacts of ACC on crop yields and food security are complex and not understood.

Yet, the adverse effects on crop yields may be more severe for resource-poor smallholders in sub-Saharan Africa, South Asia and the Caribbean than for large-scale commercial farmers in North and South America, Australia and Western Europe. It is also the tropics and subtropics of sub-Saharan Africa and South Asia where most of the billion undernourished and those prone to hunger live. These people are extremely vulnerable to ACC (HLPE-3 2012). Thus, prioritizing adaptation of agroecosystems to ACC is an essential strategy to advance global food security (Lobell et al. 2008) and to meet the challenge of feeding 9.2 billion people by 2050 (Godfray et al. 2010).

Adaptation is defined as the "initiatives and measures to reduce the vulnerability of natural and human systems against actual or expected climate change effects" (IPCC 2007a). Adaptive options are site specific, and there is a wide range of technologies to consider (Lal 2004a; Clements et al. 2011; Dinar et al. 2008; Foley et al. 2011). Adaptation measures may be anticipatory vs. reactive, private vs. public and autonomous vs. planned (HLPE-3 2012). In view of the increase in frequency and intensity of extreme events experienced since 2000, it may be prudent to adopt those anticipatory measures which enhance resilience of soils and agroecosystems to any future ACC. Thus, the focus of this chapter is improvement of soil organic carbon (SOC) concentration to above the threshold or critical limits for improving soil quality, enhancing use efficiency of inputs and reducing vulnerability of soil to climate-induced changes in soil processes. Increase in SOC concentration in soils and the terrestrial biosphere is also relevant to mitigating ACC (Lal 2004a; Hansen et al. 2008): SOC sequestration in world soils is a win-win option because it has both adaptation and mitigation effects and numerous co-benefits, like food security, which are related to soil quality.

47.2 Soil Organic Carbon and Soil Quality

The concentration of SOC in the root zone is an important determinant of soil quality. In soils with SOC concentration less than the critical limit (1.5–2.0 % by mass), several edaphic factors improve with increase in SOC concentration to the threshold level. On the contrary, depletion of SOC leads to soil degradation (Fig. 47.1). Important among the factors that are improved with increase in SOC concentration are soil physical characteristics (e.g. stable aggregates and tilth, available water capacity, bulk density, macroporosity, infiltration rate), chemical properties (e.g. cation exchange capacity, buffering, nutrient retention), biological factors (e.g. activity and species diversity of soil fauna and flora, microbial biomass carbon, earthworm activity) and

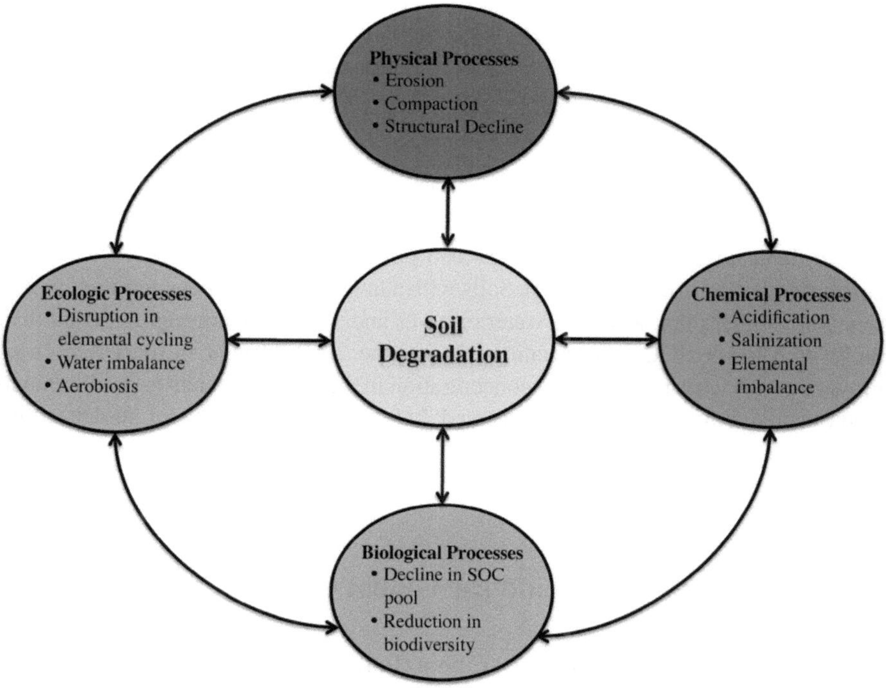

Fig. 47.1 Different processes affecting soil degradation

Table 47.1 Improvement in soil security by enhancing soil organic concentration and pool

Soil quality parameter	Favourable impact of SOC
A. Physical/mechanical quality	1. Improved soil structure and tilth
	2. Reduction in soil bulk density
	3. Decline in soil erodibility
	4. Increased aggregate strength
	5. Decline in susceptibility to crusting and compaction
B. Hydrological quality	6. Increased water infiltration capacity
	7. Decline in surface runoff
	8. Increased plant-available water capacity
C. Chemical quality	9. Increased cation/anion exchange capacity
	10. Increased buffering capacity
	11. Improved nutrient retention capacity
	12. Reduction in susceptibility to elemental imbalance
D. Biological quality	13. Increased soil biodiversity
	14. Improved bioturbation
	15. Increased microbial biomass C
E. Ecological quality	16. Improved elemental/nutrient cycling
	17. Reduced erosion and non-point-source pollution
	18. Decreased leaching of fertilizers and pesticides
	19. Increased uptake/oxidation of CH_4
	20. Reduction in nitrification/denitrification
	21. Increased efficiency of inputs
	22. Reduction in ecological footprint of production systems
F. Economic/aesthetic quality	23. Increased economic value of land
	24. Improved aesthetic value

ecological factors (e.g. elemental cycling, hydrological balance, energy budget, soil microclimate, aeration or gaseous exchange) (Table 47.1). Indeed, agronomic production in depleted and degraded soils is also positively affected by increase in SOC concentration (Lal 2006, 2010a, b). Soils with a favourable SOC concentration have a relatively higher plant-available water reserves and so are more drought tolerant than those with lower SOC concentration. During the 2012 drought in the USA, crops grown on soils with a higher SOC concentration in the topsoil (thanks to no-till, mulch farming and cover cropping) performed better than those without mulch cover (Lal et al. 2012).

47.3 Soil Carbon Pool and the Global Carbon Cycle

Soils constitute the largest terrestrial C pool (Batjes 1996; Lal 2004a; Prentice et al. 2001) and play an important role in the global C cycle (Lal 2004a). Although the exact magnitude of the C flux from land use change is difficult to estimate (Houghton 2010), emissions from the terrestrial ecosystems have been major contributors to atmospheric CO_2 enrichment for 10,000 years – since the dawn of settled agriculture (Ruddiman 2003). Soils of cropland and grazing lands have been depleted of their antecedent SOC by reduction in the quantity and quality of biomass input, alterations in soil temperature and moisture regimes and change in activity and species diversity of soil biota. In general, soil loses 30–50 % of the antecedent pool within ~50 years of cultivation in the temperate climates and ~10 years in tropical regions (Lal 2004a). The magnitude of depletion of SOC pool is greater in soils prone to accelerated erosion and other degradation processes. Drainage and cultivation of peatlands, such as those in Southeast Asia converted for production of palm oil as biofuel, can cause a large soil C debt (Fargione et al. 2008) that may take decades of C sequestration to recover.

Soils of agroecosystems, especially those that have been managed for a long period by extractive farming practices, have a large SOC sink capacity approximately equal to the amount of SOC pool depleted by historic land use and management; this can be as much as 30–50 MgC/ha, depending on soil type, climate, land use, management and the severity of depletion. The depleted SOC pool can be restored through conversion to a restorative land use and adoption of soil/crop/livestock management systems that create a positive C budget.

47.4 Soil Carbon Sequestration

Resilience of soils and agroecosystems in the face of ACC can be enhanced by improving soil quality and its physical, chemical, biological and ecological components (Fig. 47.2) through SOC sequestration. SOC sequestration implies transfer of atmospheric CO_2 into humus by application of biomass (agricultural residues, cover crops, etc.) to soils in such a way that the mean residence time (MRT) of C in soils is drastically prolonged and that the sequestered C is not re-emitted into the atmosphere. The biomass applied is converted into humus through biotic activity, especially microbial processes.

Arid and semiarid lands also contain a large amount of soil inorganic C (SIC) as carbonates and bicarbonates. These are of two types: primary or lithogenic carbonates, which originate from weathering of the parent rocks, and secondary carbonates formed through pedological processes. The latter include dissolution of CO_2 in soil air to form weak carbonic acid and its precipitation with cations (Ca^{+2}, Mg^{+2}) brought in from outside the ecosystem (e.g. deposition, application through soil amendments).

Thus, there are two broad categories of soil C sequestration: SOC and SIC (Fig. 47.3). In soils of natural (and judiciously managed) ecosystems, the MRT of C is enhanced by several mechanisms that protect SOC against microbial processes. Stability of SOC is also enhanced by transfer of C into subsoil layers (Lorenz and Lal 2005; Fontaine et al. 2007), especially because the reactive C has been moved away from the zone of natural and anthropogenic perturbations (ploughing, erosion).

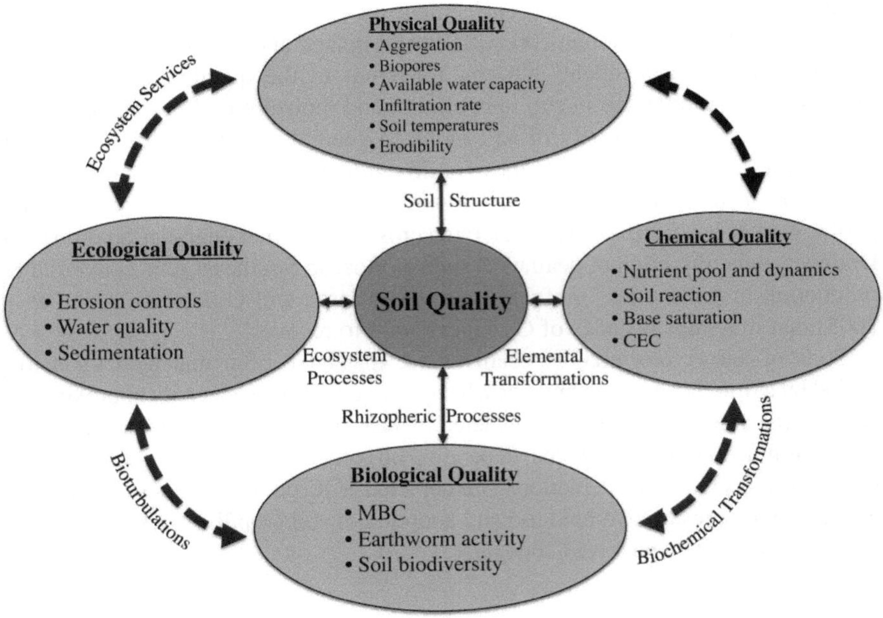

Fig. 47.2 Components of soil quality and interactive processes affecting them

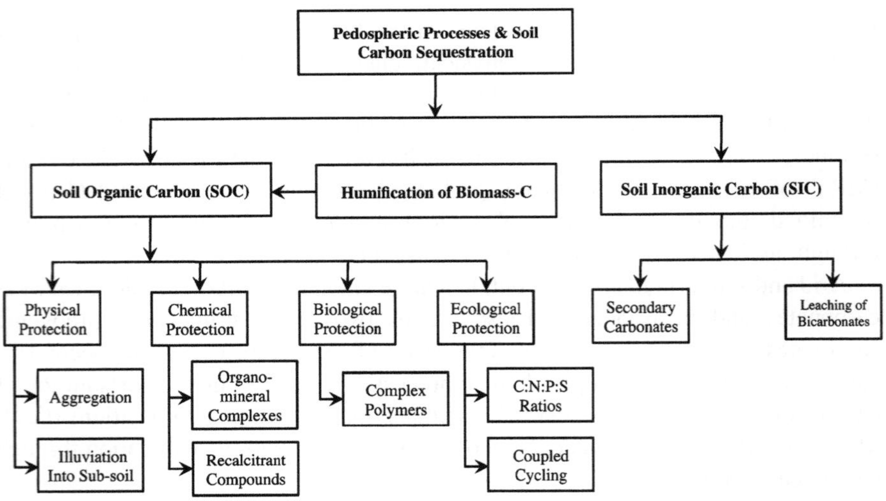

Fig. 47.3 Pedospheric processes of carbon sequestration as soil organic carbon (*SOC*) and soil inorganic carbon (*SIC*)

There is a wide range of soil, crop and water management practices that can create a positive C budget (Fig. 47.4). The use of conservation agriculture (no-till, mulching, cover cropping) in conjunction with integrated nutrient management (INM) techniques that enhance recycling of C and other elements can increase

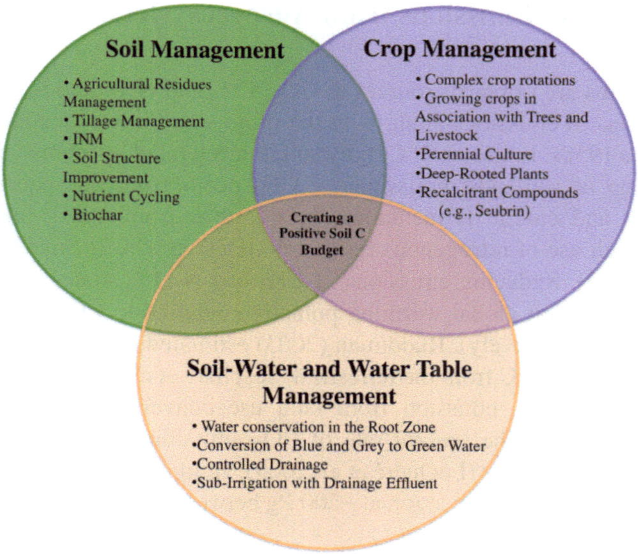

Fig. 47.4 Soil, crop and water management for creating a positive soil carbon budget (*INM* integrated nutrient management)

SOC pool over time. In general, conversion from ploughing to no-till enhances SOC sequestration (West and Marland 2002), albeit under appropriate soil, crop and climatic conditions (Blanco-Canqui and Lal 2008). Application of biochar also has a potential for enhancing the soil/ecosystem C pool (Barrow 2012), yet there is a paucity of soil-specific data on processes governing SOC dynamics with application of biochar. Soil amendments that improve soil structure by encapsulation of C within stable aggregates (e.g. zeolites, compost and manure), complex crop rotations (e.g. agroforestry) and perennial culture can also enhance SOC sequestration. Furthermore, SIC sequestration as secondary carbonates can be enhanced by the use of compost, manure and other biosolids that accentuate CO_2 concentration in soil air. Bicarbonates can be leached into the subsoil/groundwater by leaching with good quality irrigation water (low in soluble salts) (Lal 2001). Soil water conservation and effective erosion control measures are important to SOC sequestration.

Globally, restoration of degraded soils has a large potential for SOC sequestration (Lal 2001) but despite the many advantages of no-till farming, there are, also, several challenges to its adoption – especially by resource-poor farmers and smallholders (Lal 2007). Further, definite gains in SOC sequestration, while widely observed for the surface layer, may not materialize for the entire profile (Blanco-Canqui and Lal 2008). GHG budgets need to be developed at a regional and national level (Cai 2012) to assess the net C gains.

The rate of C sequestration in soils of agroecosystems ranges from 100 to 1,000 kgC/ha/year as SOC and 2–5 kgC/ha/year as SIC (Lal 2004a). The potential capacity of C sequestration in cropland is 0.4–2 PgC/year for about 50 years (Lal 2004a, 2010c). Technical mitigation potential through agriculture is estimated to be 1.5–1.6 PgC/year (IPCC 2007b).

47.5 Reducing Emissions from Agroecosystems

Agroecosystems are principal sources of CO_2, CH_4 and N_2O, and there has been a steady increase in emissions along with the progressive intensification of agriculture since the 1950s. Emission of CO_2 by soil is accelerated by intensive tillage and, also, drainage of wetlands; emissions of CH_4 occur from rice paddies, manure management and enteric fermentation by livestock (ruminants); and emissions of N_2O through use of nitrogenous fertilizers (IPCC 2007a, b; Snyder et al. 2009; Park et al. 2012). Reducing emissions of CH_4 and N_2O is strategically important because of their high global warming potentials relative to CO_2 (21 and 310 for CH_4 and N_2O, respectively). Ruddiman (2003) estimated pre-modern (8000 BC to 1750 AD) emission of C from the terrestrial biosphere at about 320 PgC. Houghton (2003) estimated CO_2 emission from land use conversion between 1850 and 2000 AD at 155 Pg. The present rate of CO_2 emission from land conversion is about 1.3 ± 0.4 PgC/year (LeQuéré et al. 2009). Holdren (2008) estimated CO_2–C emission from land use conversion at ~200 Pg between 1750 and 2002 and another 30 Pg between 2003 and 2030.

Therefore, reducing emissions from agroecosystems and land conversion is integral to any strategy of adapting to and mitigation of ACC. Reducing conversion of new land into agroecosystems is a high priority (Fig. 47.5); rationally, it is important to enhance production from existing land rather than bringing new land under agricultural management. It is estimated that adoption of high-yielding production systems has avoided emissions of 161 PgC since 1916 (Burney et al. 2010). Reducing soil degradation by erosion (and other processes) is another option; globally, accelerated erosion is responsible for emissions of about 1.1 Pg C/year (Lal 2003). Further reducing decomposition of soil organic matter (SOM) can be achieved by adopting no-till, reducing drainage of wetlands and using perennial culture (Glover et al. 2010).

Reducing emissions of CH_4 from agroecosystems presents two distinct but related opportunities. One is to reduce emissions from rice paddies, animal husbandry and manure management. The other is to enhance uptake (oxidation) of CH_4 by the soils of agroecosystems. There are some management options which can reduce CH_4 emissions from livestock and manure management (Sejian et al. 2011, 2012). Increasing productivity of livestock can also reduce emission intensity (Herrero et al. 2011). Water management in rice paddies – aerobic rice culture – is an important option to reduce emissions from rice paddies while also increasing productivity per unit consumption of water (Bouman and Tuong 2001; Bouman et al. 2005, 2006; Kreye et al. 2009).

The strategy to reduce N_2O emissions is judicious management of nitrogenous fertilizers so as to enhance N-use efficiency (Fig. 47.5). N-use efficiency varies between fertilizers (Snyder et al. 2009) and is very low in traditional farming – about 50 % in best-case scenarios (Reay et al. 2012). The use of slow-release formulations and split application are recommended management options (NRC 2009).

Mechanized farm operations that depend on fossil fuel are a major contributor to emissions of CO_2 (Lal 2004b). The energy-intensive farm operations are ploughing,

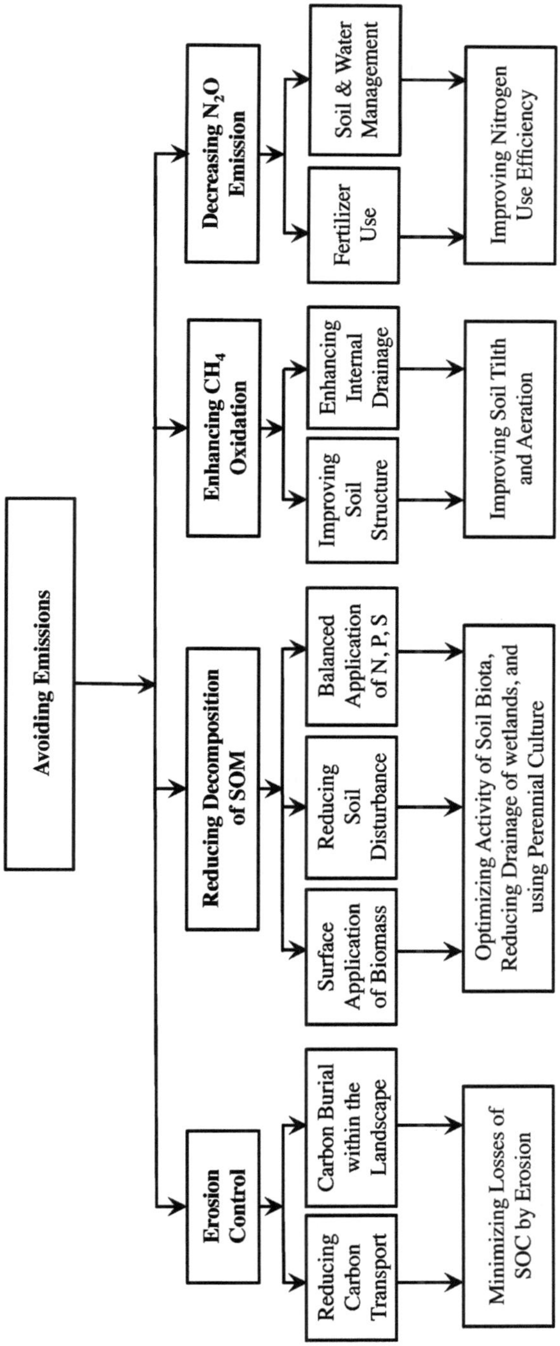

Fig. 47.5 Agronomic management systems for avoiding emissions of greenhouse gases (*GHGs*) from soils (*SOM* soil organic matter)

subsoiling, combine harvesting, use of agrochemicals (especially nitrogenous fertilizers, pesticides) and lifting groundwater for irrigation. Judicious management of these operations is critical to reducing emissions from agriculture (Lal 2004b). Harvesting crop residues for cellulosic ethanol production, or for co-combustion with coal etc., can adversely affect soil quality (Lal 2007, 2008, 2009b; Blanco-Canqui and Lal 2009a, b). Furthermore, the use of food grains for biofuel production is both inefficient (Searchinger et al. 2008, 2009) and unethical because of the competition for food availability, access and stability (Lal and Pimentel 2007). Production of ethanol from sugarcane in Brazil is more efficient than from corn in the USA (Galdos et al. 2010).

47.6 Soil Carbon Sequestration to Mitigate Food Insecurity

Food insecurity means "the inability to secure an adequate diet today and the risk of being unable to do so in the future" (HLPE-4 2012). Its dimensions include: (1) low availability arising from low production, stocks, supply and trade; (2) low access because of low income, high prices and poor market facilities; (3) low utilization by weak physiological processes of human body determined by factors such as water quality, food preparation and storage; and (4) instability over time because of volatility in production, access and utilization. Food insecurity affected 350–875 million people worldwide between 1969–1971 and 2006–2008, and FAO and WFP (2010), FAO (2011) estimated a global population vulnerable to food insecurity at 1,023 million in 2009 and 925 million in 2010.

ACC can exacerbate food insecurity by reducing crop production because of high temperatures and increasing frequency of extreme events; decreasing access by lowering income and raising prices; reducing utilization by aggravating water pollution and increasing the risks of infectious diseases; and accentuating instability by increasing risks of drought/floods and adverse trade, stocks and market conditions. This chapter focuses primarily on the effects of ACC on soil quality and agronomic/biomass productivity. The ACC-induced changes in temperature and precipitation can affect edaphic conditions (e.g. onset, end and the duration of growing season; incidence of pests and pathogens) with a profound impact on use efficiency of inputs and net primary production.

47.7 Strategies to Mitigate Climate Change and Pathways to a Low-Carbon Economy

The SOC concentration (pool) in agroecosystems is important to soil quality, agronomic productivity and global food security. In addition, sequestration of atmospheric CO_2 by soil and terrestrial ecosystems is one of the options for

Table 47.2 Soil carbon concentration and pool as indicator of past and future climate

Parameter	Usefulness as an indicator
1. A known property	1. Widely recognized and a familiar attribute
2. Assessment	2. (a) Indirect: colour (hue and chroma) (b) Direct: dry/wet combustion
3. Dimensional factor	3. Measurable in 4 dimensions: length, width, depth and position
4. Temporal variation	4. Changes over time can be assessed by repeated, synchronised measurements
5. Ecosystem services	5. Directly related to ecosystem services and functions (e.g. food security, biodiversity, water filtration)
6. Soil renewal	6. A key determinant of the rate of soil formation
7. Agronomic/biomass productivity	7. An important indicator of soil fertility and nutrient retention and supply capacity
8. Palaeoclimate	8. $\delta^{13}C$ can be used to assess palaeoclimate
9. Future climate	9. Radiative forcing can be assessed
10. Attributes of SOC/humus	10. Well-defined characteristics of SOC/humus
11. Synergy	11. Can be used in conjunction with other indicators
12. Uncertainty	12. Degree of uncertainty can be measured over temporal and spatial scales
13. Pathways	13. Its pathways can be traced over the landscape
14. History	14. Planetary and human history can be related to its fingerprints (remains of plants and animals)
15. Trading	15. It can be traded as a commodity

offsetting fossil fuel emissions, thereby mitigating anthropogenic ACC. Further, SOC is a good indicator of past and future ACC (Table 47.1). Important among these indicators are the fact that it is a known soil property which can be used synergistically with other indicators, its pathways can be traced over the landscape, and it is sensitive to ACC (Table 47.2). Furthermore, it can be traded as a commodity. In this context, the assessment of SOC pool has specific prerequisites (Table 47.3). The depth of measurement is the entire profile (or to a minimum of 1m, although short-term changes may occur only in the plough layer), and the pool (Mg) and rate of its change (MgC/ha/year) reflect the impact of land use and management (Table 47.3).

Among numerous options for offsetting anthropogenic emissions through sequestration of atmospheric CO_2 in geological and oceanic reservoirs with a long MRT, sequestration as SOC in agroecosystems is the most cost-effective (McKinsey and Co. 2009). Indeed, the cost of sequestration of CO_2 in depleted soils and degraded/desertified ecosystems can be negative because of its many collateral benefits – which include increase in agronomic productivity, decrease in risk of soil erosion, increase in biodiversity, reduction in sedimentation and non-point-source pollution and improvement in water quality. At the same time, there are numerous challenges to achieving SOC sequestration that must be addressed (Lal 2009a). At the technical level, availability of N and other nutrients (P, S) and the amount of clay and nature of clay minerals in soil are important to SOC sequestration (Christopher and

Table 47.3 Measurement of soil organic carbon pool to facilitate trading of carbon credits, carbon farming and payments for ecosystem services

Parameter	Agronomic	Climatic
1. Depth of measurement	Plough layer	2m or the whole profile
2. Units	% (by mass), g/kg	kg/m^2, Mg/ha
3. Rate of change	%/year	kgC/ha/year
4. Site	Point(s), plot	Landscape, watershed, farm, state, national, global
5. Soil bulk density	Optional	Essential
6. Objectives	Soil fertility management	Climate change adaptation and mitigations
7. Duration	Seasonal/annual changes	Decadal changes
8. Measurement of soil quality	Optional	Essential
9. Standard operating procedure	Available	Need to be developed
10. Transaction costs for trading C credits	Not applicable	Medium to high

Lal 2007). The soils of urban ecosystems also have potential for SOC sequestration (Satterthwaite et al. 2010; Zirkle et al. 2011; Selhorst and Lal 2012; Lal and Augustin 2011) and, similar to agroecosystems, there are many collateral benefits, especially with regard to water quality, microclimate and aesthetic and social/ cultural factors.

Adoption of recommended land use and management practices can be promoted through payments for ecosystem services to land managers/farmers/foresters. Widespread adoption of recommended practices needs an enabling environment that includes availability of essential inputs and credit to purchase them. Payments for ecosystem services must be based on just and fair price of soil C. Undervaluing soil C – and the same applies to water – can lead to the tragedy of the commons. Rather than subsidies, which create dependency and distort values, payments for ecosystem services can promote sustainable use of soils and other natural resources while, also, creating another income stream for land managers. Incentive payments are especially crucial for the 1.3–1.5 billion resource-poor farmers and small landholders of the tropics, who cannot adopt RMPs and invest in soil improvement because of low income.

47.8 Conclusions

Carbon sequestration in soils of agroecosystems and in degraded/desertified lands is integral to offsetting anthropogenic emissions. It is cost-effective and has numerous co-benefits. Important among these are improving soil quality, increasing use efficiency of inputs and advancing food security. The rate of C sequestration ranges from 100 to 1,000 kgC/ha/year for organic C and 2–5 kgC/ha/year for inorganic

C. The strategy is to create a positive soil C budget by adopting recommended management practices such as no-till farming, cover cropping, agroforestry, complex rotations, integrated nutrient management and application of biochar. There is a menu of options to choose from – but no silver bullet. Despite the vast potential and numerous benefits, there are several challenges of soil C sequestration – notably the competing uses of crop/animal residues and non-availability of essential inputs to resource-poor and small landholders of the developing countries. Payment for ecosystem services is, therefore, an important incentivization strategy. The way to facilitate soil C sequestration may involve measurement, monitoring and verification of the net SOC gains over time. Standard operating procedures need to be developed for site-specific management options, measurement and monitoring and establishing the fair price based on the societal value of soil C.

References

Ainsworth E, McGrath JM (2010) Direct effects of rising atmospheric carbon dioxide and ozone of crop yields. Glob Change Res 37:109–130

Barrow CJ (2012) Biochar: potential for countering land degradation and for improving agriculture. Appl Geogr 34:21–28. doi:10.1016/j.apgeog.2011.09.008

Batjes NH (1996) Total carbon and nitrogen in the soils of the world. Eur J Soil Sci 47:151–163

Blanco-Canqui H, Lal R (2008) No-tillage and soil carbon sequestration: an on-farm assessment. Soil Sci Soc Am J 72:693–701

Blanco-Canqui H, Lal R (2009a) Corn stover removal for expanded uses reduces soil fertility and structural stability. Sci Soc Am J 73(2):418–426

Blanco-Canqui H, Lal R (2009b) Indiscriminate corn stover removal reduces soil fertility, soil organic carbon and crop yields. CSA News 54:8–9

Bloom AJ, Burger M, Assensio R et al (2010) Carbon dioxide enrichment inhibits nitrate assimilation in wheat and *Arabidopsis*. Science 328:899–902

Bouman BAM, Tuong TP (2001) Field water management to save water and increase its productivity in irrigated lowland rice. Agric Water Manag 49:11–30

Bouman BAM, Peng S, Castaneda AR, Visperas RM (2005) Yield and water use of irrigated tropical aerobic rice systems. Agric Water Manag 74:87–105

Bouman BAM, Yang XG, Wang HQ et al (2006) Performance of aerobic rice varieties under irrigated conditions in North China. Field Crop Res 97:53–65

Burney JA, Davis SJ, Lobell DB (2010) Greenhouse gas mitigation by agricultural intensification. Proc Natl Acad Sci U S A 107:12052–12057

Cai Z (2012) Greenhouse gas budget for terrestrial ecosystems in China. China Earth Sci 55:173–182

Christopher S, Lal R (2007) Nitrogen limitation on carbon sequestration in North America cropland soils. Crit Rev Plant Sci 26:45–64

Clements R, Haggar J, Quezada A, Torres J (2011) Technologies for climate change adaptation – agriculture sector. UNEP Ris Centre, Roskilde

Dinar A, Somé L, Hassan R et al (2008) Climate change and agriculture in Africa: impact assessment and adaptation strategies. Earthscan/James & James, London

FAO and WFP (2010) The state of food insecurity in the world. Addressing food insecurity in protracted crises. Food and Agriculture Organization of the United Nations, Rome

FAO (2011) Hunger. Food and Agriculture Organization of the United Nations, Rome

Fargione J, Hill J, Tilman D et al (2008) Land clearing and the biofuel carbon debt. Science 319:1235–1238
Foley JA, Ramankutty N, Brauman KA et al (2011) Solutions for a cultivated planet. Nature 478:337–342
Fontaine S, Barot S, Barré P et al (2007) Stability of organic carbon in deep soil layers controlled by fresh carbon supply. Nature 450:277–280
Galdos MV, Cerri CC, Lal R et al (2010) Net greenhouse gas fluxes in Brazilian ethanol production systems. Glob Change Biol Bioenerg 2:37–44
Glover JD, Reganold JP, Bell LW et al (2010) Increased food and ecosystem security via perennial grains. Science 328:1638–1639
Godfray HCJ, Beddington JR, Crute IR et al (2010) Food security: the challenge of feeding 9 billion people. Science 327:812–818
Gornall J, Betts R, Burke E et al (2010) Implications of climate change for agricultural productivity in the early twenty-first century. Philos Trans R Soc B-Biol Sci 365:2973–2989
Hansen J, Sato M, Kharecha P et al (2008) Target atmospheric CO_2: where should humanity aim? Nat Geosci. doi:10.1038.ngeo102
Hatfield JL, Boote KJ, Kimball BA et al (2011) Climate impacts on agriculture: implications for crop production. Agron J 103:351–370
Herrero M, Thornton PK, Havlík P and Rufino M (2011) Livestock and greenhouse gas emissions: mitigation options and trade-offs. In: Wollenberg E, Nihart ML, Tapio-Bistrom, Seeberg-Elverfeldt C (eds) Climate change mitigation and agriculture. Earthscan from Routledge/CGIAR, London/Rome, pp 316–332
HLPE-3 (2012) Food security and climate change. High Level Panel of Experts, FAO, Rome
HLPE-4 (2012) Social protection for food security. High Level Panel of Experts, FAO, Rome
Holdren JP (2008) Meeting the climate change challenge. In: Eighth annual JH Chaffe memorial lecture on science and the environment. National Council for Science and the Environment (NCSE), Washington, DC
Houghton RA (2003) Revised estimates of the annual net flux of carbon to the atmosphere from changes in land use and land management 1850–2000. Tellus B 55:378–390
Houghton RA (2010) How well do we know the flux of CO_2 from land-use change? Tellus B 62:337–351
IPCC (2007a) Climate change 2007: mitigation. In: Metz B, Davidson OR, Bosch PR and others (eds) Contribution of Working Group III to the fourth assessment report of the Intergovernmental Panel on Climate Change. Cambridge University Press, Cambridge/New York
IPCC (2007b) Climate change 2007: synthesis report. In: Contribution of Working Groups I, II and III to the fourth assessment report of the Intergovernmental Panel on Climate Change. Cambridge University Press/WMO, Geneva/Cambridge
Kreye C, Bouman BAM, Castañeda AR et al (2009) Possible causes of yield failure in tropical aerobic rice. Field Crop Res 111:197–206
Lal R (2001) Potential of desertification control to sequester carbon and mitigate the greenhouse effect. Clim Change 15:35–72
Lal R (2003) Soil erosion and the global carbon budget. Environ Int 29:437–450
Lal R (2004a) Carbon emission from farm operations. Environ Int 30:981–990
Lal R (2004b) Soil carbon sequestration impacts on global climate change and food security. Science 304:1623–1627
Lal R (2006) Enhancing crop yield in the developing countries through restoration of soil organic carbon pool in agricultural lands. Land Degrad Dev 17:197–209
Lal R (2007) There is no such thing as a free biofuel from crop residues. CSA News 52:12–13
Lal R (2008) Crop residues as soil amendments and feedstock for bioethanol production. Waste Manage 28:747–758
Lal R (2009a) Challenges and opportunities in soil organic matter research. Eur J Soil Sci 60:158–169
Lal R (2009b) Soil quality impacts of residue removal for bioethanol production. Soil Tillage Res 102:233–241

Lal R (2010a) Beyond Copenhagen: mitigating climate change and achieving food security through soil carbon sequestration. Food Secur 2(2):169–177
Lal R (2010b) Enhancing eco-efficiency in agroecosystems through soil C sequestration. Crop Sci 50:S120–S131
Lal R (2010c) Managing soils and ecosystems for mitigating anthropogenic carbon emissions and advancing global food security. BioScience 6(9):708–721
Lal R, Augustin B (2011) Carbon sequestration in urban ecosystems. Springer, Dordrecht
Lal R, Pimentel D (2007) Biofuels from crop residues. Soil Tillage Res 93:237–238
Lal R, Delgado JA, Gulliford J et al (2012) Adapting agriculture to drought and extreme events. J Soil Water Conserv 67(6):153A–157A
LeQuéré C, Raupach MR, Canadell JG et al (2009) Trends in the sources and sinks of carbon dioxide. Nat Geosci. doi:10.1038/ngeo689
Lobell DB, Burke MB, Tebald C et al (2008) Prioritizing climate change adaptation needs for food security in 2030. Science 31:607–610
Lobell DB, Schlenker W, Costa-Roberts J (2011) Climate trends and global crop production since 1980. Science 333:616–620
Lorenz K, Lal R (2005) The depth distribution of soil organic carbon in relation to land use and management and the potential of carbon sequestration in sub-soil horizons. Adv Agron 88:36–66
McKinsey & Company (2009) Pathways to a low-carbon economy. Version 2 of the Global Greenhouse gas abatement cost curve. https://solutions.mckinsey.com/climatedesk/default/en-us/contact_us/fullreport.aspx
NRC (2009) Emerging technologies to benefit farmers in sub-Saharan Africa and South Asia. National Academy Press, Washington, DC
Park S, Croteau P, Boering KA et al (2012) Trends and seasonal cycles in the isotopic composition of nitrous oxide since 1940. Nat Geosci 5:261–265
Prentice IC, Farquhar GD, Le Quéré C et al (2001) Climate change 2001: working group I: the scientific basis; 3. The carbon cycle and atmospheric carbon dioxide in scientific basis. IPCC, Cambridge University Press, Cambridge
Reay DS, Davidson EA, Smith KA et al (2012) Global agriculture and nitrous oxide emissions. Nat Clim Change 2:410–416
Ruddiman WF (2003) The anthropogenic greenhouse era began thousands of years ago. Clim Change 61:261–293
Satterthwaite D, McGranahan G, Tacoli C (2010) Urbanization and its implications for food and farming. Philos Trans R Soc Lond Ser B Biol Sci 365(1554):2809–2820
Searchinger TD, Heimlich R, Houghton RA et al (2008) Use of U.S. croplands for biofuels increases greenhouse gases through emissions from land use change. Science 319:1238–1240. doi:10.1126/science.1151861
Searchinger TD, Hamburg SP, Melillo J et al (2009) Climate change. Fixing a critical climate accounting error. Science 326:527–528. doi:10.1126/science.1178797
Sejian V, Rotz CA, Lakritz J et al (2011) Forage and flax seed impact on enteric methane emission in dairy cows. Res J Vet Sci 4:1–8
Sejian V, Lakritz J, Ezeji T, Lal R (2012) Environmental stress and amelioration in livestock production. Springer, Dordrecht
Selhorst A, Lal R (2012) Effects of climate and soil properties on U.S. home lawn soil organic carbon concentration and pool. Environ Manag 50:1177–1192. doi:10.1007/s00207-012-9956-9
Snyder CS, Bruulsema TW, Jensen TL, Fixen PE (2009) Review of greenhouse gas emissions from crop production systems and fertilizer management effects. Agric Ecosyst Environ 133:247–266
West TO, Marland G (2002) A synthesis of carbon sequestration, carbon emissions, and net carbon flux in agriculture: comparing tillage practices in the United States. Agric Ecosyst Environ 91:217–232
Zirkle G, Lal R, Agustin B (2011) Modeling carbon sequestration in home lawns. Hortic Sci 46:808–814

Chapter 48
Business Case for Green Water Credits

David Dent

Abstract There is much wringing of hands about land degradation, water scarcity and climate change but not much effective action. One reason is a lack of acceptable mechanisms for translating the science into action. In the case of water scarcity, there is a missing level (at least one) between the top level ("Fix the water problem") and the information-gathering level ("What is the mechanism of water delivery and water scarcity?"). Typically, there is no one and no effective institution translating the scientific information into action.

Green Water Credits creates a market for farmers' water management activities that are at present unrecognized and unrewarded. Implementation of these payments for a specified environmental service can safeguard soil and water resources, secure rural livelihoods and combat poverty.

48.1 Context: Water Scarcity on the International Agenda

Water scarcity undermines health, food security and development. The arithmetic of diminishing freshwater resources per caput is inexorable: by 2025–2030, almost two billion people will likely be suffering absolute water scarcity, and two thirds of the world's population will be under water stress (the threshold for meeting the water requirements for agriculture, industry, domestic purposes, energy and the environment). Shortage is increasingly felt in cities (Fig. 48.1; UN Water 2007; IWMI 2007).

Water scarcity cannot be fixed in isolation. It is bound up with poverty, land degradation and climate change. This is where climate change will hit first, but the

D. Dent (✉)
Chestnut Tree Farmhouse, Forncett End, Norfolk, United Kingdom
e-mail: dentsinengland@hotmail.com

D. Dent (ed.), *Soil as World Heritage*, DOI 10.1007/978-94-007-6187-2_48,

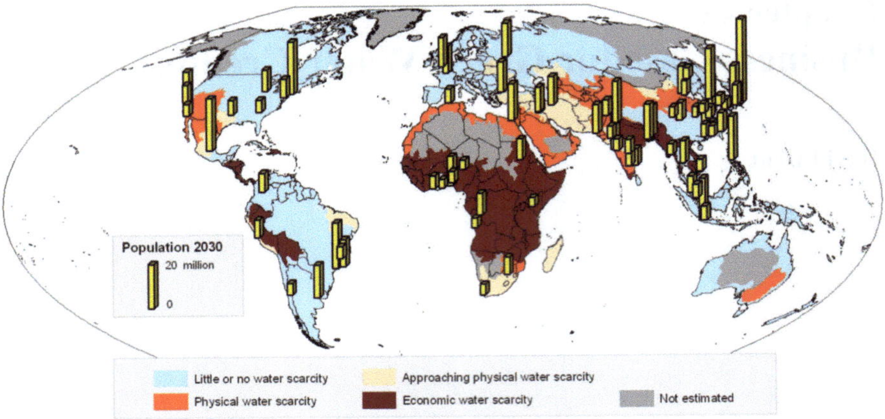

Fig. 48.1 Physical and economic water scarcity 2025 and megacities

UN Water projection does not take account of the expected increase in the variability of rainfall, which will place greater strain on water storage and distribution facilities. Most climatic models predict more extreme rainfall and increased evaporative demand that will translate into droughts and floods. Poor people in poor countries are already most afflicted and they will be hardest hit by climate change.

Soils regulate water flow – so better soil water management should be our first response to climate change. All freshwater comes from rainfall, which is received, stored and regulated by the soil – so every land use decision is a water use decision; flooding is always blamed on freak rainfall but it is actually caused by runoff from farmers' fields. Soil and groundwater are free reservoirs that hold orders of magnitude more water than all existing or conceivable man-made reservoirs – so land degradation and water scarcity are two sides of the same coin.

Burgeoning human population, economic growth and globalization are driving unprecedented land use change, and unsustainable land use is driving land degradation: soil erosion, nutrient depletion, chemical contamination, salinity and water scarcity. During the last quarter century, a quarter of all land has been degrading – mainly in Africa south of the equator, SE Asia and South China. Figure 48.2 (Bai et al. 2008) presents changes in NDVI, the greenness index, as a proxy for land degradation and improvement: and land degradation also means irretrievable loss of water resources.

Today's land degradation adds to a legacy of historical mismanagement, especially in dry lands which are home to two billion people. And two billion people, many of them the same people, live in absolute poverty; 70 % of them depend on rain-fed farming, so there is urgent need for strategies that will enable them to better manage their land, arrest land degradation and secure their livelihoods.

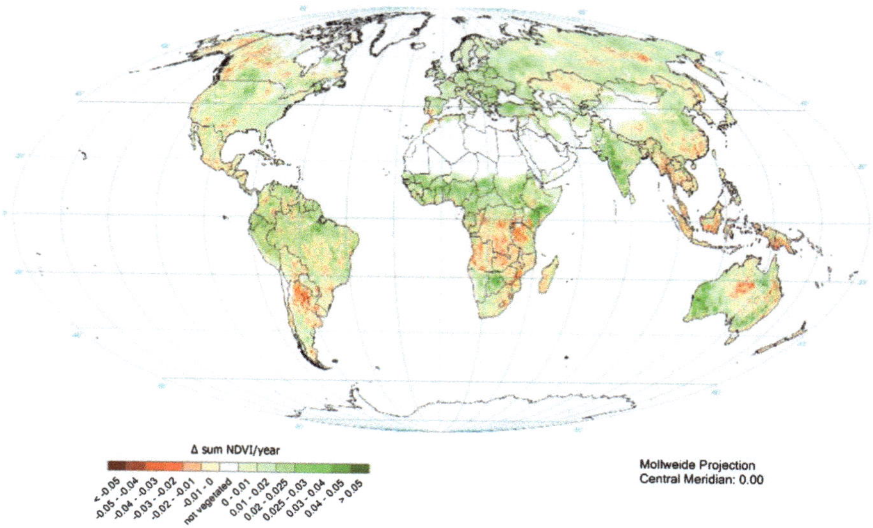

Fig. 48.2 Proxy index of global land degradation and improvement (1981–2006)

48.2 Concept of Operations

The mission is to safeguard water resources and combat poverty; *the strategy* is to create a market in water management services. There is a missing level (at least one) between the top level ("Fix the water problem") and the information-gathering level ("What is the mechanism of water delivery and water scarcity?"). Typically, there is no one and no effective institution translating the scientific information into action. Green Water Credits is a particular case of payment for environmental services. It provides a mechanism that can be taken up by any institution identified or created to make and implement action plans to deal with land and water resources.

Economies and societies are free riders on an apparently endless supply of environmental services that regulate climate, water resources and soil fertility – in short the biophysical cycles of carbon, water and nutrients. The free ride has to end because human activity is overwhelming these natural cycles (UNEP 2007). From now on, they must be purposefully managed. Conventional approaches have been tried and tried again – and found wanting – yet there is no new response. Governments are unwilling to implement their own legislation, tacitly recognizing that this would be unpopular; there is a mismatch between the responsibilities of implementing agencies and their capacity to carry them out; and over wide areas, land is actually managed by smallholders who have limited resources and their own imperatives, scarcely affected by any government initiative – good or bad.

In any case, governments do not have the capacity to manage every acre. Setting a good example is not enough and, most striking of all, local communities and any other supposed beneficiaries are absent from the formative steps of the planning process (Dalal-Clayton and Dent 2001). Poverty makes the task much harder;

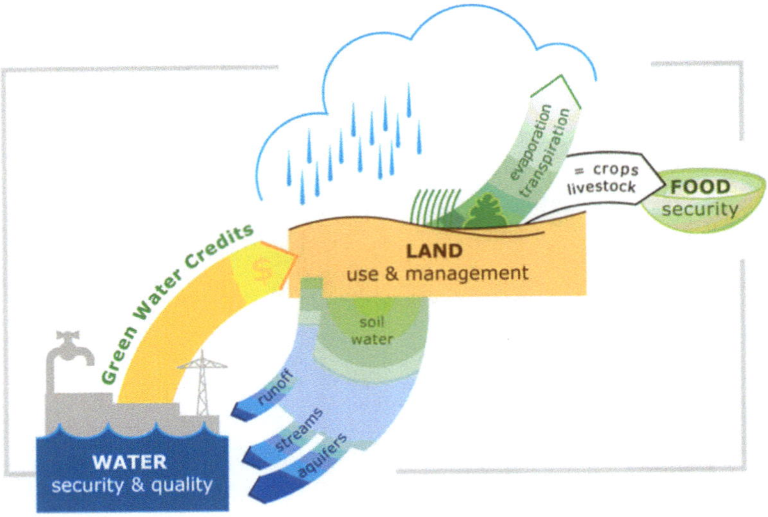

Fig. 48.3 Green Water Credits bridges the incentive gap

farmers are well aware of the benefits of good husbandry but the costs are real and higher for poor farmers because poverty imposes a short time horizon. But whereas farmers are paid for their crops and livestock, they are not paid for delivering water or for keeping the land in good shape. This is a market failure. *The alternative is to create a market in water management services* that will trigger responses by many individuals. If the price is right, cost-effective networks soon evolve to satisfy any demand, be it for environmental services or illicit drugs.

Green Water Credits addresses the market failure by bridging the incentive gap with payments for specified water management services (Fig. 48.3).

This is not a handout. Green Water Credits links upstream water managers with downstream water users who will pay for proper land and water management, safeguarding both water resources and rural livelihoods – a special case of payment for environmental services and the water equivalent of carbon credits. And the cost is no more than the marginal cost of good husbandry as opposed to bad – a fraction of the cost of conventional development projects.

48.3 Technical Aspects

Most of the rain falling on land is used by vegetation – *green water*. Worldwide, only a tenth of freshwater becomes accessible stream flow and groundwater – *blue water* (Fig. 48.4); the remainder is lost as storm runoff. The accessible proportion is much less in dry lands; for instance, only 1.7 % of rainfall in Kenya ends up as useful blue water (Fig. 48.5).

Except in deserts, absolute water shortage is not the issue. The issue is that much goes to waste; it runs off the soil surface as damaging storm water. So agricultural

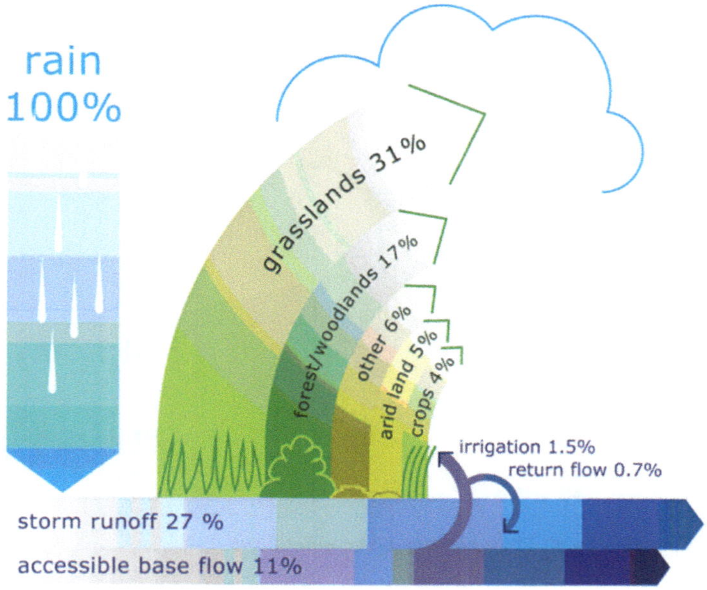

Fig. 48.4 *Green* and *blue* water global flows (Data from Falkenmark and Rockström 2004)

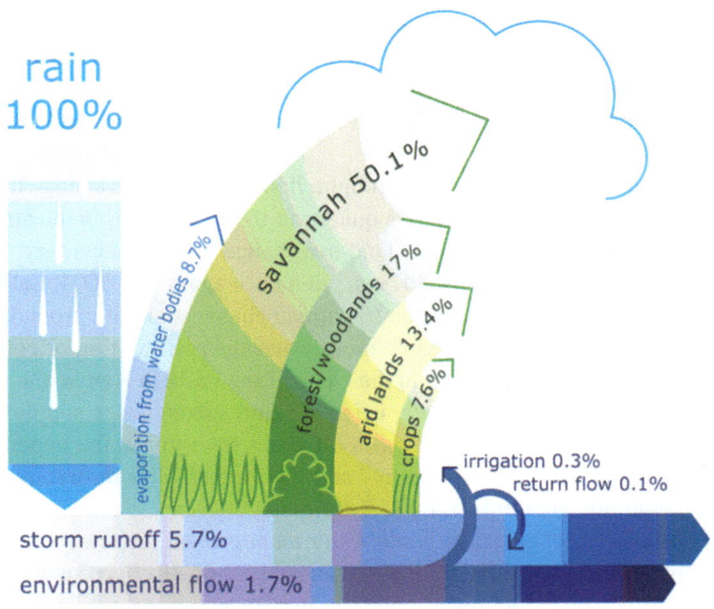

Fig. 48.5 *Green* and *blue* water flows in Kenya (Data from Rockström et al. 2005)

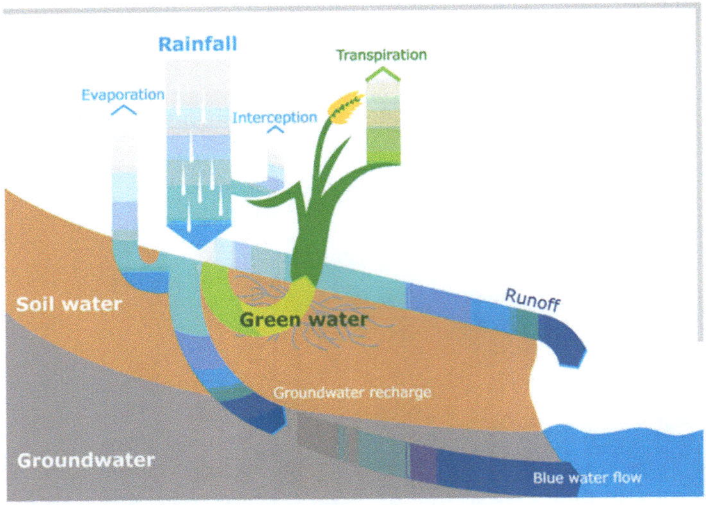

Fig. 48.6 Rainfall separated into *green* and *blue water* flows

drought, i.e. shortage of water in the soil, is much more common than meteorological drought; and political drought, where various shortcomings are attributed to drought, is commonplace. We can't make any more water but green water resources can be much increased, and downstream delivery of water can be better regulated, by increasing infiltration at the soil surface – cutting destructive storm runoff and banking this water in the soil – and by reducing unproductive evaporation. Rainwater transmitted by the soil recharges groundwater and stream base flow (Fig. 48.6).

The soil is the main buffer against drought, floods and climate change in terms of storage of water and, also, carbon; one quarter of the excess CO_2 in the atmosphere has come from the soil as a result of land use change in the last century – the best mitigation for climatic change would be to put it back again (Lal 2004; Lal Chap. 47 in this symposium). As for adaptation to climate change, soil and groundwater are free, natural reservoirs that hold much more water than all existing or conceivable man-made reservoirs; for instance, in the Upper Tana basin in Kenya, the five main reservoirs that provide hydropower, irrigation and urban water supplies have a design capacity of 2,330 million m^3; the soil water reservoir in the upper metre of their catchment holds 7,500 million m^3, and the groundwater reservoir it feeds is orders of magnitude greater (Dent and Kauffman 2007).

Good husbandry, let us call it *green water management*, increases groundwater recharge and stream base flow: mulching can achieve 65–90 % reduction in runoff and more than 25 % reduction in unproductive evaporation; conservation tillage achieves 30–90 % reduction in runoff; tied ridges, terraces and water harvesting achieve 50–100 % reduction in runoff (Ringersma et al. 2003). By arresting runoff, these practices conserve the soil and increase groundwater recharge and, hence, stream base flow.

48.4 Operational Aspects

48.4.1 Criteria for Intervention

1. Water is a national or regional issue.
2. Water is sourced in uplands with a significant rural population.
3. Profitable downstream water use can pay for upland management.
4. An enabling policy framework or the likelihood that it will be developed

A proof of concept was completed for the International Fund for Agricultural Development in 2007 in the Tana basin in Kenya (Dent and Kauffman 2007). Design and capacity building for pilot operations were completed last year in Kenya and the Sebou, in Morocco, last year. Further work is under way in the Upper Changjiang in China.

48.4.2 Who Are the Partners?

Green Water Credits is a public-private partnership. The public sector includes the government, representing the public interest. *Buyers of water services* include public utilities, most immediately water supply and hydropower generation. The private sector includes irrigators, industrial water users and reinsurance companies that cover financial losses from natural hazards like floods. The private sector will only be involved if there is a sound business case and good business practices; larger institutions are well able to look after their own interests in Green Water Credits negotiations but smaller users will need group representation.

Water services providers – the land managers – are mainly private individuals. They too will need effective group organization to participate.

Intermediaries between the users and sellers may include a regulatory authority, appointed to manage water resources, and an independent trust to manage contracts between the buyers and sellers and handle funds – this might be managed by a rural development bank. *International* partners that can provide initial investment and expertise include the International Fund for Agricultural Development (IFAD) which has experience in payments for environmental services, in particular in design of schemes to maximize benefits to the rural poor whom often lack the prerequisites for participation – such as secure land tenure and assets including human capital (Cohen 2008).

Table 48.1 is a framework of activities for implementing Green Water Credits. At the outset, supporting professional staff can be provided by the established national agencies. Start-up finance will be needed from donors and/or government budgets but, thereafter, the system should be self-financing.

Table 48.1 Framework of activities: Green Water Credits process

Item	Issues	Response	Responsibility
Establish key issues	Poverty	Integrated water management	Green Water Credits Steering group
	Water security	Green Water Credits	
	Sustainable land and water management		
Enact enabling legislation	Water to be managed as an economic and social good	Supportive legal framework	Legislature
Identify/establish responsible institutions		Responsible institution appointed	Steering group
Effective management	Responsible institution	Management structure	Responsible institution
	Buyers	Objectives	
	Sellers	Priorities	
	Intermediaries		
Specific objectives Targets	Secure rural livelihoods	Plan for optimal use of natural resources, human resources and capital	Responsible institution with participation of interested parties
	Food and water security		
	Energy security		
	Ecosystem services		
	Flood mitigation		
Identify contribution of Green Water Credits		Green Water Credits integrated within water management policy	Steering group and responsible institution
Information needs	Management information system	Design	Green Water Credits process
	Resource assessment	Capacity building	
		Information gathering	
Water management targets and water allocation		Integrated water management plans	Responsible institution in negotiation with all interested parties
		Design and capacity building	
Green water management	Green water management practices matched with local areas and targets	Design	Green Water Credits process
		Capacity building for land managers	
Establish agreed financial mechanism	Platform of negotiation between buyers and sellers of water management services	Design	Green Water Credits process
	Contracts	Capacity building for all participants	
	Collection and disbursement of Green Water Credits		
	Monitoring		
	Dispute resolution		

48.5 Measures of Effectiveness

The primary benefits of green water management are greater green water resources, increased groundwater recharge and dry-season flow of rivers, lower peak flows, fewer floods and less sediment transport and siltation of reservoirs. This does not necessarily mean any change in total downstream water delivery although this may be achieved by measures to cut unproductive evaporation; the main benefit is the regulation of flows. These benefits are not easy to quantify in ways that all beneficiaries will accept. Because of the large seasonal and variations of flow resulting from variable rainfall and abstraction, gains in groundwater recharge, river flows and cuts in siltation can only be measured over the long term by an operational gauging network. However, they can be demonstrated by well-proven basin hydrological models.

Climate change is likely to mean significant shifts in weather patterns within our lifetimes, in particular more frequent droughts and more intense storms (IPCC 2007). Without initiatives like Green Water Credits, both upstream farmers and downstream water users will be more vulnerable. Building more dams may be seen as an alternative to Green Water Credits; they also buffer floods and droughts, but siltation of reservoirs is an ever-present threat to big dams. Green Water Credits extends the life of existing reservoirs, which is always more cost-effective than building new. Moreover, a reduction in peak flows reduces the peak storage capacity needed; the capacity of the land to store water is many times greater than feasible reservoir capacity. On the other hand, reservoirs do not benefit upstream farmers – indeed a reservoir always floods a great deal of the best farmland. Finally, smallholder investments in water has often turned out to be more profitable per unit area than investments in large irrigation schemes (IWMI 2007).

More secure rural livelihoods arise directly from the diversification of income through payment of Green Water Credits to farmers for their water management activities and indirectly from more sustainable land management and better crop yields. Food security is an important collateral benefit. Green water management brings much-reduced rates of soil erosion and nutrient loss from upland farms. Enhanced green water and nutrient status bring higher and more reliable crop yields. These may be measured through increased market activity and uptake of best management practices.

48.6 Who Benefits?

Downstream water users benefit from more reliable water supplies for power generation, irrigation, domestic and industrial uses. Everyone benefits from enhanced flood protection and, hence, less risk to life and limb and damage to infrastructure. Downstream ecosystems, for instance, in wildlife conservation areas, benefit from guaranteed environmental flows.

Upstream farmers benefit directly. Green Water Credits must cover their outlay on green water management; otherwise their participation is not financially worthwhile. A reasonable rate for the job should provide dependable income in addition to

Table 48.2 Financial evaluation of green water management in the Upper Tana Basin, Kenya

	Reference	Ref 2030	Contour strips	Mulch	Tied ridges
Unmet demand (m^3 million)	247	287	244	192	194
Revenues ($ million)	182	173	179	187	186
Hydropower ($ million)	101	97	100	102	102
Irrigation ($ million)	74	69	72	77	77
Urban ($ million)	7	7	7	7	7
Hydropower (kWh million)	2,556	2,453	2,513	2,580	2,567

Reference describes the situation in 2006; the other columns the situation in 2030 – Ref 2030 without conservation measures and the following columns with conservation measures in place and assuming constant prices

ordinary farm income; they also benefit from better and more reliable crop yields and more-secure local water supplies. However, some may be excluded by clouded land tenure, some may not participate because they do not trust the government to fulfil its promises, and the poorest of the poor have no access to land – although they might benefit from increased opportunities of labour. The distribution of benefits between the various parties depends on contract negotiations between the buyers, sellers and intermediaries – but there are benefits enough to go round.

48.6.1 Costs and Benefits

The costs of green water management can easily be covered by the additional water revenues. In the Upper Tana basin in Kenya, estimated annual benefits of full implementation of Green Water Credits (at constant 1996/1997 prices) are $12–95 millions compared with annual costs of $2–20 millions. With a 20 % adoption scenario, the annual water benefits are $6–48 millions compared with costs of $0.5–4.3 millions – a tenfold return on the investment. Table 48.2 summarizes the financial evaluation of the scheme by Hoff and others (2007). Half of this benefit comes from hydropower generation; the increase in the value of that power at today's oil price makes the calculated benefits half as much again. No account was taken of the savings on sediment damage to hydropower plant, flood mitigation, higher crop yields or the environmental benefits.

The annual costs of a Green Water Credits scheme are summarized in Table 48.3 where the item loss under contour strips cautiously assumes that no cash benefits will accrue from the vegetation established on the contour strips.

48.7 Measures of Effectiveness

It is important to establish a totem by which everyone can judge the effectiveness of Green Water Credits; for instance, a river that no longer reaches the sea should reach the sea again, even in the dry season. Effectiveness might be measured more

Table 48.3 Annual costs of green water management according to the proportion of cropland covered

	Contour strips	Tied ridges	Mulch	Contour strips	Tied ridges	Mulch
	100 %	100 %	100 %	20 %	20 %	20 %
Area, ha	394,200	394,200	394,200	78,800	78,800	78,800
Construction/maintenance, $ million	1.2	19.7	9.8	0.2	3.9	*2.0*
Area loss, $ million	41.3	Nil	Nil	8.3	Nil	Nil
Total, $ million	*42.5*	*19.7*	*9.8*	*8.5*	*3.9*	*2.0*

rigorously by improving trends in river base flow, groundwater and reservoir levels, water quality and sediment loads and a reversal in land degradation. These should be monitored, but actual water levels are driven by climatic variability and the levels of water abstraction, so that improving trends will only be seen in the medium term. The impact on rural poverty is also hard to judge against an ever-changing background but, in the short term, the rates of uptake of the scheme by land users, extension of green water management and the proportion of incoming funds that is actually passed on to the farmers will be good yardsticks.

48.8 Funding a Green Water Credits Program

From the experience of the Kenyan proof of concept and current operational planning, the management costs of such an operation are $0.5 million for proof of concept and $1.5–2 million per river basin for the pilot operational design and capacity building. Implementation over a farmed area of 1/2 million acres will need annual payments of $2.5–10 million by water services buyers to water services providers.

References

Bai ZG, Dent DL, Olsson L, Schaepman ME (2008) Proxy indicator of land degradation. Soil Use and Management 24(3):223–234

Cohen A (2008) Prerequisites and semantics: the challenges of implementing pro-poor payments for watershed services. Technical background report, Rural Poverty Portal, IFAD, Rome

Dalal-Clayton DB, Dent DL (2001) Knowledge of the land. Oxford University Press, Oxford

Dent DL, Kauffman J (2007) The spark has jumped the gap. Green Water Credits Report 7, ISRIC-World Soil Information, Wageningen

Falkenmark M, Rockström J (2004) Balancing water for humans and nature. Earthscan, London

Hoff H et al (2007) Water use and demand in the Tana Basin: analysis using the water evaluation and planning tool (WEAP). Green Water Credits Report 4, ISRIC – World Soil Information, Wageningen

IPCC (2007) Climate change 2007: IPCC 4th assessment report. In: Solomon S, Qin D, Manning M and others (eds). Cambridge University Press, Cambridge, UK
IWMI (2007) Water for food water for life. A comprehensive assessment of water management in agriculture. Earthscan, London
Lal R (2004) Soil carbon sequestration impacts on global climate change and food security. Science 304:1623–1627
Ringersma J, Batjes NH, Dent DL (2003) Green water definitions and data for assessment. ISRIC report 2003/3, Wageningen
Rockström J et al (2005) Sustainable pathways to attain the Millennium Goals. Stockholm Environmental Institute, Stockholm. www.sei.se/SustMDG31Auglowers.pdf
UNEP (2007) Global environment outlook GEO4. UNEP, Nairobi
UN Water (2007) Coping with water scarcity, challenge of the 21st century. Prepared for World Water Day 2007. http://www.unwater.org/wwd07/downloads/documents/escarcity.pdf

Chapter 49
European Commission's Policy Initiatives Towards European and Global Soil Protection

Luca Montanarella

Abstract In our era of multiple crises, policymakers are exploring opportunities to shift to a green economy, seeking effective ways to implement the principles of sustainability and moving away from trade-offs to synergies between the economic, social and environmental dimensions of development. In this context, special attention should be paid to global soil resources. They are the foundation of ecosystem services but they are limited – and under pressure not only from agriculture but, increasingly, from energy production, housing and infrastructure, mining and industry. Food security for a growing population can be assured only if there is enough water and fertile soil for food production. The present legal frameworks for soil protection at national and regional levels seem unable to assure long-term sustainability: a new framework is needed – based on partnership and participatory approaches at all levels from the local to the global. The *Global Soil Partnership* proposed by the FAO and the EU could usher in a renaissance of soil protection activities to ensure the necessary availability of soil resources now and for future generations.

49.1 Introduction

Sustainable development is 'development that meets the needs of the present without compromising the ability of future generations to meet their own needs' (WCED 1987). This is an ambitious goal – given current demographic trends and a forecast global population of more than 9.3 billion by 2050 (United Nations 2011). Nonrenewable natural resources are being depleted at a rate that will certainly not allow future generations to meet their own needs. Sources of minerals, metals and energy, as well as stocks of fish, timber, water, fertile soils, clean air, biomass,

L. Montanarella (✉)
European Commission – DG JRC, Ispra, Italy
e-mail: luca.montanarella@jrc.ec.europa.eu

D. Dent (ed.), *Soil as World Heritage*, DOI 10.1007/978-94-007-6187-2_49,

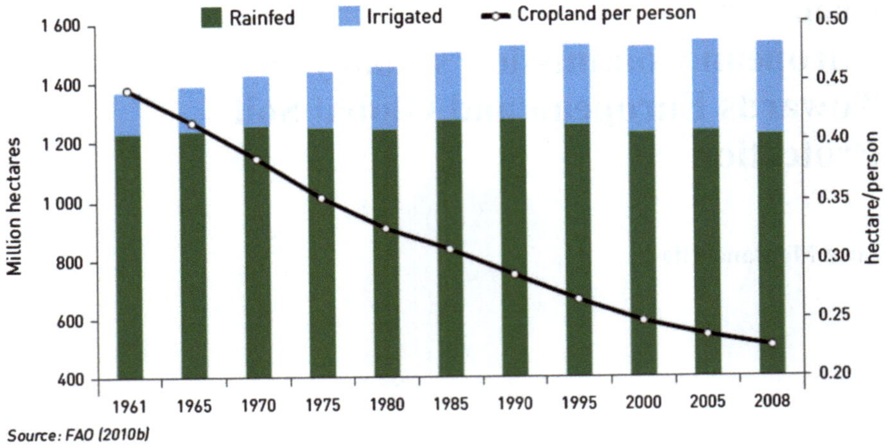

Fig. 49.1 Cropland per person 1961–2008 (FAO 2011b)

biodiversity and even the stability of the climatic system are all under pressure. Whilst world demand for food, feed and fibre is likely to increase by 70 % by 2050, 60 % of the major ecosystems that supply these resources have already been degraded or are being used unsustainably. If we carry on using resources as we do now, by 2050 we will need the equivalent of more than two planets to sustain us, and the aspirations of many for a better quality of life will not be achieved (European Commission 2011a).

Food security for the nine billion (or maybe more) humans that will populate the Earth by 2050 is becoming the main concern for development. Will there be food for all? We know that, already, we are unable to feed everyone: 850 million people are undernourished, and there is no sign of improvement of this figure (FAO 2011a). Recent assessments such as Foresight (2011) make it clear that the challenge of future food security is complex. To handle, let alone solve, the problem will require a much broader perspective than we have yet achieved to integrate the many and various social and economic factors with all the available knowledge in agricultural science and technology.

First, we must acknowledge that, without sufficient soil and water resources, no policy interventions will solve the problem of global food security (FAO 2011b). The farmed area now stands at about 4,600 million ha, of which around 1,400–1,600 million ha is arable. This area has hardly changed for the past 30 years (with a marginal increase of only 8 %, thanks to the increases in irrigated area). Given the continuous increase in the human population, we have seen a steady decrease of cropland per person, now standing at less than 0.25 ha (Fig. 49.1).

At the same time, we have seen a dramatic increase in land taken by competing land uses, especially in developing countries. Year-on-year, millions of hectares of prime cropland are lost to housing and infrastructure (European Commission 2011b), and we are losing much greater areas of fertile soil through ongoing processes of soil degradation that show no signs of relenting. The UNEP *Global*

Environmental Outlook GEO4 (2007) reckoned that 20–50,000 km^2 is lost every year through land degradation, mainly by soil erosion. Losses are two to six times higher in Africa, Latin America and Asia than in North America and Europe: erosion rates in Africa reach up to 100 t/ha/year, and in West Asia as much as one-third of the land is affected by wind erosion. Each year, we lose 24 billion tons of topsoil; what has been lost over the last two decades is equivalent to the entire cropland of the United States!

Soil is nonrenewable except in long geological time frames, so it has to be managed sustainably now if there is to be enough fertile soil for future generations. All experience shows that sustainable soil management requires adoption of legal frameworks that enforce the implementation of good practice for soil protection by landowners and major stakeholders (Hannam and Boer 2004). Successful examples of national soil protection strategies and legislation exist (Montanarella et al. 2009); probably the best known is the US Soil Protection Act of 1935 that reversed the dramatic soil degradation of the Dust Bowl in the American Midwest (Young 1994). Nevertheless, stand-alone national soil protection strategies are hard to implement in today's globalized world with pressures on resources and links between the various socio-economic factors driving land use change that go far beyond national borders.

Indirect land use changes are well documented, especially in conjunction with recent biofuel production targets adopted by developed countries (Howarth et al. 2009). In 2007, when biofuel production was still small, the EU's consumption of global cropland (domestic production plus imports minus exports) was 0.31 ha/person – one-third more than the average of 0.23 ha available per world citizen (Bringezu et al. 2012). Extensive land grabs by countries seeking more soil resources for domestic food requirements are evidence of the need for global governance of this limited resource (Cotula et al. 2009). The potential conflicts over land and water that may arise in the near future are reason enough for instruments of global governance that guarantee sustainable management of natural resources for the common good.

49.2 Towards Global Governance of Soil Resources

There has been a series of attempts to develop instruments of global governance for soils. The first, initiated by FAO in 1982, was the adoption of the *World Soil Charter* which spelt out 13 principles for sustainable soil management to be adopted by FAO member states (FAO 1982). Little if any effect can be observed on the ground, but the principles remain valid and should guide today's actions for sustainable soil management. There was renewed interest in legal frameworks in the late 1990s with the proposal of a *Convention on the Sustainable Use of Soils* elaborated by the Protestant Academy of Tutzing, Germany (Held et al. 1998), and extensively debated at stakeholder meetings and conventions (Hurni et al. 2006). But this, too, failed to win the political consensus needed for entering the intergovernmental agenda. Only following the 2008 food-price crisis have policymakers begun to perceive that land for food production is limited. As a consequence, there has been a surge of interest in a full assessment of global soil resources with major projects collecting updated soil data and information at detailed scales (Sanchez et al. 2009).

Available data already demonstrate that soil degradation is increasing and that the amount of fertile soils is shrinking rapidly (Bai et al. 2008; FAO 2011b). Quantitative targets like 'Zero Net Land Degradation' are being discussed in the framework of the United Nations Convention to Combat Desertification (UNCCD 2012) with four key actions proposed to the international community:

1. Agree on a sustainable development goal of Zero Net Land Degradation
2. Agree on a legal instrument (such as a Protocol on Zero Net Land Degradation) as a global policy and monitoring framework to focus efforts and empower the international community to act with the necessary speed and impact
3. Develop an efficient monitoring system for soil degradation based on regular collection of detailed information on the status and trends of soil properties
4. Implement an effective reporting system entailing regular assessment of global soil resources

If endorsed – *and acted upon* – by the parties, this would much improve the current situation.

49.3 Global Soil Partnership

FAO, with the support of the European Commission, launched the Global Soil Partnership (GSP) in September 2011 to raise awareness among decision makers on the vital role of soil resources in achieving food security, adapting to and mitigating climate change, and to guarantee provision of environmental services. The twin *goals* of GSP are conservation and, where possible, enhancement and restoration of world soil resources through sustainable and productive use. The *vision* is to improve global governance of soil resources – to guarantee healthy and productive soils for a food-secure world and to sustain other essential ecosystem services including water regulation and supply, climate regulation, biodiversity conservation and cultural services. The *mission* is to enhance well-ordered and applied knowledge of soil resources; develop capacity; build on the best available science and facilitate the exchange of knowledge and technologies among stakeholders, existing multilateral environmental agreements and allied technical and scientific bodies for sustainable management of soil resources at all levels. The GSP will:

- Foster awareness among decision makers and stakeholders of the key role of soil resources in productive management and sustainable development of the land
- Address critical soil issues pertaining to food security and climate change adaptation and mitigation

- Through a global platform of communication, facilitate soil knowledge management and targeted research to handle real challenges on the ground
- Establish an effective network for facing up to cross-cutting issues and ensuring synergies between relevant agricultural and environmental processes
- Develop global-governance guidelines for improving soil protection and management.

49.3.1 Composition and Governance

The Global Soil Partnership should be built upon Regional Soil Partnerships securely rooted in national and local communities. It is open to all interested parties that will join a common effort to manage soil resources sustainably by adopting the principles of the World Soil Charter.

Technical and scientific guidance to the GSP Secretariat and Partners will be provided by an Intergovernmental Technical Panel on Soils (ITPS) comprising high-level technical and scientific experts on soil-related issues. The panel will provide the urgently needed science-policy platform for land that was advocated during a recent stakeholder survey (Thomas et al. 2012) – complementing scientific advisory panels for climate change (IPCC) and biodiversity (IPBES). Its immediate tasks include launching soil-and-land issues and solutions into the wider regional and international processes, such as the achievement of the Millennium Development Goals, and interventions on integrated planning and management of land resources. In particular, it will support the World Food Summit Plan of Action and the UNCCD, UNFCCC and CBD in soil-related issues; in relation to climate change, it will advise on the urgent need to preserve soil organic carbon (UNEP 2012); for biodiversity it will throw light on the neglected world of below-ground biodiversity (Jeffery et al. 2010).

To make a difference – and avoid repetition of previous failed initiatives – there must be a new and better way of mobilizing finance and coordinating activities (Fig. 49.2) based on voluntary contributions of an 'alliance of the willing'. Many national and international authorities and organizations want to achieve sustainable soil management within a reasonable time frame. The GSP will provide the framework for them, notwithstanding the minority that are still objecting to measurable soil protection targets.

Activities at local and national levels will be coordinated through Regional Soil Partnerships (Fig. 49.3). Already there are several regional networks of soil-related institutions that may be harnessed to a common framework. For Europe, it is expected that the European Soil Bureau Network will play a leading role; similar networks may be established in the other regions to integrate existing initiatives.

It is proposed that the GSP should act on five main issues:

1. Promote sustainable management of soil resources for soil protection, conservation and sustainable outputs and services
2. Encourage awareness, education and extension, technical cooperation, policy development and investment in soils

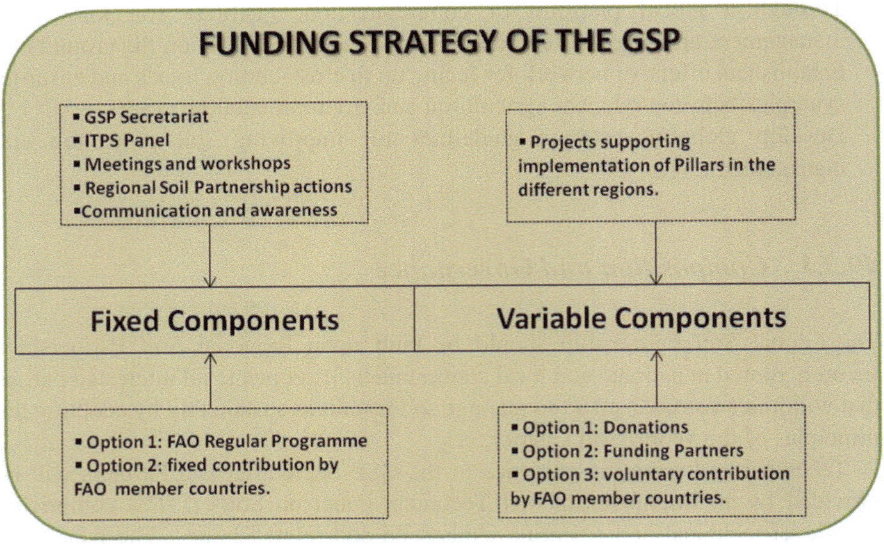

Fig. 49.2 Funding strategy for the Global Soil Partnership

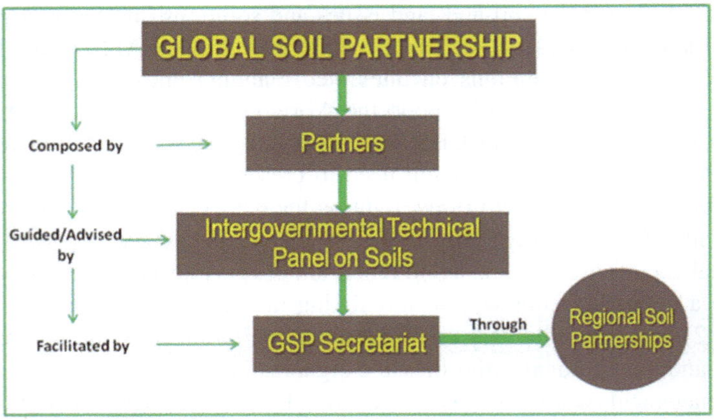

Fig. 49.3 Governance of the Global Soil Partnership

3. Promote soil research and development – focusing on identified gaps and priorities and synergies with related productive, environmental and social development actions
4. Enhance the quantity and quality of soil data and information by supporting data collection (generation), validation, analysis and reporting; monitoring the status and trends of soil resources; and integration with other disciplines
5. Harmonize methods, measurements and indicators for the sustainable management and protection of soil resources.

Drawing up an action plan to achieve these goals, together with measurable targets and indicators, will be the first task of the GSP Intergovernmental Technical Panel on Soils.

49.4 Conclusions

A road map for a green economy and improved governance of natural resources is one of the main outcomes of the Rio + 20 Conference in 2012. In this context, the role of soil resources has been well recognized by stakeholders and policymakers; the proposed Global Soil Partnership should provide a framework for the sustainable use of soil resources. After several failed attempts in the past, this should be the moment for an effective step towards global soil protection which is, surely, needed to improve the environment and livelihoods, especially in developing countries.

Sustainable development goals, agreed by all stakeholders, should include the arrest of land and soil degradation. These goals should define a clear path towards minimizing land and soil degradation and combat erosion, loss of soil organic matter and the loss of productive land to urban sprawl. Success will depend on a capacity to set ambitious and measurable targets to be achieved by an agreed date. The proposal to set a zero net soil degradation target to be achieved by 2050 is ambitious but achievable. Clear definition of indicators and criteria for such a target are necessary and will require a sound scientific foundation. By establishing a high-level Intergovernmental Technical Panel on Soils, the GSP will provide the necessary scientific foundation for this political goal. In the light of the shaky scientific foundations of the UNCCD and other international initiatives of the past, setting global soil protection on a sound scientific footing will be, in itself, a major achievement.

References

Bai ZG, Dent DL, Olsson L, Schaepman ME (2008) Proxy global assessment of land degradation. Soil Use Manag 24:223–234

Bringezu S, O'Brien M, Schütz H (2012) Beyond biofuels: assessing global land use for domestic consumption of biomass. A conceptual and empirical contribution to sustainable management of global resources. Land Use Policy 29:224–232

Cotula L, Vermeulen S, Leonard R, Keeley J (2009) Land grab or development opportunity? Agricultural investment and international land deals in Africa. IIED/FAO/IFAD, London/Rome

European Commission (2011a) Roadmap to a resource-efficient Europe. COM (2011) 571, Brussels

European Commission (2011b) Report on best practices for limiting soil sealing and mitigating its effects. Technical report 2011 – 050, Brussels

FAO (1982) World soil charter. Food and Agriculture Organisation of the United Nations, Rome

FAO (2011a) The state of food insecurity in the world. Food and Agriculture Organisation of the United Nations, Rome
FAO (2011b) The state of the world's land and water resources for food and agriculture. Food and Agriculture Organisation of the United Nations, Rome
Foresight (2011) The future of food and farming. Final project report, Government Office for Science, London
Hannam I, Boer B (2004) Drafting legislation for sustainable use of soils: a guide. IUCN, Gland/Cambridge, UK
Held M, Kummerer K, Odendahl K (1998) Preserving soils for life. The Tutzing project *Time Ecology*. In: Proposal for a convention on sustainable use of soils. Protestant Academy of Tutzing, Munich
Howarth RW, Bringezu S, Bekunda M, de Fraiture C and others (2009) Rapid assessment on bio-fuels and the environment: overview and key findings. International Council for Science (ICSU) Scientific Committee on Problems of the Environment (SCOPE). http://cip.cornell.edu/biofuels/
Hurni H, Giger M, Meyer K (eds) (2006) Soils on the global agenda. Developing international mechanisms for sustainable land management. IASUS Working Group of the International Union of Soil Sciences, Centre for Development and Environment, Bern
Jeffery S, Gardi AC, JonesA, Montanarella L et al (eds) (2010) European atlas of soil biodiversity. European Commission, Publications Office of the European Union, Luxembourg
Montanarella L, Arnold R, Micheli E (2009) Soil conservation services in the European Union and in the United States of America. Adv Geoecol 40:399–409
Sanchez PA, Ahamed S, Carré F, Hartemink AE et al (2009) Digital soil map of the world. Science 325:680–681
Thomas RJ, Akhtar-Schuster M, Stringer LC et al (2012) Fertile ground? Options for a science–policy platform for land. Environ Sci Policy 16(2):122–135
UNCCD (2012) Zero net land degradation. Policy Brief, United Nations Convention to Combat Desertification (UNCCD), Bonn
UNEP (2007) Global environmental outlook GEO-4. Chapter 3, Land. UN Environment Programme, Nairobi
UNEP (2012) Yearbook 2012. UN Environment Programme, Nairobi
United Nations Organization (2011) World population prospects: the 2010 revision. United Nations Department of Economic and Social Affairs (DESA), New York
World Commission on Environment and Development (1987) Our common future. Oxford University Press, Oxford
Young GL (1994) Soil conservation service. In: Cunningham WP et al (eds) Environmental encyclopedia. Gale Research Inc., Detroit, p 774

Chapter 50
Scientific Evidence on the Contribution of Crop Rotation to More Sustainable Agriculture

B.P. Boincean and David Dent

Abstract The *green revolution* carried production ahead of population growth but the revolution has stalled. It depended on cheap fuel, fertilizer and irrigation applied to new, responsive crop varieties but fuel and fertilizer are no longer cheap, water resources are over-committed and crop yields have levelled off – in some places they are declining. By 2050, 70 % more food will be needed than now – double in developing countries. This must come from the same land and water resources or, if present trends continue, much less: the last quarter century has seen degradation of one quarter of the land surface driven, on the one hand, by poverty and, on the other hand, by industrialised agriculture; significant areas are being turned over to the production of bio-fuel; and tracts of the best land are lost to cities and infrastructure. Climate change will bring more-severe droughts and more-intense rains: rising sea level will flood great cities and productive farmland.

The future emphasis needs to be ecological rather than technological. A sustainable alternative combines conservation and precision agriculture: no till, mulching and maintaining soil organic matter, diversification of species, and adapting farm operations to every facet of the landscape. Crop rotation is central to this strategy and data from long-term field experiments demonstrate that respecting crop rotation is cheaper than making up for its absence by application of more and more agrochemicals. Rotation means greater crop diversity and a more resilient and drought-proof ecosystem; inclusion of perennial legumes replaces nitrogen from mineral fertilizers and draws upon deep-seated nutrients and water resources. Crop rotation, alone, cannot make good the annual loss of soil organic matter in arable systems; this requires more farmyard manure which, in turn, means integration of crop production with livestock.

B.P. Boincean (✉)
Selectia Research Institute of Field Crops, Balti, Republic of Moldova
e-mail: bboincean@gmail.com.

D. Dent
Chestnut Tree Farmhouse, Forncett End, Norfolk, United Kingdom

D. Dent (ed.), *Soil as World Heritage*, DOI 10.1007/978-94-007-6187-2_50,

50.1 Context

The end of the era of cheap food and fuel has concentrated minds on food security. Between 1965 and 1980, the *green revolution* increased crop yields two- to threefold and transcended differences in soils and climate. For a generation, global food production was carried ahead of the population curve; food prices dropped and remained low; and political attention turned away from land, food and agriculture – until the food price spike of 2007/2008 (Fig. 50.1).

We can debate the contribution of speculation to the current price volatility but there are several very real reasons for financial markets to bet on steeply rising food prices. The green revolution depended on cheap fuel and derivative fertilizer and pesticides applied to new, responsive crop varieties and on the substantial extension of irrigation. Fuel and fertilizer are no longer cheap; water resources are overcommitted and rising demand from cities means less water will be available for irrigation. Crop yields have levelled off – in several areas they are declining. But burgeoning demand means that, by 2050, food production will need to be 70 % greater than now – and double in developing countries. This greater production will have to come from the same land and water resources or, if present trends continue, significantly less (FAO 2011). On top of historical land degradation going back 10,000 years in long-settled regions, the last quarter century has witnessed degradation of one quarter of the land surface – one third of forests and one quarter of the arable – and tracts of the best farmland are being lost every year to cities and connecting infrastructure (UNEP 2007; Bai et al. 2008).

Since the 1960s, the structure of agriculture and rural communities has changed profoundly. Crop rotations were simplified: perennial legumes were replaced by cheap nitrogen from mineral fertilizers, and wide spectrum fungicides and herbicides enabled second and third consecutive cereal crops to be grown before

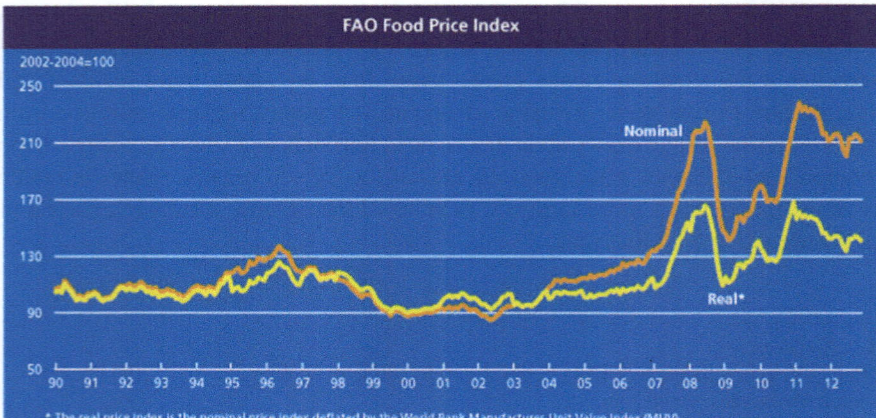

Fig. 50.1 World food price index (http://www.fao.org/worldfoodsituation/wfs-home/foodprices index/en/)

introducing a break crop. Mixed farming declined with a separation of arable and animal husbandry. Mechanization enabled more production with fewer and fewer people. The downward pressure on farm-gate prices and the increasing capital costs of modern equipment brought about a steady loss of small- and medium-sized family farms; successful commercial farms were greatly enlarged. Knowledge of the land, skills and locally adapted crop and animal breeds were extinguished.

The resulting environmental and social problems have become more and more acute. In the first place, neglect of crop rotation and of the importance of soil fertility have brought about stagnation or, even, decline of yields for most crops. Soil health and quality depend on soil biodiversity – microorganisms and microscopic animals as well as the familiar earthworm that maintain soil permeability and resilience against erosion, recycle wastes into plant nutrients and maintain a balance with pathogens. But no modern farming system is maintaining the soil organic matter that fuels soil biodiversity. A significant deterioration of soil quality has been compensated by increased inputs of mineral fertilizer – from both the energy and economic points of view, food production is becoming more expensive.

Respecting crop rotation means greater crop diversity and greater resilience of the agroecosystem. After decades of neglect by agricultural policy, the new Common Agricultural Policy is attempting to revive the practice – although we observe that a rotation comprising 75 % of one crop and 12.5 % of two others, although qualifying for subsidy, does not meet the requirements of ecology.

50.2 Ten Key Points from Long-Term Field Experiments

The economic losses from abandoning crop rotation were underestimated during the era of cheap mineral fertilizers, quite apart from the negative effects on the environment and human health. Data from the long-term field experiments on Typical chernozem soil at Selectia Research Institute of Field Crops at Balti in Moldova (and many others presented in this symposium) demonstrate real possibilities for improving soil health and cutting costs while maintaining the highest crop yields by respecting crop rotation:

1. Comparing the performance of continuous monoculture with crop rotations, mineral and organic fertilizers increase the yields of continuous monocultures much more than the same fertilizer applied to crop rotations (Tables 50.1, 50.2,

Table 50.1 Yield of winter wheat in crop rotation and in continuous culture on fertilized and unfertilized plots, 1994–2011, t/ha and %

	Fertilization		± from	± from crop rotation	
Crop sequence	Unfertilized	Fertilized	fertilization	Unfertilized	Fertilized
Crop rotation (after vetch and oats for green mass)	4.73	5.10	+0.37/8 %	+2.78/143 %	+2.26/80 %
Continuous wheat	1.95	2.84	+0.89/46 %	–	–

Table 50.2 Yields of sugar beet in crop rotation and continuous sugar beet on fertilized and unfertilized plots, 1994–2011

		Fertilization		± from fertilization		± from crop rotation	
Rotational sequence	Links of crop rotation	Unfertilized	Fertilized	t/ha	%	Unfertilized	Fertilized
Crop rotation	Spring vetch and oats for green mass > winter wheat > sugar beet	32.55	42.65	+10.1	31	+24.01/281 %	+25.82/153 %
	Maize silage > winter wheat > sugar beet	31.24	39.93	+8.69	28	+22.7/267 %	+23.1/137 %
Continuous sugar beet		8.54	16.83	+8.29	97	–	–

Table 50.3 Yield of sunflower in crop rotation and continuous culture on fertilized and unfertilized plots, 1994–2011, t/ha and %

	Fertilization			± from crop rotation	
Crop sequence	Unfertilized	Fertilized	± from fertilization	Unfertilized	Fertilized
Crop rotation	2.01	2.16	+0.15/8 %	+0.57/40 %	+0.59/38 %
Continuous culture	1.44	1.57	+0.13/9 %	–	–

Table 50.4 Yields of maize for grain in crop rotation and continuous culture on fertilized and unfertilized plots, 1994–2011, t/ha and %

	Fertilization			± from crop rotation	
Crop sequence	Unfertilized	Fertilized	± from fertilization	Unfertilized	Fertilized
Crop rotation	5.43	5.85	+0.42/8 %	+1.62/43 %	+0.38/7 %
Continuous maize	3.81	5.47	+1.66/44 %	–	–

Table 50.5 Yield of winter wheat (variety Odessa 51) after different predecessors on fertilized and unfertilized plots, 1994–2011, t/ha and %

	Treatment			Yield reduction relative to early-harvested predecessors	
Predecessors	Unfertilized	Fertilized	± from fertilization	Unfertilized	Fertilized
Spring vetch and oats for green mass	4.20	4.54	+0.34/8 %	–	–
Maize silage	3.30	4.01	+0.71/22 %	−0.90/21 %	−0.53/12 %
Maize for grain	2.57	3.59	+1.02/40 %	−1.63/39 %	−0.95/21 %

50.3 and 50.4). This is because the root systems of continuous cultures have a weak capacity to access water and nutrients compared with those of crops grown in rotation. As a corollary, the effect of crop rotation is greater on unfertilized than on fertilized plots.

2. Fertilizer reduces the effect of rotation but doesn't replace it. Even on fertilized plots, the effect of crop rotation[1] is significantly greater than the influence of fertilizers for all crops except maize. It is cheaper to respect crop rotation than to compensate for its absence with extra fertilizer.
3. In the case of winter wheat sown after a late-harvested predecessor, the positive influence of fertilization is equal to yield loss incurred by late sowing as opposed to sowing early (Table 50.5).
4. The greater the diversity of crops in a crop rotation, the higher yields of individual crops and the greater the effect of crop rotation (Table 50.6).

[1] The *effect of crop rotation* is the difference between yields in crop rotation and in continuous cultivation of the same crop, both on fertilized and unfertilized plots.

Table 50.6 Effect of 7-field and 10-field crop rotation, average for 1994–2010, t/ha and %

		10-field rotation		7-field rotation		Continuous cultures	
Crops		Unfertilized	Fertilized	Unfertilized	Fertilized	Unfertilized	Fertilized
Winter wheat	t/ha	4.64	5.06	3.96	4.29	1.95	2.84
	± t/ha/%	+2.69/138 %	+2.22/78 %	+2.01/103 %	+1.45/51 %	–	–
Sugar beet	t/ha	33.21	43.00	23.00	38.55	9.05	17.81
	± t/ha/%	+24.16/267 %	+25.19/141	+13.95/154 %	+20.74/117	–	–
Maize for grain	t/ha	5.22	5.67	5.01	5.62	3.75	5.16
	± t/ha/%	+1.47/39 %	+0.51/10 %	+1.26/34 %	+0.46/9 %	–	–
Sunflower	t/ha	1.99	2.14	1.40	1.70	1.42	1.56
	± t/ha/%	+0.57/40 %	+0.58/37 %	−0.02/0	+0.14/9 %	–	–

Table 50.7 Share of soil fertility in yield formation for different crops grown in rotation and continuous cultures, average for 1994–2011, %

		Fertilization	
Crops	Crop sequence	Fertilized	Unfertilized
Winter wheat	1	92.2	100
	2	54.4	100
Sugar beet	1	69.0	100
	2	2.90	100
Sunflower	1	92.5	100
	2	90.9	100
Maize for grain	1	92.3	100
	2	56.4	100

1 crop rotation, *2* continuous culture

5. On unfertilized plots, the yield is created entirely from inherent soil fertility; on fertilized plots, the yield is generated both from soil fertility and from fertilizers. In crop rotations, the contribution of soil fertility to yield formation *on fertilized plots* is very high for winter wheat, sunflower and maize (92–93 %) but lower for sugar beet (69 %). The contribution of soil fertility to yield formation is significantly lower in continuous cultures: 54–56 % for winter wheat and maize; 91 % for sunflower, which hardly reacts either to rotation or fertilization; and only 3 % for sugar beet (Table 50.7).

   So, fertilizer is more effective in continuous cultures than in crop rotation. *The corollary is that, by respecting crop rotation, it is possible to cut the dependence upon mineral fertilizers.* Returning perennial legumes to crop rotations replaces the nitrogen from mineral fertilizers by nitrogen from biological fixation. There are further benefits from the exploitation of deep-seated water and nutrients and from the enrichment of the soil with organic matter.

6. Crop rotations arrest infestations of disease, pests and weeds. *This means that it is possible to reduce the application of pesticides which have a negative influence on biodiversity and water quality.*
7. Crop rotations are more efficient water harvesters than continuous cultures; *they drought-proof the farming system.*
8. Soil organic matter is an integral index of soil fertility. Crop rotation alone doesn't compensate for the annual losses of soil organic matter, even with 30 % of perennial legumes and 4 tonne/ha of farmyard manure (Tables 50.8 and 50.11). More farmyard manure is needed. *Maintenance and restoration of soil organic matter, efficient recycling of nutrients and reduction of the dependence on off-farm inputs requires integration of crop and animal husbandry.* Importantly, animal husbandry provides a financial return on legumes and grass, compensating for what would otherwise be a loss of income from the cash crops forsaken.
9. Mineral and organic fertilizers influence the yield of crops and soil fertility in different ways. Both increase the yield of crops and the productivity of the whole crop rotation (Table 50.9). *Organic fertilizers are not inferior to mineral fertilizer in respect of their benefit to crop yields and the productivity of the whole crop rotation, but reliance on mineral fertilizers reduces the stocks of soil organic carbon throughout the soil profile* – to a depth of 1 m or more. By contrast, organic fertilizers like farmyard manure increase the stocks of soil carbon, especially in the deeper soil layers. All systems of fertilization deplete stocks of total nitrogen, but the negative influence of mineral fertilizers is significantly higher than that of farmyard manure or a mixture of both (Table 50.10).
10. Polyfactorial trials involving 7-field rotations with and without perennial legumes, with and without ploughing, with various combinations of farmyard manure and NPK fertilizer and without fertilizer (Table 50.11) demonstrate that farming practices drive the accumulation or decline of soil organic matter. *Soil organic matter status can be raised by crop rotations that include perennial legumes and enough farmyard manure, especially under no-till. Crop yields have responded positively to the increase in soil organic matter. Growing perennial legumes is significant both in increasing soil organic matter and in reducing dependence on artificial fertilizers.*[2]

There is one very obvious downside to crop rotation: there is more market demand for some components of rotations than others. The need to include

---

[2] A comparable picture emerges from systematic studies of the influence of agricultural management on soil organic carbon in Canada. Carbon storage increased in their black earths under no-till systems, under reduced summer fallow and in crop rotations that include grasses and perennial legumes, ploughing-in green manures and applying fertilizers. The 20-year average soil-organic-carbon-gain factors derived from field experiments and modelling were 0.1–0.14 tonne/ha/year for no-till, 0.3 tonne/ha/year for decreasing bare fallow and 0.55–0.56 tonne C/ha/year for introduction of perennials into the rotation.

Table 50.8 Annual losses of soil organic matter under crop rotations and continuous cultures, average for 47 years

	Black fallow		Permanent crops				Crop rotations (% of row crops)			
			Winter wheat		Maize for grain		40	50	60	70
Annual uncompensated loss of soil organic matter	Unfertilized	Fertilized	Unfertilized	Fertilized	Unfertilized	Fertilized	Perennial legume 30 % +4 t/ha, farmyard manure + NPK	Black fallow10 % + 0.4 t/ha, farmyard manure + NPK	8 t/ha farmyard manure + NPK	8 t/ha farmyard manure + NPK
t/ha	0.50	0.48	0.40	0.11	0.39	0.24	0.30	0.41	0.35	0.33

Table 50.9 Crop yields (t/ha) and productivity of crop rotation (t cereal units/ha) under different systems of fertilization, average for 1971–2011

Systems of fertilization in crop rotation	Crops: Winter wheat	Sugar beet	Maize for grain	Spring barley	Sunflower	Vetch and oats for green mass	Productivity of crop rotation: t c. u./ha	±t/ha and %
Without fertilization	4.22	31.1	5.9	2.4	2.1	15.4	4.5	–
130 kg a.i. NPK/ha	5.15	43.1	6.8	3.2	2.3	19.5	5.6	+1.1/ 24 %
130 kg a.i. NPK/ha + 10 t/ha of farmyard manure	5.49	43.0	6.8	3.5	2.3	21.0	5.7	+1.2/ 27 %
130 kg a.i. NPK/ha + 15 t/ha of farmyard manure	5.42	46.1	6.8	3.7	2.3	21.6	5.9	+1.4/ 31 %
15 t/ha of farmyard manure	5.36	44.9	6.8	3.5	2.3	21.3	5.8	+1.3/ 29 %

Table 50.10 Annual losses and gains of total carbon and nitrogen under different systems of fertilization in crop rotation, 1970–2009, soil layer 0–100 cm, kg/ha

Systems of fertilization	Annual losses and gains of total carbon for 0–100 cm layer	Annual losses of total nitrogen for 0–100 cm layer
Unfertilized	−600	−190
75 kg a.i. NPK/ha	−1,087	−192
130 kg a.i. NPK/ha	−1,274	−189
175 kg a.i. NPK/ha	−1,241	−195
15 t/ha farmyard manure + 75 kg a.i. NPK/ha	+256	−58
15 t/ha farmyard manure + 130kg a.i. NPK/ha	+13	−106
15 t/ha farmyard manure + 175 kg a.i. NPK/ha	+439	−60
15 t/ha farmyard manure	+46	−76

perennial legumes as a source of nitrogen means sacrificing land that might otherwise be used for oil and cereal cash crops. So far as we can see, sustainability in terms of the soil carbon, energy and nutrient budget requires that, at any one time, a quarter of the land be under perennial grasses and legumes. However, if this biomass is converted on-farm to methane and the residue used as manure, then a high degree of energy and nutrient self-sufficiency may be achieved (Triboi and Triboi, Chap. 32 in this symposium).

Table 50.11 Soil organic matter content (%) in the polyfactorial experiment

	Crop rotation with perennial crops						Crop rotation without perennial crops					
	Mouldboard + minimum tillage			Minimum tillage			Mouldboard + minimum tillage			Minimum tillage		
Soil layer	No fertilizer	Manure	Manure + NPK	No fertilizer	Manure	Manure + NPK	No fertilizer	Manure	Manure + NPK	No fertilizer	Manure	Manure + NPK
0–20 cm	Initial content in 1994											
	4.36 %						4.30 %					
	1999											
	4.24	4.14	4.51	4.13	4.32	4.41	3.95	4.10	4.05	3.96	4.13	4.20
	2009											
	3.97	4.18	4.51	4.13	4.32	4.41	3.94	4.12	4.19	4.07	4.23	4.38
	Difference between 2009 and 1999											
	−0.27	+0.04	+0.40	+0.18	+0.32	+0.32	−0.01	+0.02	+0.14	+0.11	+0.11	+0.18
20–40 cm	Initial content in 1994											
	4.07 %						4.12 %					
	3.66	3.81	3.94	3.61	3.59	3.69	3.72	3.62	3.86	3.75	3.71	3.76
	2009											
	3.32	3.82	4.36	3.82	3.94	4.04	3.51	3.73	3.98	3.51	3.90	3.82
	Difference between 2009 and 1999											
	−0.34	+0.01	+0.42	+0.21	+0.35	+0.35	−0.21	+0.11	+0.12	−0.24	+0.19	+0.06

References

Bai ZG, Dent DL, Olsson L, Schaepman ME (2008) Proxy indicator of land degradation. Soil Use Manag 24(3):223–234

FAO (2011) The state of the World's land and water resources for food and agriculture. Food and Agriculture Organisation of the United Nations/Earthscan, Rome/London

UNEP (2007) Ch 3 The land. In: Global environment outlook GEO4. UN Environment Programme, Nairobi

Recommendations of the Symposium

Rapporteur David Dent

Participants from Belarus, the Czech Republic, the European Commission, France, Germany, Italy, Moldova, the Netherlands, Romania, Russia, Switzerland, Ukraine, the United Kingdom and the USA considered the many parts played by agriculture in creating our habitat, economy and society. Our focus was the conservation and improvement of soil as a world heritage and the need to monitor the effects of management on its quality, productivity and the provision of environmental services. We reviewed the results of long-term field experiments worldwide and inspected the *Typical chernozem* and long-term field experiments at the proposed World Heritage Site at the Selectia Research Institute for Field Crops on the Balti steppe. Such experiments are a perennial spring of fundamental data, expertise, and surprises that challenge our assumptions. Unquestionably, they are cost-effective. Whether we like it or not, soils and ecosystems operate and change over longer timescales than political imperatives, and it is in all our interests to maintain a long-term outlook.

1 A World Heritage Site for Soil

Soils are truly wonderful. They sustain all life, including human life and well-being. They provide anchorage for roots and hold all the water and nutrients accessible to plants; they are home to myriad microorganisms that accomplish biochemical transformations from fixing atmospheric nitrogen to recycling wastes and to armies of microscopic animals as well as the familiar earthworms, ants and termites that create the soil's architecture – most of the world's biodiversity is in the soil, not above ground. Amongst many environmental services, soils regulate water supply, water quality and climate. Neglect of soil as a living system is degrading these functions and the soil itself. *Therefore, we are pleased to respond to the request of the President of the Republic of Moldova to support the proposal to UNESCO for listing the Typical chernozem of the Balti steppe under the Selectia long-term field experiments as the first World Heritage Site for soils and soil science. Quite simply, this is the best arable soil in the world.*

D. Dent (ed.), *Soil as World Heritage*, DOI 10.1007/978-94-007-6187-2,

2 Soil Resolution

Our analysis of the state of agriculture in Moldova supports the European Union's strategy for restructuring and reorienting farming systems. The present, all-pervading, extractive system of food and agriculture is bankrupt. It depends on nonrenewable energy and agrochemicals; it is sending the soil helter-skelter downriver – a loss that cannot be repaired on the human timescale – and hands down the consequences to future generations. We should heed the words of Ionescu Sisesti: 'The land is eternal. If the land disappears, eternity disappears'.

There is an alternative: an ecological approach respecting crop rotation with lesser demands for nonrenewable inputs and enhanced protection of soil and soil functions. But policy must change dramatically to support the necessary changes in outlook and farm practices. *Therefore, we have adopted nem. con. a Soil Resolution which may serve as a basis for legislation. On the one hand, it proposes concrete actions to support more sustainable agricultural systems and more secure rural livelihoods. On the other hand, it identifies new directions for research to underpin the new agriculture.*

2.1 Principles for Policies to Make the Best Use of Existing Knowledge

- Good soil is fundamental to the sustainability of agriculture and society. Conservation of soil and soil fertility is just as important as crop production; present knowledge enables us to do both through an *ecological approach* that is productive, economically viable and environmentally friendly. Self-sufficiency of energy may be attained by conservative use of biomass, e.g. by on-farm generation and use of methane that not only promotes sustainable farming but mitigates climate change.
- *Soil health* may be measured by indices of soil fertility such as clay content and mineralogy, bulk density, total organic carbon and nitrogen (particularly in the labile fraction of soil organic matter), earthworm population, microbial biomass and soil respiration. The impact of the farming system on the whole complex of indicators can be evaluated by calculating budgets for various elements, and energy self-sufficiency may be assessed by calculating the energy budget of the farming system.
- *Progress towards sustainability* may be demonstrated by dynamic models that are applicable to experimental plots, farms and whole landscapes. Simulation of biophysical processes enables prediction of various attributes of the farming system: its productivity, influence on the environment and sustainability. Results obtained by different methods need to be collated and maintained in a secure and accessible bank of data.

- *Regeneration of soil fertility* on each and every farm and *maintenance of environmental services* are possible only if the exigencies of the market are in harmony with agronomic and ecological laws. Such harmony appears to be achievable through lesser dependence on costly, nonrenewable inputs.

Building on these principles, policy initiatives to promote more sustainable agriculture and more secure rural livelihoods might include:

- ***A national program for food and water security and safety worked out at local, regional and state level.*** *The program should include support for or creation of markets for the required production and services such as water management and carbon sequestration.*
- ***Adoption of a soil law to secure the services provided by the soil to society and the environment.*** *This law should be the basis for allocation of payments or other incentives needed to achieve the required protection of soil services.*
- ***Application of such policy requires revitalization of state services, working in partnership with farmers, to elaborate farm and community plans for rational land use, to provide on-farm support for the adoption of best practice and to monitor the state of soil and water resources.***
- ***Support for cooperation between individual farmers for purchasing inputs, marketing produce and services, soil and water conservation at the landscape scale, mutual exchange of know-how, and support for services to cooperatives by contractors, especially for the purchase of new equipment needed for conservation farming.***
- ***Maintenance of locally based production and distribution of high-quality seeds, seedlings for orchards and vineyards, and semen for livestock.***
- ***Support for interdisciplinary research, including long-term field experiments, on rational use and management of land and water resources to avoid degradation and pollution.***

2.2 New Research Thrusts

Agriculture faces acute problems in increasing productivity, maintaining environmental services and providing secure rural livelihoods in the face of perceived contradictions between economic development and sustainability. Our knowledge is fragmented. It is applied mainly to an industrial model of production that does not meet our needs and is unsustainable. Research to support more sustainable systems should focus on:

1. *Deepening interdisciplinary research on the efficient use of land and water resources and inputs, reduction of nutrient losses through leaching and fixation as inaccessible forms, and extension of alternative technologies like allelopathy and mycorrhiza.*

2. *Self-sufficiency in energy, both for direct use and indirect consumption as nitrogen fertilizer.* Priorities include returning legumes into farming systems, reintegration of livestock and the use of biomass as a renewable energy source.
3. *Recognition and avoidance of the error of burning crop residues*; removal of crop residues means a loss of soil function through a loss of soil organic matter. Any short-term advantage of burning crop residues is much less than the losses in the long run.
4. *Extending modelling and simulation as decision-support tools* – for instance, modelling the balances of nutrients and organic matter to avoid soil degradation, and quantification of trends in soil carbon and nitrogen under the influence of different management systems and adaptation to climate change.
5. *Agro-ecological research on the whole food chain from producer to consumer* – introduction of traceability at national and European levels to identify processes of production and final products that may be a risk to health.
6. *Elaboration of incentives for farmers who practise certified, sustainable management.* This means realistic financial incentives and measurable criteria for certification (quality of products, environmental benefits like improved groundwater recharge and water quality) as well as a significant change of agriculture's financial allocation towards supporting ecosystem and societal services, and support for small- and medium-sized farms and community development. Examples include *green water credits* for practices that increase infiltration, cut damaging overland flow and prevent contamination of groundwater and *carbon credits* for accumulating soil organic matter.
7. *Reorientation of the national program of research, development and extension within the European and global network, especially intra-regional cooperation within common soil-climatic and agricultural zones (Moldova-Romania-Ukraine).* The effectiveness of research may be augmented by complementary and mutual research projects, harmonized methods and common databases. The Romanian long-term field experiments, with their unique concept of treatments established in each of the country's 16 soil-climatic regions, illustrate the advantage and need for international cooperation. As a World Heritage Site, the long-term field experiments at Balti would also offer potential for an extended program of international cooperation.
8. *Extending new concepts, techniques and indices requires education of specialists and a career structure for them to practise their skills.*

3 Soil Resolution

Participants of the International Symposium *Soil as World Heritage* in Balti, 22–23 May 2012, have adopted this resolution to be used as a legislative initiative of the President, Parliament and Government of Moldova.

Recognising:

- Soil is the main natural resource of the Republic of Moldova. It is essential to the achievement of food and water security and to the integrity of the environment.
- Services provided by the soil are indispensible to human life and the ecosystems, including agroecosystems, on which we all depend. The health of the soil determines the quality of water and air and the chain of links from soil to crops. Soil is a major buffer against climate change and the only one that we can manage effectively.
- In the face of land degradation, soil is nonrenewable on the human timescale and, therefore, needs to be protected for future generations. At present, soil is not adequately protected; it is being measurably degraded or destroyed.

1. The many parts played by the soil in assuring public health and well-being are not appreciated by society at large. Public opinion is not well informed about the role of soils in food and water security, water quality and maintaining ecosystems. *We recommend that public opinion should be so informed. This requires a concerted, well-focused, long-term program of public information and education.*
2. Improving the welfare of the people through the development of natural resources is possible only if these resources are used sustainably, preventing land degradation. Existing and future norms and actions in agriculture, rural development and use of the natural resources should be in harmony with protection and sustainable use of the soil. Parliament should recognize, as a basis for legislation, the imperative of deepening knowledge and exchanging the scientific and practical information needed to develop sustainable practices of soil management.

   Sustainable and productive farming systems can be built by taking nature as our model: structural diversity in time and space and at different hierarchical levels, land-use planning at the landscape level and respecting the principles of soil fertility maintenance including crop rotation, cover crops, shelter belts, minimum tillage and integration of crops and livestock. Preventing land degradation and maintaining soil fertility through full return of nutrients and organic matter to the soil is much more effective than controlling the consequences of bad agricultural practices. Therefore, *we recommend:*

   - *Establishment of criteria for the sustainable development of farms, irrespective of their size and ownership*
   - *An effective national agency that will rigorously monitor soil quality, stimulate farming systems in tune with the environment – for instance, through drawing up whole-farm plans in partnership with the farmers – and promote good farm practice*
   - *Training of good specialists in the universities, research stations and the private sector*

- *Assured, long-term funding of systematic, interdisciplinary research, a network of experimental and demonstration stations and, especially, long-term field experiments that comprise a national and world heritage of the skills, experience and fundamental data needed to develop and support sustainable farming systems.*

3. There is conflict between the present fragmented, under-resourced and ill-equipped ownership of land and the fundamental principles of managing the soil as a living body. *We recommend:*

   - *Social and economic reforms, including reform of land tenure, must resolve the conflict between sustainable management of the land and the economic interests of the people.*

Index

D. Dent (ed.), *Soil as World Heritage*, DOI 10.1007/978-94-007-6187-2,

M

N

O

P

MIX
Papier aus verantwortungsvollen Quellen
Paper from responsible sources
FSC® C105338

If you have any concerns about our products,
you can contact us on
ProductSafety@springernature.com

In case Publisher is established outside the EU,
the EU authorized representative is:
Springer Nature Customer Service Center GmbH
Europaplatz 3, 69115 Heidelberg, Germany

Printed by Libri Plureos GmbH
in Hamburg, Germany